Birkhäuser

Advanced Courses in Mathematics - CRM Barcelona

Since 1995 the Centre de Recerca Matemàtica (CRM) has organised a number of Advanced Courses at the post-doctoral or advanced graduate level on forefront research topics in Barcelona. The books in this series contain revised and expanded versions of the material presented by the authors in their lectures.

Bruno Vallette
Editor

Higher Structures and Operadic Calculus

Birkhäuser

Editor
Bruno Vallette
Laboratoire Analyse, Géométrie et Applications
Université Sorbonne Paris Nord
Villetaneuse, France

ISSN 2297-0304 ISSN 2297-0312 (electronic)
Advanced Courses in Mathematics - CRM Barcelona
ISBN 978-3-031-77778-3 ISBN 978-3-031-77779-0 (eBook)
https://doi.org/10.1007/978-3-031-77779-0

Mathematics Subject Classification: 18Mxx, 18Nxx, 55P62, 55P48, 14F35, 14H52, 53D55, 32S35, 14G32, 32G15

This book is published under the imprint Birkhäuser, www.birkhauser-science.com by the registered company Springer Nature Switzerland AG
The registered company address is: Gewerbestrasse 11, 6330 Cham, Switzerland

Foreword

Since their introduction 60 years ago, the notions of infinity algebras (Stasheff, Sugawara), higher categories (Boardman-Vogt), operads (May), and model categories (Quillen) have given rise to powerful new tools which made possible the resolution of open problems and prompted revolutions in many domains like algebraic topology (rational homotopy theory, faithful algebraic invariants of the homotopy type of spaces), deformation theory (formality theorems, formal moduli problems), and mathematical physics (quantization of Poisson manifolds, quantum field theories), to name but a few. This theory of higher structures using operadic calculus is currently under rapid development, and there is a need to provide the community with a modern state-of-the-art, accessible to a wide audience.

This was the primary goal of the workshop "Higher Structures and the Operadic Calculus" held at the CRM Barcelona in June 2021 during the Intensive Research Programme on "Higher Homotopical Structures." During that workshop five experts were invited to give a series of lectures on their respective domains where higher categorical or operadic structures play a crucial role.

Alexander Berglund presented a modern approach to rational homotopy theory using homotopy (aka infinity) associative, commutative, and Lie algebras, which are related to the classical models of Quillen and Sullivan. He showed the link between the Koszul property of these algebras and the (co)formality property of the spaces. He also detailed the applications to the study of the automorphisms of manifolds. Ricardo Campos gave a survey on the recent progress made in algebraic deformation theory thanks to the use of operadic structures. He emphasized the role played by the Koszul duality theory and higher categories in the study of formal moduli problems. Geoffroy Horel treated the formality property with purely concentrated weight structures. He covered recent applications in algebraic topology and algebraic geometry, either in zero or positive characteristic, made with the introduction of higher structures. Damien Calaque exposed the notion of Drinfeld's associator in a modern operadic approach. This new point of view allowed recent generalizations in higher genera and related to other groups like cyclotomic, elliptic, ellipsitomic associators for instance. Marcy Robertson covered the homotopical generalization of the notion of a modular operad called infinity modular operad and its applications to the study of the Grothendieck–Teichmüller theory.

One word about the method that lead to this book is mandatory here. Usually, speakers are enthusiastic and full of energy to prepare their lectures, which is quite natural. But they have less time and are less willing to type detailed notes, which is equally easy to understand. One interesting and fruitful method, that we would like to promote here, consists in asking an undergraduate, graduate or postdoctoral "student" (anyway we are all students), to take notes during the talks, type them, present them to the speaker, and eventually work with him or her to produce readable, detailed, and appealing texts. This process presents advantages for everybody: the speaker is pleased to see their talk typed, the "student" learns a lot on the way, and the community can enjoy neat surveys. For the present volume, Robin Stoll, Albin Grataloup, Coline Emprin, Victor Roca i Lucio, and Olivia Borghi respectively typed the aforementioned lectures. They often included extra material for the convenience of the reader and are thus to be fully considered as authors of the present volume.

We hope that the readers will enjoy this state-of-the-art that can be perceived as a door open for further developments. Indeed, these surveys are not expected to close these domains. On the opposite, they show a fruitful way to develop fields of mathematics through the introduction and the use of higher structures and operadic calculus.

We would like to express our sincere appreciation to the colleagues who read the present contributions and helped us to improve them: Pedro Boavida de Brito, Adrien Brochier, Joana Cirici, Vladimi Dotsenko, Clément Dupont, Philip Hackney, Joost Nuiten, Manuel Rivera, and Bashar Saleh. We would like to thank the other organizers of the Intensive Research Programme "Higher Homotopical Structures" Carles Casacuberta, David Gepner, and Joachim Kock and the staff of the CRM Barcelona Lluís Alsedà i Soler, Núria Hernández Martín, and David Romero. Last but not least, the editorial staff of Birkhäuser, especially Dorothy Mazlum, often had a hard time with us, so they deserve our appreciation for their ever-lasting patience and constant support.

Bures-sur-Yvette, France
September 20, 2024

Bruno Vallette

Contents

Higher Structures in Rational Homotopy Theory

Alexander Berglund and Robin Stoll

Contents

A. Berglund (✉)
Department of Mathematics, Stockholm University, Stockholm, Sweden
e-mail: alexb@math.su.se

R. Stoll
Department of Pure Mathematics and Mathematical Statistics, University of Cambridge, Cambridge, UK
e-mail: rs2348@cam.ac.uk

B. Vallette (ed.), *Higher Structures and Operadic Calculus*, Advanced Courses in Mathematics - CRM Barcelona, https://doi.org/10.1007/978-3-031-77779-0_1

1 Introduction

The purpose of these lecture notes is to give an introduction to rational homotopy theory from the perspective of higher structures. For readers already acquainted with rational homotopy theory as presented in standard textbooks, these lecture notes may serve as an introduction to the language of algebraic operads, C_∞- and L_∞-algebras, Koszul duality, etc., by seeing how it is used in a familiar context. Conversely, for readers with a background in the theory of algebraic operads, these lecture notes may serve as a concise introduction to rational homotopy theory. For readers with background in none of the theories, we hope the text can function as an introduction to both. Needless to say, it is only possible to cover a small amount of material in three lectures. We do not give full proofs of all statements but we have tried to provide references to more complete accounts throughout.

The goal of the first lecture is to explain, and outline the proof of, the result that the rational homotopy type of a simply connected space of finite type is faithfully encoded by a C_∞-algebra structure on the rational cohomology groups, or alternatively by an L_∞-algebra structure on the rational homotopy groups. We begin by reviewing the definition of C_∞- and L_∞-algebras, and then explain how the theory of algebraic operads provides a unified framework for developing the basic theory of these higher structures. We then present a short proof of the homotopy transfer theorem, which is one of the main technical tools. Next, we review the parts of Sullivan's and Quillen's theories that are necessary for proving the main result and we discuss the relation between the C_∞- and L_∞-algebra models and the minimal models of classical rational homotopy theory.

The second lecture discusses the relation between Koszul algebras and the notions of formality and coformality. We review the definition of Koszul commutative and Lie algebras, and the recent generalization to Koszul C_∞- and L_∞-infinity algebras. We emphasize that we view Koszul algebras as a calculational tool that, in favorable situations, allows one to, for example, quickly compute the rational homotopy groups of a space from a presentation of the rational cohomology ring, or quickly decide whether a given space is (co)formal or not. We give a number of examples to illustrate this point.

In the third lecture, we discuss certain higher structure in the rational cohomology of classifying spaces of automorphisms of high dimensional manifolds, discovered by Berglund–Madsen. The higher structure in question is Kontsevich's Lie graph complex and variants of it. We begin by reviewing some general results on the rational homotopy theory of spaces of self-homotopy equivalences of simply connected finite CW-complexes. Then we review the definition of modular operads and graph complexes. Finally, we sketch the results of Berglund–Madsen.

In an appendix, we discuss how the C_∞- and L_∞-models relate to Massey operations and higher order Whitehead products, respectively. We also review the nerve of an L_∞-algebra and related constructions, and we summarize some facts about rational homotopy theory of non-nilpotent spaces.

1.1 Conventions

We work over the field of rational numbers $\mathbb{Q}$, except when stated otherwise. In particular "vector space" will mean vector space over $\mathbb{Q}$, tensor products are over $\mathbb{Q}$, and so on.

We use homological grading convention, so differentials in chain complexes have degree -1, and chain complexes are unbounded unless stated otherwise. Thus, a chain complex is a collection $V = \{V_i\}_{i\in\mathbb{Z}}$ of vector spaces together with a linear morphism $d\colon V \to V$ of degree -1 such that $d^2 = 0$. We let $|v|$ denote the degree of an element $v \in V$, i.e. $|v| = i$ means $v \in V_i$. By using the convention $V^i = V_{-i}$, chain complexes can be considered as cochain complexes and vice versa, that is, subscripts refer to the homological degree and superscripts to the negated homological (i.e. cohomological) degree.

We use d as generic notation for the differential of a chain complex when it is clear from the context what is meant. For example, the differential of the tensor product $V \otimes W$ of two chain complexes can be written as

$$d(v \otimes w) = d(v) \otimes w + (-1)^{|v|} v \otimes d(w)$$

without risk of confusion. We use ∂ as generic notation for differentials in Hom-complexes $\mathrm{Hom}(V, W)$, so that

$$\partial(f) = d \circ f - (-1)^{|f|} f \circ d$$

for $f \in \mathrm{Hom}(V, W)$.

The dual of a chain complex V is denoted $V^\vee = \mathrm{Hom}(V, \mathbb{Q})$. The suspension sV is defined by $(sV)_i = V_{i-1}$ with differential $d(sv) = -sd(v)$ for $v \in V$.

2 Lecture 1: L_∞- and C_∞-Models for Spaces

The primary task of rational homotopy theory is to classify topological spaces up to rational equivalence.

Definition 2.1 A map $f\colon X \to Y$ between topological spaces is called a *rational equivalence* if the induced map in rational homology,

$$f_*\colon \mathrm{H}_*(X; \mathbb{Q}) \longrightarrow \mathrm{H}_*(Y; \mathbb{Q}),$$

is an isomorphism.

Rational equivalences need not have inverses, but rational equivalence generates an equivalence relation $\sim_{\mathbb{Q}}$ and we say that two spaces X and Y are *rationally equivalent*, or have the same *rational homotopy type*, if $X \sim_{\mathbb{Q}} Y$.

We will mainly focus on the simply connected case. If X and Y are simply connected, then $f: X \to Y$ is a rational equivalence if and only if the induced map on rational homotopy groups,

$$f_*: \pi_*(X) \otimes \mathbb{Q} \longrightarrow \pi_*(Y) \otimes \mathbb{Q},$$

is an isomorphism (see e.g. [FHT01, Theorem 8.6]). Evidently, the rational cohomology and homotopy groups are invariants of the rational homotopy type of a simply connected space. Somewhat surprisingly, a complete solution to the classification problem can be obtained by adding certain higher structure to these invariants. The aim of the first lecture is to explain the statements and proofs of the following two theorems.

Theorem 2.2 *For each simply connected space X of finite $\mathbb{Q}$-type, there exists a C_∞-algebra structure on the rational cohomology groups* $\mathrm{H}^*(X; \mathbb{Q})$ *that provides a complete invariant of the rational homotopy type, in the sense that X and Y are rationally equivalent if and only if the C_∞-structures on* $\mathrm{H}^*(X; \mathbb{Q})$ *and* $\mathrm{H}^*(Y; \mathbb{Q})$ *are C_∞-isomorphic.*

Theorem 2.3 *For each simply connected space X of finite $\mathbb{Q}$-type, there exists an L_∞-algebra structure on the shifted rational homotopy groups $\pi_{*+1}(X) \otimes \mathbb{Q}$ that provides a complete invariant of the rational homotopy type, in the sense that X and Y are rationally equivalent if and only if the L_∞-structures on $\pi_{*+1}(X) \otimes \mathbb{Q}$ and $\pi_{*+1}(Y) \otimes \mathbb{Q}$ are L_∞-isomorphic.*

Here, *finite $\mathbb{Q}$-type* means that $\mathrm{H}_k(X; \mathbb{Q})$ is finite dimensional for each k.

On one hand, both theorems are mere reformulations of results that have been known since the early days of rational homotopy theory; we will discuss the relation to Quillen's and Sullivan's theories in Sect. 2.10.

On the other hand, the perspective of higher structures has certain advantages and it has played a considerable role in recent developments of rational homotopy theory and its applications. The language of C_∞- and L_∞-algebras does not only shed new light on existing results, it also leads to new results that would be cumbersome to state and prove using only the classical theory.

Theorem 2.2 has been stated and proved in the language of C_∞-algebras by Kadeishvili [Kad09]. Statements along the lines of Theorem 2.3 can be found in various sources, for example [LM95, §1], but a treatment parallel to [Kad09] does not seem to have appeared in the literature.

2.1 A_∞-Algebras

A C_∞-algebra is a special kind of A_∞-algebra so we begin with the latter. A_∞-algebras were introduced by Stasheff [Sta63b, §2]. We will here only review a small part of the theory. For a less condensed introduction, we recommend [Kel01]. A detailed comprehensive treatment can be found in [LH03].

Definition 2.4 An A_∞*-algebra structure* on a chain complex (A, d) consists of a family of operations

$$m_n \colon A^{\otimes n} \longrightarrow A, \quad n = 2, 3, \ldots,$$

of degree $n - 2$ such that the equation

$$\sum_{r+s+t=n} (-1)^{r+st} m_{r+1+t} \circ (1^{\otimes r} \otimes m_s \otimes 1^{\otimes t}) = 0$$

holds for every $n \geq 2$, where the sum is over all positive integers r, s, t such that $r+s+t = n$, and where we set $m_1 = d$.

Note that for $n = 2$ the above equation says

$$d \circ m_2 - m_2 \circ (d \otimes 1 + 1 \otimes d) = 0.$$

In other words, m_2 is a chain map $A^{\otimes 2} \to A$. If we let ∂ denote the differential of the Hom-complex $\mathrm{Hom}(A^{\otimes k}, A)$, this equation can be rewritten as

$$\partial(m_2) = 0.$$

Similarly, for $n = 3$, the equation can be written as

$$\partial(m_3) = m_2 \circ (1 \otimes m_2) - m_2 \circ (m_2 \otimes 1). \tag{1}$$

This means that m_2 is associative up to homotopy, and that m_3 is a prescribed chain homotopy between the "associator" and the zero map. One can also think of (1) as the first step in a resolution of the associativity relation. From this perspective, the higher operations $m_4, m_5, \ldots$ take care of the syzygies in this resolution.

There is an alternative, more compact, definition of A_∞-algebras. Part of this is contained in [Sta63b, (2.4)]; the form we state can be found in [LH03, Lemma 1.2.2.1]. We write $\mathrm{T}^c(V)$ for the *tensor coalgebra* on a chain complex V, defined by

$$\mathrm{T}^c(V) = \bigoplus_{n \geq 0} V^{\otimes n}.$$

Using the "bar notation" $[v_1|\ldots|v_n] = v_1 \otimes \ldots \otimes v_n$ for $v_i \in V$, the coproduct $\Delta \colon \mathrm{T}^c(V) \to \mathrm{T}^c(V) \otimes \mathrm{T}^c(V)$ is given by

$$\Delta[v_1|\ldots|v_n] = \sum_{i=0}^{n} [v_1|\ldots|v_i] \otimes [v_{i+1}|\ldots|v_n],$$

and the counit $\eta \colon \mathbb{Q} \to \mathrm{T}^c(V)$ by the inclusion of $V^{\otimes 0} = \mathbb{Q}$. The direct sum decomposition yields an additional grading on $\mathrm{T}^c(V)$, which we call the *weight grading* and whose n-th graded piece we denote by $\mathrm{T}^c(V)^{(n)}$. We let $\mathrm{T}^c(V)^{(<n)}$ denote the sum of the components of weight $< n$. We let $\overline{\mathrm{T}}^c$ denote the sum of components in positive weight.

Proposition 2.5 *Let A be a chain complex. There is a bijection between A_∞-algebra structures on A and degree -1 self-maps b of $\mathrm{T}^c(sA)$ such that the following conditions are fulfilled:*

- *b is a coderivation, i.e. $\Delta b = (b \otimes 1 + 1 \otimes b)\Delta$.*
- *b is a perturbation of the differential d on $\mathrm{T}^c(sA)$, i.e. $(d+b)^2 = 0$.*
- *b decreases weight, i.e. $b\big(\mathrm{T}^c(sA)^{(n)}\big) \subseteq \mathrm{T}^c(sA)^{(<n)}$ for all n.*

Proof Sketch One notes that a coderivation $b\colon \mathrm{T}^c(sA) \to \mathrm{T}^c(sA)$ is uniquely determined by its linear part, i.e. the composition

$$\bigoplus_n (sA)^{\otimes n} = \mathrm{T}^c(sA) \xrightarrow{b} \mathrm{T}^c(sA) \xrightarrow{\mathrm{pr}} sA.$$

The components of this map correspond (up to shifts and signs) to the operations m_n. Under this correspondence, the A_∞-relations are equivalent to $(d+b)^2 = 0$. □

A differential graded associative algebra is the same thing as an A_∞-algebra A such that $m_n = 0$ for all $n \geq 3$. When A is a dg associative algebra, the dg coalgebra $\big(\mathrm{T}^c(sA), d+b\big)$ agrees with the classical bar construction $\mathrm{B}A$ (as defined in e.g. [FHT01, Chapter 19]). This justifies the following definition.

Definition 2.6 Let A be an A_∞-algebra. The *bar construction* of A is the dg coalgebra

$$\mathrm{B}A = \big(\mathrm{T}^c(sA), d+b\big),$$

where $d+b$ is as in Proposition 2.5.

2.2 C_∞-Algebras

C_∞-algebras go back at least to Kadeishvili [Kad88], who calls them "commutative $A(\infty)$-algebras". They are called "C_∞-algebras" by Getzler–Jones [GJ94, §5.3]. To define them we need the following auxiliary notion.

Definition 2.7 A (p,q)-*shuffle* is a permutation $\sigma \in \Sigma_{p+q}$ such that

$$\sigma(1) < \cdots < \sigma(p) \qquad \text{and} \qquad \sigma(p+1) < \cdots < \sigma(p+q)$$

hold. We denote the set of (p,q)-shuffles by $\mathrm{Sh}_{p,q}$ and we let $\tau_{p,q}$ denote the distinguished element in the group algebra $\mathbb{Q}\Sigma_{p+q}$ given by

$$\tau_{p,q} = \sum_{\sigma \in \mathrm{Sh}_{p,q}} \mathrm{sgn}(\sigma)\sigma.$$

The name "shuffle" is motivated by riffle shuffles of a deck of cards: it is cut into two parts which are subsequently mixed together without changing the order of the cards in either of the two parts. Note however that there is some inconsistency regarding the use of the term; what we call a shuffle is called an "unshuffle" in [LM95], but our usage agrees with that of e.g. [GJ94] and [LV12].

Definition 2.8 A C_∞-*algebra* is an A_∞-algebra A whose operations m_n fulfill

$$m_{p+q} \circ \tau_{p,q} = 0 \tag{2}$$

for all $p, q \geq 1$. Here $\tau_{p,q}$ acts on $A^{\otimes p+q}$ from the left by permuting the tensor factors.

Note that for $p = q = 1$ the above relation reads as

$$m_2 = m_2 \circ \tau$$

where τ is the non-trivial element of Σ_2. In particular m_2 equips A with the structure of a graded commutative (but *not* necessarily associative) binary operation and it descends to a graded commutative algebra structure on the homology of A. The higher operations m_n may be thought of as "resolving the associativity relation" of the commutative operation m_2, and (2) may be thought of as the appropriate commutativity constraint on these.

Proposition 2.5 associates to an A_∞-algebra A a differential $d + b$ on $\mathrm{B}A$. If A is a C_∞-algebra, this differential is a derivation with respect to the shuffle product $*$ of $\mathrm{B}A = \mathrm{T}^{\mathrm{c}}(\mathrm{s}A)$ (cf. [GJ94, Proposition 5.5]). In particular $d + b$ descends to a differential on the indecomposables with respect to the shuffle product,

$$\overline{\mathrm{T}}^{\mathrm{c}}(\mathrm{s}A)/(\overline{\mathrm{T}}^{\mathrm{c}}(\mathrm{s}A) * \overline{\mathrm{T}}^{\mathrm{c}}(\mathrm{s}A)).$$

The latter is isomorphic to the *cofree Lie coalgebra* $\mathbb{L}^{\mathrm{c}}(\mathrm{s}A)$ (see e.g. [LV12, Theorem 1.3.6]). This construction can also be reversed, yielding the following analog of Proposition 2.5 (cf. [LV12, Proposition 13.1.6]).

Proposition 2.9 *Let A be a chain complex. There is a bijection between C_∞-algebra structures on A and degree -1 self-maps b of $\mathbb{L}^{\mathrm{c}}(\mathrm{s}A)$ such that the following conditions are fulfilled:*

- *b is a coderivation.*
- *b is a perturbation of the differential d on $\mathbb{L}^{\mathrm{c}}(\mathrm{s}A)$, i.e. $(d + b)^2 = 0$.*
- *b decreases weight, i.e. $b\big(\mathbb{L}^{\mathrm{c}}(\mathrm{s}A)^{(w)}\big) \subseteq \mathbb{L}^{\mathrm{c}}(\mathrm{s}A)^{(<w)}$.*

Definition 2.10 Let A be a C_∞-algebra. We define the dg Lie coalgebra

$$\mathscr{L}_*(A) = \big(\mathbb{L}^{\mathrm{c}}(\mathrm{s}A), d + b\big)$$

where $d + b$ is as in Proposition 2.9. We write $\mathscr{L}^*(A) = \mathscr{L}_*(A)^\vee$ for the dual dg Lie algebra.

Remark 2.11 A commutative dg algebra (cdga) is the same thing as a C_∞-algebra A such that $m_n = 0$ for $n \geq 3$. In this case, $\mathscr{L}_*(A)$ agrees with the classical Harrison complex of [Har62] (see also [LV12, §13.1.7]). The dual $\mathscr{L}^*(A)$ agrees with the construction studied in [Tan83, I.1.(7)].

2.3 L_∞-Algebras

L_∞-algebras are a Lie algebra analog of A_∞-algebras. One of the earliest sources is Lada–Stasheff [LS93], where they are called "strongly homotopy Lie algebras".

Definition 2.12 An L_∞-*algebra* structure on a chain complex (L, d) is a family of operations

$$l_n : L^{\otimes n} \longrightarrow L, \quad n = 2, 3, \ldots$$

of degree $n - 2$ such that each l_n is anti-symmetric (i.e. invariant under the sign action of Σ_n) and such that

$$\partial(l_n) = \sum_{\substack{p,q \geq 2 \\ p+q=n+1}} \sum_{\sigma \in \mathrm{Sh}_{q,p-1}} \mathrm{sgn}(\sigma)(-1)^{p(q-1)} l_p \circ (l_q \otimes 1^{\otimes p-1}) \circ \sigma^{-1}$$

holds.[1] Here ∂ is the differential of the chain complex $\mathrm{Hom}(L^{\otimes n}, L)$.

Note that for $n = 2$ the relation above reads as

$$\partial(l_2) = 0$$

which is equivalent to l_2 being a map of chain complexes. Similarly, writing $[-, -] = l_2(-, -)$, we obtain for $n = 3$ that $\partial(l_3)(\alpha_1, \alpha_2, \alpha_3)$ is equal to

$$[[\alpha_1, \alpha_2], \alpha_3] - (-1)^{|\alpha_2||\alpha_3|}[[\alpha_1, \alpha_3], \alpha_2] + (-1)^{|\alpha_1|(|\alpha_2|+|\alpha_3|)}[[\alpha_2, \alpha_3], \alpha_1]$$

so that l_3 provides a witness for the fact that l_2 fulfills the graded Jacobi relation after taking homology. In particular l_2 equips the homology of L with a Lie algebra structure. This means that we can think of the operations l_n as "resolving the Jacobi relation".

Again there is an alternative compact definition of L_∞-algebras (cf. [LS93, §3]). We write $\Lambda^c(V)$ for the *symmetric coalgebra* on V, that is

$$\Lambda^c(V) = \bigoplus_{n \geq 0} (V^{\otimes n})_{\Sigma_n}$$

[1] There are different sign conventions in the literature. The convention used here agrees with [LS93] but differs from [LV12].

equipped with the canonical structure of a counital cocommutative coassociative coalgebra. The direct sum decomposition yields an additional grading on $\Lambda^c(V)$, which we call the *weight grading* and whose n-th graded piece we denote by $\Lambda^c(V)^{(n)}$.

Proposition 2.13 *Let L be a chain complex. There is a bijection between L_∞-algebra structures on L and degree -1 self-maps b of the cocommutative coalgebra $\Lambda^c(sL)$ such that the following conditions are fulfilled:*

- *b is a coderivation.*
- *b is a perturbation of the differential d on $\Lambda^c(sL)$, i.e. $(d+b)^2 = 0$.*
- *b decreases weight, i.e. $b\big(\Lambda^c(sL)^{(w)}\big) \subseteq \Lambda^c(sL)^{(<w)}$.*

Dg Lie algebras may be identified with L_∞-algebras L such that $l_n = 0$ for $n \geq 3$. For such L, the dg coalgebra $\big(\Lambda^c(sL), d+b\big)$ agrees with Quillen's generalization of the Chevalley–Eilenberg complex (as defined in [FHT01, Chapter 22(b)]). This justifies the following definition.

Definition 2.14 The *Chevalley–Eilenberg complex* of an L_∞-algebra L is the cocommutative dg coalgebra

$$\mathrm{C}_*(L) = \big(\Lambda^c(sL), d+b\big),$$

where $d+b$ is as in Proposition 2.13. We also write $\mathrm{C}^*(L) = \mathrm{C}_*(L)^\vee$ for the dual commutative dg algebra.

2.4 Unified Theory of ∞-Algebras via Algebraic Operads

The reader will have noticed certain patterns in the above discussion of different types of ∞-algebras. The theory of Koszul duality for operads, going back to [GK94, GJ94], provides a framework for a unified treatment. A comprehensive introduction can be found in [LV12] and in what follows we will use notation and terminology from this source. However, a reader who is unfamiliar with this theory and mainly interested the examples of A_∞-, C_∞- or L_∞-algebras can in principle read the remainder of the section simplistically by interpreting "$\mathcal{P}_\infty$" as a placeholder for A_∞, C_∞ or L_∞—the main point is that the argument follows the same pattern in each case.

For the unified treatment, we fix a dg operad $\mathcal{P}$ and a cofibrant resolution

$$\mathcal{P}_\infty \xrightarrow{\simeq} \mathcal{P}.$$

For convenience, we assume that $\mathcal{P}_\infty$ is of the form $\Omega\mathcal{C}$, i.e. that it is the cobar construction of some dg cooperad $\mathcal{C}$. Not every resolution is of this form, but resolutions of this form always exist. In general, one can take $\mathcal{C}$ to be the bar construction $\mathrm{B}\mathcal{P}$. If $\mathcal{P}$ is a Koszul operad, a smaller (in fact minimal) resolution is obtained by taking $\mathcal{C}$ to be the Koszul dual cooperad $\mathcal{P}^{¡}$.

The main examples are the operads

$$\mathcal{A}\mathrm{ss}, \quad \mathcal{C}\mathrm{om}, \quad \mathcal{L}\mathrm{ie},$$

governing associative, commutative and Lie algebras, respectively. These operads are Koszul and their Koszul dual cooperads may be identified with

$$\mathcal{A}\mathrm{ss}^{\text{i}} = (\mathbb{S}\mathcal{A}\mathrm{ss})^\vee, \quad \mathcal{C}\mathrm{om}^{\text{i}} = (\mathbb{S}\mathcal{L}\mathrm{ie})^\vee, \quad \mathcal{L}\mathrm{ie}^{\text{i}} = (\mathbb{S}\mathcal{C}\mathrm{om})^\vee,$$

respectively, where $\mathbb{S}$ denotes the operadic suspension and $(-)^\vee$ denotes linear dual. The resulting cofibrant resolutions are the dg operads

$$\mathcal{A}\mathrm{ss}_\infty = \Omega(\mathbb{S}\mathcal{A}\mathrm{ss})^\vee, \quad \mathcal{C}\mathrm{om}_\infty = \Omega(\mathbb{S}\mathcal{L}\mathrm{ie})^\vee, \quad \mathcal{L}\mathrm{ie}_\infty = \Omega(\mathbb{S}\mathcal{C}\mathrm{om})^\vee,$$

that govern A_∞-, C_∞-, and L_∞-algebras, respectively (cf. [LV12, §10.1.5 f. §13.1.8]).

We will now discuss a common generalization of the constructions T^{c}, $\mathbb{L}^{\mathrm{c}}$, and Λ^{c}. The *cofree conilpotent* $\mathcal{C}$-coalgebra on a chain complex A is the chain complex

$$\mathcal{C}[A] = \bigoplus_{n=0}^{\infty} \mathcal{C}(n) \otimes_{\Sigma_n} A^{\otimes n}.$$

The cooperad structure of $\mathcal{C}$ induces a natural $\mathcal{C}$-coalgebra structure on $\mathcal{C}[A]$. We call $\mathcal{C}[A]^{(w)} = \mathcal{C}(w) \otimes_{\Sigma_w} A^{\otimes w} \subseteq \mathcal{C}[A]$ the *weight w* part of $\mathcal{C}[A]$. The following can be found in [GJ94, Proposition 2.15].

Proposition 2.15 *Let A be a chain complex. There is a bijection between $\mathcal{P}_\infty$-algebra structures on A and degree -1 self-maps b of $\mathcal{C}[A]$ such that the following conditions are fulfilled:*

- *b is a coderivation of the $\mathcal{C}$-coalgebra $\mathcal{C}[A]$.*
- *b is a perturbation of the differential d on $\mathcal{C}[A]$, i.e. $(d+b)^2 = 0$.*
- *b decreases weight, i.e. $b\big(\mathcal{C}[A]^{(w)}\big) \subseteq \mathcal{C}[A]^{(<w)}$.*

Definition 2.16 Let A be a $\mathcal{P}_\infty$-algebra. The *bar construction* of A is the dg $\mathcal{C}$-coalgebra

$$\mathrm{B}_{\mathcal{P}} A = \big(\mathcal{C}[A], d + b\big),$$

where $d + b$ is as in Proposition 2.15.

Remark 2.17 Proposition 2.15 subsumes Propositions 2.5, 2.9, and 2.13. Furthermore, we recover the constructions discussed in the previous section, up to suspension. Indeed, if $\mathcal{P}$ is equal to $\mathcal{A}\mathrm{ss}$, $\mathcal{C}\mathrm{om}$, or $\mathcal{L}\mathrm{ie}$, and if $\mathcal{C} = \mathcal{P}^{\text{i}}$, then the suspension of the bar construction $\mathrm{B}_{\mathcal{P}} A$ may be identified with $\mathrm{B}A$, $\mathscr{L}_*(A)$, or $\mathrm{C}_*(A)$, respectively.

2.5 ∞-Morphisms

As in the previous section, we consider a dg operad $\mathcal{P}$ together with a resolution

$$\mathcal{P}_\infty \xrightarrow{\simeq} \mathcal{P}$$

of the form $\mathcal{P}_\infty = \Omega\mathcal{C}$ for a dg cooperad $\mathcal{C}$. We will assume that $\mathcal{C}$ is connected in the sense that $\mathcal{C}(0) = 0$ and $\mathcal{C}(1) = \mathbb{Q}$.

A morphism of $\mathcal{P}_\infty$-algebras $A \to A'$ is a morphism of chain complexes that commutes with the $\mathcal{P}_\infty$-algebra structure maps. By relaxing this requirement up to coherent homotopy, one arrives at the notion of ∞-morphisms, or strongly homotopy maps (sometimes referred to as "shmaps" [SH70]). The following definition can be compared to [LV12, §10.2.2].

Definition 2.18 Let A and A' be $\mathcal{P}_\infty$-algebras. A $\mathcal{P}_\infty$*-morphism*, or ∞*-morphism*, from A to A', written

$$f\colon A \rightsquigarrow A',$$

is a morphism $f\colon \mathrm{B}A \to \mathrm{B}A'$ of $\mathcal{C}$-coalgebras.

The linear part of an ∞-morphism f is the chain map

$$f_1\colon A \longrightarrow A'$$

given by extracting the weight one component of the map

$$f\colon \bigoplus_{n\geq 1} \mathcal{C}(n) \otimes_{\Sigma_n} A^{\otimes n} \longrightarrow \bigoplus_{n\geq 1} \mathcal{C}(n) \otimes_{\Sigma_n} (A')^{\otimes n}.$$

Since the coderivation b on BA decreases weight, one sees that f_1 is a chain map. We say that f is an ∞*-isomorphism* if f_1 is an isomorphism, and that f is a ∞*-quasi-isomorphism* if f_1 is a quasi-isomorphism.

It should be noted that the notion of ∞-isomorphism is much more flexible than the usual notion of an isomorphism. See Remark 3.5 below for an example that illustrates this point.

Remark 2.19 Note that the definition of ∞-morphisms makes sense, and is interesting, also in the case when A and A' are ordinary $\mathcal{P}$-algebras. For instance, the category **DASH** of associative dg algebras and strongly homotopy multiplicative maps has been studied in e.g. [GM74].

Lemma 2.20 *Assume that $\mathcal{C}$ is equipped with a coaugmention. Then two $\mathcal{P}$-algebras A and A' are ∞-quasi-isomorphic if and only if they are quasi-isomorphic.*

Proof The "if" direction is clear. For the "only if" direction it is enough to consider the case where there is a ∞-quasi-isomorphism $f: A \rightsquigarrow A'$, i.e. a quasi-isomorphism of $\mathcal{C}$-coalgebras $f: \mathrm{B}A \to \mathrm{B}A'$. The bar–cobar adjunction

$$\mathbf{Coalg}_{\mathcal{C}} \underset{\mathrm{B}}{\overset{\Omega}{\rightleftarrows}} \mathbf{Alg}_{\mathcal{P}} \quad (\Omega \dashv \mathrm{B})$$

yields a zig-zag of maps of $\mathcal{P}$-algebras

$$A \xleftarrow{\simeq} \Omega\mathrm{B}A \xrightarrow{\Omega f} \Omega\mathrm{B}A' \xrightarrow{\simeq} A'$$

where the counit maps are quasi-isomorphism (cf. [LV12, §11.3]). To see that Ωf is a quasi-isomorphism as well, we note that, since $\mathcal{C}$ is coaugmented, there is a commutative square

$$\begin{array}{ccc} \Omega\mathrm{B}A & \xrightarrow{\Omega f} & \Omega\mathrm{B}A' \\ \uparrow & & \uparrow \\ A & \underset{f_1}{\xrightarrow{\simeq}} & A' \end{array}$$

where f_1 is the linear part of f, such that each of the two vertical maps is a section of the respective counit map, and hence a quasi-isomorphism itself. □

2.6 Homotopy Transfer Theorem

A principal advantage that $\mathcal{P}_\infty$-algebras have over $\mathcal{P}$-algebras is that they are homotopy invariant. The basic principles for homotopy invariant algebraic structures in topology were laid down by Boardman–Vogt [BV73]. For an adaptation to the differential graded context, see [Mar04]. One of the defining properties of homotopy invariant algebraic structures is that they can be transferred along homotopy equivalences. This is the main content of the homotopy transfer theorem.

There is an abundance of different approaches to, and variants of, the homotopy transfer theorem for $\mathcal{P}_\infty$-algebras and we will not attempt to give a complete account of all of these here. Instead, we will give a short proof that uses homological perturbation theory, following [Ber14a]. A feature of this approach is that it works entirely with the compact description of $\mathcal{P}_\infty$-algebras as coderivation differentials. Moreover, it yields explicit formulas that are easier to work with than the tree formulas (as found in e.g. [LV12, §10.3]).

Homological perturbation theory is a tool that arose in the study of chain models for fibrations [Bro65, Gug72]. The idea of using homological perturbation theory to transfer algebraic structures encoded by (co)derivation differentials was successfully realized for

A_∞-algebras early on, see e.g. [GLS91, HK91]. The hurdles for realizing this idea for algebras over general operads, allowing the unified treatment presented here, were overcome in [Ber14a]. The formulation and proof we give here are taken from [Ber14a, Theorem 1.3].

Theorem 2.21 (Homotopy Transfer Theorem) *Let $\mathcal{C}$ be a dg cooperad which is connected in the sense that $\mathcal{C}(0) = 0$ and $\mathcal{C}(1) = \mathbb{Q}$ and let $\mathcal{P}_\infty = \Omega\mathcal{C}$. Let*

$$h \circlearrowright A \underset{g}{\overset{f}{\rightleftarrows}} B$$

be a contraction of chain complexes, i.e. f and g are degree 0 chain maps, h is a map of degree 1, and the following equations hold: $fg = 1$, $gf = 1 + \partial(h)$, $fh = 0$, $hg = 0$, and $h^2 = 0$. For every $\mathcal{P}_\infty$-algebra structure b on A, there is an induced $\mathcal{P}_\infty$-algebra structure b' on B and extensions of f and g to ∞-quasi-isomorphisms.

Proof The first step is to extend the given contraction to a contraction

$$H \circlearrowright \mathcal{C}[A] \underset{G}{\overset{F}{\rightleftarrows}} \mathcal{C}[B].$$

Here, $\mathcal{C}[A]$ and $\mathcal{C}[B]$ are equipped with the internal differentials (which are induced by the ones of $\mathcal{C}$, as well as A and B, respectively). The maps F and G are defined in the obvious way, $F = \mathcal{C}[f]$ and $G = \mathcal{C}[g]$. It is less obvious how to define H, but the following works: Define a self-map of $A^{\otimes n}$ by

$$h_n^\Sigma = \sum_{j=1}^{n} \sum_{\substack{\epsilon \in \{0,1\}^n \\ \epsilon_j = 0}} a_{n,|\epsilon|+1} \pi^{\epsilon_1} \otimes \ldots \otimes \pi^{\epsilon_{j-1}} \otimes h \otimes \pi^{\epsilon_{j+1}} \otimes \ldots \otimes \pi^{\epsilon_n},$$

where $\pi = gf$, $|\epsilon| = \epsilon_1 + \ldots + \epsilon_n$, and

$$a_{n,k} = \frac{1}{\binom{n}{k} k}.$$

Since h_n^Σ is symmetric, it induces a self-map of $\mathcal{C}(n) \otimes_{\Sigma_n} A^{\otimes n}$ and H is defined by taking the sum of h_n^Σ over all n. One can in principle check by hand that this yields a contraction, but for a more conceptual proof we refer to [Ber14a, §5].

Next, the "basic perturbation lemma" (cf. [Bro65] or [Gug72, §3]), where the perturbation b is the input, produces a new contraction,

$$H' \circlearrowright (\mathcal{C}[A], d + b) \underset{G'}{\overset{F'}{\rightleftarrows}} (\mathcal{C}[B], d + b'),$$

given by the explicit formulas

$$
\begin{aligned}
b' &= F\Sigma G, \\
F' &= F(1+\Sigma H), \\
H' &= H(1+\Sigma H), \\
G' &= (1+H\Sigma)G,
\end{aligned}
$$

where

$$\Sigma = \sum_{k=0}^{\infty} b(Hb)^k.$$

Note that the above series is finite when evaluated on an element since b decreases weight.

It is a non-obvious fact that b' is a coderivation and that F' and G' are morphisms of $\mathcal{C}$-coalgebras. Again, this can in principle be checked by hand, but a more conceptual explanation can be found in [Ber14a].

Lastly we note that in weight 1 the maps F' and G' are given by f and g, respectively, so that the former actually extend the latter. This also implies that F' and G' are ∞-quasi-isomorphisms. □

Remark 2.22 The coefficients $a_{n,k}$ in the formula for h_n^{Σ} are the entries in "Leibniz's harmonic triangle" (see Fig. 1): they can be defined by the recursive formulas

$$
\begin{aligned}
a_{n,1} &= \frac{1}{n}, \\
a_{n,k} &= a_{n-1,k-1} - a_{n,k-1},
\end{aligned}
$$

for $1 < k \leq n$. For example,

$h_1^{\Sigma} = h,$

$$
\begin{array}{ccccccccccccc}
 & & & & & & 1 & & & & & & \\
 & & & & & \frac{1}{2} & & \frac{1}{2} & & & & & \\
 & & & & \frac{1}{3} & & \frac{1}{6} & & \frac{1}{3} & & & & \\
 & & & \frac{1}{4} & & \frac{1}{12} & & \frac{1}{12} & & \frac{1}{4} & & & \\
 & & \frac{1}{5} & & \frac{1}{20} & & \frac{1}{30} & & \frac{1}{20} & & \frac{1}{5} & & \\
 & \frac{1}{6} & & \frac{1}{30} & & \frac{1}{60} & & \frac{1}{60} & & \frac{1}{30} & & \frac{1}{6} & \\
\frac{1}{7} & & \frac{1}{42} & & \frac{1}{105} & & \frac{1}{140} & & \frac{1}{105} & & \frac{1}{42} & & \frac{1}{7}
\end{array}
$$

Fig. 1 Leibniz's harmonic triangle

$$
\begin{aligned}
h_2^{\Sigma} &= \frac{1}{2}\big(h \otimes 1 + 1 \otimes h\big) + \frac{1}{2}\big(h \otimes \pi + \pi \otimes h\big), \\
h_3^{\Sigma} &= \frac{1}{3}\big(h \otimes 1 \otimes 1 + 1 \otimes h \otimes 1 + 1 \otimes 1 \otimes h\big) \\
&\quad + \frac{1}{6}\big(h \otimes \pi \otimes 1 + h \otimes 1 \otimes \pi + \pi \otimes h \otimes 1 + 1 \otimes h \otimes \pi + \pi \otimes 1 \otimes h + 1 \otimes \pi \otimes h\big) \\
&\quad + \frac{1}{3}\big(h \otimes \pi \otimes \pi + \pi \otimes h \otimes \pi + \pi \otimes \pi \otimes h\big).
\end{aligned}
$$

2.7 *Minimal* $\mathcal{P}_\infty$*-Algebra Model*

A $\mathcal{P}_\infty$-algebra is called *minimal* if its differential is trivial. The property that justifies the terminology is that an ∞-quasi-isomorphism between minimal $\mathcal{P}_\infty$-algebras is an ∞-isomorphism. Also, as we will discuss in Sect. 2.10 below, minimal L_∞-algebras correspond to minimal Sullivan models and minimal C_∞-algebras correspond to minimal Lie models. The minimality theorem for A_∞-algebras goes back Kadeishvili [Kad82]. The proof is an application of the homotopy transfer theorem.

Theorem 2.23 *Let A and A' be P-algebras.*

- *There is a $\mathcal{P}_\infty$-algebra structure on* $\mathrm{H}^*(A)$ *and ∞-quasi-isomorphisms from A to* $\mathrm{H}^*(A)$ *and vice versa.*
- *The $\mathcal{P}$-algebras A and A' are quasi-isomorphic if and only if* $\mathrm{H}^*(A)$ *and* $\mathrm{H}^*(A')$ *are ∞-isomorphic.*

Proof For the first part we use the homotopy transfer theorem: over a field, one can always find a contraction between a cochain complex A and its cohomology $\mathrm{H}^*(A)$ (this is well-known, see e.g. [Wei94, Exercise 1.4.4] or [BM20, Lemma B.1] and [Ber14a, Remark 2.1]).

For the second part we first prove the "only if" direction. To this end it is enough to consider the case where there is a quasi-isomorphism $f : A \to A'$. Combined with the ∞-quasi-isomorphisms from the first part, we obtain ∞-quasi-isomorphisms

$$\mathrm{H}^*(A) \overset{\simeq}{\rightsquigarrow} A \xrightarrow{f} A' \overset{\simeq}{\rightsquigarrow} \mathrm{H}^*(A')$$

the composite of which is an ∞-quasi-isomorphism from $\mathrm{H}^*(A)$ to $\mathrm{H}^*(A')$. Since the source and target are minimal, this is necessarily an ∞-isomorphism.

For the "if" direction we use that an ∞-isomorphism from $\mathrm{H}^*(A)$ to $\mathrm{H}^*(A')$, combined with the ∞-quasi-isomorphisms from the first part,

$$A \overset{\simeq}{\rightsquigarrow} \mathrm{H}^*(A) \overset{\cong}{\rightsquigarrow} \mathrm{H}^*(A') \overset{\simeq}{\rightsquigarrow} A'$$

yield an ∞-quasi-isomorphism $f : A \rightsquigarrow A'$. By Lemma 2.20, this implies that A and A' are quasi-isomorphic as desired. □

2.8 Sullivan's Rational Homotopy Theory

There are many good sources on Sullivan's approach to rational homotopy theory. In addition to the original [Sul77], comprehensive accounts can be found, for example, in any of [BG76, FHT01, Ber12, GM13]. We will here give a condensed summary pointing out the key facts needed for proving Theorems 2.2 and 2.3, referring to the literature for proofs and further details.

Definition 2.24 We denote by $\Omega_\bullet$ the simplicial cdga given by

$$\Omega_n = \frac{\Lambda\big(t_0, \ldots, t_n, dt_0, \ldots, dt_n\big)}{\big(t_0 + \cdots + t_n - 1, dt_0 + \cdots + dt_n\big)}$$

with t_k in degree 0 and dt_k in cohomological degree 1 (i.e. homological degree -1); here Λ denotes the free graded commutative algebra on the given generators. The differential d is, as the notation suggests, determined by $d(t_k) = dt_k$. For a morphism $\varphi\colon [m] \to [n]$ in the simplex category Δ, the cdga morphism $\varphi^*\colon \Omega_n \to \Omega_m$ is determined by

$$\varphi^*(t_i) = \sum_{j \in \varphi^{-1}(i)} t_j.$$

We can think of Ω_n as de Rham forms on an n-simplex that are "polynomial". We can extend this construction to a general simplicial set X by taking one copy of Ω_n for each n-simplex and gluing them together according to the simplicial structure. The resulting cdga, which we denote by $\Omega^*(X)$, is called *polynomial de Rham forms* on X.

More abstractly, we obtain from $\Omega_\bullet$, via the general nerve/realization construction, an adjunction

$$\mathbf{sSet} \underset{\langle - \rangle}{\overset{\Omega^*(-)}{\rightleftarrows}} \mathbf{CDGA}^{\mathrm{op}} \quad (\perp) \tag{3}$$

where the functors are explicitly given by

$$\Omega^*(X) = \operatorname{colim}_{\Delta^n \to X} \Omega_n$$
$$\langle A \rangle = \operatorname{Hom}_{\mathbf{CDGA}}(A, \Omega_\bullet)$$

We call $\langle A \rangle$ the *realization* of the cdga A.

Since $\Omega^*(X)$ is a simplicial analogue of the de Rham complex, one would expect it to compute the (rational) cohomology of X. This is the content of the following theorem; see [FHT01, Theorem 10.15] or [BG76, Theorem 2.2] for a proof.

Theorem 2.25 (Polynomial de Rham Theorem) *Let X be a simplicial set. Integration of polynomial differential forms defines a natural quasi-isomorphism of cochain complexes*

$$I\colon \Omega^*(X) \xrightarrow{\simeq} C^*(X; \mathbb{Q})$$

from polynomial de Rham forms to singular cochains. Explicitly,

$$I(\omega)(\sigma) = \int_{\Delta^n} \omega_\sigma,$$

for $\omega \in \Omega^*(X)$ *and* $\sigma\colon \Delta^n \to X$.

Remark 2.26 The quasi-isomorphism I in the preceding theorem is not a morphism of dg algebras, but [BG76, Proposition 3.3] shows that it extends to an A_∞-quasi-isomorphism. In particular, this implies that the induced map in cohomology,

$$\mathrm{H}^*(I)\colon\ \mathrm{H}^*(\Omega^*(X)) \to \mathrm{H}^*(X;\mathbb{Q}),$$

is an isomorphisms of algebras. By Theorem 2.23, this also implies that $\Omega^*(X)$ is quasi-isomorphic to $\mathrm{C}^*(X;\mathbb{Q})$ as an associative dg-algebra. For an alternative proof of this statement, see [FHT01, Corollary 10.10].

Remark 2.27 In his proof of the de Rham theorem, Dupont [Dup78, §2] constructs a contraction

$$s \circlearrowright \Omega^*(X) \underset{E}{\overset{I}{\rightleftarrows}} \mathrm{C}^*(X;\mathbb{Q}) \tag{4}$$

that is *natural* in X. As observed by Cheng–Getzler [CG08], an application of the homotopy transfer theorem for C_∞-algebras then shows that the cochain complex $\mathrm{C}^*(X;\mathbb{Q})$ carries a natural C_∞-algebra structure, which is naturally C_∞-quasi-isomorphic to $\Omega^*(X)$. The existence of such a C_∞-algebra structure on $\mathrm{C}^*(X;\mathbb{Q})$ has also been noticed by Sullivan using different methods, see the appendix of [TZ07].

Next, we review the definition of Sullivan algebras. These play the role of CW-complexes in the category of cdgas: they are cofibrant and every connected cdga can be replaced by a Sullivan algebra up to quasi-isomorphism.

Definition 2.28 A *Sullivan algebra* is a cdga of the form $(\Lambda V, d)$, where V is a graded vector space concentrated in positive cohomological degrees and the differential d satisfies the following *nilpotence condition*: there exists an exhaustive filtration

$$0 = F_{-1}V \subseteq F_0V \subseteq F_1V \subseteq \cdots \subseteq V$$

such that $d(F_pV) \subseteq \Lambda(F_{p-1}V)$ for all $p \geq 0$.

A Sullivan algebra $(\Lambda V, d)$ is called *minimal* if $d(V) \subseteq \Lambda^{\geq 2}V$. It is called *of finite type* if V (or, equivalently, ΛV) is of finite type.

Every quasi-isomorphism between minimal Sullivan algebras is an isomorphism (see [FHT01, Theorem 14.11]). This justifies the terminology "minimal". The reader may compare with minimal Kan complexes in simplicial homotopy theory (see e.g. [GJ99, Chapter I.10]) or minimal resolutions in homological algebra (see e.g. [Avr98, Proposition 1.1.2]).

Remark 2.29 Sullivan algebras are cofibrant. In particular this implies that, for any quasi-isomorphism $q\colon A \to B$ of cdgas and any map $f\colon (\Lambda V, d) \to B$ there exists a lift $\tilde{f}$ such that the diagram

$$\begin{array}{ccc} & & A \\ & \overset{\tilde{f}}{\nearrow} & \downarrow {\scriptstyle q}\ \simeq \\ (\Lambda V, d) & \xrightarrow{f} & B \end{array}$$

commutes up to homotopy, and this lift is unique up to homotopy (see e.g. [FHT01, Proposition 12.9]).

The following theorem states that (connected) spaces can always be modeled by minimal Sullivan algebras. A proof can be found, for example, in [FHT01, Corollary, p.191] or [BG76, Proposition 7.7].

Theorem 2.30 (Existence of Minimal Models) *Let X be a connected space. Then there exists a unique, up to non-canonical isomorphism, minimal Sullivan algebra $\mathcal{M}_X = (\Lambda V, d)$ that is quasi-isomorphic to $\Omega^*(X)$.*

Definition 2.31 The Sullivan algebra $\mathcal{M}_X$ of Theorem 2.30 is called the *minimal Sullivan model* of X.

Remark 2.32 If X is nilpotent and of finite $\mathbb{Q}$-type, then the minimal model $\mathcal{M}_X$ is of finite type, see [BG76, Theorem 10.1].

The next theorem states that the cohomology of the realization of a finite type Sullivan algebra agrees with the cohomology of the algebra. See [Sul77, Theorem 8.1], [FHT01, Theorem 17.10], or [BG76, Theorem 10.1].

Theorem 2.33 (Cohomology of the Spatial Realization) *Let $(\Lambda V, d)$ be a Sullivan algebra of finite type. Then the unit $(\Lambda V, d) \to \Omega^*\langle(\Lambda V, d)\rangle$ of the adjunction* (3) *is a quasi-isomorphism.*

Exercise 2.34 Using the facts stated above, show that two nilpotent spaces X and Y of finite $\mathbb{Q}$-type are rationally equivalent if and only if $\Omega^*(X)$ and $\Omega^*(Y)$ are quasi-isomorphic as cdgas.

Granted this, the proof of Theorem 2.2 is straightforward.

Proof of Theorem 2.2 By Exercise 2.34 two simply connected spaces X and Y are rationally equivalent if and only if $\Omega^*(X)$ and $\Omega^*(Y)$ are quasi-isomorphic as cdgas. By Theorems 2.23 and 2.25 the latter condition is equivalent to $\mathrm{H}^*(X;\mathbb{Q})$ and $\mathrm{H}^*(Y;\mathbb{Q})$ being C_∞-isomorphic. □

2.9 *Quillen's Rational Homotopy Theory*

Quillen [Qui69] defined a functor

$$\lambda\colon \mathbf{Top}_1 \longrightarrow \mathbf{DGL}_{\geq 1}$$

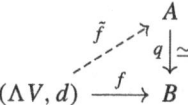

from the category of simply connected pointed topological spaces to the category of positively graded dg Lie algebras over $\mathbb{Q}$, and showed that it induces an equivalence of categories after formally inverting the rational equivalences in $\mathbf{Top}_1$ and the quasi-isomorphisms in $\mathbf{DGL}_{\geq 1}$. In particular, X and Y are rationally equivalent if and only if λX and λY are quasi-isomorphic. Another feature of Quillen's functor is the existence of a natural isomorphism

$$\mathrm{H}_*(\lambda X) \cong \pi_{*+1}(X) \otimes \mathbb{Q}.$$

Granted these facts, the proof of Theorem 2.3 is a simple application of the homotopy transfer theorem for L_∞-algebras.

Proof of Theorem 2.3 Applied to λX, the homotopy transfer theorem yields an L_∞-algebra structure on $\mathrm{H}_*(\lambda X) \cong \pi_{*+1}(X) \otimes \mathbb{Q}$. Next, by Quillen's theory, the spaces X and Y are rationally equivalent if and only if λX and λY are quasi-isomorphic, which happens if and only if $\pi_{*+1}(X) \otimes \mathbb{Q}$ and $\pi_{*+1}(Y) \otimes \mathbb{Q}$ are L_∞-isomorphic by Theorem 2.23. □

Counterparts of Sullivan's minimal models for dg Lie algebras were developed by Baues–Lemaire [BL77] and Neisendorfer [Nei78]. We will stick to the simply connected case here for simplicity, but we should mention that generalizations to non-nilpotent spaces have been developed recently by Buijs et al. [BFMT20].

A dg Lie algebra is called *minimal* if it is of the form $(\mathbb{L}V, \delta)$, where $\mathbb{L}V$ is the free Lie algebra on a positively graded vector space V and δ is decomposable in the sense that

$$\delta(V) \subseteq \mathbb{L}^{\geq 2}V.$$

Every positively graded dg Lie algebra admits a unique, up to isomorphism, minimal model, see e.g. [BL77, Theorem 2.3], [Nei78, Proposition 5.6(c)] or [FHT01, Theorem 22.13]. Also see [BFMT20, Theorem 3.19] for a generalization to non-negatively graded complete dg Lie algebras. Applied to Quillen's λX, this yields the following result.

Theorem 2.35 *For each simply connected space X, there is a unique, up to non-canonical isomorphism, minimal dg Lie algebra $\mathbb{L}_X$ such that*

$$\mathbb{L}_X \simeq \lambda X.$$

Definition 2.36 We will call the minimal dg Lie algebra $\mathbb{L}_X$ in the above theorem the *minimal Quillen model* of X.

2.10 Nothing New Under the Sun: ∞-Algebras vs. Minimal Models

As hinted in the introduction, the homotopy transfer and minimality theorems were known long before ∞-algebras became part of mainstream mathematics, but in a different language. Finite type nilpotent L_∞-algebras are essentially the same thing as Sullivan algebras of finite type. More formally we have the following statement, a proof of which can be found e.g. in [Ber15, Theorem 2.3]. Recall, from Sect. 2.7, that an L_∞-algebra is called minimal if its differential is trivial. We say that an L_∞-algebra L is *nilpotent* if its lower central series terminates degreewise, see Definition B.1.

Proposition 2.37 *The functor* $C^*(-)$ *restricts to an equivalence of categories*

$$\left\{ \begin{array}{c} \text{Nilpotent } L_\infty\text{-algebras } L \\ \text{of finite type} \end{array} \right\} \longrightarrow \left\{ \begin{array}{c} \text{Sullivan algebras } (\Lambda V, d) \\ \text{of finite type} \end{array} \right\},$$

where the morphisms are ∞*-morphisms of* L_∞*-algebras and morphisms of cdgas, respectively. Moreover, a nilpotent* L_∞*-algebra* L *is minimal if and only if the Sullivan algebra* $C^*(L)$ *is minimal.*

Remark 2.38 Under this dictionary, the minimality theorem for L_∞-algebras (Theorem 2.23) is nothing but the classical result that a Sullivan algebra has a minimal model, cf. [Sul77, Theorem 2.2] or [FHT01, Theorem 14.9].

A key feature of minimal Sullivan models is that the rational homotopy groups can be read off easily. This is the content of the following statement, a proof of which can be found in, for example, [Sul77, Theorem 10.1(i)] or [FHT01, Theorem 15.11]. For a proof using nerves of L_∞-algebras, see Remark B.7.

Proposition 2.39 *If* X *is a simply connected space of finite* $\mathbb{Q}$*-type with minimal Sullivan model*

$$\mathcal{M}_X = (\Lambda V, d) \simeq \Omega^*(X),$$

then V^k *is dual to* $\pi_k(X) \otimes \mathbb{Q}$ *for every* k*.*

The preceding two propositions together imply that the differential d of the minimal Sullivan model $\mathcal{M}_X$ corresponds to a minimal nilpotent L_∞-algebra structure on $L_X = \pi_{*+1}(X) \otimes \mathbb{Q}$ such that $C^*(L_X) \cong \mathcal{M}_X$. We can use this to give an alternative proof of Theorem 2.3.

Second Proof of Theorem 2.3 By Theorem 2.30 and Exercise 2.34, two simply connected spaces X and Y of finite $\mathbb{Q}$-type are rationally equivalent if and only if their minimal Sullivan models $\mathcal{M}_X$ and $\mathcal{M}_Y$ are isomorphic. By the preceding discussion this is equivalent to $\pi_{*+1}(X) \otimes \mathbb{Q}$ and $\pi_{*+1}(Y) \otimes \mathbb{Q}$ being L_∞-isomorphic. □

Remark 2.40 It is not a priori clear that the L_∞-structure on $\pi_{*+1}(X) \otimes \mathbb{Q}$ obtained from the minimal Sullivan model via the equivalence of categories in Proposition 2.37 is L_∞-isomorphic to the one obtained from Quillen's dg Lie algebra λX via the homotopy transfer theorem. The statement that they are is equivalent to a conjecture formulated by Baues–Lemaire [BL77, Conjecture 3.5], proved by Majewski [Maj00, Theorem 4.90]. See also [Ber24] for a short proof.

The following analog of Proposition 2.37 is a straightforward consequence of Proposition 2.9. We say that a non-unital C_∞-algebra is *simply connected* if it is concentrated in cohomological degrees ≥ 2.

Proposition 2.41 *The functor $\mathscr{L}^*(-)$ restricts to an equivalence of categories*

$$\left\{ \begin{array}{c} \text{Finite type, simply} \\ \text{connected } C_\infty\text{-algebras } A \end{array} \right\} \longrightarrow \left\{ \begin{array}{c} \text{Finite type dg Lie algebras} \\ (\mathbb{L}V, \delta) \text{ with } V = V_{\geq 1} \end{array} \right\},$$

where the morphisms are ∞-morphisms of C_∞-algebras and morphisms of dg Lie algebras, respectively. Moreover, a C_∞-algebra A is minimal if and only if the dg Lie algebra $\mathscr{L}^*(A)$ is minimal.

The following is dual to Proposition 2.39. For a proof, see [FHT01, Proposition 24.4].

Proposition 2.42 *If X is a simply connected space with minimal Quillen model*

$$\mathbb{L}_X = (\mathbb{L}V, \delta) \simeq \lambda X,$$

then $V_k \cong \widetilde{\mathrm{H}}_{k+1}(X; \mathbb{Q})$ for all k.

In particular, Proposition 2.42 together with Proposition 2.41 imply that the differential in the minimal Quillen model $\mathbb{L}_X$ of a simply connected space X of finite $\mathbb{Q}$-type corresponds to a minimal C_∞-algebra structure on the cohomology $\widetilde{\mathrm{H}}^*(X; \mathbb{Q})$ and, moreover, $\mathbb{L}_X$ may be identified with the Harrison cochains $\mathscr{L}^*(\widetilde{\mathrm{H}}^*(X; \mathbb{Q}))$ on this C_∞-algebra. This leads to an alternative proof of Theorem 2.2.

Second Proof of Theorem 2.2 By Theorem 2.35, two simply connected spaces X and Y of finite $\mathbb{Q}$-type are rationally equivalent if and only if their minimal Quillen models $\mathbb{L}_X$ and $\mathbb{L}_Y$ are isomorphic. Since we may identify $\mathbb{L}_X$ with the construction $\mathscr{L}^*(\widetilde{\mathrm{H}}^*(X; \mathbb{Q}))$, this is equivalent to $\mathrm{H}^*(X; \mathbb{Q})$ and $\mathrm{H}^*(Y; \mathbb{Q})$ being C_∞-isomorphic. □

We can summarize the situation in a diagram

$$\begin{array}{ccc} \text{Sullivan model } (\Lambda V, d) & \xrightarrow{\text{HTT}} & C_\infty\text{-algebra } \widetilde{\mathrm{H}}^*(X; \mathbb{Q}) \\ {\scriptstyle C^*(-)}\uparrow & & \downarrow{\scriptstyle \mathscr{L}^*(-)} \\ L_\infty\text{-algebra } \pi_{*+1}(X) \otimes \mathbb{Q} & \xleftarrow{\text{HTT}} & \text{Quillen model } (\mathbb{L}(W), \delta) \end{array}$$

of different models of a simply connected space X.

3 Lecture 2: Koszul Algebras, Formality, and Coformality

The notion of formality has long been a central one in rational homotopy theory. A landmark paper featuring the notion is [DGMS75]. The notion of coformality, introduced in [NM78], has not been studied to the same extent, perhaps because it is not as clearly linked to geometric structures on the space, but examples of coformal spaces abound.

Koszul algebras were introduced by Priddy [Pri70] as a tool for computing cohomology algebras $\mathrm{Ext}^*_A(k,k)$ of augmented associative algebras $A \to k$.

In this lecture, we will review the interplay between the notions of formality, coformality and Koszul algebras that was elucidated in [Ber14b]. We will also adapt the notion of Koszul A_∞-algebras that was introduced in [BB20] to C_∞- and L_∞-algebras and establish analogs of the results obtained there.

3.1 Formality and Coformality

We begin by defining the titular notions.

Definition 3.1 Let X be a simply connected space of finite $\mathbb{Q}$-type.

- X is called *formal* if $\Omega^*(X)$ is quasi-isomorphic to $\mathrm{H}^*(X;\mathbb{Q})$ as a cdga.
- X is called *coformal* if λX is quasi-isomorphic to $\pi_{*+1}(X)\otimes\mathbb{Q}$ as a dg Lie algebra.

Example 3.2

(1) A compact Kähler manifold is formal by the work of Deligne et al. [DGMS75]. In particular, smooth complex projective varieties are examples of formal spaces.
(2) A theorem of Miller [Mil79] says that for all $k > 1$, every $(k-1)$-connected Poincaré duality space of dimension at most $4k-2$ is formal. In particular, all simply connected closed manifolds of dimension at most 6 are formal.
(3) A simply connected space X is called *rationally elliptic* if $\mathrm{H}^*(X;\mathbb{Q})$ and $\pi_*(X)\otimes\mathbb{Q}$ are finite dimensional. Elliptic spaces with positive Euler characteristic are formal. In fact, the cohomology ring of such a space has the form $\mathbb{Q}[x_1,\ldots,x_n]/(f_1,\ldots,f_n)$, where $x_1,\ldots,x_n$ are even dimensional classes and $f_1,\ldots,f_n$ is a regular sequence [FHT01, Proposition 32.16(ii)]. A space whose cohomology ring has this form is automatically formal, see [BG76, §16]. Examples of elliptic spaces with positive Euler characteristic include complex projective spaces and Grassmannians, or more generally homogeneous spaces G/H for H a closed subgroup of maximal rank in a connected compact Lie group G.
(4) The arguably simplest example of a non-formal Sullivan algebra is
$$\big(\Lambda(x,y,z),d\big),$$
where x and y are cocycles of odd degree k and z is a generator of degree $2k-1$ such that $dz = xy$. Formality is obstructed by the existence of non-trivial Massey products, e.g. $\langle [x],[x],[y]\rangle \neq 0$. This Sullivan algebra represents an elliptic Poincaré duality space of dimension $4k-1$ with Euler characteristic zero, so it shows that the bound in Miller's theorem is sharp, and it shows that the non-vanishing of the Euler characteristic is necessary in the above example. This space is however coformal. See Example 3.41 for a further discussion.
(5) Kähler manifolds are examples of symplectic manifolds, but symplectic manifolds need not be formal in general, see [BT00, FM08] and [FHT01, p.163].

Whether a space is (co)formal is reflected by certain algebraic properties of its various rational models. This is made precise by the following two propositions.

We say that the differential in a minimal Quillen model $(\mathbb{L}V, d)$ is quadratic if it satisfies $d(V) \subseteq \mathbb{L}^2 V$. Similarly, the differential of a minimal Sullivan model $(\Lambda V, d)$ is said to be quadratic if $d(V) \subseteq \Lambda^2 V$.

Proposition 3.3 *Let X be a simply connected space of finite $\mathbb{Q}$-type. Then the following are equivalent*

(1) The space X is formal.
(2) The C_∞-algebra $\mathrm{H}^*(X; \mathbb{Q})$ is C_∞-isomorphic to one with $m_n = 0$ for $n \neq 2$.
(3) The space X admits a minimal Quillen model with quadratic differential.

Proposition 3.4 *Let X be a simply connected space of finite $\mathbb{Q}$-type. Then the following are equivalent*

(1) The space X is coformal.
(2) The L_∞-algebra $\pi_{*+1}(X) \otimes \mathbb{Q}$ is L_∞-isomorphic to one with $\ell_n = 0$ for $n \neq 2$.
(3) The space X admits a minimal Sullivan model with quadratic differential.

Proof We prove Proposition 3.3. The proof of Proposition 3.4 is entirely analogous. An application of the second part of Theorem 2.23 to the cdgas $A = \Omega^*(X)$ and $A' = \mathrm{H}^*(X; \mathbb{Q})$ shows that X is formal if and only if $\mathrm{H}^*(X; \mathbb{Q})$, with the transferred C_∞-algebra structure, is C_∞-isomorphic to $\mathrm{H}^*(X; \mathbb{Q})$ with the trivial C_∞-algebra structure. For the equivalence with the last condition, one simply notes that under the equivalence of categories in Proposition 2.41, a C_∞-algebra A has $m_n = 0$ for $n \neq 2$ precisely when the differential of $\mathscr{L}^*(A)$ is quadratic. □

Remark 3.5 One should be careful when talking about "the" C_∞-structure on $\mathrm{H}^*(X; \mathbb{Q})$, because it is only defined up to C_∞-isomorphism, a notion which is much more flexible than the usual notion of isomorphism.

For example, consider the space

$$X = S^2 \vee \left(S^2 \times S^3\right).$$

It has minimal Quillen model

$$\left(\mathbb{L}(\alpha, \beta, \gamma, \xi), \delta\right),$$

where $|\alpha| = |\beta| = 1$, $|\gamma| = 2$, and $|\xi| = 4$. The differential δ is trivial on α, β, and γ, but

$$\delta(\xi) = [\alpha, \gamma].$$

Another minimal Quillen model is given by

$$\left(\mathbb{L}(\alpha, \beta, \gamma, \xi), \delta'\right),$$

where δ' agrees with δ except

$$\delta'(\xi) = [\alpha, \gamma] + [\alpha, [\alpha, \beta]].$$

An isomorphism $\varphi\colon (\mathbb{L}, \delta) \to (\mathbb{L}, \delta')$ is determined by

$$\varphi(\gamma) = \gamma + [\alpha, \beta]$$

and by letting φ act by the identity on the other generators.

Under the equivalence of Proposition 2.41, the differentials δ and δ' correspond to two different C_∞-structures m and m' on $\mathrm{H}^*(X; \mathbb{Q})$ and φ corresponds to a C_∞-isomorphism between them. In the first $m_n = 0$ for all $n \neq 2$, but the presence of the cubic term in $\delta'(\xi)$ means that $m'_3 \neq 0$.

Remark 3.6 As discussed in Remark 2.26, the dg-algebra $\Omega^*(X)$ is quasi-isomorphic to $\mathrm{C}^*(X; \mathbb{Q})$. In particular, if X is formal, then $\mathrm{C}^*(X; \mathbb{Q})$ is formal as an associative dg algebra. Since $\mathrm{C}^*(X; \mathbb{Q})$ is in general non-commutative, it is not obvious that the converse should hold, but this has been proved by Saleh [Sal17]. Similarly, Saleh shows that X is coformal if and only if the associative dg algebra $\mathrm{C}_*(\Omega X; \mathbb{Q})$ is formal. A generalization was obtained by Campos et al. [CPRNW19], who proved that two cdgas are quasi-isomorphic if and only if their underlying associative dg algebras are quasi-isomorphic.

3.2 *Koszul Algebras and Simultaneous Formality and Coformality*

In this section, we will examine how the notions of formality, coformality, and Koszul algebras interact, following [Ber14b].

In what follows, A will be a graded commutative non-unital algebra (e.g. $\widetilde{\mathrm{H}}^*(X; \mathbb{Q})$ for some space X) and L a graded Lie algebra (e.g. $\pi_{*+1}(X) \otimes \mathbb{Q}$).

Definition 3.7 A *weight grading* on a graded commutative non-unital algebra A is a decomposition

$$A = \bigoplus_{w=1}^{\infty} A^{(w)}$$

such that $A^{(p)} \cdot A^{(q)} \subseteq A^{(p+q)}$. We consider it to be a cohomological grading (note that it is, accordingly, written as a superscript).

A weight grading on A induces a weight grading on

$$\mathscr{L}_*(A) = \big(\mathbb{L}^{\mathrm{c}}(\mathrm{s}A), b\big)$$

by letting s have weight -1 (here b is as in Proposition 2.9; note that the d occurring there is trivial since A has no differential). We obtain a cochain complex

$$\mathscr{L}_*(A)^{(0)} \xrightarrow{b} \mathscr{L}_*(A)^{(1)} \xrightarrow{b} \dots \tag{5}$$

since b increases weight by 1.

Definition 3.8 A graded commutative non-unital algebra A is *Koszul* if it admits a weight grading such that the cochain complex (5) is exact, i.e. if $\mathrm{H}^*(\mathscr{L}_*(A)) = \ker b \cap \mathscr{L}_*(A)^{(0)}$.

Next, we will see that Koszul algebras admit presentations of the following special form.

Definition 3.9 A *quadratic presentation* of a graded commutative non-unital algebra A is a graded vector space V together with a surjective map of graded algebras

$$\pi : \Lambda V \longrightarrow A$$

such that its kernel $I = \ker \pi$ is generated as an ideal by the kernel R of the restriction $\Lambda^2 V \subseteq \Lambda V \to A$.

A graded commutative non-unital algebra A is *quadratic* if it has some quadratic presentation.

Remark 3.10 The quadratic presentation of a quadratic algebra A is unique up to isomorphism. More precisely we have that, given two quadratic presentations $p\colon \Lambda V \to A$ and $q\colon \Lambda W \to A$ of A, there exists an isomorphism $f\colon V \to W$ of graded vector spaces and an automorphism g of the algebra A such that $q \circ \Lambda f = g \circ p$.

To prove this, choose a lift $\tilde{p}\colon \Lambda V \to \Lambda W$ of p along q and check that its homogeneous part $f = \tilde{p}_0\colon V \to W$ is an isomorphism of graded vector spaces such that Λf descends to an algebra isomorphism g of A.

Remark 3.11 A quadratic presentation of A induces a weight grading on A by setting $A^{(w)}$ to be $\pi(\Lambda^w V)$.

The following proposition is a special case of [Ber14b, Theorem 2.11].

Proposition 3.12 *Let A be a graded commutative non-unital algebra. If A is Koszul, then A is quadratic.*

More explicitly, let $A = \bigoplus_{w=1}^{\infty} A^{(w)}$ be a weight grading such that the cochain complex (5) is exact. Then, writing $V = A^{(1)}$, the multiplication map $\Lambda V \to A$ is a quadratic presentation of A, i.e. we have two short exact sequences

$$0 \longrightarrow I \longrightarrow \Lambda V \longrightarrow A \longrightarrow 0$$

$$0 \longrightarrow R \longrightarrow \Lambda^2 V \longrightarrow A^{(2)} \longrightarrow 0 \tag{6}$$

such that I is the ideal generated by $R = I^{(2)}$.

We will now discuss the Koszul property for Lie algebras.

Definition 3.13 A *weight grading* on a graded Lie algebra L is a decomposition

$$L = \bigoplus_{w=1}^{\infty} L^{(w)}$$

such that $[A^{(p)}, A^{(q)}] \subseteq A^{(p+q)}$. We consider it to be a cohomological grading (note that it is, accordingly, written as a superscript).

A weight grading on L induces a weight grading on

$$\mathrm{C}_*(L) = \big(\Lambda^{\mathrm{c}}(\mathrm{s}L), b\big)$$

by letting s have weight -1 (here b is as in Proposition 2.13; note that the d occurring there is trivial since L has no differential). We obtain a cochain complex

$$\mathrm{C}_*(L)^{(0)} \xrightarrow{b} \mathrm{C}_*(L)^{(1)} \xrightarrow{b} \dots \tag{7}$$

since b increases weight by 1.

Definition 3.14 A graded Lie algebra L is *Koszul* if it has some weight grading such that the cochain complex (7) is exact, i.e. if $\mathrm{H}^*(\mathrm{C}_*(L)) = \ker b \cap \mathrm{C}_*(L)^{(0)}$.

Next, we will see that Koszul graded Lie algebras admit presentations of the following special form.

Definition 3.15 A *quadratic presentation* of a graded Lie algebra L is a graded vector space W together with a surjective map of graded Lie algebras

$$\pi : \mathbb{L}W \longrightarrow L$$

such the kernel $J = \ker \pi$ is generated by the kernel S of the restriction $\mathbb{L}^2 W \subseteq \mathbb{L}W \to L$.

A graded Lie algebra L is *quadratic* if it has some quadratic presentation.

Remark 3.16 As in Remark 3.10, the quadratic presentation of a quadratic Lie algebra L is unique up to isomorphism.

Remark 3.17 A quadratic presentation of L induces a weight grading on L by setting $L^{(w)}$ to be $\pi(\mathbb{L}^w W)$.

The following proposition is a special case of [Ber14b, Theorem 2.11].

Proposition 3.18 *Let L be a graded Lie algebra. If L is Koszul, then L is quadratic.*

More explicitly, let $L = \bigoplus_{w=1}^{\infty} L^{(w)}$ be a weight grading such that the cochain complex (7) *is exact. Then, writing $W = L^{(1)}$, the induced Lie algebra map $\mathbb{L}W \to L$*

is a quadratic presentation of L, i.e. we have two short exact sequences

$$0 \longrightarrow J \longrightarrow \mathbb{L}W \longrightarrow L \longrightarrow 0$$

$$0 \longrightarrow S \longrightarrow \mathbb{L}^2 W \longrightarrow L^{(2)} \longrightarrow 0$$

such that J is the Lie ideal generated by $S = J^{(2)}$.

We now discuss the Koszul dual of a quadratic commutative algebra or Lie algebra.

Definition 3.19 We say that a graded commutative non-unital algebra A and a graded Lie algebra L are *Koszul dual* if there exist some quadratic presentations $\Lambda V \to A$ and $\mathbb{L}W \to L$ such that there is a nondegenerate pairing of degree -1

$$\langle -, - \rangle : V \otimes W \longrightarrow \mathbb{Q}$$

such that under the induced degree -2 pairing $\Lambda^2 V \otimes \mathbb{L}^2 W \to \mathbb{Q}$ given by

$$\begin{aligned} \langle x \wedge y, [\alpha, \beta] \rangle =& (-1)^{|y||\alpha|+|x|+|\alpha|} \langle x, \alpha \rangle \langle y, \beta \rangle \\ & - (-1)^{|\alpha||\beta|+|y||\beta|+|x|+|\beta|} \langle x, \beta \rangle \langle y, \alpha \rangle \end{aligned} \tag{8}$$

we have $R = S^{\perp}$ with R and S as in Propositions 3.12 and 3.18, respectively.

When A and L are Koszul dual, we write $L^! = A$ and $A^! = L$.

The following is the more traditional way of introducing Koszul duals, which justifies the notations $L^!$ and $A^!$. It is a straightforward reformulation of the preceding definition (using Remarks 3.10 and 3.16).

Proposition 3.20 *Let A be a graded commutative non-unital algebra and L a graded Lie algebra. The following are equivalent:*

(1) A and L are Koszul dual.
(2) For any quadratic presentation $\pi : \Lambda V \to A$ of A we have $L \cong \mathbb{L}(sV^\vee)/(S)$ where S is the orthogonal complement of $R = \ker(\pi|_{\Lambda^2 V})$ under the pairing (8) *associated to the pairing $V \otimes sV^\vee \to \mathbb{Q}$ of degree -1.*
(3) For any quadratic presentation $\pi : \mathbb{L}W \to L$ of L we have that $A \cong \Lambda(sW^\vee)/(R)$ where R is the orthogonal complement of $S = \ker(\pi|_{\mathbb{L}^2 W})$ under the pairing (8) *associated to the pairing $sW^\vee \otimes W \to \mathbb{Q}$ of degree -1.*

The following theorem is proven in [Ber14b, Theorem 1.2 and Theorem 1.3].

Theorem 3.21 (Berglund) *The following are equivalent for a simply connected space X:*

(1) X is formal and coformal.
(2) X is formal and $\widetilde{\mathrm{H}}^*(X; \mathbb{Q})$ is a Koszul algebra.
(3) X is coformal and $\pi_{*+1}(X) \otimes \mathbb{Q}$ is a Koszul Lie algebra.

If these conditions are satisfied, then $\widetilde{\mathrm{H}}^*(X;\mathbb{Q})$ *and* $\pi_{*+1}(X)\otimes\mathbb{Q}$ *are Koszul dual via the restriction of the Hurewicz pairing*

$$\begin{aligned} \widetilde{\mathrm{H}}^*(X;\mathbb{Q})\otimes\pi_{*+1}(X) &\longrightarrow \mathbb{Q} \\ x\otimes\alpha &\longmapsto \langle x, \mathrm{hur}(\alpha)\rangle \end{aligned}$$

to indecomposables. Here hur *denotes the Hurewicz homomorphism.*

We now give a number of examples to illustrate how Theorem 3.21 can be used to make computations or to deduce non-formality results. Many of these are taken from [Ber14b, §5].

Example 3.22 (*H*-Spaces) Let X be a simply connected H-space (for example the loop space ΩY of a 2-connected space Y) and assume X has finite $\mathbb{Q}$-type. Then the cohomology ring $\mathrm{H}^*(X;\mathbb{Q})$ is a free graded commutative algebra (cf. [FHT01, p.143]), and hence it admits a quadratic presentation

$$\mathrm{H}^*(X;\mathbb{Q})\cong\Lambda V,$$

with no non-trivial relations, for some graded vector space V. Free algebras are intrinsically formal and Koszul. By Theorem 3.21, X is also coformal, and the homotopy Lie algebra is isomorphic to the Koszul dual of the cohomology ring. Thus,

$$\pi_{*+1}(X)\otimes\mathbb{Q}\cong(\Lambda V)^!\cong\mathbb{L}W/\big(\mathbb{L}^2W\big)\cong W,$$

for $W=(sV)^\vee$. In particular, the Koszul duality between homotopy and cohomology in this case boils down to the statement that the indecomposables of the cohomology ring of an H-space may be identified with the dual of $\pi_*(X)\otimes\mathbb{Q}$.

Example 3.23 (Co-*H*-Spaces) Let X be a simply connected co-H-space, e.g. the suspension ΣY of a connected space Y, and assume X has finite $\mathbb{Q}$-type. By [Ber61], X is rationally equivalent to a wedge of spheres. In particular, X formal and coformal. Moreover, since all cup products in the reduced cohomology of a co-H-space are trivial, the cohomology ring admits the quadratic presentation

$$\mathrm{H}^*(X;\mathbb{Q})\cong\Lambda V/(\Lambda^2V)$$

where $V=\widetilde{\mathrm{H}}^*(X;\mathbb{Q})$. This is a Koszul algebra. The homotopy Lie algebra is isomorphic to the Koszul dual Lie algebra, which is free in this case;

$$\pi_{*+1}(X)\otimes\mathbb{Q}\cong\Big(\Lambda V/(\Lambda^2V)\Big)^!\cong\mathbb{L}W$$

where $W=(sV)^\vee\cong\widetilde{\mathrm{H}}_{*+1}(X;\mathbb{Q})$.

Example 3.24 (Highly Connected Manifolds) Let $n\geq 2$ and suppose that M is an $(n-1)$-connected closed oriented manifold of dimension $\leq 3n-2$. Neisendorfer–

Miller [NM78] observed that such manifolds are both formal and coformal provided $\dim \mathrm{H}^*(M; \mathbb{Q}) \geq 4$. A description of the homotopy Lie algebra was stated in [Nei79, §5]. We will here revisit these results from the point of view of higher structures, following [BB17, §4.1].

Denote by m_k the operations of the C_∞-algebra structure on $\mathrm{H}^*(M; \mathbb{Q})$. Suppose that $x_1, \ldots, x_k$ are non-trivial cohomology classes of positive degree. By the connectivity assumption, we must have $|x_i| \geq n$ for all i. If $k \geq 3$, this implies

$$\begin{aligned} |m_k(x_1, \ldots, x_k)| &= |x_1| + \cdots + |x_k| + 2 - k \\ &\geq nk + 2 - k \\ &\geq 3(n-1) + 2 = 3n - 1, \end{aligned}$$

so that $m_k = 0$ for $k \neq 2$. By Proposition 3.3, this implies that M is formal.

By Poincaré duality the cohomology of M is of the form $\mathbb{Q}1 \oplus V \oplus \mathbb{Q}z$ where V consists of indecomposable cohomology classes and z is a fundamental cohomology class. When $\dim V > 1$, one can deduce that $\mathrm{H}^*(M; \mathbb{Q})$ is Koszul (cf. [BB17, Theorem 4.2]) and admits the quadratic presentation

$$\mathrm{H}^*(M; \mathbb{Q}) \cong \Lambda V/(R),$$

where R is the kernel of the map induced by the cup product pairing defined as $\mu(x \wedge y) = \langle x \cup y, [M] \rangle$:

$$0 \longrightarrow R \longrightarrow \Lambda^2 V \xrightarrow{\mu} \mathbb{Q} \longrightarrow 0.$$

Hence, by Theorem 3.21, M is coformal and the homotopy Lie algebra admits the presentation

$$\pi_{*+1}(M) \otimes \mathbb{Q} \cong \mathbb{L}(W)/(\omega),$$

where $W = (sV)^\vee$ and $\omega \in \mathbb{L}^2(W)$ is dual to μ.

Example 3.25 (Symplectic Manifolds) As pointed out above, symplectic manifolds need not be formal, but Lupton–Oprea [LO94, Corollary 2.7] show that every coformal simply connected compact symplectic manifold M is formal. Moreover, the cohomology ring $\mathrm{H}^*(M; \mathbb{Q})$ is generated in degree 2 in this case. This result also shows that non-formal symplectic manifolds are examples of spaces that are neither formal nor coformal.

Example 3.26 (Moduli Spaces of Genus 0 Curves) The Deligne–Mumford compactification $\overline{\mathcal{M}}_{0,n}$ of the moduli space of genus 0 curves with n marked points is known to be formal, and Dotsenko has shown that $\mathrm{H}^*(\overline{\mathcal{M}}_{0,n}; \mathbb{Q})$ is Koszul, see [Dot22].

Example 3.27 (Configuration Spaces) Consider the configuration space of n (ordered) points in $\mathbb{R}^m$,

$$\mathrm{F}(\mathbb{R}^m, n) = \left\{ (x_1, \ldots, x_n) \in (\mathbb{R}^m)^n \mid x_i \neq x_j \text{ when } i \neq j \right\}.$$

Let $a_{i,j} \in \mathrm{H}^{m-1}\big(\mathrm{F}(\mathbb{R}^m, n); \mathbb{Q}\big)$ be the class obtained by pulling back the fundamental cohomology class of S^{m-1} along the map

$$\xi_{i,j} \colon \mathrm{F}(\mathbb{R}^m, n) \longrightarrow \mathrm{S}^{m-1}$$

that sends a point $(x_1, \ldots, x_n)$ to the unit vector $\frac{x_i - x_j}{|x_i - x_j|}$. The classes $a_{i,j}$ for $1 \le i < j \le n$ generate the cohomology ring, and the kernel I of the map

$$\Lambda(a_{i,j}) \longrightarrow \mathrm{H}^*\left(\mathrm{F}(\mathbb{R}^m, n); \mathbb{Q}\right)$$

is generated by the so-called *Arnold relations*:

$$\begin{aligned} a_{i,j}^2 &= 0, \\ a_{i,j}a_{j,k} + a_{j,k}a_{k,i} + a_{k,i}a_{i,j} &= 0, \end{aligned}$$

for i, j, k distinct (here we set $a_{i,j} = (-1)^m a_{j,i}$ if $i > j$); this is due to Arnold [Arn69] for $m = 2$ and Cohen [Coh76, Theorem 11.7] in the general case. This algebra can be seen to be Koszul (cf. [Bez94, Theorem 1.2] in the case $m = 2$, or [Ber14b, Example 5.5]). Next, it is well-known that the space $\mathrm{F}(\mathbb{R}^m, n)$ is formal. For $m = 2$ this was observed in [Arn69]. For $m > 2$, this statement is contained in the much more general result [LV14, Theorem 1.2].

For $m > 2$, it follows that $\mathrm{F}(\mathbb{R}^m, n)$ is coformal and that its homotopy Lie algebra may be computed as the Koszul dual Lie algebra. One calculates that the Koszul dual is given by

$$\pi_{*+1}\big(\mathrm{F}(\mathbb{R}^m, n)\big) \otimes \mathbb{Q} \cong \mathbb{L}(\alpha_{i,j})/(S)$$

for certain classes $\alpha_{i,j}$ of degree $m - 2$ for $1 \le i < j \le n$ and (S) the ideal generated by

$$\begin{aligned} &[\alpha_{i,j}, \alpha_{k,l}], && \text{for distinct } i, j, k, l, \\ &[\alpha_{i,j}, \alpha_{i,k} + \alpha_{j,k}], && \text{for distinct } i, j, k. \end{aligned}$$

Here we interpret $\alpha_{j,i} = (-1)^m \alpha_{i,j}$ if $i < j$. This is the *Drinfeld–Kohno Lie algebra*. In other words, the Arnold relations are orthogonal to the Drinfeld–Kohno relations. A computation of the homotopy Lie algebra using different methods can be found in [CG02].

Remark 3.28 If X is a connected, not necessarily simply connected, space of finite $\mathbb{Q}$-type, then one can show that Theorem 3.21 remains true upon replacing $\pi_{*+1}(X) \otimes \mathbb{Q}$ by $\pi_{*+1}(\mathbb{Q}_\infty X)$, where $\mathbb{Q}_\infty X$ denotes the Bousfield–Kan $\mathbb{Q}$-completion of X (see Appendix C). Quillen's functor λX is only defined for simply connected spaces, but an appropriate notion of coformality is obtained by replacing λX in Definition 3.1 with the complete dg Lie algebra $\mathscr{L}^*(\widetilde{\mathrm{H}}^*(X; \mathbb{Q}))$, the Harrison cochains on the cohomology C_∞-algebra. This agrees with the notion of coformality discussed in [BFMT20, §10.3].

A special case of Theorem 3.21 is the following result of Papadima–Yuzvinsky [PY99]. This was observed in [Ber14b, Corollary 1.9] in the nilpotent case, but by using Remark 3.28 one can upgrade this observation to cover the general case.

Theorem 3.29 (Papadima–Yuzvinsky) *Let X be a connected space of finite type. If X is formal, then X is a rational $K(\pi, 1)$-space if and only if the cohomology ring* $\mathrm{H}^*(X; \mathbb{Q})$ *is Koszul with weight equal to cohomological degree.*

Example 3.30 The space $\mathrm{F}(\mathbb{R}^2, n)$ is not simply connected. In fact, it is a model for the classifying space of the pure braid group on n strands. Formality and Koszulness of the cohomology ring then imply that the Bousfield–Kan $\mathbb{Q}$-completion of $\mathrm{F}(\mathbb{R}^2, n)$ is a rational $K(\pi, 1)$ and that its fundamental group, which may be identified with the Malcev completion of the pure braid group, is the Malcev group associated to the completion of the Drinfeld–Kohno Lie algebra. See Remark 3.28 and Theorem 3.29 as well as [FHT15, §8.6].

Remark 3.31 Salvatore [Sal20] has shown that $\mathrm{F}(\mathbb{R}^m, n)$ is intrinsically formal over any commutative ring provided $n \leq m$, but $\mathrm{F}(\mathbb{R}^2, n)$ is not formal over $\mathbb{F}_2$ when $n \geq 4$.

3.3 *Koszul C_∞- and L_∞-Algebras*

Berglund–Börjeson [BB20] extended the notion of Koszul algebras to A_∞-algebras and linked it to formality. The following is an adaptation of the results of [BB20] to C_∞- and L_∞-algebras. It leads to a generalization of the results of [Ber14b] discussed in the previous section.

Definition 3.32 A C_∞-algebra A is *Koszul* if it is equipped with a weight grading

$$A = \bigoplus_{w=1}^{\infty} A^{(w)}$$

such that the structure operations m_n are homogeneous of weight $2 - n$ and such that the cochain complex

$$\mathscr{L}_*(A)^{(0)} \xrightarrow{b} \mathscr{L}_*(A)^{(1)} \xrightarrow{b} \dots$$

is exact, where $\mathscr{L}_*(A) = (\mathbb{L}^c(sA), b)$ is weight graded by letting s have weight -1.

Similarly an L_∞-algebra L is *Koszul* if it is equipped with a weight grading

$$L = \bigoplus_{w=1}^{\infty} L^{(w)}$$

such that the structure operations l_n are homogeneous of weight $2 - n$ and such that the cochain complex

$$\mathrm{C}_*(L)^{(0)} \xrightarrow{b} \mathrm{C}_*(L)^{(1)} \xrightarrow{b} \ldots$$

is exact, where $\mathrm{C}_*(L) = (\Lambda^c(\mathrm{s}L), b)$ is weight graded by letting s have weight -1.

We consider these weight gradings to be cohomological (note that they are, accordingly, written as superscripts).

In contrast to the preceding subsection, here we allow A and L to have a non-trivial differential. Basic examples of Koszul L_∞-algebras are provided by free L_∞-algebras.

Example 3.33 Let W be a graded vector space. There is a weight grading on the free L_∞-algebra $\mathbb{L}_\infty(W)$ determined by letting l_n have weight $2 - n$ and putting W in weight 1. With this weight grading $\mathbb{L}_\infty(W)$ is Koszul.

Proposition 3.34 *Every minimal Koszul L_∞-algebra L is generated in weight* 1*, i.e., letting $W = L^{(1)}$, the canonical (strict) morphism of L_∞-algebras*

$$\mathbb{L}_\infty(W) \longrightarrow L$$

is surjective.

Proof By definition of Koszulness, there is a quasi-isomorphism $C \to \mathrm{C}_*(L)$ of dg coalgebras, where C denotes the weight zero homology of $\mathrm{C}_*(L)$. We then obtain weight homogeneous strict quasi-isomorphisms of L_∞-algebras

$$\mathscr{L}^*_\infty(C) \xrightarrow{\simeq} \mathscr{L}^*_\infty(\mathrm{C}_*(L)) \xrightarrow{\simeq} L.$$

Since L has trivial differential, the composite map must be surjective. Since $\mathscr{L}^*_\infty(C) = \mathbb{L}_\infty(s^{-1}C)$ is generated in weight 1, this shows that L is generated in weight 1. □

We now turn to the definition of the Koszul dual commutative algebra of a Koszul L_∞-algebra.

Lemma 3.35 *If* $\mathrm{s}W$ *is dual to* V*, then* $\mathbb{L}_\infty(W)^{(2)}$ *is dual to* $\mathrm{s}^2\Lambda V$.

Proof Note that $\mathbb{L}_\infty(W)^{(2)}$ is spanned by terms of the form $l_n(w_1, \ldots, w_n)$, so we may identify

$$\mathbb{L}_\infty(W)^{(2)} = \bigoplus_{n\geq 1} E(n) \otimes_{\Sigma_n} W^{\otimes n},$$

where $E(n)$ is the sign representation of Σ_n concentrated in degree $n - 2$. The right hand side can be rewritten as $s^{-2}\Lambda s W$, which is dual to $s^2\Lambda V$. □

Definition 3.36 Let L be a Koszul L_∞-algebra, let $W = L^{(1)}$ and consider the short exact sequence in weight 2

$$0 \longrightarrow S \longrightarrow \mathbb{L}_\infty(W)^{(2)} \longrightarrow L^{(2)} \longrightarrow 0$$

arising from the surjective map $\mathbb{L}_\infty(W) \to L$. We let V be dual to sW and denote by $S^\perp \subseteq \Lambda V$ the linear subspace obtained (up to shifts) as the orthogonal complement of S under the pairing of Lemma 3.35. We define the *Koszul dual algebra of* L to be the algebra

$$L^! = \Lambda V/(S^\perp)$$

i.e. the quotient of ΛV by the ideal generated by $S^\perp$.

The Koszul dual Lie algebra of a Koszul C_∞-algebra is defined similarly.

Definition 3.37 Let A be a Koszul C_∞-algebra and let $V = A^{(1)}$. The *Koszul dual Lie algebra* is defined by

$$A^! = \mathbb{L}(W)/(R^\perp),$$

where W is dual to sV and where $R^\perp$ is the orthogonal complement to the kernel R in the exact sequence

$$0 \longrightarrow R \longrightarrow \mathbb{C}_\infty(V)^{(2)} \longrightarrow A^{(2)} \longrightarrow 0$$

of graded vector spaces.

If L is a Koszul L_∞-algebra, then the Koszul dual commutative algebra $L^!$ is exactly the weight zero cohomology of weight graded cochain algebra $\mathrm{C}^*(L)$, and there is an exact sequence

$$\ldots \longrightarrow \mathrm{C}^*(L)_{(1)} \longrightarrow \mathrm{C}^*(L)_{(0)} \longrightarrow L^! \longrightarrow 0.$$

Similarly, if A is a Koszul C_∞-algebra, then the Koszul dual Lie algebra $A^!$ is the weight zero cohomology of $\mathscr{L}^*(A)$.

Theorem 3.38 *Let X be a simply connected space of finite $\mathbb{Q}$-type.*

(1) X is formal if and only if the L_∞-algebra $\pi_{*+1}(X) \otimes \mathbb{Q}$ is Koszul. In this case $\widetilde{\mathrm{H}}^*(X;\mathbb{Q})$ is the Koszul dual algebra.
(2) X is coformal if and only if the C_∞-algebra $\widetilde{\mathrm{H}}^*(X;\mathbb{Q})$ is Koszul. In this case $\pi_{*+1}(X) \otimes \mathbb{Q}$ is the Koszul dual Lie algebra.

Proof See the proof of [BB20, Theorem 2.9]. □

Remark 3.39 That $L = \pi_{*+1}(X) \otimes \mathbb{Q}$ is Koszul implies that the minimal model $\mathrm{C}^*(L)$ has an extra homological grading such that the chain complex

$$\ldots \longrightarrow \mathrm{C}^*(L)_{(1)} \longrightarrow \mathrm{C}^*(L)_{(0)} \longrightarrow \widetilde{\mathrm{H}}^*(X;\mathbb{Q}) \longrightarrow 0 \tag{9}$$

is exact. This recovers the "bigraded model" of Halperin–Stasheff [HS79].

Example 3.40 Complex projective space $\mathbb{C}\mathrm{P}^n$ is formal, for example since it is a compact Kähler manifold, but the cohomology algebra

$$\mathrm{H}^*(\mathbb{C}\mathrm{P}^n; \mathbb{Q}) \cong \mathbb{Q}[x]/(x^{n+1})$$

is not Koszul for $n > 1$, so $\mathbb{C}\mathrm{P}^n$ is not coformal by Theorem 3.21. The homotopy L_∞-algebra may be presented as

$$\pi_{*+1}(\mathbb{C}\mathrm{P}^n) \otimes \mathbb{Q} \cong \mathbb{Q}\langle \alpha, \beta \rangle,$$

where $|\alpha| = 1$, $|\beta| = 2n$ and the only non-trivial operation is given by

$$l_{n+1}(\alpha, \ldots, \alpha) = (n+1)!\beta.$$

This can be seen by observing that the minimal Sullivan model is given by

$$\big(\Lambda(x, y), d\big), \quad dx = 0, \ dy = x^{n+1}, \quad |x| = 2, \ |y| = 2n+1,$$

and then taking the corresponding L_∞-algebra as in Proposition 2.37. Alternatively, one can use the homotopy fiber sequence

$$\mathrm{S}^1 \longrightarrow \mathrm{S}^{2n+1} \longrightarrow \mathbb{C}\mathrm{P}^n$$

to argue that $\pi_*(\mathbb{C}\mathrm{P}^n) \otimes \mathbb{Q}$ is spanned by the homotopy classes of the evident maps

$$\alpha\colon \mathrm{S}^2 \cong \mathbb{C}\mathrm{P}^1 \longrightarrow \mathbb{C}\mathrm{P}^n,$$
$$\pi\colon \mathrm{S}^{2n+1} \longrightarrow \mathbb{C}\mathrm{P}^n,$$

and then observe, following Porter [Por67], that $(n+1)!\pi$ may be identified with the Whitehead product $\langle \alpha, \ldots, \alpha \rangle$ of order $n+1$. Using the connection between higher order Whitehead products and the L_∞-algebra structure on $\pi_{*+1}(X) \otimes \mathbb{Q}$ (see Proposition A.2), this implies the result.

The homotopy L_∞-algebra is Koszul if α and β are assigned weight 1 and 2, respectively. The sequence (9) assumes the form

$$0 \longrightarrow \mathbb{Q}[x]y \xrightarrow{d} \mathbb{Q}[x] \longrightarrow \mathbb{Q}[x]/(x^{n+1}) \longrightarrow 0$$

in this case.

The Koszul duality between the homotopy L_∞-algebra and the cohomology algebra, as expressed in Definition 3.36, takes the following form: in weight 2, we have the exact sequence

$$0 \longrightarrow S \longrightarrow \mathbb{L}_\infty(\alpha)^{(2)} \longrightarrow \mathbb{Q}\langle \beta \rangle \longrightarrow 0$$

where S is spanned by $l_k(\alpha, \ldots, \alpha)$ for all $k \geq 2$ except $k = n + 1$. Since we have that

$$\langle l_k(\alpha, \ldots, \alpha), x^m \rangle \neq 0$$

if and only if $k = m$, we see that $S^{\perp} = \mathbb{Q}\langle x^{n+1} \rangle$ and thus

$$\mathrm{H}^*(\mathbb{C}\mathrm{P}^n; \mathbb{Q}) \cong \left(\pi_{*+1}(\mathbb{C}\mathrm{P}^n) \otimes \mathbb{Q}\right)^! = \mathbb{Q}[x]/(x^{n+1}),$$

as expected.

Example 3.41 Let $n \geq 2$ and let

$$\mathrm{S}^{2n-1} \longrightarrow M \longrightarrow \mathrm{S}^n \times \mathrm{S}^n \tag{10}$$

be a smooth sphere bundle with non-trivial Euler class

$$0 \neq e \in \mathrm{H}^{2n}(\mathrm{S}^n \times \mathrm{S}^n; \mathbb{Q}).$$

For example, we can take M to be the sphere bundle associated to the pullback $c^*(\mathrm{TS}^{2n})$ of the tangent bundle of S^{2n} along the collapse map

$$c \colon \mathrm{S}^n \times \mathrm{S}^n \longrightarrow \mathrm{S}^{2n}.$$

The total space M is an $(n-1)$-connected closed manifold of dimension $4n - 1$ and, as we will see, it is coformal but not formal.

The long exact sequence of rational homotopy groups associated to the fibration (10) is easily seen to split, which yields a decomposition

$$\pi_{*+1}(M) \otimes \mathbb{Q} \cong \left(\pi_{*+1}(\mathrm{S}^{2n-1}) \otimes \mathbb{Q}\right) \oplus \left(\pi_{*+1}(\mathrm{S}^n \times \mathrm{S}^n) \otimes \mathbb{Q}\right).$$

The left summand is one-dimensional, spanned by a class γ in degree $2n-2$, and the right summand is spanned by classes α and β in degree $n-1$ and, if n is even, by $[\alpha, \alpha]$ and $[\beta, \beta]$ in degree $2n-2$. There is no room for non-trivial L_∞-operations l_k for $k \geq 3$, so M is necessarily coformal. The only possible non-trivial Lie bracket not already accounted for is $[\alpha, \beta]$. Since it vanishes in $\pi_{*+1}(S^n \times S^n)$ we must have $[\alpha, \beta] = a\gamma$ for some $a \in \mathbb{Q}$ and one can check that $a \neq 0$ since the Euler class is non-trivial. This implies that the homotopy Lie algebra admits the following cubic presentation:

$$\pi_{*+1}(M) \otimes \mathbb{Q} \cong \mathbb{L}(\alpha, \beta)/\left([\alpha, [\alpha, \beta]], [\beta, [\alpha, \beta]]\right).$$

It is easily seen that this graded Lie algebra does not admit any quadratic presentation, so it cannot be Koszul. Since M is coformal, this implies that M is not formal by Theorem 3.21.

We obtain the minimal Sullivan model by applying Chevalley–Eilenberg cochains. For a suitable choice of generators, it assumes the form

$$\left(\Lambda(x, y, z), d\right), \quad dx = dy = 0, \ dz = xy,$$

where $|x| = |y| = n$ and $|z| = 2n - 1$, if n is odd. If n is even, it assumes the form

$$\big(\Lambda(x, y, s, t, z), d\big), \quad dx = dy = 0, \; ds = x^2, \; dt = y^2, \; dz = xy,$$

where $|x| = |y| = n$ and $|s| = |t| = |z| = 2n - 1$.

The cohomology C_∞-algebra of M is easily computed from the Sullivan model by using the homotopy transfer theorem (cf. [BB20, Corollary 4.12]). For a suitable choice of basis it is given by

$$\mathrm{H}^*(M; \mathbb{Q}) \cong \mathbb{Q}\langle x, y, u, v, w\rangle,$$

where $|x| = |y| = n$, $|u| = |v| = 3n - 1$, $|w| = 4n - 1$, and the only non-trivial C_∞-operations are given by

$$xv = yu = w,$$
$$m_3(x, x, y) = -m_3(y, x, x) = -u,$$
$$m_3(x, y, y) = -m_3(x, y, y) = v.$$

This is a Koszul C_∞-algebra if x, y are assigned weight 1, the classes u, v are assigned weight 2 and w is assigned weight 3.

4 Lecture 3: Graph Complexes and Automorphisms of Manifolds

In this lecture, we will discuss another type of higher structure in rational homotopy theory. It arises in the study of the rational cohomology of automorphisms of high dimensional manifolds and was discovered by Berglund–Madsen [BM20]. The higher structure in question is the homology of a certain graph complex in the sense of Kontsevich [Kon93, Kon94].

We begin by reviewing a few results, some classical and some more recent, on the rational homotopy theory of classifying spaces of homotopy automorphisms of simply connected CW-complexes. Then we review some basics about *modular operads*, which is an efficient tool for handling graph complexes. Finally, we discuss the results of Berglund–Madsen [BM20].

4.1 Classifying Spaces of Homotopy Automorphisms

Let X be a finite CW-complex and denote by

$$\mathrm{aut}(X)$$

the topological monoid of homotopy equivalences from X to itself, equipped with the compact-open topology. Let $\mathrm{Baut}(X)$ be its classifying space, defined, for example, using the geometric bar construction [May75]. This space classifies fibrations with fiber X. More precisely, for every CW-complex B, there is a natural bijection

$$[B, \mathrm{Baut}(X)] \cong \{\pi : E \to B \text{ fibration} \mid \pi^{-1}(b) \simeq X \text{ for all } b \in B\}/\sim$$

where the right-hand side denotes the set of fiber homotopy equivalence classes of fibrations over B whose fiber is homotopy equivalent to X, cf. [Sta63a, May75]. In particular, this means that the cohomology ring $\mathrm{H}^*(\mathrm{Baut}(X))$ may be identified with the ring of characteristic classes of fibrations with fiber X. It is a fundamental problem in homotopy theory to compute this ring.

To compute it rationally, one may attempt to use tools from rational homotopy theory. An issue is that the space $\mathrm{Baut}(X)$ is in general not nilpotent, so the standard methods are not directly applicable, but one can study the homotopy fiber sequence

$$\mathrm{Baut}(X)\langle 1\rangle \longrightarrow \mathrm{Baut}(X) \longrightarrow \mathrm{B}\mathscr{E}(X).$$

Here, the base is the classifying space of the discrete group

$$\mathscr{E}(X) = \pi_1(\mathrm{Baut}(X)) = \pi_0(\mathrm{aut}(X))$$

of homotopy classes of self–homotopy equivalences of the space X. The fiber $\mathrm{Baut}(X)\langle 1\rangle$ is the 1-connected cover of $\mathrm{Baut}(X)$, i.e. the unique (up to weak homotopy equivalence) simply connected space admitting a map to $\mathrm{Baut}(X)$ that induces an isomorphism on $\pi_k(-)$ for $k > 1$.[2] Since $\mathrm{Baut}(X)\langle 1\rangle$ is simply connected, the standard methods of rational homotopy theory are in principle applicable, and in fact tractable Lie models for this space can be written down (see Theorem 4.4 below).

The following result, due independently to Sullivan [Sul77] and Wilkerson [Wil76], puts strong constraints on the group $\mathscr{E}(X)$ when X is a simply connected finite CW-complex. This was one of the primary applications of Sullivan's rational homotopy theory.

Theorem 4.1 (Sullivan, Wilkerson) *If X is a simply connected finite CW-complex, then $\mathscr{E}(X)$ is an arithmetic group.*

Typical examples of arithmetic groups are the $\mathbb{Z}$-points in linear algebraic groups defined over $\mathbb{Q}$, such as $\mathrm{GL}_n(\mathbb{Z})$ or $\mathrm{SL}_n(\mathbb{Z})$. See [Ser79] for an introduction.

Example 4.2 It is an exercise[3] to check that

$$\mathscr{E}\left(\bigvee^n \mathrm{S}^k\right) \cong \begin{cases} \mathrm{Out}(\mathrm{F}_n), & k = 1, \\ \mathrm{GL}_n(\mathbb{Z}), & k > 1. \end{cases}$$

---

[2] The universal cover, when it exists, is a model for the 1-connected cover.

[3] In fact, this is [Hat02, §4.A, Exercises 3 and 4].

The group $\mathrm{Out}(\mathrm{F}_n)$ of outer automorphisms of a free group on n generators is finitely presented, but it is not arithmetic when $n \geq 3$ (cf. [Vog02, §2.8.1]).

Remark 4.3 Arithmetic groups Γ are known to be of *finite type* in the sense that $\mathrm{B}\Gamma$ has the homotopy type of a CW-complex with finitely many cells in each dimension. Groups of finite type are in particular finitely presented.

Dror–Dwyer–Kan have shown that $\mathscr{E}(X)$ is of finite type whenever X is virtually nilpotent [DDK81] (they do not establish the stronger property of arithmeticity in this generality). If X is not virtually nilpotent, then $\mathscr{E}(X)$ is not necessarily of finite type. An example where $\mathscr{E}(X)$ is not even finitely generated is given by

$$X = S^1 \vee S^2 \vee S^3,$$

see [FK77].

Rational models for the simply connected cover of $\mathrm{Baut}(X)$ are known. We here review a version for $\mathrm{Baut}_*(X)$, where $\mathrm{aut}_*(X)$ denotes the topological monoid of pointed self–homotopy equivalences of X.

For a dg Lie algebra L with differential δ, we let $\mathrm{Der}(L)$ denote the dg Lie algebra of derivations of L. Its elements of degree n are maps $\theta\colon L \to L$ of degree n that satisfy

$$\theta[x, y] = [\theta(x), y] + (-1)^{n|x|}[x, \theta(y)]$$

for all $x, y \in L$. The commutator

$$[\theta, \eta] = \theta \circ \eta - (-1)^{|\theta||\eta|}\eta \circ \theta$$

makes $\mathrm{Der}(L)$ into a graded Lie algebra, and equipped with the differential $[\delta, -]$, it becomes a dg Lie algebra. The truncation $\mathrm{Der}(L)\langle 1\rangle$ is defined as the subalgebra which is zero in non-positive degrees, agrees with $\mathrm{Der}(L)$ in degrees > 1 and consists of the derivations θ such that $[\delta, \theta] = 0$ in degree 1.

The following is due to Tanré [Tan83, Corollaire VII.4.(4)] and Schlessinger–Stasheff [SS12].

Theorem 4.4 *Let X be a simply connected finite CW-complex. The simply connected cover* $\mathrm{Baut}_*(X)\langle 1\rangle$ *has Lie model* $\mathrm{Der}(\mathbb{L}_X)\langle 1\rangle$, *where* $\mathbb{L}_X$ *is the minimal Quillen model of* X.

Let us briefly mention some recent generalizations and extensions of this result:

- Berglund–Saleh [BS20] construct Lie models for $\mathrm{Baut}_A(X)\langle 1\rangle$, where $\mathrm{aut}_A(X)$ is the topological monoid of homotopy automorphisms of X that restrict to the identity on a given non-empty subcomplex $A \subseteq X$.
- Félix et al. [FFM22] construct Lie models for nilpotent covers $\mathrm{Baut}_G(X) \to \mathrm{Baut}(X)$ associated to subgroups $G \subseteq \pi_0(\mathrm{aut}_*(X))$ that act nilpotently on $\mathrm{H}_*(X)$ (that such covers are nilpotent is a result due to Dror–Zabrodsky [DZ79]).

- Berglund–Zeman [BZ22] incorporate the action of the deck transformation group Γ into Lie models $\mathfrak{g}$ for certain nilpotent covers of the form $\mathrm{Baut}_G(X)$ and use this to express the rational homotopy type of $\mathrm{Baut}(X)$ as a homotopy orbit space $\mathrm{MC}_\bullet(\mathfrak{g})_{h\Gamma}$.

Next, let M be a simply connected compact m-dimensional manifold with boundary $\partial M = \mathrm{S}^{m-1}$. We will let

$$\mathrm{aut}_\partial(M)$$

denote the topological monoid of self–homotopy equivalences of M that restrict to the identity on the boundary. Fix a minimal Quillen model $\mathbb{L}_M$ for M. There is a distinguished cycle $\omega \in \mathbb{L}_M$ of degree $m-2$ that represents the inclusion $\mathrm{S}^{m-1} = \partial M \to M$ under the isomorphism

$$\pi_{m-1}(M) \otimes \mathbb{Q} \cong \mathrm{H}_{m-2}(\mathbb{L}_M).$$

Let $\mathrm{Der}_\omega(\mathbb{L}_M)$ denote the dg Lie algebra of derivations $\theta\colon \mathbb{L}_M \to \mathbb{L}_M$ such that $\theta(\omega) = 0$. Note that $\mathrm{Der}_\omega(\mathbb{L}_M)$ is closed under the differential $[\delta, -]$ since $\delta(\omega) = 0$. The following theorem is proven in [BM20, Theorem 3.12]. (It simplifies the model provided by Berglund–Saleh [BS20].)

Theorem 4.5 (Berglund–Madsen) *Let M be a simply connected compact m-manifold with boundary $\partial M = S^{m-1}$. The simply connected cover* $\mathrm{Baut}_\partial(M)\langle 1\rangle$ *has Lie model* $\mathrm{Der}_\omega(\mathbb{L}_M)\langle 1\rangle$.

Example 4.6 Consider the $2d$-dimensional manifold

$$W_g = \#_g(\mathrm{S}^d \times \mathrm{S}^d)$$

and let $W_{g,1}$ denote the result of removing the interior of an embedded disk $\mathrm{D}^{2g} \subset W_g$. For $d = 1$, the manifold $W_{g,1}$ is an orientable surface of genus g and one boundary component. Now assume that $d > 1$. Then $W_{g,1} \simeq \bigvee^{2g} \mathrm{S}^d$ has Lie model $\mathbb{L}V_g$ with trivial differential $\delta = 0$, where V_g is the graded vector space $s^{-1}\widetilde{\mathrm{H}}_*(W_{g,1})$, which is concentrated in degree $d-1$. For a suitable choice of basis $\alpha_1, \beta_1, \ldots, \alpha_g, \beta_g$ for V_g, the element ω assumes the form

$$\omega = [\alpha_1, \beta_1] + \cdots + [\alpha_g, \beta_g].$$

It follows that $\mathrm{Baut}_\partial(W_{g,1})\langle 1\rangle$ has Lie model $\mathrm{Der}_\omega(\mathbb{L}V_g)\langle 1\rangle$. The differential $[\delta, -]$ is trivial, so in particular $\mathrm{Baut}_\partial(W_{g,1})\langle 1\rangle$ is coformal.

4.2 Graph Complexes and Modular Operads

Graph complexes, invented by Kontsevich [Kon93], were a major impetus for the development of the theory of algebraic operads and Koszul duality for operads. A spectacular application is that the cohomology of various groups of central importance to geometric topology, such as mapping class groups of surfaces or automorphism groups of free groups, can be expressed in terms of graph homology (see Theorem 4.12 below for a precise statement of one such result). However, the computation of graph homology is a difficult problem the complete solution of which remains an open problem.

Modular operads and the Feynman transform of Getzler–Kapranov [GK98] provide a convenient framework for handling Kontsevich's graph complexes and generalizations thereof. The Feynman transform may be thought of as a (dualized) bar construction for modular operads. To define modular operads, we first need to review cyclic operads [GK95].

Definition 4.7 A *cyclic operad* is an operad $\mathcal{P}$ together with an extension of the action of Σ_n on $\mathcal{P}(n)$ to an action of $\Sigma_{n+} = \mathrm{Bij}(\{0, \dots, n\})$ (one should think of this as also permuting the output with the inputs) such that

$$(a \circ_m b) \,.\, t_{m+n-1} = (a \,.\, t_m) \circ_1 (b \,.\, t_n)$$

holds for all $a \in \mathcal{P}(m)$ and $b \in \mathcal{P}(n)$. Here t_k denotes the permutation $(01 \dots k) \in \Sigma_{k+}$ and $c \circ_l d$ is the result of piping the output of d into the l-th input of c.

For a cyclic operad $\mathcal{P}$ and $n \geq 0$, we set $\mathcal{P}(\!(n+1)\!) = \mathcal{P}(n)$ as a Σ_{n+}-module.

Remark 4.8 The condition in the definition of a cyclic operad is equivalent to requiring composition along *un*rooted trees to be well-defined, analogously to operads and rooted trees.

Definition 4.9 A *modular operad* is a cyclic operad $\mathcal{P}$ equipped with a grading

$$\mathcal{P}(\!(n)\!) = \bigoplus_{g=0}^{\infty} \mathcal{P}(\!(g, n)\!)$$

(called the *genus* grading) and for each pair $i, j \in \{0, \dots, n\}$ with $i \neq j$ a contraction operation

$$\xi_{i,j} \colon \mathcal{P}(\!(g, n)\!) \longrightarrow \mathcal{P}(\!(g+1, n-2)\!)$$

(which one should think of as connecting the i-th and j-th in/output to each other) such that composition along arbitrary graphs is well-defined. An explicit list of axioms equivalent to this condition can be found in [GK98, Theorem 3.7].

Remark 4.10 It is customary to impose the so called "stability condition"

$$\mathcal{P}(\!(g, n)\!) \cong 0 \text{ when } 2g + n \leq 2$$

in the definition of a modular operad. We will not need this restriction however, and so we drop it from the definition.

Example 4.11 The operads $\mathcal{L}ie$ and C_∞ are cyclic operads. Any cyclic operad can be considered as a modular operad by putting everything in genus 0 and letting all contractions act trivially.

There is another way of producing a modular operad from C_∞ (or any other cyclic operad). Namely we let $\mathcal{F}$ be the modular operad freely generated by the cyclic operad C_∞. In genus 0 this has $\mathcal{F}(\!(0,n)\!) \cong C_\infty(\!(n)\!)$ and in higher genera it is freely generated from genus 0 under applications of the contraction operations. The modular operad $\mathcal{F}$ is the "modular envelope" of the cyclic operad C_∞ and isomorphic to the "Feynman transform" of the cyclic operad $\mathcal{L}ie$ considered as a modular operad, see [War19, Corollary 9.3].

The differential of C_∞ induces a differential ∂ on $\mathcal{F}$. For example we have

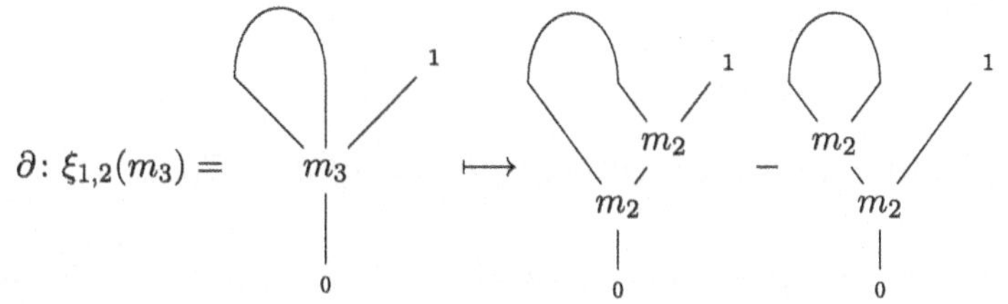

in $\mathcal{F}(\!(1,2)\!)$. Hence $\mathcal{F}(\!(g,n)\!)$ is a finite chain complex

$$\mathcal{F}(\!(g,n)\!)_{2g-3+n} \xrightarrow{\partial} \mathcal{F}(\!(g,n)\!)_{2g-4+n} \xrightarrow{\partial} \cdots \xrightarrow{\partial} \mathcal{F}(\!(g,n)\!)_0$$

(which degrees can be non-trivial is an easy combinatorial consequence of m_k having degree $k-2$). This is one example of a "graph complex". In genus 0 its homology is easy to describe, being given by

$$\mathrm{H}_k(\mathcal{F}(\!(0,n)\!)) \cong \begin{cases} \mathbb{Q}, & k=0 \\ 0, & k>0 \end{cases}$$

but in higher genera this has a very interesting, and complicated, structure which is not yet fully understood.

The homology of the graph complex $\mathcal{F}$ turns out to be related to the homology of certain automorphism groups of free groups. As discussed above (see Example 4.2), the group of homotopy classes of self–homotopy equivalences of a bouquet of g circles may be identified with the group of outer automorphisms of the free group on g generators,

$$\mathscr{E}\left(\bigvee^g S^1\right) \cong \mathrm{Out}(\mathrm{F}_g).$$

More generally, define $A_{g,n}$ to be the group of homotopy classes of self–homotopy equivalences of $\bigvee^g S^1$ relative to n marked points. Then

$$A_{g,0} \cong \mathrm{Out}(\mathrm{F}_g), \quad A_{g,1} \cong \mathrm{Aut}(\mathrm{F}_g), \quad A_{g,n} \cong \mathrm{Aut}(\mathrm{F}_g) \ltimes \mathrm{F}_g^{n-1}.$$

The following is due to Kontsevich [Kon93, Kon94] for $n = 0$ and Conant et al. [CKV13, Theorem 11.1] for $n > 0$.

Theorem 4.12 (Kontsevich, Conant–Kassabov–Vogtmann) *For all $g + n \geq 2$ and all k, there is an isomorphism*

$$\mathrm{H}_k(\mathcal{F}(\!(g,n)\!)) \cong \mathrm{H}_k(A_{g,n};\mathbb{Q}).$$

The homology of either side is still largely unknown. See [CKV13, CHKV16] for some computations and a discussion.

4.3 Graph Homology and Automorphisms of Manifolds

We will now discuss the results of Berglund–Madsen [BM20] that relate the cohomology of automorphisms of high dimensional manifolds to Kontsevich graph homology. Certain key parts of the proof [BM20] have been simplified by Berglund–Zeman [BZ22, §5.2], and the outline presented in this section will follow the latter.

Recall that

$$W_g = \#_g(\mathrm{S}^d \times \mathrm{S}^d)$$

and that $W_{g,1}$ denotes what is left after removing the interior of a small embedded disk $\mathrm{D}^{2d} \subset W_g$.

Let Γ_g be the group of automorphisms of $\mathrm{H}_d(W_g)$ that are induced by a self–homotopy equivalance of W_g. Under the identification of the group of automorphisms of the abelian group $\mathrm{H}_d(W_g) \cong \mathbb{Z}^{2g}$ with $\mathrm{GL}_{2g}(\mathbb{Z})$, one can check that

$$\Gamma_g = \begin{cases} \mathrm{O}_{g,g}(\mathbb{Z}), & d \text{ even}, \\ \mathrm{Sp}_{2g}(\mathbb{Z}), & d = 1,3,7, \\ \mathrm{Sp}_{2g}^q(\mathbb{Z}), & d \text{ odd}, d \neq 1,3,7, \end{cases}$$

where $\mathrm{Sp}_{2g}^q(\mathbb{Z}) \subseteq \mathrm{Sp}_{2g}(\mathbb{Z})$ is the finite-index subgroup of symplectic $(2g \times 2g)$-matrices

$$\begin{pmatrix} A & B \\ C & D \end{pmatrix}$$

such that the diagonal entries of the $(g \times g)$-matrices $C^t A$ and $D^t B$ are even. It can be shown that each element of Γ_g can be realized by a self–homotopy equivalence (in fact diffeomorphism) of $W_{g,1}$ that fixes the boundary pointwise, cf. [BM20, §5.2].

Next, one considers the homotopy fiber sequence

$$\mathrm{Btor}_\partial(W_{g,1}) \longrightarrow \mathrm{Baut}_\partial(W_{g,1}) \longrightarrow \mathrm{B}\Gamma_g, \tag{11}$$

where $\mathrm{tor}_\partial(W_{g,1}) \subset \mathrm{aut}_\partial(W_{g,1})$ denotes the "Torelli submonoid", i.e. the submonoid of self–homotopy equivalences that act trivially on $\mathrm{H}_d(W_g)$. The space $\mathrm{Btor}_\partial(W_{g,1})$ is nilpotent and rationally equivalent to $\mathrm{Baut}_\partial(W_{g,1})\langle 1\rangle$ and, as discussed in Example 4.6, the latter space has Lie model

$$\mathfrak{g}_g = \mathrm{Der}_\omega(\mathbb{L}V_g)\langle 1\rangle.$$

In particular, this means that there is an isomorphism of algebras

$$\mathrm{H}^*\big(\mathrm{Btor}_\partial(W_{g,1}); \mathbb{Q}\big) \cong \mathrm{H}^*_{CE}(\mathfrak{g}_g).$$

The spectral sequence of the homotopy fiber sequence (11) turns out to collapse at the E_2-page. This was proved stably (i.e. for g large compared to the cohomological degree) in [BM20], and later without any restrictions in [BZ22, Theorem 5.32]. This yields an isomorphism ([BZ22] even shows there is an isomorphism of algebras)

$$\mathrm{H}^*(\mathrm{Baut}_\partial(W_{g,1}); \mathbb{Q}) \cong \mathrm{H}^*\big(\Gamma_g; \mathrm{H}^*_{CE}(\mathfrak{g}_g)\big).$$

The action of Γ_g on the Chevalley–Eilenberg cohomology of $\mathfrak{g}_g = \mathrm{Der}_\omega(\mathbb{L}V_g)\langle 1\rangle$ is induced by the evident action on V_g. By using vanishing results for the cohomology of arithmetic groups due to Borel [Bor81], one can show that the right-hand side reduces to

$$\mathrm{H}^*(\Gamma_g; \mathbb{Q}) \otimes \mathrm{H}^*_{CE}(\mathfrak{g}_g)^{\Gamma_g}$$

in the stable range. In this range, the left factor $\mathrm{H}^*(\Gamma_g; \mathbb{Q})$ is well-understood; results of Borel [Bor74] show that it is a polynomial ring on certain classes x_i for $i \geq 1$.

The right factor $\mathrm{H}^*_{CE}(\mathfrak{g}_g)^{\Gamma_g}$ can be expressed in terms of graph homology, as we now will explain.

Forgetting the homological grading, the Lie algebra $\mathfrak{g}_g$ is identical to the positive part of the Lie algebra of symplectic derivations considered by Kontsevich [Kon93]. He showed that the Chevalley–Eilenberg cohomology of it is related the homology of the Lie graph complex $\mathcal{F}$. To account for the homological grading, we define a regraded version $\mathcal{F}^d$ of $\mathcal{F}$ by

$$\mathcal{F}^d\,(\!(g, n)\!) = s^{2d(1-g)}\mathcal{F}\,(\!(g, n)\!)\,.$$

For a graded vector space P, we let

$$\mathcal{F}^d[P] = \bigoplus_{g,n} \mathcal{F}^d \left(\!\left(g, n \right)\!\right) \otimes_{\Sigma_n} P^{\otimes n},$$

and note that

$$\mathcal{F}^d[0] = \bigoplus_{g} \mathcal{F}^d \left(\!\left(g, 0 \right)\!\right).$$

The following is essentially dual to [BM20, Theorem 9.1].

Theorem 4.13 *There is an isomorphism of chain complexes*

$$\lim_g C^*(\mathfrak{g}_g)^{\Gamma_g} \cong \Lambda \mathcal{F}^d[0]$$

where the right hand side denotes the free graded commutative algebra on the chain complex $\mathcal{F}^d[0]$.

The above (together with homological stability results and semisemplicity of the Γ_g-representation $C^*(\mathfrak{g}_g)$) lead to

Theorem 4.14 (Berglund–Madsen) *There is an isomorphism*

$$\lim_g \mathrm{H}^*(\mathrm{Baut}_\partial(W_{g,1}); \mathbb{Q}) \cong \mathbb{Q}[x_1, x_2, \ldots] \otimes \Lambda \mathrm{H}_*(\mathcal{F}^d)[0].$$

Remark 4.15 Stoll [Sto22] recently proved similar results where W_g is replaced by manifolds of the form

$$\#_g (\mathrm{S}^k \times \mathrm{S}^l),$$

for $3 \leq k < \ell \leq 2k - 2$. An interesting new phenomenon is that a twisted version of the Lie graph complex $\mathcal{F}$ appears in the odd dimensional case.

Berglund–Madsen also consider the block diffeomorphism group $\widetilde{\mathrm{Diff}}_\partial(W_{g,1})$. Interestingly, the full graph complex $\mathcal{F}$, and not just the "vacuum" part $\mathcal{F}[0]$, appears in the description of its stable cohomology. Considerations similar in outline to the above imply the following result, where $P \subset \mathrm{H}^*(BO; \mathbb{Q})$ denotes the graded vector space spanned by all Pontryagin classes p_i of degree $4i > d$.

Theorem 4.16 (Berglund–Madsen) *There is an isomorphism*

$$\mathrm{H}^*(\mathrm{B}\widetilde{\mathrm{Diff}}_\partial(W_{\infty,1}); \mathbb{Q}) \cong \mathbb{Q}[x_1, x_2, \ldots] \otimes \Lambda \mathrm{H}_*(\mathcal{F}^d)[P].$$

In view of the fact that the Lie graph complex $\mathcal{F}$ originated as a device to compute the homology of automorphism groups of free groups [Kon93, Kon94], it is surprising that it also appears in the cohomology of automorphisms of high dimensional manifolds. Whether a more direct connection between the homology of automorphism groups of free groups and the cohomology of automorphism of high dimensional manifolds can be found is an open problem.

Appendix A: Massey Products and Higher Order Whitehead Products

The C_∞-algebra structure on cohomology $\mathrm{H}^*(X; \mathbb{Q})$ and the L_∞-algebra structure on the rational homotopy groups $\pi_{*+1}(X) \otimes \mathbb{Q}$ are closely related to Massey products and higher order Whitehead products, respectively. The goal of this section is to make these statements more precise and collect some relevant references.

A.1 Massey Products

The discussion of Massey products does not require commutativity, so let us fix an associative dg algebra (A, d). The n-fold Massey product of $n \geq 2$ homology classes $x_1, \ldots, x_n \in \mathrm{H}_*(A)$ is a subset

$$\langle x_1, \ldots, x_n \rangle \subseteq \mathrm{H}_*(A)$$

that is defined whenever $\langle x_i, \ldots, x_j \rangle$ is defined and contains 0 for all $1 \leq i < j \leq n$ such that $(i, j) \neq (1, n)$. Assume this condition holds; then there exists a family of chains $a_{i,j} \in A$, indexed by all $(i, j) \neq (0, n)$ with $0 \leq i < j \leq n$, such that $a_{i-1,i}$ is a representative of x_i and such that

$$d(a_{i,j}) = \sum_{i<k<j} \overline{a}_{i,k} a_{k,j}$$

where $\overline{a} = (-1)^{|a|+1} a$. Now $\langle x_1, \ldots, x_n \rangle$ is defined to be the set of all homology classes associated to cycles of the form

$$\sum_{0<k<n} \overline{a}_{0,k} a_{k,n}$$

where $a_{i,j}$ is some family of chains as above. Note that $\langle x_1, x_2 \rangle$ is given by the singleton $\{(-1)^{|x_1|+1} x_1 x_2\}$.

The following proposition summarizes some relations between Massey products and the A_∞-algebra structure on $\mathrm{H}_*(A)$ obtained via the homotopy transfer theorem. When specialized to a cdga model A for a space X, this gives relations between the minimal C_∞-algebra structure on $H^*(X; \mathbb{Q})$ and Massey products. We refer the reader to [BMFM20] and [GM13, §18.3] for more details.

Proposition A.1 *Let* $x_1, \dots, x_n \in \mathrm{H}_*(A)$ *and set* $e = \sum_{i=1}^{n}(n-i)|x_i|$.

(1) Assume that $\langle x_1, \dots, x_n \rangle$ *is defined. Then for all* $x \in \langle x_1, \dots, x_n \rangle$ *we have* $[x] = [(-1)^e m_n(x_1, \dots, x_n)]$ *in the quotient* $\mathrm{H}_*(A)/\sum_{k<n} \mathrm{im}(m_k)$.
(2) Assume that $m_k = 0$ *for all* $k \leq n-2$ *and that* $\langle x_1, \dots, x_n \rangle$ *is defined. Then* $(-1)^e m_n(x_1, \dots, x_n) \in \langle x_1, \dots, x_n \rangle$.
(3) Each $x \in \langle x_1, \dots, x_n \rangle$ *can be realized as* $\epsilon m_n(x_1, \dots, x_n)$ *as above for some sign* $\epsilon \in \{\pm 1\}$, *after possibly passing to an* A_∞*-isomorphic* A_∞*-algebra structure on* $\mathrm{H}_*(A)$.

A.2 Lie–Massey Products and Higher Order Whitehead Products

The analogs of Massey products for dg Lie algebras are called *Lie–Massey products* and they can be defined as follows (cf. [Ret93]). For a dg Lie algebra (L, d), the n-fold Lie–Massey product of $n \geq 2$ homology classes $\alpha_1, \dots, \alpha_n \in \mathrm{H}_*(L)$ is a subset

$$\langle \alpha_1, \dots, \alpha_n \rangle \subseteq \mathrm{H}_*(L)$$

that is defined whenever $\langle \alpha_{i_1}, \dots, \alpha_{i_k} \rangle$ is defined and contains 0 for all proper subsets $\{i_1, \dots, i_k\} \subsetneq \{1, \dots, n\}$ of cardinality at least 2. Assume this condition holds; then there exists a family of chains $u_S \in L$, indexed by all proper subsets $S \subsetneq \{1, \dots, n\}$, such that $u_{\{i\}}$ is a representative of α_i and such that

$$d(u_S) = \sum (-1)^{e_{V,W}} [u_V, u_W]$$

where the sum is over all partitions $S = V \sqcup W$ such that the smallest element of S is contained in V. Here $e_{V,W}$ is defined to be $\sum_{v \in V}(|\alpha_v| + 1)$ plus the Koszul sign incurred by permuting $\bigotimes_{s \in S} s\alpha_s$ into $\bigotimes_{v \in V} s\alpha_v \otimes \bigotimes_{w \in W} s\alpha_w$. Now $\langle \alpha_1, \dots, \alpha_n \rangle$ is defined to be the set of all homology classes associated to cycles of the form

$$\sum (-1)^{e_{V,W}} [u_V, u_W]$$

where u_S is a family as above, and the sum is over all partitions $\{1, \dots, n\} = V \sqcup W$ with $1 \in V$. Note that $\langle \alpha_1, \alpha_2 \rangle$ is given by the one element set $\{(-1)^{|\alpha_1|+1}[\alpha_1, \alpha_2]\}$.

The following proposition summarizes some results that connect Lie–Massey brackets to the L_∞-algebra structure on $\mathrm{H}_*(L)$ obtained from the homotopy transfer theorem, see [BBMFM17]. Note that there is no analogue of the third statement of Proposition A.1.

Proposition A.2 *Let* $\alpha_1, \dots, \alpha_n \in \mathrm{H}_*(A)$ *and set* $e = \sum_{i=1}^{n}(n-i)|\alpha_i|$.

(1) Assume that $\langle \alpha_1, \dots, \alpha_n \rangle$ *is defined. Then, for all* $\alpha \in \langle \alpha_1, \dots, \alpha_n \rangle$, *we have* $[\alpha] = [(-1)^e l_n(\alpha_1, \dots, \alpha_n)]$ *in the quotient* $\mathrm{H}_*(L)/\sum_{k<n} \mathrm{im}(l_k)$.
(2) Assume that $l_k = 0$ *for all* $k \leq n-2$ *and that* $\langle \alpha_1, \dots, \alpha_n \rangle$ *is defined. Then* $(-1)^e l_n(\alpha_1, \dots, \alpha_n) \in \langle \alpha_1, \dots, \alpha_n \rangle$.

The topological significance of Lie–Massey products is that they are related to *higher order Whitehead products*. Higher order Whitehead products are certain partially defined operations on homotopy groups that were introduced in the 1960s, cf. [Har61, Por65]. Allday [All73, All77] showed that higher order Whitehead products in the rational homotopy groups of a simply connected space X can be computed in terms of Lie–Massey brackets in the homology of a dg Lie model for X. Andrews–Arkowitz [AA78] computed higher order Whitehead products in terms of Sullivan models. Under the dictionary in Proposition 2.37, this implicitly gives a relation between the higher order Whitehead products and the minimal L_∞-algebra structure on the rational homotopy groups. More explicit statements formulated directly in terms of L_∞-algebras can be found in [BBMFM17].

Appendix B: The Nerve of an L_∞-Algebra

In this appendix we will review the *Maurer–Cartan space*, or *nerve*, of an L_∞-algebra, and related constructions. The nerve of a positively graded L_∞-algebra of finite type is nothing but the spatial realization of the associated Sullivan algebra, but the advantage of the Maurer–Cartan space perspective becomes apparent when one works with unbounded L_∞-algebras and models for non-connected spaces such as mapping spaces. The nerve also provides a more direct link between spaces and dg Lie algebras than Quillen's theory [Qui69].

The nerve of a dg Lie algebra was introduced by Hinich [Hin97] and was further brought into the spotlight by Getzler [Get09]. The nerve $\mathrm{MC}_\bullet(\mathfrak{g})$, and Getzler's small model for it, $\gamma_\bullet(\mathfrak{g})$, have played an important role in recent developments of rational homotopy theory (see e.g. [BM13, Ber15, LM15, BFMT20, RNV20]) and we will briefly discuss some of these in this appendix.

Definition B.1 The *lower central series* of an L_∞-algebra $\mathfrak{g}$ is by definition the smallest filtration

$$\mathfrak{g} = \Gamma^1\mathfrak{g} \supseteq \Gamma^2\mathfrak{g} \supseteq \dots$$

that is compatible with the L_∞-structure in the sense that

$$\left[\Gamma^{i_1}\mathfrak{g}, \dots, \Gamma^{i_r}\mathfrak{g}\right] \subseteq \Gamma^{i_1+\dots+i_r}\mathfrak{g}$$

for all $i_1, \dots, i_r$.

We say that $\mathfrak{g}$ is *nilpotent* if for every n, there exists a k such that

$$\Gamma^k\mathfrak{g}_n = 0.$$

The *completion* of $\mathfrak{g}$ is the L_∞-algebra

$$\widehat{\mathfrak{g}} = \varprojlim \mathfrak{g}/\Gamma^k\mathfrak{g}$$

and we say that $\mathfrak{g}$ is *complete* if the canonical map

$$\mathfrak{g} \longrightarrow \widehat{\mathfrak{g}}$$

is an isomorphism.

Remark B.2 Clearly, every positively graded L_∞-algebra is nilpotent and every nilpotent L_∞-algebra is complete. Moreover, a finite type L_∞-algebra is complete if and only if it is nilpotent [Ber15, Proposition 5.2]. (Note that the notion of nilpotence considered here coincides with the notion of "degree-wise nilpotence" of [Ber15, Definition 2.1].) The definition of completeness given here is slightly more restrictive than the one given in [Ber15] in that we only consider the lower central series filtration.

If $\mathfrak{g}$ is a complete L_∞-algebra, then the series

$$\mathcal{F}(\tau) = \sum_{n\geq 0} \frac{1}{n!} \ell_n(\tau, \ldots, \tau)$$

converges for every $\tau \in \mathfrak{g}_{-1}$. If

$$\mathcal{F}(\tau) = 0,$$

then τ is called a *Maurer–Cartan element*. The set of Maurer–Cartan elements in $\mathfrak{g}$ is denoted $\mathrm{MC}(\mathfrak{g})$.

Definition B.3 If A is a cdga and $\mathfrak{g}$ is a complete L_∞-algebra, the tensor product of the underlying chain complexes $A \otimes \mathfrak{g}$ admits an L_∞-algebra structure where

$$[a_1 \otimes x_1, \ldots, a_n \otimes x_n] = \pm a_1 \ldots a_n \otimes [x_1, \ldots, x_n]$$

for all $n \geq 2$ and all $a_1, \ldots, a_n \in A$, $x_1, \ldots, x_n \in \mathfrak{g}$. The completion of this L_∞-algebra is called the *completed tensor product* and it is denoted

$$A \mathbin{\widehat{\otimes}} \mathfrak{g}.$$

This construction is functorial in A, so if we form the completed tensor product with the simplicial cdga $\Omega_\bullet$ (see Definition 2.24), we obtain a simplicial complete L_∞-algebra $\Omega_\bullet \mathbin{\widehat{\otimes}} \mathfrak{g}$.

Definition B.4 The *Maurer–Cartan space*, or *nerve*, of a complete L_∞-algebra $\mathfrak{g}$ is the simplicial set

$$\mathrm{MC}_\bullet(\mathfrak{g}) = \mathrm{MC}(\Omega_\bullet \mathbin{\widehat{\otimes}} \mathfrak{g}).$$

The vertices of the simplicial set $\mathrm{MC}_\bullet(\mathfrak{g})$ are the Maurer–Cartan elements of $\mathfrak{g}$ and the connected components are the so-called gauge equivalence classes of Maurer–Cartan elements,

$$\pi_0(\mathrm{MC}_\bullet(\mathfrak{g})) = \mathrm{MC}(\mathfrak{g})/\sim.$$

The higher homotopy groups can be computed in terms of the homology of twists of $\mathfrak{g}$ by Maurer–Cartan elements. For a Maurer–Cartan element $\tau \in \mathrm{MC}(\mathfrak{g})$, the twisted L_∞-algebra $\mathfrak{g}^\tau$ is the complete L_∞-algebra with the same underlying graded vector space as $\mathfrak{g}$ and structure maps ℓ_n^τ given by

$$\ell_n^\tau(x_1, \ldots, x_n) = \sum_{k\geq 0} \frac{1}{k!}\ell_{k+n}(\tau, \ldots, \tau, x_1, \ldots, x_n).$$

The following is [Ber15, Theorem 1.1].

Theorem B.5 *For every Maurer–Cartan element τ in $\mathfrak{g}$ and every $k \geq 0$, there is an isomorphism*

$$\pi_{k+1}(\mathrm{MC}_\bullet(\mathfrak{g}), \tau) \cong \mathrm{H}_k(\mathfrak{g}^\tau).$$

For $k > 0$, this is an isomorphism of abelian groups. For $k = 0$, this is an isomorphism of groups where the complete Lie algebra $\mathrm{H}_0(\mathfrak{g}^\tau)$ is given a group structure via the Baker–Campbell–Hausdorff formula.

Under the correspondence in Proposition 2.37, the Maurer–Cartan space corresponds to Sullivan's simplicial realization. The following is [Ber15, Proposition 6.1].

Proposition B.6 *Let L be a non-negatively graded nilpotent L_∞-algebra of finite type and let*

$$\mathrm{C}^*(L) = (\Lambda V, d)$$

be the corresponding Sullivan algebra. The Maurer–Cartan space of L is isomorphic to the simplicial realization of $(\Lambda V, d)$,

$$\mathrm{MC}_\bullet(L) \cong \langle \Lambda V, d\rangle.$$

Remark B.7 The graded vector space V is dual to sL, so Proposition B.6 together with Theorem B.5 can be used to prove the fact (Proposition 2.39) that the rational homotopy groups of a simply connected space are dual to the indecomposables of the minimal Sullivan model,

$$\pi_k\big(\langle V, d\rangle\big) \cong \mathrm{Hom}(V^k, \mathbb{Q})$$

for all k.

Let us give an example of an application of Maurer–Cartan spaces to rational homotopy theory. They are very useful for studying mapping spaces. The following is [Ber15, Theorem 1.4].

Theorem B.8 *If A is a cdga model for a connected space X and L is a complete L_∞-model for a $\mathbb{Q}$-local nilpotent space Y, then $A \mathbin{\widehat{\otimes}} L$ is a model for the space of maps from X to Y in the sense that*

$$\operatorname{map}(X, Y) \simeq \mathrm{MC}_\bullet(A \mathbin{\widehat{\otimes}} L).$$

In particular, there is a bijection

$$[X, Y] \cong \mathrm{MC}(A \mathbin{\widehat{\otimes}} L)/\!\sim \tag{12}$$

and for all $k \geq 0$ there is an isomorphism

$$\pi_{k+1}\big(\operatorname{map}(X, Y), f\big) \cong \mathrm{H}_k\left((A \mathbin{\widehat{\otimes}} L)^\tau\right),$$

whenever the homotopy class of the map $f\colon X \to Y$ corresponds to the gauge equivalence class of the Maurer–Cartan element $\tau \in \mathrm{MC}(A \mathbin{\widehat{\otimes}} L)$ under the bijection (12).

Remark B.9 While it is in principle possible to write down models for mapping spaces in the classical language of Sullivan algebras—in fact this has been done by Brown–Szczarba [BS97]—the model formulated using the language of L_∞-algebras is arguably more perspicuous.

We end this appendix with a brief review of Getzler's functor $\gamma_\bullet(\mathfrak{g})$, introduced in [Get09], and some recent alternative constructions of it due to Buijs et al. [BFMT20] and Robert-Nicoud–Vallette [RNV20].

Using Dupont's contraction (4) specialized to the n-simplex $X = \Delta^n$, Getzler defines

$$\gamma_n(\mathfrak{g}) = \mathrm{MC}_n(\mathfrak{g}) \cap \ker(s \otimes 1)$$

and shows that this defines a simplicial subset $\gamma_\bullet(\mathfrak{g})$ of $\mathrm{MC}_\bullet(\mathfrak{g})$ that has a number of remarkable properties, including:

- $\gamma_\bullet(\mathfrak{g})$ is a Kan complex and the inclusion $\gamma_\bullet(\mathfrak{g}) \to \mathrm{MC}_\bullet(\mathfrak{g})$ is a homotopy equivalence.
- For $\mathfrak{g}$ abelian, $\gamma_\bullet(\mathfrak{g})$ is isomorphic to the simplicial vector space associated to the chain complex $(s\mathfrak{g})\langle 0\rangle$ under the Dold–Kan correspondence.
- For a nilpotent Lie algebra $\mathfrak{g}$ concentrated in degree zero, $\gamma_\bullet(\mathfrak{g})$ is isomorphic to the nerve of the nilpotent group $\exp(\mathfrak{g})$ associated to $\mathfrak{g}$, i.e., the group with underlying set $\mathfrak{g}$ and multiplication given by the Baker–Campbell–Hausdorff formula.

Recently, alternative constructions of Getzler's functor have been found. This is a nice application of the technology discussed in the first lecture, so let us sketch the main ideas. Buijs–Félix–Murillo–Tanré observe that one can use the natural C_∞-algebra structure on

$C^*(X;\mathbb{Q})$ (see Remark 2.27) to define a cosimplicial complete dg Lie algebra

$$\mathfrak{L}^\bullet = \mathscr{L}^*\big(C^*(\Delta^\bullet;\mathbb{Q})\big).$$

Buijs et al. [BFMT20] and Robert-Nicoud [RN19] proved that this cosimplicial object represents Getzler's functor.

Theorem B.10 *There is a natural isomorphism*

$$\gamma_\bullet(\mathfrak{g}) \cong \mathrm{Hom}_{\mathbf{cDGL}}(\mathfrak{L}^\bullet, \mathfrak{g})$$

for complete dg Lie algebras $\mathfrak{g}$.

In particular, this means that the functor $\gamma_\bullet$ admits a well-behaved left adjoint when restricted to complete dg Lie algebras; the cosimplicial complete dg Lie algebra $\mathfrak{L}^\bullet$ gives rise to an adjunction

$$\mathbf{sSet} \underset{\langle - \rangle}{\overset{\mathfrak{L}(-)}{\rightleftarrows}} \mathbf{cDGL} \qquad (13)$$

where

$$\mathfrak{L}(X) = \mathrm{colim}_{\Delta^n \to X}\mathfrak{L}^n$$
$$\langle \mathfrak{g} \rangle = \mathrm{Hom}_{\mathbf{cDGL}}(\mathfrak{L}^\bullet, \mathfrak{g}).$$

The homotopy theoretical properties of this adjunction are worked out in the recent book [BFMT20].

Next, one might wonder what happens if we do not restrict $\gamma_\bullet$ to complete dg Lie algebras. Is the functor $\gamma_\bullet$ representable by a cosimplicial complete L_∞-algebra? Recently, Robert-Nicoud–Vallette [RNV20] gave a positive answer to this question. The key is to work with the canonical resolution of the commutative operad,

$$\Omega \mathrm{B}\mathcal{C}\mathrm{om} \xrightarrow{\simeq} \mathcal{C}\mathrm{om}, \qquad (14)$$

instead of the minimal resolution

$$\mathcal{C}\mathrm{om}_\infty = \Omega(\mathbb{S}\mathcal{L}\mathrm{ie})^\vee \xrightarrow{\simeq} \mathcal{C}\mathrm{om}.$$

Note that the cooperad $\mathrm{B}\mathcal{C}\mathrm{om}$ may be identified with the dual of the shifted L_∞-operad $\mathbb{S}\mathcal{L}\mathrm{ie}_\infty$. This means in particular that one can associate a complete L_∞-algebra,

$$\mathscr{L}^*_\infty(A),$$

to every $\Omega\mathrm{B}\mathcal{C}\mathrm{om}$-algebra A. This could be viewed as an L_∞-version of Harrison cochains. Viewing $\Omega^*(X)$ as a $\Omega\mathrm{B}\mathcal{C}\mathrm{om}$-algebra by pullback along (14), an application of the homotopy transfer theorem to Dupont's contraction (4) shows that there is a natural $\Omega\mathrm{B}\mathfrak{C}\mathrm{om}$-algebra structure on $\mathrm{C}^*(X; \mathbb{Q})$ for every simplicial set X. In particular, there is a cosimplicial complete L_∞-algebra

$$\mathfrak{L}_\infty^\bullet = \mathscr{L}_\infty^*\big(\mathrm{C}^*(\Delta^\bullet; \mathbb{Q})\big).$$

Robert-Nicoud–Vallette [RNV20] then prove the following.

Theorem B.11 *There is a natural isomorphism*

$$\gamma_\bullet(\mathfrak{g}) \cong \mathrm{Hom}_{\mathbf{cL}_\infty}(\mathfrak{L}_\infty^\bullet, \mathfrak{g}),$$

for complete L_∞-algebras $\mathfrak{g}$.

As before, a consequence of this result is that Getzler's functor admits a left adjoint that can be described in quite concrete terms; there is an adjunction

$$\mathbf{sSet} \underset{\langle - \rangle_\infty}{\overset{\mathfrak{L}_\infty(-)}{\rightleftarrows}} \mathbf{cL}_\infty \qquad (\perp) \tag{15}$$

where

$$\mathfrak{L}_\infty(X) = \mathrm{colim}_{\Delta^n \to X} \mathfrak{L}_\infty^n,$$
$$\langle \mathfrak{g} \rangle_\infty = \mathrm{Hom}_{\mathbf{cL}_\infty}(\mathfrak{L}_\infty^\bullet, \mathfrak{g}).$$

We refer to [RNV20] for a detailed study of this construction. The cosimplicial objects $\mathfrak{L}^\bullet$ and $\mathfrak{L}_\infty^\bullet$, and the adjunctions they give rise to, will likely play an important role in future developments of rational homotopy theory.

Appendix C: Rational Homotopy Theory for Non-nilpotent Spaces

In this appendix we summarize some aspects of rational homotopy theory for spaces that are not necessarily nilpotent. We discuss Bousfield's $\mathrm{H}_*(-; \mathbb{Q})$-localization and its relation to the Bousfield–Kan $\mathbb{Q}$-completion and the Malcev completion for groups, and we discuss what Sullivan's minimal model can say about a non-nilpotent space.

Bousfield's $\mathrm{H}_*(-; \mathbb{Q})$-localization [Bou75] is a natural generalization of the rationalization of a nilpotent space (see e.g. [HMR75] or [FHT01, Chapter 9] for accounts of the latter). The $\mathrm{H}_*(-; \mathbb{Q})$-localization is a functor $L_\mathbb{Q}$ from the homotopy category of pointed spaces to itself, equipped with a natural transformation $\eta_X \colon X \to L_{\mathbb{Q}}(X)$. It is characterized by the following properties:

(1) The map $\eta_X \colon X \to L_{\mathbb{Q}}(X)$ is a rational homology isomorphism.
(2) Every rational homology isomorphism $f \colon X \to Y$ factors uniquely in the homotopy category as $f = r\eta_X$ for some $r \colon L_{\mathbb{Q}}(X) \to Y$.

The Bousfield–Kan $\mathbb{Q}$-completion $X \to \mathbb{Q}_\infty X$ [BK72] provides a concrete model for the $\mathrm{H}_*(-;\mathbb{Q})$-localization in many, but not all, cases. We summarize some of its properties (see [BK72]):

- A map $X \to Y$ is a rational homology isomorphism if and only if the induced map $\mathbb{Q}_\infty X \to \mathbb{Q}_\infty Y$ is a weak homotopy equivalence.
- The canonical map $X \to \mathbb{Q}_\infty X$ may or may not be a rational homology isomorphism. The space X is called $\mathbb{Q}$-*good* if it is and $\mathbb{Q}$-*bad* if it is not.
- If X is $\mathbb{Q}$-good, then the Bousfield–Kan $\mathbb{Q}$-completion $\mathbb{Q}_\infty X$ is a model for the $\mathrm{H}_*(-;\mathbb{Q})$-localization.
- Nilpotent spaces, and more generally virtually nilpotent spaces, are $\mathbb{Q}$-good [DDK77, Proposition 3.4]. In particular, simply connected spaces or spaces with finite fundamental group are $\mathbb{Q}$-good.
- If X is $\mathbb{Q}$-bad, then so is $\mathbb{Q}_\infty X$. An example of a $\mathbb{Q}$-bad space is the wedge of n circles $\vee^n S^1$ for $n > 1$ [IM19].

The following key result links the Bousfield–Kan completion to the minimal Sullivan model.

Theorem C.1 *Consider a connected simplicial set X of finite $\mathbb{Q}$-type with minimal Sullivan model $(\Lambda V, d) \to \Omega^*(X)$. The adjoint map*

$$X \longrightarrow \langle \Lambda V, d \rangle$$

is weakly equivalent to the Bousfield–Kan $\mathbb{Q}$-completion $X \to \mathbb{Q}_\infty X$. Furthermore, there is a bijection for every $k \geq 1$

$$\pi_k(\mathbb{Q}_\infty X) \cong \mathrm{Hom}(V^k, \mathbb{Q}).$$

For $k \geq 2$ this is an isomorphism of abelian groups.

Proof See Theorem [BG76, Theorem 12.2] and [BG76, Theorem 12.8(iii)]. □

If X is nilpotent, then there is a simple relation between the homotopy groups of X and the homotopy groups of $\mathbb{Q}_\infty X$, namely

$$\pi_k(\mathbb{Q}_\infty X) \cong \pi_k(X) \otimes \mathbb{Q}$$

for all k, where $\pi_1(X) \otimes \mathbb{Q}$ should be interpreted as the rationalization of the nilpotent group $\pi_1(X)$.

For general X, the relation is not as simple, but some things can be said. If X is connected and $\mathrm{H}_1(X;\mathbb{Q})$ is finite-dimensional, then $\pi_1(\mathbb{Q}_\infty X)$ may be identified with the Malcev completion $\pi_1(X)^\wedge_\mathbb{Q}$ (cf. [FHT15, Corollary 7.4]).

The Malcev completion of a group G can be defined by

$$G^\wedge_\mathbb{Q} = \varprojlim \big((G/\Gamma_n G)\otimes\mathbb{Q}\big),$$

i.e. the inverse limit of the rationalizations of the nilpotent groups $G/\Gamma_n G$, where $\{\Gamma_n G\}_n$ is the lower central series of G, defined by $\Gamma_1 G = G$ and $\Gamma_{n+1}G = [G,\Gamma_n G]$. Alternatively, one can define $G^\wedge_\mathbb{Q}$ as the grouplike elements in the complete Hopf algebra $\widehat{\mathbb{Q}}[G]$, see [Qui69, Appendix A3].

In general, the homotopy groups of $\mathbb{Q}_\infty X$ can be wildly different from the rational homotopy groups of X. For example, if $X = \mathrm{B}\Gamma$ for a perfect group Γ of finite type, then the higher homotopy groups of X are trivial, whereas $\mathbb{Q}_\infty X$ is a model for the rationalization of the Quillen plus construction $\mathrm{B}\Gamma^+$, so that

$$\pi_k(\mathbb{Q}_\infty \mathrm{B}\Gamma) \cong \pi_k(\mathrm{B}\Gamma^+)\otimes\mathbb{Q}$$

for all k, cf. [FHT01, pp. 212f.].

References

[AA78] Peter Andrews and Martin Arkowitz. Sullivan's minimal models and higher order Whitehead products. *Canadian J. Math.*, 30(5):961–982, 1978.

[All73] Christopher Allday. Rational Whitehead products and a spectral sequence of Quillen. *Pacific J. Math.*, 46:313–323, 1973.

[All77] Christopher Allday. Rational Whitehead products and a spectral sequence of Quillen. II. *Houston J. Math.*, 3(3):301–308, 1977.

[Arn69] V. I. Arnol'd. The cohomology ring of the group of dyed braids. *Mat. Zametki*, 5:227–231, 1969.

[Avr98] Luchezar L. Avramov. Infinite free resolutions. In *Six lectures on commutative algebra*, volume 166 of *Progress in Mathematics*. Birkhäuser, 1998.

[BB17] Alexander Berglund and Kaj Börjeson. Free loop space homology of highly connected manifolds. *Forum Math.*, 29(1):201–228, 2017.

[BB20] Alexander Berglund and Kaj Börjeson. Koszul A_∞-algebras and free loop space homology. *Proc. Edinb. Math. Soc. (2)*, 63(1):37–65, 2020.

[BBMFM17] Francisco Belchí, Urtzi Buijs, José M. Moreno-Fernández, and Aniceto Murillo. Higher order Whitehead products and L_∞ structures on the homology of a DGL. *Linear Algebra Appl.*, 520:16–31, 2017.

[Ber61] Israël Berstein. Homotopy mod. C of spaces of category 2. *Comment. Math. Helv.*, 35:9–14, 1961.

[Ber24] Alexander Berglund. On exponential groups and Maurer-Cartan spaces. *Proc. Amer. Math. Soc. Ser. B*, 11:358–370, 2024.

[Ber12] Alexander Berglund. Rational homotopy theory. https://staff.math.su.se/alexb/rathom2.pdf, 2012. Lecture notes.

[Ber14a] Alexander Berglund. Homological perturbation theory for algebras over operads. *Algebr. Geom. Topol.*, 14(5):2511–2548, 2014.

[Ber14b] Alexander Berglund. Koszul spaces. *Trans. Amer. Math. Soc.*, 366(9):4551–4569, 2014.

[Ber15] Alexander Berglund. Rational homotopy theory of mapping spaces via Lie theory for L_∞-algebras. *Homology Homotopy Appl.*, 17(2):343–369, 2015.

[Bez94] R. Bezrukavnikov. Koszul DG-algebras arising from configuration spaces. *Geom. Funct. Anal.*, 4(2):119–135, 1994.

[BFMT20] Urtzi Buijs, Yves Félix, Aniceto Murillo, and Daniel Tanré. *Lie models in topology*, volume 335 of *Progress in Mathematics*. Birkhäuser, 2020.

[BG76] A. K. Bousfield and V. K. A. M. Gugenheim. On PL de Rham theory and rational homotopy type. *Mem. Amer. Math. Soc.*, 8(179), 1976.

[BK72] A. K. Bousfield and D. M. Kan. *Homotopy limits, completions and localizations*, volume 304 of *Lecture Notes in Mathematics*. Springer-Verlag, 1972.

[BL77] H. J. Baues and J.-M. Lemaire. Minimal models in homotopy theory. *Math. Ann.*, 225(3):219–242, 1977.

[BM13] Urtzi Buijs and Aniceto Murillo. Algebraic models of non-connected spaces and homotopy theory of L_∞ algebras. *Adv. Math.*, 236:60–91, 2013.

[BM20] Alexander Berglund and Ib Madsen. Rational homotopy theory of automorphisms of manifolds. *Acta Math.*, 224(1):67–185, 2020.

[BMFM20] Urtzi Buijs, José M. Moreno-Fernández, and Aniceto Murillo. A_∞ structures and Massey products. *Mediterr. J. Math.*, 17(1):Article 31, 2020.

[Bor74] Armand Borel. Stable real cohomology of arithmetic groups. *Ann. Sci. École Norm. Sup.*, 7:235–272, 1974.

[Bor81] Armand Borel. Stable real cohomology of arithmetic groups. II. In *Manifolds and Lie groups (Notre Dame, Ind., 1980)*, volume 14 of *Progr. Math.*, pages 21–55. Birkhäuser, Boston, Mass., 1981.

[Bou75] A. K. Bousfield. The localization of spaces with respect to homology. *Topology*, 14:133–150, 1975.

[Bro65] R. Brown. The twisted Eilenberg-Zilber theorem. In *Simposio di Topologia (Messina, 1964)*, pages 33–37. Edizioni Oderisi, Gubbio, 1965.

[BS97] Edgar H. Brown, Jr. and Robert H. Szczarba. On the rational homotopy type of function spaces. *Trans. Amer. Math. Soc.*, 349(12):4931–4951, 1997.

[BS20] Alexander Berglund and Bashar Saleh. A dg Lie model for relative homotopy automorphisms. *Homology Homotopy Appl.*, 22(2):105–121, 2020.

[BT00] I. K. Babenko and I. A. Taĭmanov. On nonformal simply connected symplectic manifolds. *Sibirsk. Mat. Zh.*, 41(2):253–269, 2000.

[BV73] John M. Boardman and Rainer M. Vogt. *Homotopy invariant algebraic structures on topological spaces*, volume 347 of *Lecture Notes in Mathematics*. Springer-Verlag, 1973.

[BZ22] Alexander Berglund and Tomas Zeman. Algebraic models for classifying spaces of fibrations. arXiv: `2203.02462v4`, 2022. To appear in Geometry & Topology.

[CG02] F. R. Cohen and S. Gitler. On loop spaces of configuration spaces. *Trans. Amer. Math. Soc.*, 354(5):1705–1748, 2002.

[CG08] Xue Zhi Cheng and Ezra Getzler. Transferring homotopy commutative algebraic structures. *J. Pure Appl. Algebra*, 212(11):2535–2542, 2008.

[CHKV16] James Conant, Allen Hatcher, Martin Kassabov, and Karen Vogtmann. Assembling homology classes in automorphism groups of free groups. *Comment. Math. Helv.*, 91(4):751–806, 2016.

[CKV13] James Conant, Martin Kassabov, and Karen Vogtmann. Hairy graphs and the unstable homology of $\mathrm{Mod}(g, s)$, $\mathrm{Out}(F_n)$ and $\mathrm{Aut}(F_n)$. *J. Topol.*, 6(1):119–153, 2013.

[Coh76] Frederick R. Cohen. The homology of $\mathscr{C}_{n+1}$-spaces. In Frederick R. Cohen, Thomas J. Lada, and J. Peter May, editors, *The homology of iterated loop spaces*, volume 533 of *Lecture Notes in Mathematics*. Springer-Verlag, 1976.

[CPRNW19] Campos, Ricardo and Petersen, Dan and Robert-Nicoud, Daniel and Wierstra, Felix. Lie, associative and commutative quasi-isomorphism. *Acta Math.*, 233(2):195–238 (2024).

[DDK77] E. Dror, W. G. Dwyer, and D. M. Kan. An arithmetic square for virtually nilpotent spaces. *Illinois J. Math.*, 21(2):242–254, 1977.

[DDK81] E. Dror, W. G. Dwyer, and D. M. Kan. Self-homotopy equivalences of virtually nilpotent spaces. *Comment. Math. Helv.*, 56(4):599–614, 1981.

[DGMS75] Pierre Deligne, Phillip Griffiths, John Morgan, and Dennis Sullivan. Real homotopy theory of Kähler manifolds. *Invent. Math.*, 29(3):245–274, 1975.

[Dot22] Vladimir Dotsenko. Homotopy invariants for $\overline{\mathscr{M}}_{0,n}$ via Koszul duality. *Invent. Math.*, 228(1):77–106, 2022.

[Dup78] Johan L. Dupont. *Curvature and characteristic classes*, volume 640 of *Lecture Notes in Mathematics*. Springer-Verlag, 1978.

[DZ79] Emmanuel Dror and Alexander Zabrodsky. Unipotency and nilpotency in homotopy equivalences. *Topology*, 18(3):187–197, 1979.

[FFM22] Yves Félix, Mario Fuentes, and Aniceto Murillo. Lie models of homotopy automorphism monoids and classifying fibrations. *Adv. Math.*, 402:108359, 2022.

[FHT01] Yves Félix, Stephen Halperin, and Jean-Claude Thomas. *Rational homotopy theory*, volume 205 of *Graduate Texts in Mathematics*. Springer-Verlag, 2001.

[FHT15] Yves Félix, Steve Halperin, and Jean-Claude Thomas. *Rational homotopy theory II*. World Scientific, 2015.

[FK77] David Frank and Donald W. Kahn. Finite complexes with infinitely-generated groups of self-equivalences. *Topology*, 16(2):189–192, 1977.

[FM08] Marisa Fernández and Vicente Muñoz. An 8-dimensional nonformal, simply connected, symplectic manifold. *Ann. of Math.*, 167(3):1045–1054, 2008.

[Get09] Ezra Getzler. Lie theory for nilpotent L_∞-algebras. *Ann. of Math.*, 170(1):271–301, 2009.

[GJ94] E. Getzler and J. D. S. Jones. Operads, homotopy algebra and iterated integrals for double loop spaces. arXiv: `hep-th/9403055`, 1994. Preprint.

[GJ99] Paul G. Goerss and John F. Jardine. *Simplicial homotopy theory*, volume 174 of *Progress in Mathematics*. Birkhäuser, 1999.

[GK94] V. Ginzburg and M. Kapranov. Koszul duality for operads. *Duke Math. J.*, 76(1):203–272, 1994.

[GK95] E. Getzler and M. M. Kapranov. Cyclic operads and cyclic homology. In *Geometry, Topology and Physics for Raoul Bott*, volume 4 of *Conference Proceedings and Lecture Notes in Geometry and Topology*, pages 167–201. International Press, 1995.

[GK98] E. Getzler and M. M. Kapranov. Modular operads. *Compositio Math.*, 110(1):65–126, 1998.

[GLS91] V. K. A. M. Gugenheim, L. A. Lambe, and J. D. Stasheff. Perturbation theory in differential homological algebra. II. *Illinois J. Math.*, 35(3):357–373, 1991.

[GM74] V. K. A. M. Gugenheim and H. J. Munkholm. On the extended functoriality of Tor and Cotor. *J. Pure Appl. Algebra*, 4:9–29, 1974.

[GM13] Phillip Griffiths and John Morgan. *Rational homotopy theory and differential forms*, volume 16 of *Progress in Mathematics*. Springer, second edition, 2013.

[Gug72] V. K. A. M. Gugenheim. On the chain-complex of a fibration. *Illinois J. Math.*, 16:398–414, 1972.

[Har61] K. A. Hardie. Higher Whitehead products. *Quart. J. Math. Oxford Ser.*, 12:241–249, 1961.

[Har62] D. K. Harrison. Commutative algebras and cohomology. *Trans. Amer. Math. Soc.*, 104:191–204, 1962.

[Hat02] Allen Hatcher. *Algebraic topology*. Cambridge University Press, 2002.

[Hin97] Vladimir Hinich. Descent of Deligne groupoids. *Internat. Math. Res. Notices*, (5):223–239, 1997.

[HK91] Johannes Huebschmann and Tornike Kadeishvili. Small models for chain algebras. *Math. Z.*, 207(2):245–280, 1991.

[HMR75] Peter Hilton, Guido Mislin, and Joe Roitberg. *Localization of nilpotent groups and spaces*, volume 15 of *North-Holland Mathematics Studies*. North-Holland Publishing Company, 1975.

[HS79] Stephen Halperin and James Stasheff. Obstructions to homotopy equivalences. *Adv. in Math.*, 32(3):233–279, 1979.

[IM19] Sergei O. Ivanov and Roman Mikhailov. A finite $\mathbb{Q}$-bad space. *Geom. Topol.*, 23(3):1237–1249, 2019.

[Kad82] T. V. Kadeishvili. The algebraic structure in the homology of an $A(\infty)$-algebra. *Soobshch. Akad. Nauk Gruzin. SSR*, 108(2):249–252, 1982.

[Kad88] T. V. Kadeishvili. The structure of the $A(\infty)$-algebra, and the Hochschild and Harrison cohomologies. *Trudy Tbiliss. Mat. Inst. Razmadze Akad. Nauk Gruzin. SSR*, 91:19–27, 1988.

[Kad09] Tornike Kadeishvili. Cohomology C_∞-algebra and rational homotopy type. In *Algebraic topology—old and new*, volume 85 of *Banach Center Publications*, pages 225–240. Institute of Mathematics, Polish Academy of Sciences, 2009.

[Kel01] B. Keller. Introduction to A-infinity algebras and modules. *Homology Homotopy Appl.*, 3(1):1–35, 2001.

[Kon93] Maxim Kontsevich. Formal (non)commutative symplectic geometry. In *The Gel'fand Mathematical Seminars, 1990–1992*, pages 173–187. Birkhäuser, 1993.

[Kon94] Maxim Kontsevich. Feynman diagrams and low-dimensional topology. In *First European Congress of Mathematics, Volume II*, volume 120 of *Progress in Mathematics*, pages 97–121. Birkhäuser, 1994.

[LH03] K. Lefevre-Hasegawa. Sur les A-infini catégories. arXiv: `math/0310337`, 2003. Preprint.

[LM95] T. Lada and M. Markl. Strongly homotopy Lie algebras. *Comm. Algebra*, 23(6):2147–2161, 1995.

[LM15] Andrey Lazarev and Martin Markl. Disconnected rational homotopy theory. *Adv. Math.*, 283:303–361, 2015.

[LO94] Gregory Lupton and John Oprea. Symplectic manifolds and formality. *J. Pure Appl. Algebra*, 91(1–3):193–207, 1994.

[LS93] Tom Lada and Jim Stasheff. Introduction to SH Lie algebras for physicists. *Internat. J. Theoret. Phys.*, 32(7):1087–1103, 1993.

[LV12] Jean-Louis Loday and Bruno Vallette. *Algebraic operads*, volume 346 of *Grundlehren der Mathematischen Wissenschaften*. Springer-Verlag, 2012.

[LV14] Pascal Lambrechts and Ismar Volić. Formality of the little N-disks operad. *Mem. Amer. Math. Soc.*, 230(1079), 2014.

[Maj00] Martin Majewski. Rational homotopical models and uniqueness. *Mem. Amer. Math. Soc.*, 143(682), 2000.

[Mar04] Martin Markl. Homotopy algebras are homotopy algebras. *Forum Math.*, 16(1):129–160, 2004.

[May75] J. Peter May. Classifying spaces and fibrations. *Mem. Amer. Math. Soc.*, 1(155), 1975.

[Mil79] Timothy James Miller. On the formality of $(k-1)$-connected compact manifolds of dimension less than or equal to $4k-2$. *Illinois J. Math.*, 23(2):253–258, 1979.

[Nei78] Joseph Neisendorfer. Lie algebras, coalgebras and rational homotopy theory for nilpotent spaces. *Pacific J. Math.*, 74(2):429–460, 1978.

[Nei79] Joseph Neisendorfer. The rational homotopy groups of complete intersections. *Illinois J. Math.*, 23(2):175–182, 1979.

[NM78] Joseph Neisendorfer and Timothy Miller. Formal and coformal spaces. *Illinois J. Math.*, 22(4):565–580, 1978.

[Por65] Gerald J. Porter. Higher-order Whitehead products. *Topology*, 3:123–135, 1965.

[Por67] Gerald J. Porter. Higher order Whitehead products and Postnikov systems. *Illinois J. Math.*, 11:414–416, 1967.

[Pri70] S.B. Priddy. Koszul resolutions. *Trans. Amer. Math. Soc.*, 152:39–60, 1970.

[PY99] Stefan Papadima and Sergey Yuzvinsky. On rational $K[\pi, 1]$ spaces and Koszul algebras. *J. Pure Appl. Algebra*, 144(2):157–167, 1999.

[Qui69] D. Quillen. Rational homotopy theory. *Ann. of Math.*, 90:205–295, 1969.

[Ret93] Vladimir S. Retakh. Lie-Massey brackets and n-homotopically multiplicative maps of differential graded Lie algebras. *J. Pure Appl. Algebra*, 89(1–2):217–229, 1993.

[RN19] Daniel Robert-Nicoud. Representing the deformation ∞-groupoid. *Algebr. Geom. Topol.*, 19(3):1453–1476, 2019.

[RNV20] Daniel Robert-Nicoud and Bruno Vallette. Higher lie theory. arXiv: 2010.10485, 2020. Preprint.

[Sal17] Bashar Saleh. Noncommutative formality implies commutative and Lie formality. *Algebr. Geom. Topol.*, 17(4):2523–2542, 2017.

[Sal20] Paolo Salvatore. Non-formality of planar configuration spaces in characteristic 2. *Int. Math. Res. Not. IMRN*, (10):3100–3129, 2020.

[SH70] James Stasheff and Steve Halperin. Differential algebra in its own rite. In *Proceedings of the Advanced Study Institute on Algebraic Topology (August 10–23, 1970), Vol. III*, volume 113 of *Various Publications Series*, pages 567–577. Aarhus Universitet, 1970.

[Ser79] J.-P. Serre. Arithmetic groups. In *Homological Group Theory*, pages 105–136. Cambridge University Press, 1979.

[SS12] Mike Schlessinger and Jim Stasheff. Deformation theory and rational homotopy type. arXiv: 1211.1647, 2012. Preprint.

[Sta63a] James Stasheff. A classification theorem for fibre spaces. *Topology*, 2:239–246, 1963.

[Sta63b] James D. Stasheff. Homotopy associativity of H-spaces. II. *Trans. Amer. Math. Soc.*, 108:293–312, 1963.

[Sto22] Stoll, Robin. The stable cohomology of self-equivalences of connected sums of products of spheres. *Forum Math. Sigma*, 12: Article e1 (2024).

[Sul77] D. Sullivan. Infinitesimal computations in topology. *Inst. Hautes Études Sci. Publ. Math.*, (47):269–331 (1978), 1977.

[Tan83] Daniel Tanré. *Homotopie rationnelle: modèles de Chen, Quillen, Sullivan*, volume 1025 of *Lecture Notes in Mathematics*. Springer-Verlag, 1983.

[TZ07] Thomas Tradler and Mahmoud Zeinalian. Infinity structure of Poincaré duality spaces. *Algebr. Geom. Topol.*, 7:233–260, 2007. Appendix A by Dennis Sullivan.

[Vog02] Karen Vogtmann. Automorphisms of free groups and outer space. In *Proceedings of the Conference on Geometric and Combinatorial Group Theory, Part I (Haifa, 2000)*, volume 94, pages 1–31, 2002.

[War19] Benjamin C. Ward. Six operations formalism for generalized operads. *Theory Appl. Categ.*, 34(6):121–169, 2019.

[Wei94] C.A. Weibel. *An introduction to homological algebra*, volume 38 of *Cambridge Studies in Advanced Mathematics*. Cambridge University Press, 1994.

[Wil76] Clarence W. Wilkerson. Applications of minimal simplicial groups. *Topology*, 15(2):111–130, 1976.

Operadic Deformation Theory

Ricardo Campos and Albin Grataloup

Contents

1 Introduction

Deformation theory is classically understood to be the study of variations of an object X, to infinitesimally nearby objects X_ϵ. One of the most salient examples, dating back

R. Campos (✉)
Ricardo Campos, Institut de Mathématiques de Toulouse, UMR5219, Université de Toulouse, CNRS, UPS, Toulouse Cedex 9, France
e-mail: ricardo.campos@math.univ-toulouse.fr

A. Grataloup
Albin Grataloup, IMAG, University of Montpellier, CNRS, Montpellier, France
e-mail: albin.grataloup@bfh.ch

B. Vallette (ed.), *Higher Structures and Operadic Calculus*, Advanced Courses in Mathematics - CRM Barcelona, https://doi.org/10.1007/978-3-031-77779-0_2

to the 50's, are the works of Kodaira–Spencer [KS58, KS60] and Fröhlicher–Nijenhuis [FN57] on the deformation theory of complex manifolds and algebraic varieties where it is shown that "first-order" deformations of an algebraic variety are related to the tangent cohomology.

The approach to deformation theory by Grothendieck, Mumford, Schlessinger (see [AST76]) consists of associating to a deformation problem, a "deformation functor"

$$F\colon \text{Parametrizations} \to \text{Sets}$$
$$\text{Infinitesimal } \epsilon \mapsto \{X_\epsilon\}$$

such that when $\epsilon = 0$ we get back the original object $X_0 = X$. To make some sense of the association above, we can replace the source by the category of local Artinian commutative algebras (or their spectra). The prototypical example are the algebras $k[\epsilon]/(\epsilon^n)$. From the point of view of algebraic geometry, these represent infinitesimal thickenings of a point.

If we want to capture the essence of a deformation problem, we should have a way to encode when and how two deformations can be equivalent, which suggests that we should take groupoids instead of sets to be the target of F.

It turns out that in characteristic zero one can often associate to a deformation problem a differential graded (dg) Lie algebra $\mathfrak{g}$ encoding the problem in the sense that the associated deformation functor takes the form

$$R \mapsto \text{Maurer–Cartan elements of } \mathfrak{g} \otimes \mathfrak{m}_R, \tag{1}$$

where $\mathfrak{m}_R$ is the unique maximal ideal of R. These are degree 1 solutions of the Maurer–Cartan equation $d\mu + \frac{1}{2}[\mu, \mu] = 0$ and in fact there is a natural *gauge group* acting on them such that the functor naturally lands in groupoids.

To exemplify, let us consider a very classical algebraic situation:

Example 1.1 Let A be an associative algebra. A first-order deformation of A is given by a map $\mu\colon A \otimes A \to A$, such that the formula, $a \cdot b = ab + \mu_1(a, b)\epsilon$, defines an ϵ-linear associative product on $A[\epsilon]/(\epsilon^2)$. The requirement that the product be associative is equivalent to

$$\mu_1(a, b)c + \mu_1(ab, c) = a\mu_1(b, c) + \mu_1(a, bc), \quad \forall a, b, c \in A$$

More generally, a higher order deformation along $k[\epsilon]/(\epsilon^{n+1})$ is an associative product on $A[\epsilon]/(\epsilon^{n+1})$, which amounts to the choice of $\mu_1, \ldots, \mu_n$ satisfying generalisations of the equation above.

The dg Lie algebra encoding this deformation problem is the Hochschild Lie algebra given in degree n by $\mathfrak{g}^n = \mathrm{Hom}(A^{\otimes n+1}, A)$ (with appropriate differential and bracket). A deformation of order n, $\mu = \epsilon\mu_1 + \cdots + \epsilon^n\mu_n$, is precisely a Maurer–Cartan element of $g \otimes \epsilon k[\epsilon]/(\epsilon^{n+1})$. Two noteworthy takeaways are that the Maurer–Cartan equation only involves dg Lie algebraic data and that the dg Lie algebra is the same regardless of the type of deformation.

Examples such as the previous one led Deligne, Drinfeld and Quillen [Dri14] to formulate the following informal slogan:

Any deformation problem over a field k of characteristic zero is controlled by a dg Lie k-algebra.

This slogan was fruitfully used in works of Kontsevich [Kon03], Hinich [Hin01], Goldman–Millson [GM88], Manetti [Man02]. This leads us to the central questions that we wish to address:

Question

(1) Given a deformation problem, how to obtain the dg Lie algebra encoding it?
(2) Which deformations are "equivalent"?
(3) What is the structure of the moduli space of deformations up to equivalence?

The notion of equivalence tends to come naturally associated to the problem. For instance, take two deformations $\widetilde{X}$ and $\widetilde{X}'$ of a complex manifold or more generally a k-scheme, X. Then they are called equivalent if there is an isomorphism (or more generally a weak equivalence) that reduces to the identity on X (up to homotopy). In the case of associative algebras, two deformations are equivalent if there is an isomorphism intertwining μ and μ'.

The main purpose of this text is to present how the theory of algebraic operads gives us very efficient tools to address these questions. Let us consider again the illustrative example of deforming associative algebras. An object of interest is the set (ideally a "space") of associative algebra structures on a vector space, or more generally a cochain complex, A. These are precisely all possible structures of left modules on A (with respect to a monoidal product $\circ$) over the operad **Ass** of associative algebras $\mathbf{Ass} \circ A \to A$.

Crucially, the symmetric monoidal category of cochain complexes is closed which allows us to see A as a representation of the operad **Ass**: There is an endomorphisms operad of A, $\mathbf{End}_A$, such that the set of associative algebra structures on A is

$$\{\text{Associative algebra structures on } A\} \cong \mathrm{Hom}_{\mathrm{Operads}}(\mathbf{Ass}, \mathbf{End}_A).$$

It follows that to understand deformations of a particular algebraic structure on A, we can equivalently study deformations of the corresponding map of operads $f : \mathbf{Ass} \to \mathbf{End}_A$.

A homotopical perspective gives us some insight in how to progress further. Indeed, using a model structure on operads, this Hom-set can be naturally upgraded to a mapping space. The deformations we care about live in an infinitesimal neighbourhood of the point f in the mapping space. Here, Koszul duality plays a key role in constructing for us a simple cofibrant replacement of **Ass**, which is the operad governing A_∞-algebras (algebras only associative up to homotopy).

Furthermore, this procedure expresses the mapping space naturally as the Maurer–Cartan space of a dg Lie algebra $\mathfrak{g}$, seen as the *deformation complex of* f. In the present case $\mathfrak{g}$ is exactly the Hochschild dg Lie algebra of Example 1.1.

In fact, the slogan itself that dg Lie algebras control deformation problems should be interpreted in an appropriate homotopical sense. On one hand, quasi-isomorphic dg Lie algebras encode the same deformation problem. On the other hand, rather than a set or a groupoid, the target of a deformation functor should be a space, concretely an ∞-groupoid. A formalisation of a deformation functor was only achieved in the 2010s in the works of Lurie [Lur11] which he called a *formal moduli problem*, which is an ∞-categorical.

A formalisation of the slogan was by Lurie [Lur11] and Pridham [Pri10] in the form an equivalence between the ∞-categories of formal moduli problems and dg Lie algebras. One of the equivalences is, informally speaking, given by the assignment (1). In the other direction, the equivalence is given by the tangent complex functor $\mathbb{T}$ whose dg Lie algebra structure can be quite inexplicit.

Here too, the theory of algebraic operads provides a conceptual reason for why deformations are encoded by Lie algebras. It is an instance of Koszul duality between dg commutative algebras (the local Artinian algebras) and dg Lie algebras.

This survey splits into three chapters. In the first one, we review the theory of algebraic operads, with the goal of understanding Koszul duality and the structure of the model category of operads and of algebras over a fixed operad. In the second chapter we focus on the tools to treat deformation problems associated to a given dg Lie algebra, namely the Deligne groupoid associated to a Lie algebra and the more general problem of integrating homotopy Lie algebras. We review ∞-algebras and ∞-morphisms under this perspective as well as the homotopy transfer theorem. We will see how these methods apply to specific problems, such as Kontsevich's formality theorem or to rational homotopy theory. In the final chapter we address the more (derived) geometrical approach to deformation theory, namely the notion of formal moduli problems and the formalization of the deformation slogan, under the algebraic operadist perspective.

We will review the construction of the (co)tangent complex and understand problems that can be infinitesimally deformed along non-commutative algebras. At the very end, we will review recent developments, that fall outside of the framework of this such as deformation theory in positive characteristic.

1.1 Conventions

Throughout the text, unless otherwise specified we work over a field k of characteristic zero. We always work over cochain complexes and use cohomological conventions, regardless of whether a subscript or a superscript is used for the degrees. For example, both Hochschild homology and cohomology arise from a complex whose differential has degree $+1$.

All algebraic objects are considered differential graded by assumption and we will thus not use the prefix "dg". For instance, what we will call an associative algebra is what is more commonly called a differential graded algebra elsewhere. The few instances we will work with non-dg objects, we will explicitly use the phrasing "graded" to mean that there is no differential and "concentrated in degree 0" for objects such as vector spaces, having nothing in non-zero degree. In particular, k-module and cochain complex are synonyms.

Whenever applicable, we distinguish the internal Hom from the Hom set by using $\underline{\mathrm{Hom}}(V, W)$ for the first and $\mathrm{Hom}(V, W)$ for the latter.

A list of notations is present at the end.

2 Algebraic Operad Theory

The goal of this section is to introduce the important elements of the theory of algebraic symmetric operads. In Sect. 2.1, we introduce the definitions of symmetric operads, cooperads and the main examples of (co)operads we are going to use in this review. In Sect. 2.2, we discuss the notion of algebra and coalgebra over operads and cooperads. We also give the construction of the (co)free (co)algebra adjunctions. In Sect. 2.3 we describe the convolution operad, the bar-cobar constructions and the notion of twisting morphisms. These three sections are meant to be a shortcut for the reader to the content of [LV12] up to Section 6, which is generally the reference we recommend for details. Finally, Sects. 2.4, 2.5, and 2.6 are devoted to model categorical aspects of operad theory, describing the model structure on operads and on algebras over an operad. We also discuss the notion of Koszul duality as a means to obtain simple cofibrant replacements of operads.

2.1 Algebraic Operads and Cooperads

Let us start with a motivating example. A (differential graded or dg) associative algebra is a cochain complex (A, d) together with a bilinear product $\cdot$ such that for all $a, b, c \in A$

1. $(a \cdot b) \cdot c = a \cdot (b \cdot c)$ (associativity),
2. $d(a \cdot b) = da \cdot b + (-1)^{\deg a} a \cdot db$ (compatibility with the differential).

The operadic approach consists in removing the spotlight from the elements a, b, c (which are arbitrary) and rather focus on the multiplication operation itself. One could instead define an associative algebra structure on a cochain complex A to be a map $\mu \colon A \otimes A \to A$, such that $\mu(\mu, \mathrm{id}_A) = \mu(\mathrm{id}_A, \mu)$. Notice that this is an equality of maps $A \otimes A \otimes A \to A$. An immediate advantage of this perspective is that we no longer need to encode property (2.1). Indeed, it is hidden in the assumption that $\mu \colon A \otimes A \to A$ is a map of complexes. Property 2.1 is "categorical", it follows formally from working with the underlying category of cochain complexes and this is the reason why we choose to drop the prefix dg: a dg associative algebra is the same as an associative algebra in cochain complexes. From this perspective it also follows immediately from applying the homology functor to μ that $H(A)$ is an associative algebra (in graded vector spaces). Property 2.1 on the other hand is "structural" and is the one suitable to be studied with the operadic machinery.

Even if we start with an operation taking two inputs, by composing it, we can write identities that take place in $\mathrm{End}_A(n) := \underline{\mathrm{Hom}}(A^{\otimes n}, A)$. We think of these as multiendomorphisms or *arity n* endomorphisms of A. We could have considered instead *unital* associative algebras, which would amount to furthermore choosing an arity 0 endomorphism $1\colon k \to A$, such that $\mu(1, \mathrm{id}_A) = \mu(\mathrm{id}_A, 1) = \mathrm{id}_A\colon A \to A$.

An operad $\mathscr{P}$ will be the kind of object such that its representations, i.e. maps of operads $\mathscr{P} \to \mathrm{End}_A$, correspond precisely to $\mathscr{P}$-algebra structures on A. The important properties to retain are the existence of an *arity*, representing the number of inputs of the operation, as well as a way to *compose* operations.

Furthermore, a way to encode symmetries is necessary: if we want to consider commutative algebras this amounts to require $\mu\colon A \otimes A \to A$ to factor through the Σ_2-coinvariants of the source. Or in a Lie algebra, even ignoring the anti-symmetry of the bracket, if we wish to encode the Jacobi identity, we need to permute the inputs of $[a, [b, c]]$.

The objects encoding the collection of n-ary operations of an operad will be taken in a fixed category. We will use the category Mod_k of cochain complexes over k in this survey, although much of what is discussed can be extended to any closed symmetric monoidal model category such as the category of simplicial sets sSet or A-modules Mod_A for example. The first thing we define is a category whose objects are given by collections of n-ary operations.

Definition 2.1 A *symmetric sequence* is a collection of elements in Mod_k, $M := \{M(n)\}_{n\geq 0}$ together with right actions of the symmetric groups $M(n) \curvearrowleft \Sigma_n$ for all $n \geq 0$. A morphism $f\colon M \to N$ between symmetric sequences is the data of a collection of maps in Mod_k, $f_n : M(n) \to N(n)$ that are invariant with respect to the actions of Σ_n on $M(n)$ and $N(n)$. The category of symmetric sequences is denoted $\mathbf{Seq}_\Sigma$.

Remark 2.2 Let $\mathbb{N}^\sim$ denote the category with objects the sets $\{1, \ldots, n\}$ for all $n \in \mathbb{N}$ and morphisms given by the permutations of these sets.

The category of symmetric sequences in Mod_k is isomorphic to the category of functors, $\mathrm{Fun}\,(\mathbb{N}^\sim, \mathrm{Mod}_k)$, that is, we have an equivalence[1] of categories:

$$\mathbf{Seq}_\Sigma \simeq \mathrm{Fun}\,(\mathbb{N}^\sim, \mathrm{Mod}_k)\,.$$

Remark 2.3 The most common variations on the previous definition is to consider:

- $\mathbb{N}^o \subset \mathbb{N}^\sim$ the sub-category with the same objects but with only identities as morphisms.
- Given a set S of *colors*, denote $\mathbb{N}_S$ the category whose objects are k-tuples of elements on S for all $k \in \mathbb{N}$, and whose morphisms are given by all permutations of these tuples, $(a_1, \cdots, a_k) \to (a_{\sigma_1}, \cdots, a_{\sigma_k})$ that preserve the colors.

[1] Sending a functor F to the collection $(F(n))_{n\in\mathbb{N}}$ and the permutation of $\{1, \cdots, n\}$ are the morphisms in $\mathbb{N}^\sim$ that induce the action of Σ_n on $F(n)$.

We can use the functorial construction with these alternative categories, we obtain:

- For $\mathbb{N}^o$, the notion of *non-symmetric operads* (with no action of Σ_n).
- For $\mathbb{N}_S$, the notion of *colored operad*.

Remark 2.4 We think of the cochain complex $M(n)$ as the space of all n-ary operations of our operad. An operation $m \in M(n)$ will be depicted by a rooted planar tree as follows:

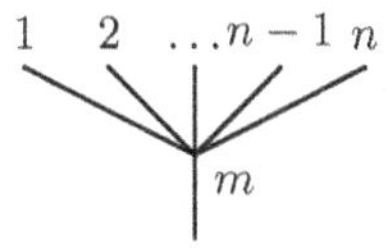

(2)

The action of Σ_n is the action that permutes the n leaves of the tree and sends m to an other operation $\sigma.m$.

The idea behind algebraic operads is that these trees correspond to operations with n inputs and 1 output. We will need to make sense of how to compose these trees by concatenating them as depicted by the following graph:

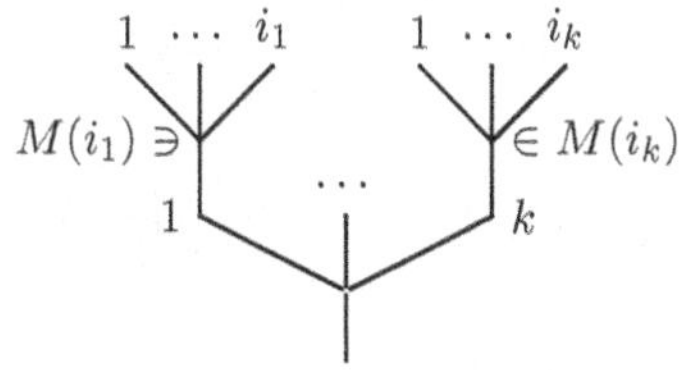

To make sense of having the set of all k-tuples of trees that we want as input for an other k-ary tree, we define a symmetric tensor product on $\mathbf{Seq}_\Sigma$. In the tree above, the upper level should be interpreted as an arity $i_1 + \cdots + i_k$ element of $M(i_1) \otimes \cdots \otimes M(i_k)$.

Definition 2.5 Given $\mathscr{D}$, a $\mathscr{V}$-enriched category and $\mathscr{V}$, a closed symmetric monoidal category with small limits and colimits. Then the *Day convolution* of two functors $F, G : \mathscr{D} \to \mathscr{V}$ is defined by the coend construction (see [Lor15, Proposition 6.2.1]):

$$F \otimes_{\mathrm{Day}} G(-) := \int^{(c,c')\in \mathscr{D}\times\mathscr{D}} \mathscr{D}(c \otimes_{\mathscr{D}} c', -) \otimes_{\mathscr{V}} F(c) \otimes_{\mathscr{V}} G(c')$$

Will will now describe the Day convolution using concrete formulas in the context we are interested in, that is when $\mathscr{D} = \mathbb{N}^{\sim}$ and $\mathscr{V} = \mathrm{Mod}_k$.

We then have the following formula for $\otimes_{\mathrm{Day}}$ which can equally been taken as a definition for the reader less confortable with coend calculus.

Proposition 2.6 *For $\mathscr{D} = \mathbb{N}^{\sim}$ and $\mathscr{V} = \mathrm{Mod}_k$, we obtain:*

$$F \otimes_{\mathrm{Day}} G(n) = \bigoplus_{p+q=n} \mathrm{Ind}_{\Sigma_p \times \Sigma_q}^{\Sigma_n} (F(p) \otimes G(q)).$$

where $\mathrm{Ind}_{\Sigma_p\times\Sigma_q}^{\Sigma_n}$ *denotes the induction*[2] *from* $\Sigma_p \times \Sigma_q$ *representations to* Σ_n-*representations.*

Proof For $\mathscr{D} = \mathbb{N}^{\sim}$ we get:[3]

$$F \otimes_{\mathrm{Day}} G(d) = \int^{p,q\in\mathbb{N}} \underline{\mathrm{Hom}}(p \otimes q, d) \otimes F(p) \otimes G(q)$$

We obtain that $F \otimes_{\mathrm{Day}} G(d)$ is equal to:

$$\mathrm{coeq}\left(\coprod_{p,q\in\mathbb{N}} \underline{\mathrm{Hom}}(p+q,d)\otimes F(p)\otimes G(q) \leftleftarrows \coprod_{f:p\to p',\ g:q\to q'} \underline{\mathrm{Hom}}(p+q,d)\otimes F(p')\otimes G(q') \right)$$

$\underline{\mathrm{Hom}}(p+q, d)$ is non-zero only if $p+q=d$, $p=p'$ and $q=q'$. Therefore we get:

$$\mathrm{coeq}\left(\coprod_{p+q=d} k[\Sigma_d]\otimes F(p)\otimes G(q) \leftleftarrows \coprod_{p+q=d}\coprod_{\Sigma_p\times\Sigma_q} k[\Sigma_d]\otimes F(p)\otimes G(q) \right)$$

$$= \coprod_{p+q=d} \mathrm{coeq}\left(k[\Sigma_d]\otimes F(p)\otimes G(q) \leftleftarrows \coprod_{\Sigma_p\times\Sigma_q} k[\Sigma_d]\otimes F(p)\otimes G(q) \right)$$

We now need to relate this coequalizer to the tensor product over $k[\Sigma_p \times \Sigma_q]$. To do that we notice that for any $p, q \in \mathbb{N}$ with $p+q=d$, $N \in k[\Sigma_d]-\mathrm{Mod}$, a map

$$\mathrm{Hom}_{k[\Sigma_d]}\left(\mathrm{coeq}\left(k[\Sigma_d]\otimes F(p)\otimes G(q) \leftleftarrows \coprod_{\Sigma_p\times\Sigma_q} k[\Sigma_d]\otimes F(p)\otimes G(q) \right), N \right)$$

is exactly given by a map $f : k[\Sigma_d] \otimes F(p) \otimes G(q) \to N$ such that for all $(\alpha_p, \alpha_q) \in \Sigma_p \times \Sigma_q, \sigma \in \Sigma_d, f_p \in F(p)$ and $g_q \in G(q)$ we have:

$$f(\sigma.(\alpha_p,\alpha_q)\otimes f_p\otimes g_q) = f(\sigma\otimes\alpha_p.f_p\otimes\alpha_q.g_q)$$

[2] Recall that given a morphism of associative algebras $f: A \to B$, we have an induction/extension of scalars functor $B \otimes_A -: \mathrm{Mod}_A \to \mathrm{Mod}_B$, left adjoint to the restriction of scalars along f. When f arises from a group morphism, such as $\Sigma_p \times \Sigma_q \hookrightarrow \Sigma_n$ for $p+q=n$, we refer to the corresponding induction as the induced representation

$$k[\Sigma_n]\otimes_{k[\Sigma_p\times\Sigma_q]}(-) = \mathrm{Ind}_{\Sigma_p\times\Sigma_q}^{\Sigma_n} : (\Sigma_p\times\Sigma_q)\text{-Rep} \to \Sigma_n\text{-Rep}.$$

[3] With $p \otimes q := p+q$.

But this is exactly the data of a map:

$$\mathrm{Hom}_{\mathrm{Mod}_{k[\Sigma_d]}}\left(k[\Sigma_d] \otimes_{k[\Sigma_p \times \Sigma_q]} F(p) \otimes G(q), N\right)$$

This shows that:

$$\begin{aligned} F \otimes_{\mathrm{Day}} G(d) &= \coprod_{p+q=d} (F(p) \otimes G(q)) \otimes_{k[\Sigma_p \times \Sigma_q]} k[\Sigma_d] \\ &= \coprod_{p+q=d} \mathrm{Ind}_{\Sigma_p \times \Sigma_q}^{\Sigma_d} (F(p) \otimes G(q)) \end{aligned}$$

□

Remark 2.7 Taking successive Day tensor powers of a functor leads to the following formula:

$$G^{\otimes_{\mathrm{Day}} k}(n) = \bigoplus_{i_1+\cdots+i_k=n} \mathrm{Ind}_{\Sigma_{i_1} \times \cdots \times \Sigma_{i_k}}^{\Sigma_n} (F(i_1) \otimes \cdots \otimes F(i_k))$$

Objects of arity n in this symmetric sequence can be graphically understood as having k trees in parallel with total arity $i_1 + \cdots + i_k = n$, with labels of different trees potentially permuted among them:

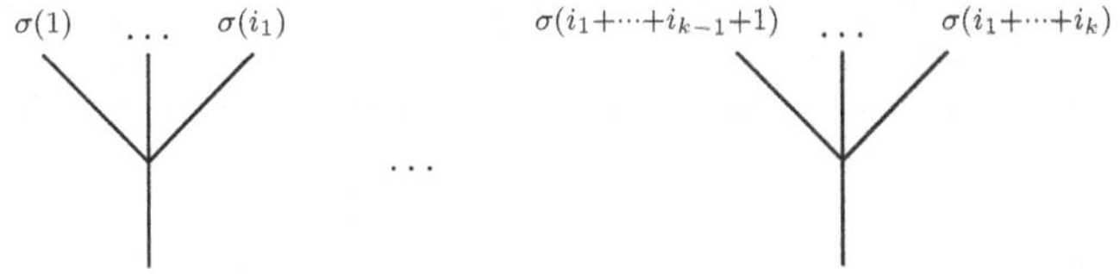

The induction, $\mathrm{Ind}_{\Sigma_{i_1} \times \cdots \times \Sigma_{i_k}}^{\Sigma_n}$, describes the action of Σ_n on the n entries in a way that is compatible with the actions given on each individual tree, $\Sigma_{i_1} \times \cdots \times \Sigma_{i_k}$. This consists in adding the $i_1, \ldots, i_k$ shuffles $\mathrm{Sh}(i_1, \ldots, i_k)$ which are canonical representatives of the cosets $\Sigma_n / \Sigma_{i_1} \times \cdots \times \Sigma_{i_k}$.

We want now to describe all the ways we can put k trees (elements in $M^{\otimes_{\mathrm{Day}} k}$) as input for a k-ary tree so that we can compose them. This defines a new symmetric sequence described by the following non-symmetric monoidal product:

Definition 2.8 We define the *composition of symmetric sequences* to be the monoidal structure (see [Lor15, Section 6.3]) on $\mathbf{Seq}_\Sigma$ given by:

$$F \circ G := \int^m F(m) \otimes G^{\otimes_{\mathrm{Day}} m}$$

We obtain the formula:

$$F \circ G(n) = \bigoplus_{k \geq 0} F(k) \otimes_{k[\Sigma_k]} \left(\bigoplus_{i_1+\cdots+i_k=n} \mathrm{Ind}_{\Sigma_{i_1} \times \cdots \times \Sigma_{i_k}}^{\Sigma_n} (G(i_1) \otimes \cdots \otimes G(i_n)) \right)$$

Graphically $F \circ G$ corresponds to the symmetric sequence of all possible concatenation of a F-tree with k-entries with k G-trees on top on it:

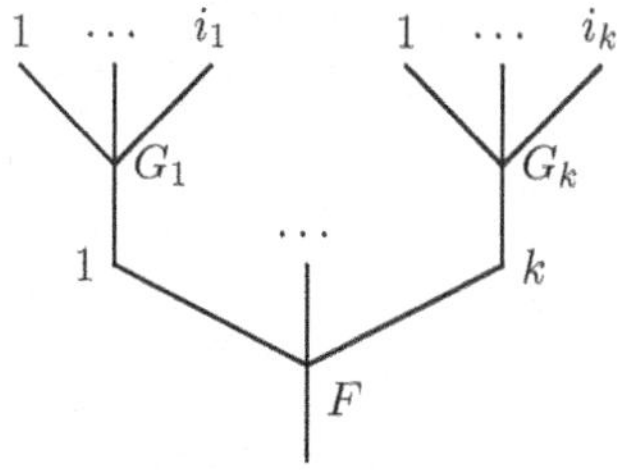

$\in (F \circ G)(i_1 + \cdots + i_k)$ (3)

Such concatenation will be called a 2-leveled tree. In the picture above we display the unconcatenated labels, but keep in mind that the bottom labels disappear and the top level is now labeled by a permutation of $i_1 + \cdots + i_k$. In general, a n-fold product for $\circ$ will give a n-leveled tree.

The unit of this monoidal product will be denoted by I and is the symmetric sequence given by $I(1) = k$ and $I(n) = 0$ for $n \neq 1$.

The notion of composition of an operad amounts to sending a 2-leveled tree, given by an element in $F \circ F$, to a new operation in F viewed as the "composition" of these leveled trees.

Definition 2.9 The category of *operads* $\mathscr{P}$ valued in Mod_k, denoted by **Op**, is the category of monoid objects in $\mathbf{Seq}_\Sigma$ for $\circ$. Concretely an object $\mathscr{P} \in \mathbf{Op}$ is a symmetric sequence $\mathscr{P}$ together with an associative multiplication and a unit morphisms:

$$\mu : \mathscr{P} \circ \mathscr{P} \to \mathscr{P} \qquad \eta : I \to \mathscr{P}$$

μ corresponds to composing the operations by sending any concatenation of trees as depicted in 3 (with $G = F = \mathscr{P}$) to a new element in $\mathscr{P}$. In particular for each $n \in \mathbb{N}$ and $i_1, \cdots, i_n \in \mathbb{N}$, we get the composition map:

$$\mathscr{P}(n) \otimes (\mathscr{P}(i_1) \otimes \cdots \otimes \mathscr{P}(i_k)) \to \mathscr{P}(i_1 + \cdots + i_k)$$

Which pictorially gives:

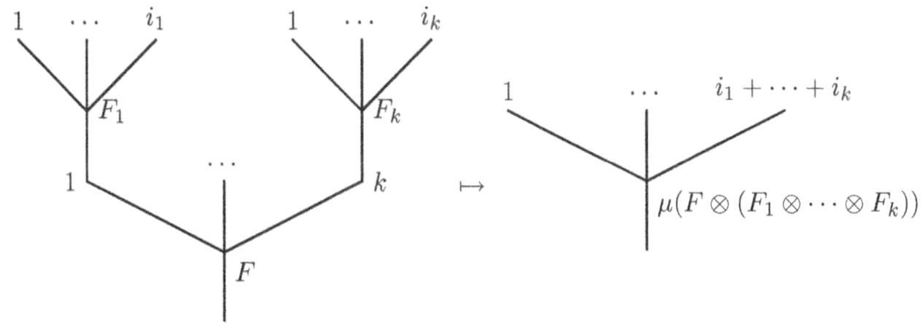

These maps satisfy the natural associativity and unitality conditions coming from the definition of monoids. In particular this means that given a n-level tree in $\mathscr{P}^{\circ n}$, it is canonically sent to a tree $\mathscr{P}$ no matter in which order we chose to compose each levels.

Definition 2.10 Equivalently, an operadic structure can be completely described via the *partial compositions*. For $\alpha \in \mathscr{P}(n)$, $\beta \in \mathscr{P}(m)$ and for all $1 \leq i \leq n$ we defined $\circ_i$ as:

$$\alpha \circ_i \beta = \mu_{\mathscr{P}}(\alpha, (I, \cdots, I, \beta, I, \cdots, I)) \in \mathscr{P}(n+m-1)$$

where β is in the i-th position in the n-tuple $(I, \cdots, I, \beta, I, \cdots, I)$. This can be depicted as:

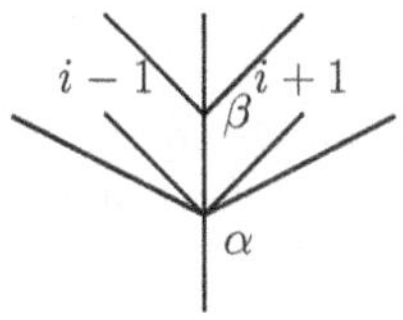

In the picture above we display the original label, but keep in mind that relabeling after composition will make the $i+1$ label into $i+m$.

As a consequence of this equivalent description, one can interpret an operad structure on a symmetric sequence $\mathscr{P}$, to be a rule to assign to any tree T (rooted and planar, but without any leveling), with internal vertices decorated by elements of $\mathscr{P}$ in appropriate arity, a single element of $\mathscr{P}(l)$, where l is the number of leaves of T.

Construction 2.11

- **Free operad:** The forgetful functor $\mathbf{Op} \to \mathbf{Seq}_\Sigma$ has a left adjoint $\mathfrak{T}$. Given a symmetric sequence E, we denote by $\mathfrak{T}E$ the *free operad* generated by E.

  As a symmetric sequence, $\mathfrak{T}E$ is spanned by all possible trees whose internal vertices are decorated by elements of E of the appropriate arity and by I of arity 1 (which we interpret as having no internal vertices). The (cohomological) degree of the tree is the sum of the degrees of all internal vertices and the differential acts vertex by vertex. The operadic structure is done by grafting as depicted in the following example:

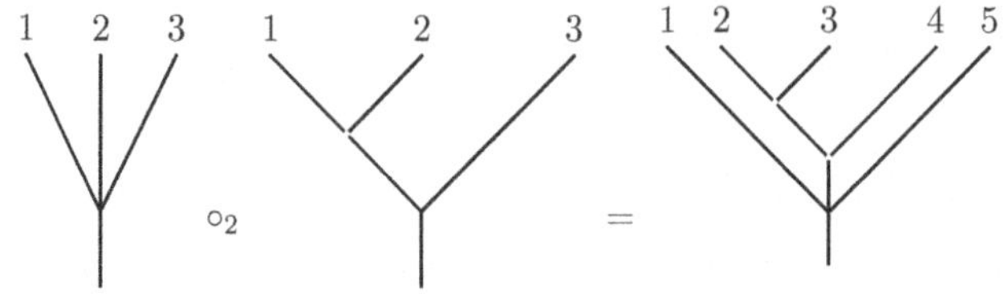

We will denote by $\mathfrak{T}^{(n)}E \subset \mathfrak{T}E$ the subcomplex spanned by trees with exactly n internal vertices.

- **Operad defined by generators and relations:** Consider an operadic ideal[4] $\mathscr{I} \subset \mathscr{P}$. In this situation, the cokernel of the map $\mathscr{I} \to \mathscr{P}$, denoted by ${}^{\mathscr{P}}\!/_{\mathscr{I}}$, is called the *quotient operad.*

  If we consider $E \in \mathbf{Seq}_{\Sigma}$, $R \subset \mathfrak{T}E$ and (R) the operadic ideal generated by R in $\mathfrak{T}E$. Then the operad generated by E with relations R is defined by:

$$\mathscr{P}(E, R) := {}^{\mathfrak{T}E}\!/_{(R)}.$$

Definition 2.12 An operad presented in terms of generators and relations $\mathscr{P}(E, R)$ as above is said to be *binary* if the generating symmetric sequence is concentrated in arity 2, i.e. $E(\neq 2) = 0$. It is said to be *quadratic* if the $R \subset \mathfrak{T}^{(n)}E$, i.e. if relations always involve two operations.

Loosely speaking, operads are naturally given in terms of generators and relations, whenever the algebras they describe are defined in terms of their defining operations and the properties they must satisfy.

Remark 2.13 Let E be a symmetric sequence such that $E(\neq 2) = 0$. While there are only two possible arity 3 trees one can write using two binary vertices, due to the symmetries, $\mathfrak{T}^{(2)}E \cong (E(2) \otimes E(2))^{\oplus 3}$. Indeed, there are exactly three $(2, 1)$ shuffles. Pictorially the three types of elements of $\mathfrak{T}^{(2)}E$ are:

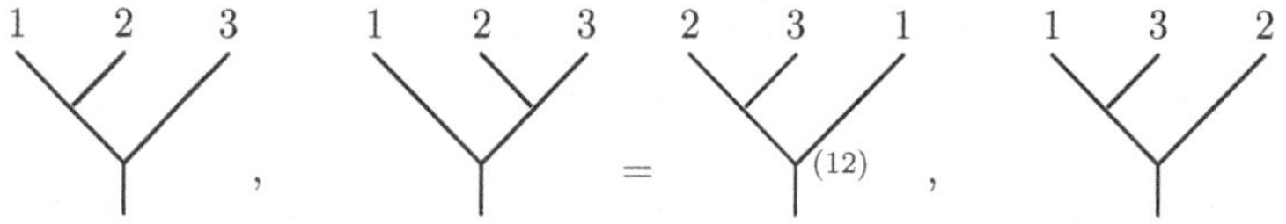

with internal vertices labeled by elements of $E(2)$.

[4] $\mathscr{I} \subset \mathscr{P}$ is called an *operadic ideal* if $\mu(l \otimes (l_1 \otimes \cdots \otimes l_k)) \in \mathscr{I}$ as soon as one of $l, l_1, \cdots, l_k$ is in $\mathscr{I}$.

Example 2.14

- **Trivial operad:** The symmetric sequence I defined by $I(1) = k$ and $I(k) = 0$ for $k \neq 1$ is the unit for $\circ$ and possesses a natural operad structure given by the identity as monoidal product and unit:

$$I = I \circ I \to I \qquad I \to I$$

- **Associative operad:** The associative operad is the operad generated by two binary operations, m and its permutation $m' = m \cdot (12)$, subject to the quadratic associativity relations which is the smallest Σ_3-submodule of $\mathfrak{T}^{(2)}E$ containing:

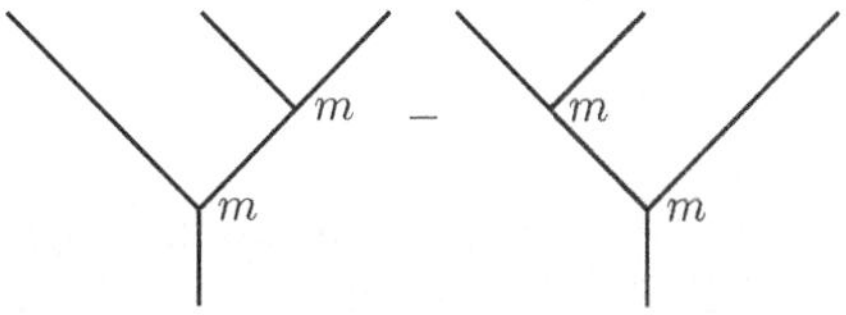

The complex $\mathfrak{T}^{(2)}E = (E(2) \otimes E(2))^{\oplus 3}$ is generated by all permutations of the partial compositions $m \circ_1 m$ and $m \circ_2 m$. The relations imposed for the associative algebra are generated by all the permutations of $m \circ_1 m - m \circ_2 m$:

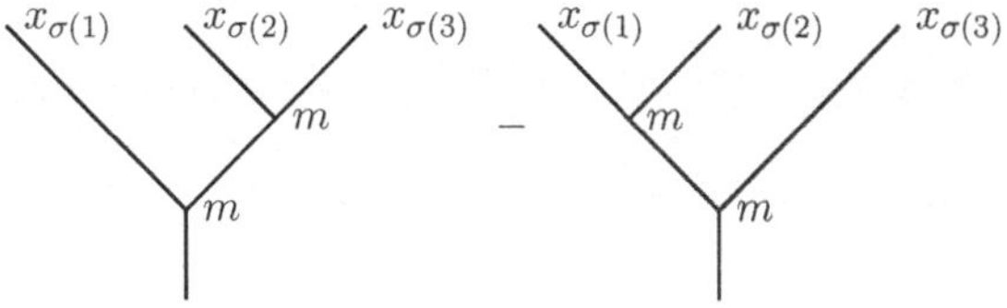

The quotient vector space we obtain in arity 3 is therefore 6-dimensional and is isomorphic to $k[\Sigma_3]$. In general, we have that **Ass**$(0) = 0$ and for $n \geq 1$, **Ass**$(n) = k[\Sigma_n]$ together with the natural action of Σ_n.

- **Commutative operad:** The commutative operad is defined as the operad generated by a single binary operation, invariant by the Σ_2 action (i.e. commutative) with quadratic relations encoding associativity. The free operad satisfies $\mathfrak{T}^{(2)}E = (E(2) \otimes E(2))^{\oplus 3}$ and is therefore 3-dimensional. If we write xy for the product in $E(2)$, $\mathfrak{T}^{(2)}E$ is freely generated by $x(yz)$, $(xz)y$ and $(xy)z$.

  So we only have to add the associativity relations $x(yz) - y(zx)$, $y(zx) - z(xy) = (zx)y - z(xy)$. We give a pictorial description for the first relation:

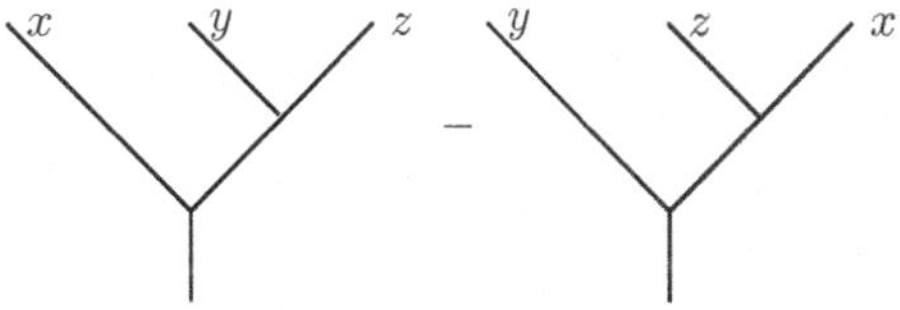

Since there are 3 generators and 2 relations, the commutative operad is therefore 1-dimensional in arity 3. In general, we have that $\mathbf{Com}(n) = k$ for all $n \geq 1$ with the trivial action.

- **Unital commutative operad:**

  The unital commutative operad **uCom** is an example of an operad which is neither binary nor quadratic. Besides the binary generator there is one additional generator, corresponding to the unit, which is an arity zero element $E(0) = k$. The relation involving the unit involves a quadratic and a constant term

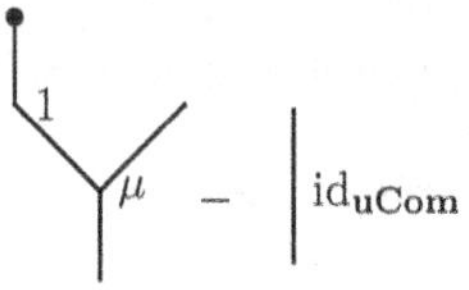

  which lives in $\mathfrak{T}^{(2)}E \oplus \mathfrak{T}^{(0)}E$. The operad **uCom** differs from **Com** only in arity 0, $\mathbf{uCom}(0) = k$. Similarly, there is an operad of unital associative algebras which only differs from **Ass** in that $\mathbf{uAss}(0) = k$.
- **Lie operad:** The Lie operad, denoted by **Lie**, is generated by a binary operation [-, -] acted upon by Σ_2 via $[\text{-}, \text{-}] \cdot \tau = -[\text{-}, \text{-}]$. It satisfies the associativity relation given by the Jacobiator. We have $\mathbf{Lie}(2) = k$ but the description of $\mathbf{Lie}(n)$ for $n > 2$ is more complicated. We refer to [LV12, Section 13.2.3 and 13.2.4] for more details on the description of **Lie**.
- **Endomorphisms operad:** We will now see the first operad not naturally presented in terms of generators and relations. Take $V \in \mathrm{Mod}_k$. We define the endomorphism operad of V, denoted $\mathbf{End}_V$, by:

$$\mathbf{End}_V(n) := \underline{\mathrm{Hom}}_k \left(V^{\otimes n}, V\right)$$

  where the action of Σ_n on $\mathbf{End}(n)$ is obtained from the natural action of Σ_n on $V^{\otimes n}$. The composition is naturally given by the composition for $\underline{\mathrm{Hom}}$ defined by $\mu(f, f_1, \cdots, f_m) = f \circ (f_1 \otimes \cdots \otimes f_m)$ with $f \in \mathbf{End}_V(m)$.

  Loosely speaking, $\mathbf{End}_V$ is the operad in which all possible operations on V live. We will see in Sect. 2.2 that we can define a $\mathscr{P}$-algebra structure on V as a morphism of operads $\mu_V : \mathscr{P} \to \mathbf{End}_V$.
- **Coendomorphisms operad:** Take $V \in \mathrm{Mod}_k$. We define the coendomorphism operad of V to be $\mathbf{coEnd}_V$ where:

$$\mathbf{coEnd}_V(n) := \underline{\mathrm{Hom}}(V, V^{\otimes n})$$

  and the action of Σ_n on $\mathbf{coEnd}(n)$ is obtained from the natural action of Σ_n on $V^{\otimes n}$. The composition is naturally given by the composition for $\underline{\mathrm{Hom}}$ given by $\mu(f, f_1, \cdots, f_m) = f_1 \otimes \cdots \otimes f_m \circ f$ with $f \in \mathbf{coEnd}_V(m)$.

  Similar to the previous example, $\mathbf{coEnd}_V$ can be used to define $\mathscr{P}$-coalgebra structures on V.

Now that we have discussed the notion of operad, which amounts to describing how to compose n-ary operations, we will describe the notion of cooperad, where instead of composing we will decompose an n-ary operation into a sum of i-leveled trees. In a similar fashion to Definition 2.8, we define the *complete composition* of symmetric sequences to be

$$F\hat{\circ}G = \int_m F(m) \otimes G^{\otimes_{\mathrm{Day}} m}.$$

Proposition 2.15 *For F and G in* $\mathbf{Seq}_\Sigma$, *then we have the following formula:*

$$F\hat{\circ}G(n) = \prod_{k\geq 0}\left(F(k)\otimes_k\left(\bigoplus_{i_1+\cdots+i_k=n}\mathrm{Ind}_{\Sigma_{i_1}\times\cdots\Sigma_{i_k}}^{\Sigma_n}(G(i_1)\otimes\cdots\otimes G(i_k))\right)\right)^{\Sigma_k}$$

Compared with $\circ$, $\hat{\circ}$ carries some disadvantages, one of which being that it is not monoidal (since direct products do not distribute along tensor products).

In order to circumvent that problem, we define another monoidal product $\bar{\circ}$ as the sub-symmetric sequence of $\int_m F(m) \otimes G^{\otimes_{\mathrm{Day}} m}$ given by the direct sum instead. Using the formula obtained in Proposition 2.15 we can define the following monoidal product by replacing the product by a direct sum:

$$F\bar{\circ}G(n) = \bigoplus_{k\geq 0}\left(F(k)\otimes_k\left(\bigoplus_{i_1+\cdots+i_k=n}\mathrm{Ind}_{\Sigma_{i_1}\times\cdots\Sigma_{i_k}}^{\Sigma_n}(G(i_1)\otimes\cdots\otimes G(i_k))\right)\right)^{\Sigma_k} \tag{4}$$

Remark 2.16 The normalization morphism $V^G \to V_G$ from invariants to coinvariants induces a natural transformation $\bar{\circ} \to \circ$ sending the direct sum of invariants to the direct sum of coinvariants. Since we work in characteristic 0 this is a natural isomorphism and therefore we will not worry too much about the distinction. In particular, $\bar{\circ}$ is indeed monoidal.

Definition 2.17 The category of *cooperads in* Mod_k, denoted **coOp** is defined a the category of comonoids in $\mathbf{Seq}_\Sigma$ for $\bar{\circ}$. Concretely an object $\mathscr{C} \in$ **coOp** is a symmetric sequence $\mathscr{C}$ together with cocomposition and counit morphisms:

$$\Delta : \mathscr{C} \to \mathscr{C}\bar{\circ}\mathscr{C} \qquad \epsilon : \mathscr{C} \to I$$

Remark 2.18 This pictorial description also involves trees for the n-ary operations in $M(n)$. This does not change compared to operad because of the natural isomorphism of Remark 2.16. However, the difference with cooperads is that the coalgebra structure will not compose trees but rather decompose them into a sum of concatenations of trees built via $\bar{\circ}$.

For example:

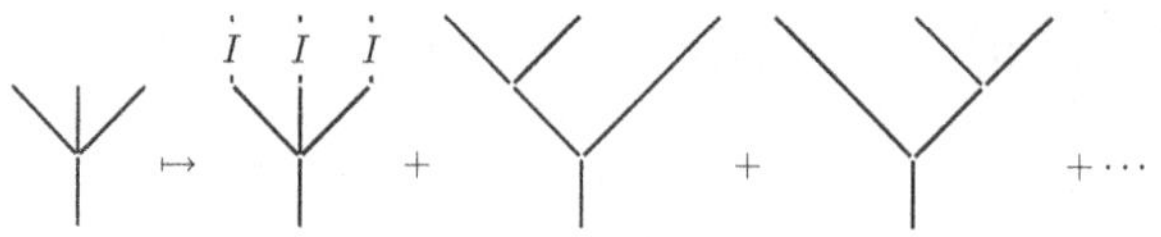

Concretely, it is the data of maps:

$$\mathscr{C}(n) \to \bigoplus_{k\geq 0} \left(\mathscr{C}(k) \otimes \left(\bigoplus_{i_1+\cdots+i_k=n} \mathrm{Ind}_{\Sigma_{i_1}\times\cdots\Sigma_{i_k}}^{\Sigma_n} (\mathscr{C}(i_1) \otimes \cdots \otimes \mathscr{C}(i_k)) \right)\right)^{\Sigma_k}$$

together with a counit, satisfying some coassociativity and counitality conditions derived from the comonoid axioms.

Similar to the case of operads, cooperads also have a notion of partial cocomposition, which consists in cocomposing along a single edge i into two vertex trees. This is given by maps $\Delta_i : \mathscr{C}(n) \to \mathscr{C}(n-k+1) \otimes \mathscr{C}(k)$, for all $n \geq 0$, $i = 1, \ldots, n$ and $k \leq n$.

Definition 2.19 A cooperad $\mathscr{C}$ is *conilpotent* if for every $m \in \mathscr{C}(n)$, there exists an N such that the iteration of any choice of N partial cocomposition maps is equal to 0. The category of conilpotent cooperads will be denoted by $\mathbf{coOp}^{\mathrm{conil}}$

All cooperads we will consider in this text will be conilpotent. Notice that any cooperad such that $\mathscr{C}(0) = 0$ and $\mathscr{C}(1) = k$ is conilpotent (N can be taken to be $n-1$ for a given $m \in \mathscr{C}(n)$).

Construction 2.20

- **Cofree conilpotent cooperad:** The cofree conilpotent cooperad associated to a symmetric sequence E is denoted $\mathfrak{T}^{\mathrm{co}}E$ is the image on E via the cofree functor defined as the right adjoint of the forgetful functor $F : \mathbf{coOp}^{\mathrm{conil}} \to \mathbf{Seq}_\Sigma$. The underlying symmetric sequence of $\mathfrak{T}^{\mathrm{co}}E$ is isomorphic to the one of $\mathfrak{T}E$ and the cocomposition of a tree is obtained by "cutting" along an internal edge. Such a cooperad is conilpotent since a given tree with e edges can be only non-trivially partially cocomposed up to e times.
- **Cooperad by generators and relations:** The cooperad cogenerated by a symmetric sequence E with corelations $R \subset \mathfrak{T}^{\mathrm{co}}E$, denoted $\mathscr{C}(E, R)$, is defined to be the smallest subcooperad of $\mathfrak{T}^{\mathrm{co}}E$ containing R. In the quadratic binary case, we have $\mathscr{C}(E, R)(2) = E$ and $\mathscr{C}(E, R)(3) = R$.
- **Dual of an operad:** The linear dual of a symmetric sequence M is a symmetric sequence $M^\vee$ defined in each arity by the linear dual $M^\vee(n)$. For any cooperad $\mathscr{C}$, $\mathscr{C}^\vee$ has the structure of an operad. Since $\bar{\circ}$ is not exactly the dual version of $\circ$, the converse is not always true. For an operad $\mathscr{P}$ we only get a map, $\mathscr{P}^\vee \to \mathscr{P}^\vee \hat{\circ} \mathscr{P}^\vee$ which might not factor through a comonoid $\mathscr{P}^\vee \to \mathscr{P}^\vee \bar{\circ} \mathscr{P}^\vee$. If we further suppose that $\mathscr{P}(0) = 0$ and that the pre-image of each element in $\mathscr{P}$ by $\mu_{\mathscr{P}}$ is finite dimensional then we get a cooperad structure on $\mathscr{P}^\vee$ (see [LV12, Section 5.8.2]).

2.2 Algebraic Structures Over an Operad or Cooperad

In this section we are going to define the notion of algebra and coalgebra both over operads and cooperads. We already gave some definitions in the case of operads in Example 2.14, but there are definitions in terms of modules and comodules over the operads viewed as monoids. Proposition 2.26 shows that both notions coincide. More generally, in this section will define several adjunctions involving the monoidal products defined earlier. This will allow us to describe the free $\mathscr{P}$-algebra and cofree $\mathscr{P}$-coalgebra functors associated to an operad $\mathscr{P}$ (see Proposition 2.27). Finally we will discuss the naturality of the category of algebras with respect to morphisms of operads.

Definition 2.21 Let $A \in \mathrm{Mod}_k$ and $\mathscr{P} \in \mathbf{Op}$.

- The structure of a *$\mathscr{P}$-algebra* on A is the structure of a $\mathscr{P}$-module on A with respect to $\circ$, where we interpret A as a symmetric sequence concentrated in arity 0:

  $$\mu\colon \mathscr{P} \circ A \to A$$

  Unraveling the definition, this amounts to associating to every $p \in \mathscr{P}(n)$ and $a_1, \ldots, a_n$ a single element in A, which we can shorten to $p(a_1, \ldots, a_n)$ if the map μ is implicit. The category of $\mathscr{P}$-algebras is denoted

  $$\mathbf{Alg}_{\mathscr{P}} := \mathscr{P}\text{-}\mathrm{Mod}_{\circ},$$

  using the convention that modules will always refer to sequences concentrated in arity 0. We will see in Proposition 2.26 that a $\mathscr{P}$-algebra structure on A is equivalent to the choice of a morphism of operads $\mathscr{P} \to \mathbf{End}_A$.
- Perhaps the expected way to define a $\mathscr{P}$-coalgebra would be in terms of a map $A \to \mathscr{P}^{\vee} \hat{\circ} A$. However, as $\hat{\circ}$ is not monoidal, it makes no sense to ask for this map to be a comodule. As such, a *$\mathscr{P}$-coalgebra* structure on A is defined as a map of operads:

  $$\mathscr{P} \to \mathbf{coEnd}_A$$

  The category of $\mathscr{P}$-coalgebras is denoted

  $$\mathbf{coAlg}_{\mathscr{P}} := \mathrm{Hom}_{\mathbf{Op}}\left(\mathscr{P}, \mathbf{coEnd}_A\right)$$

- We will see with Lemma 2.25 that a coalgebra $\mathscr{P} \to \mathbf{coEnd}_A$ induces a map $A \to \mathscr{P}^{\vee} \hat{\circ} A$, assuming $\mathscr{P}$ is arity-wise dualizable[5] and $\mathscr{P}^{\vee}$ is a cooperad.

[5] A symmetric sequence M is called arity-wise dualizable if for each $n \in \mathbb{N}$, $M(n)$ is dualizable, which in our setting corresponds to bounded and finite dimensional in every degree.

Therefore we can define a *conilpotent coalgebra* as a comodule:[6]

$$A \to \mathscr{P}^{\vee}\bar{\circ}A$$

The category of conilpotent $\mathscr{P}$-coalgebras is denoted by $\mathbf{coAlg}_{\mathscr{P}}^{\mathrm{conil}}$ and we have:

$$\mathbf{coAlg}_{\mathscr{P}}^{\mathrm{conil}} = \mathscr{P}^{\vee} - \mathrm{coMod}_{\bar{\circ}}$$

Definition 2.22 Let $A \in \mathrm{Mod}_k$ and $\mathscr{C}$ be a cooperad. The structure of a *$\mathscr{C}$-(co)algebra on A* is defined as a $\mathscr{C}^{\vee}$-(co)algebra structure on A. We denote by $\mathbf{Alg}_{\mathscr{C}}$ and $\mathbf{coAlg}_{\mathscr{C}}$ the categories of all $\mathscr{C}$-algebras and $\mathscr{C}$-coalgebras respectively. We have:

$$\mathbf{Alg}_{\mathscr{C}} := \mathscr{C}^{\vee}\text{-}\mathrm{Mod}_{\circ} \qquad \mathbf{coAlg}_{\mathscr{C}} := \mathrm{Hom}_{\mathbf{Op}}\left(\mathscr{C}^{\vee}, \mathbf{coEnd}_A\right)$$

Moreover, a conilpotent $\mathscr{C}$-coalgebra can be defined as a comodule in $\mathscr{C} - \mathrm{coMod}_{\bar{\circ}}$. We get:

$$\mathbf{coAlg}_{\mathscr{C}}^{\mathrm{conil}} := \mathscr{C} - \mathrm{coMod}_{\bar{\circ}}$$

Remark 2.23 If we replace Mod_k by Mod_R (with R a commutative algebra), to obtain (co)operads valued in Mod_R, we obtain the notion of (co)algebra in R-modules, that is R-linear algebraic structures.

In the Example 2.14, we described the endomorphism operad, $\mathbf{End}_A$, and defined a $\mathscr{P}$-algebra structure on A as a morphism of operad $\mathscr{P} \to \mathbf{End}_A$. It turns out that this coincides with the previous definition as we will explain with Proposition 2.26. We start with a fundamental Lemma expressing some adjunctions between the monoidal product and the (co)endomorphism symmetric sequences.

Definition 2.24 Given $M, N \in \mathrm{Mod}_k$, we define the following symmetric sequences:

$$\mathbf{End}_N^M(n) = \underline{\mathrm{Hom}}_k\left(M^{\otimes n}, N\right)$$

$$\mathbf{coEnd}_N^M(n) = \underline{\mathrm{Hom}}_k\left(M, N^{\otimes n}\right).$$

An application of some end-coend gymnastics, together with the hom-tensor adjunction, give the following lemma.

Lemma 2.25 *For all $\mathscr{P} \in \mathbf{Seq}_{\Sigma}$ and for all $M, N \in \mathrm{Mod}_k$, there is an isomorphism:*

$$\mathrm{Hom}_k(\mathscr{P} \circ M, N) \cong \mathrm{Hom}_{\mathbf{Seq}_{\Sigma}}\left(\mathscr{P}, \mathbf{End}_N^M\right)$$

---

[6] This comodule extends to a map $A \to \mathscr{P}^{\vee}\hat{\circ}A$ which is equivalent, thanks to Lemma 2.25, to a map of symmetric sequences $\mathscr{P} \to \mathbf{End}_A$. This turns out to also be a map of operad that defines the underlying coalgebra structure (by forgetting the conilpotency).

Morover if $\mathscr{P}$ is arity-wise dualizable then we have an isomorphism:

$$\mathrm{Hom}_k\left(M, \mathscr{P}^\vee \hat{\circ} N\right) \cong \mathrm{Hom}_{\mathbf{Seq}_\Sigma}\left(\mathscr{P}, \mathbf{coEnd}_N^M\right)$$

Proposition 2.26 *Given $A \in \mathrm{Mod}_k$ and $\mathscr{P} \in \mathbf{Op}$, we have:*

$$\mathscr{P}\text{-}\mathrm{Mod}_\circ(A) \cong \mathrm{Hom}_{\mathbf{Op}}(\mathscr{P}, \mathbf{End}_A)$$

Proof Following Lemma 2.25 it is a straightforward verification that the condition of $\mathscr{P} \circ A \to A$ being a $\mathscr{P}$-module structure on A is equivalent to the map of symmetric sequences $\mathscr{P} \to \mathbf{End}_A$ being a map of operads. □

The previous results and Lemma 2.25 give us some adjunctions using the monoidal product $\circ$ and the **End** constructions. In particular, we obtain the following constructions:

Proposition 2.27

- *Given $\mathscr{P}$ an operad, we have the following free-forgetful adjunction:*

$$\mathscr{P}(-) : \mathrm{Mod}_k \rightleftarrows \mathbf{Alg}_{\mathscr{P}} : (-)^\sharp$$

  The left adjoint $\mathscr{P}(-) : \mathrm{Mod}_k \to \mathbf{Alg}_{\mathscr{P}}$ is the free $\mathscr{P}$-algebra *functor that sends $A \in \mathrm{Mod}_k$ to $\mathscr{P}(A) := \mathscr{P} \circ A$.*
- *Given $\mathscr{C}$ an arity-wise dualizable cooperad, we have the following adjunction:*

$$(-)^\sharp : \mathbf{coAlg}_{\mathscr{C}}^{\mathrm{conil}} \rightleftarrows \mathrm{Mod}_k : \mathscr{C}(-)$$

  The right adjoint $\mathscr{C}(-) : \mathrm{Mod}_k \to \mathbf{coAlg}_{\mathscr{C}}^{\mathrm{conil}}$ is the cofree conilpotent coalgebra *functor that sends $A \in \mathrm{Mod}_k$ to $\mathscr{C}(A) := \mathscr{C}\bar{\circ}A$.*

  In particular, if $\mathscr{P}$ is arity-wise dualizable we get:

$$(-)^\sharp : \mathbf{coAlg}_{\mathscr{P}^\vee}^{\mathrm{conil}} \simeq \mathbf{coAlg}_{\mathscr{P}}^{\mathrm{conil}} \rightleftarrows \mathrm{Mod}_k : \mathscr{P}^\vee(-)$$

  this gives us the cofree $\mathscr{P}$-coalgebra.

Proof Let us consider the first adjunction. Notice that $\mathscr{P}(A)$ is naturally a $\mathscr{P}$-algebra via the structure maps of $\mathscr{P}$: $(\mathscr{P} \circ \mathscr{P}) \circ A \to \mathscr{P} \circ A$.

There is an inclusion of chain complexes $A = I \circ A \to \mathscr{P} \circ A$ induced by the unit $I \to \mathscr{P}$. Pulling back along this inclusion yields the "restriction to generators" map:

$$\mathrm{Hom}_{\mathbf{Alg}_{\mathscr{P}}}(\mathscr{P}(A), B) \to \mathrm{Hom}_k\left(A, B^\sharp\right)$$

We shall show that this map is a bijection by constructing its inverse. Let $f \in \mathrm{Hom}_k\left(A, B^\sharp\right)$. We consider

$$\mathscr{P} \circ A \overset{\mathrm{id} \circ f}{\to} \mathscr{P} \circ B \to B,$$

where the second map is the $\mathscr{P}$-algebra structure on B. It is easy to check that this defines a map of $\mathscr{P}$-algebras and establishes the inverse map.

The proof of the second bullet point is dual. □

Example 2.28

- For **uAss**, the free associative algebra functor is the tensor algebra functor. Indeed, given $V \in \mathrm{Mod}_k$, we have that:

$$\mathbf{Ass}(V) = \bigoplus_{n \geq 0} k[\Sigma_n] \otimes_{k[\Sigma_n]} V^{\otimes n} \simeq \bigoplus_{n \geq 0} V^{\otimes n} = TV$$

- For **Com**, the free commutative algebra functor is the non-unital symmetric algebra functor:

$$\mathbf{Com}(V) = \bigoplus_{n \geq 1} k \otimes_{k[\Sigma_n]} V^{\otimes n} \simeq \bigoplus_{n \geq 1} (V^{\otimes n})_{\Sigma_n} = \mathrm{Sym}_k^{\geq 1} V$$

Proposition 2.29 *Given $f : \mathscr{P} \to \mathscr{Q}$ a morphism of operads we obtain an adjunction:*

$$f_! : \mathbf{Alg}_{\mathscr{P}} \rightleftarrows \mathbf{Alg}_{\mathscr{Q}} : f^*$$

where f^* sends the $\mathscr{Q}$-structure $\mathscr{Q} \to \mathbf{End}_A$ to the composition $\mathscr{P} \to \mathscr{Q} \to \mathbf{End}_A$ and $f_!(B)$, with $B \in \mathbf{Alg}_{\mathscr{P}}$, is defined as the reflexive coequalizer:

$$\mathscr{Q} \circ_{\mathscr{P}} B := \mathrm{coeq}\left(\mathscr{Q} \circ \mathscr{P} \circ B \overset{\mu_{\mathscr{Q}} \circ f}{\underset{\mu_B}{\rightrightarrows}} \mathscr{Q} \circ B \right)$$

Remark 2.30 The notation $\circ_{\mathscr{P}}$ is suggestive. It is a relative version of the usual operation $\circ$. In fact for $f : I \to \mathscr{P}$ this adjunction recovers the free-forget adjunction, we get $f^* A = A^\sharp$ and $f_! V = \mathscr{P} \circ_I V = \mathscr{P} \circ V$.

Remark 2.31 When $\mathscr{P}$ is an augmented operad, that is an operad together a retract $\mathscr{P} \to I$ of the unit, we can define the trivial algebra functor given as the right adjoint $\mathrm{Triv} : \mathbf{Alg}_I = \mathrm{Mod}_k \to \mathbf{Alg}_{\mathscr{P}}$ coming from Proposition 2.29 applied to the augmentation.

2.2.1 A Short Digression on the Universal Enveloping Algebra

Recall that the universal enveloping algebra of a Lie algebra $\mathfrak{g}$ is classically defined to be $\mathfrak{U}\mathfrak{g} := T(\mathfrak{g})/[x, y] = x \otimes y - y \otimes x$. This is exactly the induction functor along the operad morphism $f\colon \mathbf{Lie} \to \mathbf{Ass}$,

$$\mathbf{Alg}_{\mathbf{Lie}} \underset{f^*}{\overset{\mathfrak{U}}{\rightleftarrows}} \mathbf{Alg}_{\mathbf{Ass}}$$

where $f^*(A)$ is the Lie algebra structure on A given by the Lie bracket $[x, y] = xy - yx$.

It will be important later on to notice that there is a natural coproduct on $\mathfrak{U}\mathfrak{g}$ given by

$$\Delta(x) = 1 \otimes x + x \otimes 1 \in \mathfrak{U}\mathfrak{g} \otimes \mathfrak{U}\mathfrak{g}, \quad \text{for } x \in \mathfrak{g},$$

which can be extended to the whole $\mathfrak{U}\mathfrak{g}$ by making $\Delta(xy) = \Delta(x)\Delta(y)$. Defining an antipodal map $S(x) = -x$, it follows that the universal enveloping algebra functor extends to a functor into (cocommutative) Hopf algebras.

Definition 2.32 Given a Hopf algebra H, we say that $g \in H \setminus \{0\}$ is *grouplike* if $\Delta(g) = g \otimes g$. We say that $x \in H$ is *primitive* if $\Delta(x) = 1 \otimes x + x \otimes 1$.

One can observe that for any Hopf algebra, the grouplike elements form a group, with inverse given by the antipode. As we will make precise in Sect. 3.2.1, under some completion assumptions, when H is a universal enveloping algebra $\mathfrak{U}\mathfrak{g}$, the grouplike elements can be interpreted as the exponential group of $\mathfrak{g}$.

It is generally difficult to explicitly describe the cochain complex (in other words provide a basis) underlying the induction $f_!$ along an arbitrary morphism of operads $f\colon \mathscr{P} \to \mathscr{Q}$. In this particular case $f\colon \mathbf{Lie} \to \mathbf{Ass}$, the PBW (Poincaré–Birkoff–Witt) theorem gives a complete answer to this problem (in characteristic zero).

Theorem 2.33 (PBW, [Qui69, Theorem B.2.3]) *There is an isomorphism of cochain complexes:*

$$\mathfrak{U}\mathfrak{g} \cong \operatorname{Sym} \mathfrak{g}.$$

Furthermore, interpreting $\operatorname{Sym} \mathfrak{g}$ *as the cofree cocommutative coalgebra generated by* $\mathfrak{g}$, *this is an isomorphism of coalgebras.*

2.3 Classical Constructions for Algebraic Operads

This section is a recollection of classical constructions and definitions on operads. We start by defining the convolution operad (Definition 2.34) and the totalization of an operad (Definition 2.36). These will be used to define the notion of (Koszul) twisting morphism (Definition 2.55), which provides a convenient alternative description for the bar-cobar constructions (Definition 2.39). The goal of this section is to give the main definitions around the bar-cobar adjunction, which will be used later in order to take good cofibrant replacements (Proposition 2.56), describe homotopy algebras (Sect. 3.1.1) and study deformations of algebraic structures (Sect. 3.2).

Definition 2.34 Given an operad $\mathscr{P}$ and a cooperad $\mathscr{C}$, we can construct the *convolution operad*:

$$\mathrm{Conv}(\mathscr{C}, \mathscr{P})(n) := \underline{\mathrm{Hom}}_k\,(\mathscr{C}(n), \mathscr{P}(n))$$

The symmetric sequence structure on $\mathrm{Conv}(\mathscr{C}, \mathscr{P})$ is given, for each $n \geq 0$, by the action by conjugation: $\sigma.f := x \in \mathscr{C}(n) \mapsto \sigma.f(\sigma^{-1}.x)$.

The operadic structure is given by the map that sends

$$\phi := f \otimes (g_1 \otimes \cdots \otimes g_k) \in \mathrm{Conv}(\mathscr{C}, \mathscr{P}) \circ \mathrm{Conv}(\mathscr{C}, \mathscr{P})(n),$$

with $g_l \in \mathrm{Conv}(\mathscr{C}, \mathscr{P})(i_l)$ and $i_1 + \cdots + i_k = n$, to the element in $\mathrm{Conv}(\mathscr{C}, \mathscr{P})$ of arity n defined by the following composition:

$$\mathscr{C} \xrightarrow{\Delta} \mathscr{C}\hat{\circ}\mathscr{C} \to \mathscr{C}(k) \otimes \mathscr{C}(i_1) \otimes \cdots \otimes \mathscr{C}(i_k) \xrightarrow[\phi]{} \mathscr{P}(k) \otimes \mathscr{P}(i_1) \otimes \cdots \otimes \mathscr{P}(i_k) \hookrightarrow \mathscr{P} \circ \mathscr{P} \xrightarrow{\mu} \mathscr{P}$$

The object we are going to be interested in is not the convolution operad itself, but rather its totalization. The totalization of an operad is a pre-Lie algebra whose associated Lie algebra will be important in defining twisting morphisms and describing deformations of algebraic structures.

Definition 2.35 (Pre-Lie and Permutative Operads)

- A (right) *permutative algebra* is an associative algebra A with a binary operation satisfying:

$$x \cdot (y \cdot z) = (-1)^{|y||z|} x \cdot (z \cdot y)$$

  We denote by **Perm** the quadratic operad encoding permutative algebras.
- A (right) *pre-Lie algebra* is given by $\mathfrak{g} \in \mathrm{Mod}_k$ with a bilinear operation $\star$ such that:

$$(x \star y) \star z - x \star (y \star z) = (-1)^{|y||z|} \big((x \star z) \star y - x \star (z \star y)\big)$$

  We denote by **preLie** the quadratic operad encoding pre-Lie algebras. **Perm** and **preLie** are related by the fact that they are Koszul dual of each other (Definition 2.60).

Pre-Lie algebras are so called since, similarly to associative algebras, the formula $[p, q] = p \star q - q \star p$ defines a Lie algebra structure on $\mathfrak{g}$. Moreover, there is a map **preLie** $\to$ **Ass** so that all associative algebras are pre-Lie algebras. In this text, the fundamental instances of pre-Lie algebras that do not come from an associative algebra will arise as totalizations of operads.

Definition 2.36 Given $\mathscr{P}$ an operad we can produce a pre-Lie algebra called the *totalization of* $\mathscr{P}$:

$$\mathrm{Tot}(\mathscr{P}) := \prod_{n \geq 0} \mathscr{P}(n)^{\Sigma_n} \in \mathbf{Alg}_{\mathbf{preLie}}$$

The pre-Lie product $\alpha \star \beta$ is defined as the sum of all possible ways to compose α and β. Pictorially, if $\alpha \in \mathscr{P}(3)^{\Sigma_3}$ and $\beta \in \mathscr{P}(2)^{\Sigma_2}$,

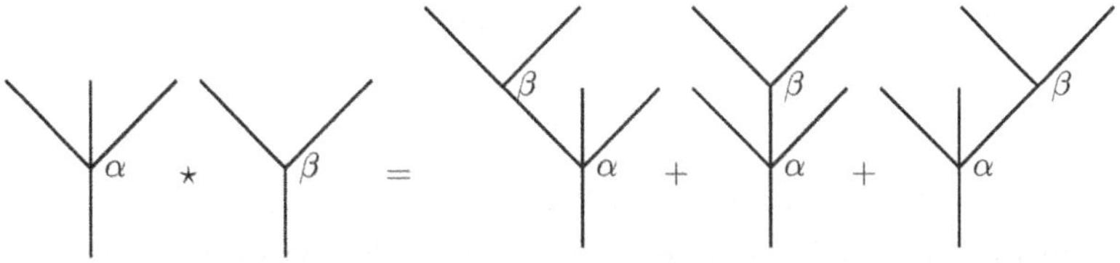

To be precise, the incoming edges should be labeled by all possible shuffles (this is similar to Remark 2.13), we refer to [LV12, Section 5.4.3] for details.

For non-symmetric operads the formula is exactly the sum of partial compositions:

$$\alpha \star \beta = \sum_{i=1}^{n} \alpha \circ_i \beta$$

It should not be surprising that one can construct pre-Lie algebras out of "tree-shaped compositions". Indeed, the pre-Lie operad is isomorphic to the operad of rooted trees [CL01].

Definition 2.37 Given a Lie algebra $\mathfrak{g}$, we define its *set of Maurer–Cartan elements* of $\mathfrak{g}$ to be:

$$\mathrm{MC}(\mathfrak{g}) := \left\{ x \in \mathfrak{g}_1 \mid dx + \frac{1}{2}[x, x] = 0 \right\}$$

Moreover, MC defines a functor from Lie algebras to sets.

Definition 2.38 (Twisting Morphisms) Given a cooperad $\mathscr{C}$ and an operad $\mathscr{P}$, the set of *twisting morphisms* is the set of Maurer–Cartan elements in $\mathrm{Tot}(\mathrm{Conv}(\mathscr{C}, \mathscr{P}))$:

$$\mathrm{Tw}(\mathscr{C}, \mathscr{P}) := \mathrm{MC}(\mathrm{Tot}(\mathrm{Conv}(\mathscr{C}, \mathscr{P})))$$

It corresponds to the set of morphisms $\alpha : \mathscr{C} \dashrightarrow \mathscr{P}$ of symmetric sequences of degree 1 satisfying the Maurer–Cartan equation, $d\alpha + \alpha \star \alpha = 0$.

Definition 2.39 ([LV12, Section 6.5.3]) There is an adjunction called the *Bar-cobar adjunction*:

$$\Omega : \mathbf{coOp}^{\mathrm{conil}} \rightleftarrows \mathbf{Op}^{\mathrm{aug}} : \mathbf{B}$$

between coaugmented conilpotent cooperads and augmented operads defined by $\Omega(\mathscr{C}) = \left(\mathfrak{T}(\overline{\mathscr{C}}[-1]), d_{\mathscr{C}} + d_{\Omega}\right)$ with $\overline{\mathscr{C}} := {}^{\mathscr{C}}/_{I}$ for Ω, and with $\mathbf{B}\mathscr{P} := \left(\mathfrak{T}^{co}(\overline{\mathscr{P}}[1]), d_{\mathscr{P}} + d_{\mathbf{B}}\right)$.

The underlying symmetric sequence of $\Omega(\mathscr{C})$ depends only on the underlying symmetric sequence of $\mathscr{C}$ and the cooperad structure only plays a role for the differential d_{Ω}. The differential d_{Ω} is defined on the generators, i.e. 1-vertex trees labeled by $c \in \overline{\mathscr{C}}[-1]$, as the sum over all possible partial cocompositions of c, and then extended to an arbitrary tree by derivations, which amounts to sum this formula for all vertices. The "coassociativity" property of partial cocomposition, together with the signs produced by the degree shifts, guarantees that d_{Ω} squares to zero (see [LV12, Section 6.5.2] for more details).

Similarly, $d_{\mathbf{B}}$ sums over all possible ways of contracting an edge on a tree using the partial composition of $\mathscr{P}$, see [LV12, Section 6.5.1]. We will see in Sects. 2.4 and 2.5 that this adjunction can provide a cofibrant replacement of an operad and will be used to define homotopy algebras.

In fact, maps $\Omega\mathscr{C} \to \mathscr{P}$ (or equivalently $\mathscr{C} \to \mathbf{B}\mathscr{P}$) can be described in terms of twisting morphisms. This point of view will prove to be very important when it comes to the description of the deformations of algebraic structures (see Sect. 3.2.1).

Proposition 2.40 (Rosetta Stone) *There are natural equivalences:*

$$\mathrm{Hom}_{\mathbf{Op}}(\Omega\mathscr{C}, \mathscr{P}) \simeq \mathrm{Tw}(\mathscr{C}, \mathscr{P}) \simeq \mathrm{Hom}_{\mathbf{coOp}^{\mathrm{conil}}}(\mathscr{C}, \mathbf{B}\mathscr{P})$$

Proof Take $\alpha : \Omega\mathscr{C} \to \mathscr{P}$. Ignoring differentials, this corresponds to a map from the free operad on $\overline{\mathscr{C}}[-1]$ and is thus equivalent to having a (co)unit preserving map $\tilde{\alpha} : \mathscr{C} \dashrightarrow \mathscr{P}$ of degree 1 of symmetric sequences. Moreover α is compatible with the differentials if and only if $\tilde{\alpha}$ is a Maurer–Cartan element.

Indeed, we have the following equivalences:

$$\begin{aligned}
\alpha(d_{\Omega\mathscr{C}}) = d_{\mathscr{P}} \circ \alpha &\Leftrightarrow \alpha(d_{\Delta} + d_{\mathscr{C}}) = d_{\mathscr{P}} \circ \alpha \\
&\Leftrightarrow (\alpha \circ d_{\Delta} + \alpha \circ d_{\mathscr{C}}) - d_{\mathscr{P}} \circ \alpha = 0 \\
&\Leftrightarrow \alpha \star \alpha + \partial\alpha = 0 \\
&\Leftrightarrow \frac{1}{2}[\alpha, \alpha] + \partial\alpha = 0
\end{aligned}$$

□

Remark 2.41 (A Remark on Augmentations) In Definition 2.39 we consider (co)operads that are (co)augmented which forces us to remove the (co)unit. Instead, we could have worked from the beginning with the equivalent categories of non-(co)unital (co)operads, but this makes the descriptions of (co)algebras, namely the (co)free one, more complicated.

While we will mostly hide such details under the rug, in practice this means that strictly speaking for propositions such as the Rosetta Stone to hold, one should require the (co)augmentation ideals the twisting morphisms to be maps compatible with the (co)augmentation, should they exist.

2.4 Model Categorical Aspects

Given an operad $\mathscr{P}$, we are interested in understanding the category of $\mathscr{P}$-algebras up to homotopy, i.e. together with the notion of quasi-isomorphisms as "weak equivalences". The theory of model categories gives us a way to handle these weak equivalences. In this section we will see that algebras over operads, and operads themselves form model categories.

2.4.1 Model Categories and Homotopical Algebra

Model categories are a tool introduced by Quillen in [Qui67] in order to study homotopy theory. The main problem is to understand objects of a category up to a notion of weak equivalences which are typically non-invertible morphisms.

A naive way to proceed is to formally invert all the weak equivalences (see [Hov07, Definition 1.2.1]). For a category $\mathscr{C}$ with a class of morphisms which we call weak equivalences $\mathscr{W}$, this formal localization will be denoted $\mathrm{h}\mathscr{C} := \mathscr{C}[\mathscr{W}^{-1}]$. But in general, descriptions of $\mathscr{C}[\mathscr{W}^{-1}]$ are very difficult to deal with. The category $\mathrm{h}\mathscr{C}$ can however be described via a universal property: There is a natural functor $\mathscr{C} \to \mathrm{h}\mathscr{C}$ such that any functor $\mathscr{C} \to \mathscr{D}$ sending morphisms in $\mathscr{W}$ to isomorphisms in $\mathscr{D}$ factors uniquely through $\mathrm{h}\mathscr{C}$. This universal property characterizes $\mathrm{h}\mathscr{C}$ (see [Hov07, Lemma 1.2.2]) up to equivalence of categories. Model structures are a tool to get workable descriptions of $\mathrm{h}\mathscr{C}$. In our framework, we will consider $\mathscr{C}$ a category having all small limits and small colimits which, besides a class of *weak equivalences* is equipped with two other classes of morphisms called *fibrations* and *cofibrations* and satisfying certain axioms making $\mathscr{C}$ a *model category*. For a detailed account we refer to [Hov07]. We use the phrase *trivial (co)fibration* to refer to a (co)fibration which is also a weak equivalence. Let us present the main features of model categories:

- Any morphism $A \to B$ can be factorized both as a cofibration followed by a trivial fibration or as a fibration followed by a trivial cofibration:

$$A \xrightarrow{\text{cof.}} B' \xrightarrow{\text{triv.fib.}} B$$

$$A \xrightarrow{\text{triv.cof.}} A' \xrightarrow{\text{fib.}} B$$

- An object A is called *fibrant* (resp. *cofibrant*) if the map to (resp. from) the terminal (resp. initial) object is a fibration (resp. cofibration). The factorization axioms ensure that all objects are weakly equivalent to a fibrant-cofibrant object (such an object is called a fibrant-cofibrant replacement).
- The class of all cofibrations (respectively fibrations) is completely determined by the classes of weak equivalences and fibrations (respectively weak equivalences and cofibrations) (see [Hov07, Lemma 1.1.10]). For example, a morphism $f : A \to B$ is a cofibration if and only if it has the *left lifting property* with respect all trivial fibrations.

This means that for all trivial fibrations $g : A' \to B'$ and all commutative squares:

$$\begin{array}{ccc} A & \longrightarrow & A' \\ {\scriptstyle f}\downarrow & {\scriptstyle h}\nearrow & \downarrow {\scriptstyle g} \\ B & \longrightarrow & B' \end{array}$$

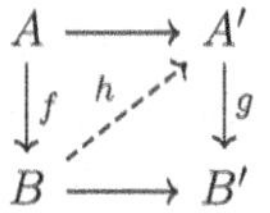

there exists a lift $h : B \dashrightarrow A'$ making the diagram commute. In particular, in many cases in this survey we will describe a model structure by describing the class of weak equivalences and fibrations, knowing that cofibrations are determined by them, even though they are usually more difficult to describe.

Note that similarly, fibrations and trivial fibrations are determined by trivial cofibrations and cofibrations via the right lifting property.

A priori, a weak equivalence may not have an inverse or even a quasi-inverse.[7] This means that to define the notion of "being weak equivalent", the only sensible to do is to ask for two objects, A and B, to be weakly equivalent if there is a zig-zag of weak equivalences connecting them:

$$\begin{array}{ccccccc} & A_1 & & \cdots & & A_n & \\ \swarrow & & \searrow & & \swarrow & & \searrow \\ A & \cdots & & & \cdots & & B \end{array}$$

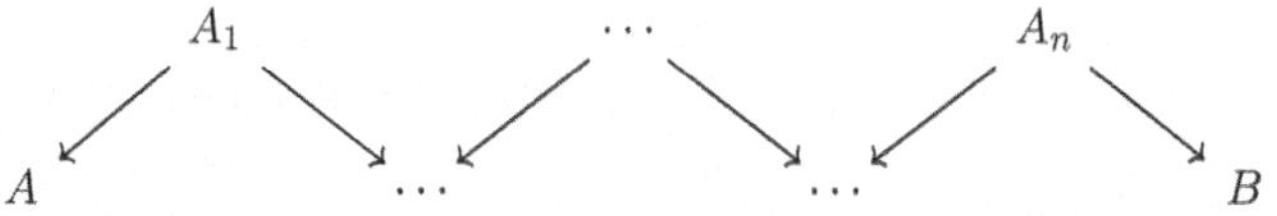

We denote this equivalence by $A \sim B$. On the other hand, a consequence of the axioms of a model category is that a weak equivalence between objects which are both fibrant and cofibrant admits a "homotopy-inverse" which makes "being weakly equivalent" a simpler equivalence relation, and it means that we have a nicer description of the homotopy category, [Hov07, Theorem 1.2.10] where $\mathscr{C}^{\mathrm{cf}}$ denote the full sub-category of $\mathscr{C}$ given by objects that are both fibrant and cofibrant.

The most basic example in this survey is the model structure on cochain complexes.

Theorem 2.42 ([Hov07, Theorem 2.3.11]) *Let R be a ring. There is a model structure on Mod_R such that a map is:*

- *a weak-equivalence if it is a quasi-isomorphism.*
- *a fibration if it is degree-wise surjective.*

In Mod_k, all objects are both fibrant and cofibrant (because all objects are projective) and therefore all quasi-isomorphisms are quasi-invertible. Later on, we will transfer this model structure into other categories (such as the category of algebras over an operad) which *will not* satisfy this property.

Even though fibrant and cofibrant objects have good homotopical properties, they can be quite big and complicated. For an arbitrary object in a model category, we may want to

---

[7] A quasi-inverse (or homotopy inverse) is a map in the opposite direction inducing an inverse in the homotopy category.

consider a class of objects that are *minimal*, in which any weak equivalence, $f : A \to B$, between two minimal objects must be an isomorphism. For Mod_k we will consider the class of objects given by the cohomologies of the complexes, that is the minimal objects are graded complexes with no differentials.

Definition 2.43 Take $\mathscr{C}$ a model category and a class of minimal objects. Then we say that an object A is *formal* if there is a weak equivalence between A and a minimal object $\tilde{A}$.

We did not make the class of minimal objects explicit in that definition since we will always work with model categories where the class of minimal objects is clear.

Proposition 2.44 *Every object in* Mod_k *is formal. Moreover, every cochain complex has a* deformation retract *to its cohomology, i.e. there exists a diagram*

$$h \circlearrowright (A, d_A) \underset{i}{\overset{p}{\rightleftarrows}} (H(A), 0)$$

such that $pi = \mathrm{id}_{H(A)}$ *and* h *is a* homotopy *between* Id_A *and* $i \circ p$ *of degree* -1, *meaning that* $\mathrm{Id}_A - ip = d_A h - h d_A$.

Proof Using the fact that k is a field we can non-canonically decompose A as a direct sum $A \cong B[1] \oplus \underbrace{B \oplus H(A)}_{\ker d_A}$, where the only non-zero piece of the differential is the "identity" shift: $B[1] \to B$. The maps p, i are defined in the obvious way and $h := \mathrm{shift}^{-1}$. □

The natural notion of morphisms between different model categories is the notion of *Quillen adjunctions*. Given $\mathscr{C}$ and $\mathscr{D}$ two model categories, a Quillen adjunction is an adjunction:

$$L : \mathscr{C} \rightleftarrows \mathscr{D} : R$$

such that either L preserves cofibrations and trivial cofibrations or equivalently R preserves fibrations and trivial fibrations. This condition is enough to ensure that L and R induce an adjunction between the homotopy categories. Moreover a Quillen adjunction is called a *Quillen equivalence* if it induces an equivalence between the homotopy categories. We refer to [Hov07, Corollary 1.3.16] for a more practical description of Quillen equivalences.

Given an ordinary adjunction:

$$L : \mathscr{C} \rightleftarrows \mathscr{D} : R$$

such that $\mathscr{D}$ (respectively $\mathscr{C}$) is a model category. We ask whether we can *transfer* the model structure on $\mathscr{C}$ (respectively $\mathscr{D}$) such that the adjunction is Quillen. To do so, we define the following classes on $\mathscr{C}$ (respectively $\mathscr{D}$):

- f is a cofibration in $\mathscr{C}$ if and only if $L(f)$ is a cofibration in $\mathscr{D}$ (respectively f is a fibration in $\mathscr{D}$ if and only if $R(f)$ is a fibration in $\mathscr{C}$).

- f is a weak equivalence in $\mathscr{C}$ if and only if $L(f)$ is a weak equivalence in $\mathscr{D}$ (respectively f is a weak equivalence in $\mathscr{D}$ if and only if $R(f)$ is a weak equivalence in $\mathscr{C}$).
- The fibrations are determined by the lifting properties.

If these classes define a model structure (which might not always be the case), then the model structure on $\mathscr{C}$ is called the *left transferred model structure* (respectively the model structure on $\mathscr{D}$ is called the *right transferred model structure*). Note that for these choices of classes of maps, the adjunction is automatically Quillen.

Even given a Quillen adjunction $L \dashv R$ the functors L and R might not have good homotopical properties (such as preserving weak equivalences). However, the properties of a Quillen adjunction ensure that these functors can be "derived" to a new adjunction

$$\mathbb{L}(L) : \mathscr{C} \rightleftarrows \mathscr{D} : \mathbb{R}(R)$$

where $\mathbb{L}(L)$ is the *left derived functor of* L and $\mathbb{R}(R)$ is the *right derived functor of* R. These derived functors preserve weak-equivalences and are obtained by precomposing L and R by an appropriate weakly equivalent replacement (a cofibrant replacement for L and a fibrant replacement for R). In the rest of this survey, whenever a functorial construction is "derived" or when taking a "homotopy" pullback or pushout, the reader may keep in mind that it is the classical construction applied to an appropriate replacement.

Although model structures are very convenient tools to handle weak equivalences, a model structure is not always available. In such situations, namely those coming up in Sect. 4, we need to use ∞-categories (see [Lur09]).

2.4.2 Model Structures for (Co)algebras

Theorem 2.45 ([Hin97, Theorem 4.1.1]) *For an operad* $\mathscr{P} \in \mathbf{Op}$, *consider the adjunction of Proposition 2.27,*

$$\mathscr{P}(-) : \ \mathrm{Mod}_k \rightleftarrows \mathbf{Alg}_{\mathscr{P}} \ : (-)^\sharp$$

Then we can right transfer the model structure on Mod_k *(Theorem 2.42) to a model structure on* $\mathbf{Alg}_{\mathscr{P}}$ *such that a map* f *is weak equivalence (respectively a fibration) in* $\mathbf{Alg}_{\mathscr{O}}(\mathrm{Mod}_k)$ *if and only if* $f^\sharp$ *is a quasi-isomorphism (respectively a fibration). With this model structure this adjunction is Quillen.*

Remark 2.46 Let us note that if $\mathscr{P}$ has no differential, then $H(A)$ naturally has a $\mathscr{P}$-algebra structure. While every cochain complex is formal, we can ask whether a $\mathscr{P}$-algebra is formal (Definition 2.43), that is, do we have a zig-zag of quasi-isomorphisms from A to its homology *in the category of* $\mathscr{P}$*-algebras*. It turns out that the answer is that not all $\mathscr{P}$-algebras are formal. On the other hand, we will see that the homotopy transfer theorem (Sect. 3.1.2) will provide a *homotopy* $\mathscr{P}$-algebra structure on $H(A)$, which in a way quantifies the defect of formality to be satisfied. This extended structure on $H(A)$ will be weakly equivalent to A.

The model structures on algebras over operads are compatible with change of operads in the following sense: Recall from Proposition 2.29 that a morphism of operads $f\colon \mathscr{P} \to \mathscr{Q}$ induces an adjunction:

$$f_! : \mathbf{Alg}_{\mathscr{P}} \rightleftarrows \mathbf{Alg}_{\mathscr{Q}} : f^*$$

Theorem 2.47 ([Hin97, Theorem 4.6.4]) *The adjunction above is Quillen. Furthermore, if $f\colon \mathscr{P} \to \mathscr{Q}$ is a quasi-isomorphism, this adjunction is a Quillen equivalence.*

The first assertion follows from the fact that the right adjoint preserves the underlying complex. Therefore it sends fibrations to fibrations and weak equivalences to weak equivalences.

Warning 2.48 While the right adjoint will always preserve quasi-isomorphisms, the left adjoint has no reason to do so.

As an example, take the canonical map $f\colon \mathbf{Ass} \to \mathbf{Com}$. The induction $f_!$ takes an associative algebra to its abelianisation. Let A be the quasi-free[8] associative algebra,

$$A = T_k(x, y, z), dx = dy = 0, dz = xy - yx.$$

The algebra A is quasi-isomorphic to its own homology $H(A) = \mathrm{Sym}_k(x, y)$. But the (homology of the) abelianization of A is $\mathrm{Sym}_k(x, y, z)$.

2.4.3 Model Structure on Operads

In this Section, we introduce a model structure on operads. This model structure is obtained via transfer from the model structure on $\mathbf{Seq}_\Sigma$. This discussion would also extend when replacing Mod_k by a "good enough" model category such as $\mathbf{sSet}$ or Mod_A for example (see [BM03]).

Theorem 2.49 ([BM03, Theorem 3.2]) *There is a cofibrantly generated model structure on the category* $\mathbf{Op}$ *such that a morphism $\mathscr{P} \to \mathscr{Q}$ is a weak equivalence (respectively fibration) if for all $n \in \mathbb{N}$ the maps $\mathscr{P}(n) \to \mathscr{Q}(n)$ are weak-equivalences (respectively fibrations) in the projective model structure on* Mod_k.

Remark 2.50 The free operad functor is left adjoint to the forgetful functor $\mathbf{Op} \to \mathbf{Seq}_\Sigma$ where the model structure $\mathbf{Seq}_\Sigma$ is the projective model structure on the functor category $\mathrm{Fun}\,(\mathbb{N}^\sim, \mathrm{Mod}_k)$, where fibrations and weak-equivalences are defined object-wise. This forgetful functor clearly preserves fibrations and trivial fibrations which makes it a right Quillen functor. In fact the model structure on $\mathbf{Op}$ is left transferred from the model structure on $\mathbf{Seq}_\Sigma$ via the free-forget adjunction.

[8] Free as a graded algebra, if we forget the differential.

Remark 2.51 The original version on Theorem 3.2 in [BM03] is a more general version of this theorem giving a model structure on the category of operads valued in any closed model category (with some additional technical assumptions).

Remark 2.52 There is no natural model structure on cooperads having weak equivalences given by all quasi-isomorphisms and making the bar-cobar adjunction Quillen. This comes from the fact that the cobar construction does not preserve quasi-isomorphisms.

Despite this last remark, the bar-cobar adjunction is well behaved with respect to the model structure on operads.

Theorem 2.53 ([LV12, Theorem 6.6.3]) *Both the unit* $\epsilon : \Omega\mathbf{B}\mathscr{P} \to \mathscr{P}$ *and the counit* $\eta : \mathscr{C} \to \mathbf{B}\Omega\mathscr{C}$ *of the bar-cobar adjunction are quasi-isomorphisms.*

Sketch of Proof Basis elements of $\Omega\mathbf{B}\mathscr{P}$ can be seen as trees whose vertices are themselves ("inner") trees whose vertices are labeled by $\mathscr{P}$. Filtering by the number of inner edges we recover at the level of the associated graded only the piece of the differential corresponding to the one from $\mathscr{P}$ and a second one making an inner edges into an outer edge. One can check that the associated graded retracts into $\mathscr{P}$ by constructing a homotopy that makes an outer edge into an inner edge. A similar argument shows that $\mathscr{C} \to \mathbf{B}\Omega\mathscr{C}$ is a quasi-isomorphism. □

Proposition 2.54 ([LV12, Proposition 6.5.3]) *The functor* **B** *preserve quasi-isomorphisms.*

In general Ω will not preserve all quasi-isomorphisms. It preserves quasi-isomorphisms between non-negatively graded cooperads such that $\mathscr{C}(0) = 0$ and $\mathscr{C}(1) = k$ (see [LV12, Proposition 6.5.6]).

Definition 2.55

- A twisting morphism (Definition 2.38) $\alpha : \mathscr{C} \dashrightarrow \mathscr{P}$ is said to be *Koszul* if the induced map $\Omega\mathscr{C} \to \mathscr{P}$ is a quasi-isomorphism. The full sub-category of **Tw** generated by Koszul twisting morphisms will be denoted by **Kos**.
- A twisting morphism $\alpha : \mathscr{C} \dashrightarrow \mathscr{P}$ is said to be *weakly Koszul* if the induced map $\mathscr{C} \to \mathbf{B}\mathscr{P}$ is a quasi-isomorphism.
- The counit morphism $\Omega\mathbf{B}\mathscr{P} \to \mathscr{P}$ is a quasi-isomorphism (see Theorem 2.53) and induces the *universal twisting morphism* $\mathbf{B}\mathscr{P} \to \mathscr{P} \in \mathbf{Tw}$ (see Sect. 2.6).

Since the bar construction preserves quasi-isomorphisms (see Proposition 2.54), Koszul morphisms induce a quasi-isomorphism $\mathbf{B}\Omega\mathscr{C} \to \mathbf{B}\mathscr{P}$ and are therefore weakly Koszul. There is also a version of the bar-cobar adjunction between (co)algebras associated to a given twisting morphism.

Proposition 2.56 *If* $\mathscr{C} \in$ **coOp** *satisfies* $\mathscr{C}(0) = 0$ *and* $\mathscr{C}(1) = k$ *then* $\Omega\mathscr{C}$ *is cofibrant. In particular, if* $\mathscr{P} \in$ **Op** *satisfies the same conditions, the unit map* $\Omega\mathbf{B}\mathscr{P} \to \mathscr{P}$ *is a cofibrant resolution of* $\mathscr{P}$.

2.5 Koszul Duality

In Theorem 2.47 we saw that from the homotopical point of view it is the same to consider algebras over an operad $\mathscr{P}$ or over a cofibrant replacement $\mathscr{Q}$ of such an operad. If $\mathscr{P}$ has no differential, algebras over $\mathscr{Q}$ have a typically more complicated structure. On the other hand, algebras over cofibrant operads, especially those of the form $\Omega\mathscr{C}$, have very nice properties that we will discuss in the next section. One of such properties was foreshadowed in the Rosetta Stone Proposition 2.40, where we saw that maps out of $\Omega\mathscr{C}$ amount to solving the Maurer–Cartan equation.

This section is all about finding a "simple" cofibrant replacement of operads using the methods of Koszul duality, originally introduced by Ginzburg and Kapranov [GK94]. This is inspired in the more classical Koszul duality for associative algebras, introduced by Priddy [Pri70] as a tool to produce small resolutions for a large class of algebras. We will assume in this Section that $\mathscr{P}(0) = 0$ and $\mathscr{P}(1) = k$.

Definition 2.57 Given an operad $\mathscr{P}$, we say that a cooperad $\mathscr{C}$ is the *generalized Koszul dual cooperad* of $\mathscr{P}$ if there is a quasi-isomorphism $\Omega\mathscr{C} \to \mathscr{P}$. In particular it is given by a Koszul twisting morphism $\mathscr{C} \dashrightarrow \mathscr{P}$ (Definition 2.55).

Remark 2.58 If $\mathscr{P}(0) = 0$ and $\mathscr{P}(1) = k$, a generalised Koszul dual always exists since, thanks to Proposition 2.56, $\mathbf{B}\mathscr{P}$ is a generalized Koszul dual of $\mathscr{P}$ (given by the universal twisting morphism from Definition 2.55).

Moreover, the generalized Koszul dual is unique up to weak equivalence. Given the cofibrant resolution $\Omega\mathscr{C} \to \mathscr{P}$, we can always find the following sequence of quasi-isomorphisms (thanks to Theorem 2.53 and Proposition 2.54):

$$\mathscr{C} \xrightarrow{\text{counit}} \mathbf{B}\Omega\mathscr{C} \longrightarrow \mathbf{B}\mathscr{P}$$

The main problem is that $\mathbf{B}\mathscr{P}$ is typically a very big cooperad. We will therefore be interested in a class of operads who admit Koszul dual cooperads which are easy to compute. This is the class of Koszul operads, and these operads are in particular quadratic operads (Definition 2.12).

Using the notation from Constructions 2.11 and 2.20, notice that the degree shift isomorphism $\mathfrak{T}^{(1)}E[1] \to \mathfrak{T}^{(1)}E$ induces a canonical twisting morphism $\kappa : \mathscr{C}(E[1], R[2]) \dashrightarrow \mathscr{P}(E, R)$ (see [LV12, Section 7.4.1]).

Definition 2.59 A quadratic operad $\mathscr{P} = \mathscr{P}(E, R)$ is *Koszul* if

$$\Omega\mathscr{C}(E[1], R[2]) \to \mathscr{P}$$

is a quasi-isomorphism. Equivalently, $\mathscr{P}$ is Koszul if $\mathscr{C}(E[1], R[2])$ is a generalized Koszul dual of $\mathscr{P}$. In that case, we define the *Koszul dual cooperad* of $\mathscr{P}$ to be $\mathscr{P}^{\text{i}} := \mathscr{C}(E[1], R[2])$.

Classically, Koszul duality is a duality between operads and operads (rather than operads and cooperads). To recover such a duality we could in principle simply define the Koszul dual operad of a Koszul operad $\mathscr{P}$ to be $(\mathscr{P}^{\text{i}})^\vee$, which is equivalent data if the generators are arity-wise finite dimensional.

In practice, since many important Koszul operads (for example **Ass**, **Com** and **Lie**) are binary and concentrated in degree zero, one typically introduces a degree shift so that their Koszul duals are also concentrated in degree zero. A way to achieve this is to tensor it in each arity with the endomorphisms operad of $k[1]$, which has the effect of raising the degree of the arity n piece by $n - 1$:

$$\mathscr{Q}\{-n\} := \mathscr{Q} \otimes \mathbf{End}_{k[n]}$$

This is mostly for psychological reasons, since the categories of algebras are equivalent via the map shifting the degree of the underlying vector space.

Definition 2.60 Let $\mathscr{P} = \mathscr{P}(E, R)$ be a Koszul operad. The *operad Koszul dual* to $\mathscr{P}$ is:

$$\mathscr{P}^! := (\mathscr{P}^{\text{i}})^\vee\{-1\}$$

If $\mathscr{P}$ is binary, this takes the form $\mathscr{P} = \mathscr{P}\left(\operatorname{sgn} \otimes E^\vee, R^\perp\right)$, where sgn denotes the sign representation of Σ_2 and $R^\perp$ denotes the orthogonal complement $R^\perp \cong \left(\mathfrak{T}^{(2)}E/_R\right)^\vee$.

Proposition 2.61 *We have that* $(\mathscr{P}^!)^! = \mathscr{P}$ *for* $\mathscr{P}$ *an arity-wise dualizable Koszul operad. Furthermore,* $\mathscr{P}$ *is Koszul if and only if* $\mathscr{P}^!$ *is Koszul.*

Proof The first part is immediate from the definition. For the second statement, if $\Omega\mathscr{P}^{\text{i}} \to \mathscr{P}$ is a quasi-isomorphism, taking **B** on both sides yields a quasi-isomorphism $\mathscr{P}^{\text{i}} \to \mathbf{B}\mathscr{P}$. Dualizing and shifting gives the result. □

While not all quadratic operads are Koszul, this should be taken as a very common property for all operads appearing in practice. In particular, we have:

Theorem 2.62 *The operads* **Ass**, **Com**, **Lie**, **preLie** *and* **Perm** *are Koszul. Furthermore, we have the following Koszul dualities:*

$$\mathbf{Ass}^! = \mathbf{Ass} \qquad \mathbf{Com}^! = \mathbf{Lie} \qquad \mathbf{preLie}^! = \mathbf{Perm}$$

When $\mathscr{P}$ is a Koszul operad with no differential, it is easy to convince oneself that $\Omega\mathscr{P}^{\text{i}}$ is the smallest possible quasi-free resolution of $\mathscr{P}$. This motivates the following.

Convention 2.63 *From here onwards, we will tacitly suppose that a Koszul operad* $\mathscr{P}$ *has no differential, even if it might be concentrated in non-zero degrees. We will use the notation*

$$\mathscr{P}_\infty := \Omega\mathscr{P}^{i}.$$

Algebras over $\mathscr{P}_\infty$ are sometimes called homotopy $\mathscr{P}$-algebras. *For the associative, Lie and commutative operads, we will rather use the classical notations A_∞, L_∞ and C_∞ respectively.*

2.6 *Bar-Cobar Adjunction for Algebras*

Definition 2.64 ([LV12, Section 11.3]) Given a twisting morphism $\alpha : \mathscr{C} \dashrightarrow \mathscr{P}$, we can define a *bar-cobar adjunction associated to α*:

$$\Omega_\alpha : \mathbf{coAlg}_{\mathscr{C}}^{\mathrm{conil}} \rightleftarrows \mathbf{Alg}_{\mathscr{P}} : \mathbf{B}_\alpha$$

where Ω_α is defined as the quasi-free $\mathscr{P}$-algebra, given on objects by $\Omega_\alpha(C) = (\mathscr{P} \circ C, d_C + d_{\mathscr{P}} + d_\Omega)$ together with the cobar differential extending the differential on $\mathscr{P}$, on C and with d_Ω the unique differential extending the following composition:

$$C \xrightarrow{\Delta_C} \mathscr{C} \circ C \xrightarrow{\alpha \circ \mathrm{id}} \mathscr{P} \circ C$$

$\mathbf{B}_\alpha$ is the quasi-cofree $\mathscr{C}$-coalgebra $\mathbf{B}_\alpha(A) = (\mathscr{C} \circ A, d_A + d_{\mathscr{C}} + d_{\mathbf{B}})$ with $d_{\mathbf{B}}$ the unique codifferential on $\mathscr{C} \circ A$ extending the following composition:

$$\mathscr{C} \circ A \xrightarrow{\alpha \circ \mathrm{id}} \mathscr{P} \circ A \xrightarrow{\mu_A} A$$

Example 2.65

- For $\mathfrak{g}$ a Lie algebra, the bar construction associated to the canonical twisting morphism gives the Chevalley–Eilenberg chain complex up to a degree shift (which is in fact a shifted cocommutative coalgebra):

$$\mathbf{B}_\kappa \mathfrak{g} = \left(\mathrm{Sym}_k^{\geq 1}(\mathfrak{g}[1])[-1], \delta_{\mathrm{CE}}\right) = \mathrm{CE}_*(\mathfrak{g}).$$

- For A an associative algebra, we obtain the Hochschild chain complex with coefficients in itself (a shifted coassociative coalgebra):

$$\mathbf{B}_\kappa A = \bigoplus_{n \geq 1} A^{\otimes n}[n-1] = \mathrm{CH}_*(A).$$

Remark 2.66 There is also a notion of twisting morphism from a $\mathscr{C}$-coalgebra C to a $\mathscr{P}$-algebra A, given by [LV12, Definition 11.1.1]. This leads to a version of the Rosetta Stone for algebras (see [LV12, Proposition 11.3.1]).

Remark 2.67 The bar-cobar adjunction is natural with respect to morphisms of twisting morphism (Definition 2.38) in the following sense. Given a morphism of twisting morphisms i.e., a commutative square:

$$\begin{array}{ccc} \mathscr{C} & \overset{\alpha}{\dashrightarrow} & \mathscr{P} \\ \downarrow{\scriptstyle f} & & \downarrow{\scriptstyle g} \\ \mathscr{D} & \overset{\beta}{\dashrightarrow} & \mathscr{Q} \end{array}$$

we have the following commutative diagrams of left and right adjoint functors.

$$\begin{array}{ccc} \mathbf{coAlg}_{\mathscr{C}} & \xleftarrow[\mathbf{B}_\alpha]{} & \mathbf{Alg}_{\mathscr{P}} \\ {\scriptstyle f^!}\uparrow & & {\scriptstyle g_*}\uparrow \\ \mathbf{coAlg}_{\mathscr{D}} & \xleftarrow[\mathbf{B}_\beta]{} & \mathbf{Alg}_{\mathscr{Q}} \end{array} \qquad \begin{array}{ccc} \mathbf{coAlg}_{\mathscr{C}} & \xrightarrow{\Omega_\alpha} & \mathbf{Alg}_{\mathscr{P}} \\ \downarrow{\scriptstyle f^*} & & \downarrow{\scriptstyle g_!} \\ \mathbf{coAlg}_{\mathscr{D}} & \xrightarrow{\Omega_\beta} & \mathbf{Alg}_{\mathscr{Q}} \end{array}$$

If we ignore the differentials, this is just follows from the fact that the composition of adjoints is an adjoint. One can then check that the differentials agree. See [CT20, Lemma 3.25] for a proof of the commutativity the first diagram (which is equivalent to saying that the diagram of right adjoints is also commutative).

The condition of being Koszul gives extra properties to the adjunction, in particular Theorems 11.3.3 and 11.3.4 in [LV12] relate the criteria of being Koszul to having a unit and counit of the bar-cobar adjunction that are quasi-isomorphisms.

Proposition 2.68 ([LV12, Corollary 11.3.5]) *For $\mathscr{P}$ as Koszul operad, the counit $\Omega_\kappa \mathbf{B}_\kappa A \to A$ is a quasi-free resolution of A and in particular $\Omega_\kappa \mathbf{B}_\kappa$ gives a canonical cofibrant replacement on the category of $\mathscr{P}$-algebras. Furthermore, if C is a conilpotent $\mathscr{P}^{i}$-coalgebra, the unit $C \to \mathbf{B}_\kappa \Omega_\kappa C$ is a quasi-isomorphism.*

Notice that the bar-cobar adjunction give us a direct connection between algebras over a Koszul operad $\mathscr{P}$ and algebras over it's Koszul dual $\mathscr{P}^!$. We may define the *Koszul dual algebra* of a $\mathscr{P}$-algebra A to be $\mathfrak{D}(A)$, where $\mathfrak{D}$ is given by:

$$\mathfrak{D}\colon \mathbf{Alg}_{\mathscr{P}} \xrightarrow{\mathbf{B}_\kappa} \mathbf{coAlg}_{\mathscr{P}^{i}} \xrightarrow{\vee} \mathbf{Alg}_{\mathscr{P}^!\{-1\}}{}^{\mathrm{op}} \xrightarrow{A \mapsto A[-1]} \mathbf{Alg}_{\mathscr{P}^!}{}^{\mathrm{op}}$$

Example 2.69

- The Koszul dual commutative algebra of a Lie algebra $\mathfrak{g}$ is

$$\mathfrak{D}(\mathfrak{g}) = \left(\mathrm{Sym}_{k}^{\geq 1}\left((\mathfrak{g}[1])^\vee\right), d\right),$$

  where the differential d is the unique derivation extending the dual of the Lie bracket $\ell^\vee\colon \mathfrak{g}^\vee \to \mathfrak{g}^\vee \otimes \mathfrak{g}^\vee$. When $\mathfrak{g}$ is a finite dimensional Lie algebra in degree 0, up to the term $\mathrm{Sym}_k^0(\mathfrak{g}^\vee) = 0$, this is precisely the classical Chevalley–Eilenberg cochain complex of $\mathfrak{g}$ with coefficients in the trivial module k, usually written:

$$\mathrm{CE}^*(\mathfrak{g}) = \big(\mathrm{Hom}(\Lambda^*\mathfrak{g}, k), d_{\mathrm{CE}}\big)$$

- Similarly, in the finite dimensional and degree 0 case and up to a term k, the dual algebra of an associative algebra A is its Hochschild cochain complex with coefficients in the trivial module k:

$$\mathfrak{D}(A) = \mathrm{CH}^*(A) = \big(\mathrm{Hom}(A[1]^{\otimes *}, k), d_{\mathrm{Hoch}}\big)$$

Warning 2.70 Suppose the space of generators of $\mathscr{P}$ is finite dimensional. Despite the terminology "dual" there is no direct natural transformation between $\mathfrak{D}^2$ and $\mathrm{id}_{\mathbf{Alg}_{\mathscr{P}}}$ in any direction. However, if A is a degreewise finite dimensional $\mathscr{P}$-algebra, we have $\mathbf{B}(A)^\vee = \Omega(A^\vee)$, and in that case, there is a map $\mathfrak{D}^2 A \to A$ which is equivalent to the bar-cobar resolution of A. In general, $\mathfrak{D}^2 A$ and A might not even have isomorphic homology.

The bar-cobar adjunction has nice properties with respect to a model structure on coalgebras.

Theorem 2.71 ([DCH16, Theorem 3.11] or [Val20, Theorem 2.1]) *Given any twisting morphism $\alpha : \mathscr{C} \dashrightarrow \mathscr{P}$, we can define the α-model structure on $\mathbf{coAlg}_{\mathscr{C}}$ as the model structure obtain via the left transfer along the bar-cobar adjunction for $\mathscr{P}$-algebras (see Definition 2.64):*

$$\Omega_\alpha : \mathbf{coAlg}_{\mathscr{C}} \rightleftarrows \mathbf{Alg}_{\mathscr{P}} : \mathbf{B}_\alpha$$

This construction gives us a Quillen adjunction which is an equivalence if the twisting morphism is Koszul. This adjunction is natural in the choice of α, $\mathscr{P}$ and $\mathscr{C}$ (see [DCH16, Remark 3.12]).

Remark 2.72 If we take $\alpha : \mathscr{C} \to I$ the augmentation on $\mathscr{C}$, we obtain a model structure transferred from $\mathbf{Alg}_I = \mathrm{Mod}_k$, where weak equivalences are quasi-isomorphisms. However we will not make use of this model structure, since it is not compatible with the one on $\mathscr{P}$-algebras. In fact, we will only be interested in the case where α is Koszul.

Corollary 2.73 *Given a Koszul operad $\mathscr{P}$, there are Quillen equivalences:*

$$\mathbf{coAlg}_{\mathscr{P}^{¡}} \rightleftarrows \mathbf{Alg}_{\mathscr{P}} \leftrightarrows \mathbf{Alg}_{\mathscr{P}_\infty}$$

Proof The first equivalence follows from Theorem 2.71 and the second one from Theorem 2.47. □

2.6.1 A Comment on Operads as Algebras

The reader might wonder what is the relation between the bar-cobar adjunction for algebras and for operads.

In a roundabout way, non-symmetric operads can themselves be seen as algebras over a colored operad $\mathcal{O}^{ns}$ with set of colors $S = \mathbb{N}$ (that is, $\mathcal{O}^{ns}$ is a monoid in $\mathrm{Fun}(\mathbb{N}_{\mathbb{N}}, \mathrm{Mod}_k)$, see Remark 2.3). The colored operad $\mathcal{O}^{ns}$ is generated by binary operations consisting of the partial compositions $\circ_i$, see [vdL04]. This observation can be used to show that operads form a model category by applying a colored version of Theorem 2.45. There is also an operad $\mathcal{O}$ encoding symmetric operads which is colored in groupoids, to take properly into account the symmetric group actions, see [War21].

Furthermore, the operad $\mathcal{O}$ is quadratic and Koszul self dual $\mathcal{O}^! = \mathcal{O}$. The (colored enhancement of the) bar-cobar construction from Definition 2.64 with respect to the canonical $\mathcal{O}^{¡} \dashrightarrow \mathcal{O}$ corresponds precisely to the operadic bar-cobar construction.

Example 2.74 It is a curious observation that given a permutative algebra A, we can define an operad $\mathcal{P}$ where $\mathcal{P}(n) = A$ for all n and setting all partial compositions to be the multiplication on A.[9]

Using that $\mathbf{preLie}^! = \mathbf{Perm}$ we can use this observation to recover conceptually (albeit in an overcomplicated way) the pre-Lie structure on the totalization $\mathrm{Tot}(\mathcal{P})$ of an operad as follows:

There is an adjunction between operads and groupoid colored operads

$$L : \mathbf{Op} \rightleftarrows \mathbf{gcOp} : R$$

in which $L(\mathcal{Q})(c_1, \ldots, c_n; c_0) = \mathcal{Q}(n)$ for any choice of colors. The observation above manifests itself as a map of operads $\mathcal{O} \to L(\mathbf{Perm})$, which, by taking Koszul duals yields a map $L(\mathbf{preLie}) \to \mathcal{O}^! = \mathcal{O}$.

The data of an operad structure on $\mathcal{P}$ is equivalent to a map $\mathcal{O} \to \mathbf{End}_{\mathcal{P}}$. Composing these maps and applying the adjunction, we obtain a map

$$\mathbf{preLie} \to R(\mathbf{End}_{\mathcal{P}}),$$

One can show that $R(\mathbf{End}_{\mathcal{P}}) = \mathbf{End}_{\mathrm{Tot}(\mathcal{P})}$ so the map above recovers precisely the pre-Lie structure on $\mathrm{Tot}(\mathcal{P})$.

For more details, see [CCN22, Section 3.3].

[9] In fact, this produces a non-unital operad, since A is a priori non-unital. This is consistent with the detail hidden under the rug that $\mathcal{O}$ encodes in fact non-unital operads.

3 Algebraic Deformation Theory

As mentioned in the introduction, it is an old heuristic that deformation problems are in correspondence with Lie algebras. We will not define precisely for now what we mean by deformation problem, a formalisation of this correspondence will be treated in detail in Sect. 4. Nevertheless we point out that there is no general explicit procedure to produce a Lie algebra out of a deformation problem.

In the other direction, though, this heuristic tells us that if, for some reason, we have access to the Lie algebra $\mathfrak{g}$ controlling a deformation problem, then the set Maurer–Cartan elements of $\mathfrak{g}$, $\mathrm{MC}(\mathfrak{g})$, corresponds to such deformations. Furthermore, a deformation problem should come with some notion of equivalence between two deformations which corresponds on the Lie algebra side, to an action of the "gauge group" or "exponential group" of $\mathfrak{g}$ on $\mathrm{MC}(\mathfrak{g})$.

The goal of this section is to present how the operadic machinery developed in the previous section give us very efficient tools to produce Lie algebras associated to certain algebraic deformation problems. Throughout we will work with a cooperad $\mathscr{C}$ such that $\mathscr{C}(0) = 0$ and $\mathscr{C}(1) = k$ so that $\Omega\mathscr{C}$ is cofibrant. This condition implies in particular that $\mathscr{C}$ is conilpotent.

We will start, in Sect. 3.1, by recalling the definitions of homotopy $\mathscr{P}$-algebras, ∞-morphisms and the homotopy transfer theorem. It turns out that these homotopy algebras provide a good context in which we can deform algebraic structures. In Sect. 3.2, we discuss the construction of spaces of deformations as the Maurer–Cartan space of the convolution Lie algebra. In that section, we will define two groupoids, one given by some Maurer–Cartan element together with gauge action between them, and the other given by deformations of algebraic structures together with equivalences of such deformations. The main result, Theorem 3.33, says that those groupoids are in fact equivalent. Then, in Sect. 3.3, we will focus on the interplay between commutative and associative structures.

3.1 Algebraic Structures Up to Homotopy

As we mentioned in Sect. 2.4, the non-invertibility of quasi-isomorphisms of $\mathscr{P}$-algebras makes the study of the question of "A being quasi-isomorphic to B" rather unpleasant since it, in principle, forces us to consider all possible zig-zags of quasi-isomorphisms of algebras between A and B.

On the other hand, the category of $\mathscr{P}$-algebras forms a model category in which every object is fibrant, so this already reduces that question to the existence of a single quasi-isomorphism $Q(A) \xrightarrow{\sim} B$, where $Q(A)$ is any cofibrant replacement of A. This is not always a very efficient way to address this question or more generally to study the homotopy category of $\mathscr{P}$-algebras.

In this section, we will always assume that $\mathscr{P}$ is a Koszul operad. We will see that the class of morphisms of $\mathscr{P}$-algebras can be enlarged to an explicit notion of ∞-*morphism*, in which ∞-quasi-isomorphisms admit quasi-inverses, which gives us great control over the homotopy category. This is summarized by the following theorem whose proof we will sketch at the end of the section.

Theorem 3.1 *The faithful inclusion from $\mathscr{P}$-algebras with ordinary morphisms to $\mathscr{P}$-algebras with ∞-morphisms induces an equivalence at the level of the homotopy categories.*

In particular, two $\mathscr{P}$-algebras are weakly equivalent if and only if there exists a direct ∞-quasi-isomorphism between them.

3.1.1 Homotopy Algebras and Morphisms

We will in fact start by enlarging the objects of the category of $\mathscr{P}$-algebras rather than the morphisms, by going to the category of $\mathscr{P}_\infty$-algebras, which recovers the classical notions of A_∞- and L_∞-algebras. The real advantages of doing so will only appear in Sect. 3.1.2, but for the moment notice that from the homotopical point of view, nothing is gained or lost thanks to Corollary 2.73.

There are several different equivalent ways to define $\mathscr{P}_\infty$-algebras. Given our presentation, the most natural definition is to define a $\mathscr{P}_\infty$-structure on A to be a morphism $\mathscr{P}_\infty := \Omega\mathscr{P}^{\mathrm{i}} \to \mathbf{End}_A$. But using Proposition 2.40, we have the equivalences:

$$\mathrm{Hom}_{\mathbf{Op}}\left(\mathscr{P}_\infty, \mathbf{End}_A\right) \simeq \mathrm{Tw}(\mathscr{P}^{\mathrm{i}}, \mathbf{End}_A) \simeq \mathrm{Hom}_{\mathbf{coOp}^{\mathrm{conil}}}\left(\mathscr{P}^{\mathrm{i}}, \mathbf{BEnd}_A\right)$$

Lemma 3.2 *A $\mathscr{P}_\infty$-structure on A is exactly given by a codifferential on the cofree $\mathscr{P}^{\mathrm{i}}$-algebra $\mathscr{P}^{\mathrm{i}}(A)$. We denote by* codiff($\mathscr{P}^{\mathrm{i}}(A)$) *the set of such codifferentials.*

By codifferential on $\mathscr{P}^{\mathrm{i}}(A)$, we mean a degree 1 square zero coderivation D, such that the underlying map $A \to A$ is just the differential on A:

$$D = d_A + D^{\geq 2}$$

Proof This is essentially the content of [LV12, Proposition 10.1.11]. In general a coderivation on $\mathscr{P}^{\mathrm{i}}(A)$ is completly determined (because of cofreeness of $\mathscr{P}^{\mathrm{i}}(A)$) by a map in $\mathrm{Hom}_{\mathbf{Seq}_\Sigma}\left(\mathscr{P}^{\mathrm{i}}, \mathbf{End}_A\right)$ of degree 1 (see [LV12, Proposition 6.3.8] together with Lemma 2.25). For this map to correspond to a $\mathscr{P}_\infty$-structure on A, it needs to be a twisting morphism via the Rosetta Stone. It turns out that to satisfy the Maurer–Cartan equation in $\mathrm{Hom}_{\mathbf{Seq}_\Sigma}\left(\mathscr{P}^{\mathrm{i}}, \mathbf{End}_A\right)$ is equivalent to saying that the associated coderivation on $\mathscr{P}^{\mathrm{i}}(A)$ squares to zero. □

Remark 3.3 Notice that for $\mathscr{P} = \mathscr{P}(E, R)$ Koszul, the Lie algebra Tot(Conv $(\mathscr{P}^{\mathrm{i}}, \mathbf{End}_A)$) is filtered complete, with filtration given by the vertex filtration on $\mathfrak{T}^c(E)$, which descends to $\mathscr{P}^{\mathrm{i}}$.

This filtration splits as a weight grading by the number of vertices:

$$\mathrm{Tot}(\mathrm{Conv}(\mathscr{P}^{\mathrm{i}}, \mathbf{End}_A)) = \prod_{n\geq 1} \mathrm{Hom}_\Sigma(\mathscr{P}^{\mathrm{i}(n)}, \mathbf{End}_A))$$

A twisting morphism then decomposes in $\phi = \phi_1 + \phi_2 + \cdots$ and the Maurer–Cartan equation becomes:

$$-\sum_{\substack{p+q=n \\ p<n,q<n}} \phi_k \star \phi_l = \partial \phi_n$$

where ∂ and $\star$ give the pre-Lie structure of the convolution pre-Lie algebra.

Example 3.4

- $\mathbf{As}_\infty$ and A_∞-algebras:

  For simplicity reasons let us consider **As** be the *non-symmetric* version of the associative operad and let us describe homotopy algebras over it. An A_∞-algebra structure is given by a twisting morphism in $\mathrm{Tw}(\mathbf{As}^{¡}, \mathbf{End}_A)$. Forgetting for now the Maurer–Cartan condition, this is given by a set of maps from $\mathbf{As}^{¡}(n)$ to $\mathbf{End}_A(n)[-1]$ for each $n \in \mathbb{N}$. We have that:

$$\mathbf{As}^{!}(n) = (\mathbf{As}^{¡}(n))^{\vee} \otimes \mathbf{End}_{k[-1]}(n) \simeq (\mathbf{As}^{¡}(n))^{\vee}[n-1]$$

  Therefore we get $\mathbf{As}^{¡}(n) = (\mathbf{As}^{!}(n))^{\vee}[1-n] = \mathbf{As}(n)^{\vee}[1-n]$ (since $\mathbf{As}^{!} = \mathbf{As}$). Therefore we have an isomorphism $\mathbf{As}^{¡}(n) \cong k[1-n]$ for $n \geq 1$ and such a morphism is given for each n by an element in $\mathbf{End}_A(n)$ of degree $2-n$, that is an element of degree 0 in $\underline{\mathrm{Hom}}_k\left(A^{\otimes n}, A\right)[n-2]$. This means that for each $n \geq 2$, we get an n-ary operation of degree $2-n$:

$$\mu_n : A^{\otimes n} \to A$$

  satisfying some conditions corresponding to the Maurer–Cartan equation satisfied by the twisting morphism:

$$\partial\alpha + \alpha \star \alpha = 0,$$

  where ∂ is the differential on $\mathrm{Tw}(\mathbf{As}^{¡}, \mathbf{End}_A)$ given by $\partial\alpha = \alpha \circ d_{\mathbf{As}^{¡}} + \partial_A \circ \alpha$. But we have $\partial_{\mathbf{As}^{¡}} = 0$ and we get:

$$\partial_A \circ \alpha + \alpha \star \alpha = 0$$

  From the codifferential point of view, $\mathscr{P}^{¡}(A)$ is simply the shifted cofree tensor coalgebra on the graded module $A[-1]$ given by the tensor algebra $(T(A[-1]))[1]$. Then the maps μ_n correspond to a map, D, on $\mathscr{P}^{¡}(A)$ with

$$D = d_A + \mu_2 + \cdots : \mathscr{P}^{¡}(A) \to A$$

defining a derivation D on $\mathscr{P}^{\text{i}}(A)$. Then, if we write $\mu_1 := d_A$, the condition $D^2 = 0$ recovers the classical equations encoding the associativity up to homotopy of the A_∞-algebras:

$$\sum_{p+q+r=n} (-1)^{p+qr} \mu_{p+r+1} \circ (\mathrm{id}^{\otimes p} \otimes \mu_q \otimes \mathrm{id}^{\otimes r}) = 0, \quad \forall n \geq 1$$

- $\mathbf{Lie}_\infty$ and L_∞-algebras:
  Similarly to the associative situation, we can see that $\mathbf{Lie}^{\text{i}}(n) \cong k[1-n]$ with the sign action of Σ_n. An L_∞-algebra structure on A is given by a sequence of graded skew-symmetric maps for each $n \geq 2$:

$$\ell_n : A^{\otimes n} \to A[n-2]$$

The cofree $\mathbf{Lie}^{\text{i}}$-coalgebra generated by A is given by $(\mathrm{Sym}\, A[-1])[1]$ together with the natural decomposition product. If $D : \mathbf{Lie}^{\text{i}}(A) \to A$ is the degree 1 codifferential associated to the sequence of ℓ_n, and d_A the differential on A, then $D^2 = 0$ is equivalent to the following conditions:

$$\sum_{\substack{p+q=n \\ p>1, q>1}} \sum_{\sigma \in \mathrm{Unsh}(p,q)} \mathrm{sgn}(\sigma)(-1)^{(p-1)q} (\ell_p \circ \ell_q)^\sigma = \partial_A(\ell_n)$$

Such homotopy algebraic structures, being algebras over an operad $\mathscr{P}_\infty$ have a natural notion of morphism, which commutes with all the $\mathscr{P}_\infty$ operations. Such morphisms are sometimes called *strict* morphisms, in opposition with a weaker notion of morphisms called ∞-*morphism*. Even when the $\mathscr{P}_\infty$ structure arises from a strict $\mathscr{P}$-algebra structure, this is a novel notion. These new type of morphisms give us a better control of the homotopy theory of $\mathscr{P}_\infty$-algebras, have very nice invertibility properties and naturally appear in the homotopy transfer theorem (Theorem 3.10).

Definition 3.5 Let A and B be two $\mathscr{P}_\infty$-algebras given by codifferentials $\left(\mathscr{P}^{\text{i}}(A), \Delta_A\right)$ and $\left(\mathscr{P}^{\text{i}}(B), \Delta_B\right)$ (see Lemma 3.2). An ∞-*morphism* from A to B, denoted $f : A \rightsquigarrow B$, is defined to be a morphism of $\mathscr{P}^{\text{i}}$-coalgebras preserving the codifferential:

$$f : \left(\mathscr{P}^{\text{i}}(A), \Delta_A\right) \to \left(\mathscr{P}^{\text{i}}(B), \Delta_B\right)$$

Let us unravel a bit the definition above. By virtue of being a map into a cofree conilpotent coalgebra, an ∞-morphism $f : A \rightsquigarrow B$ is in fact completely determined by the projection $\mathscr{P}^{\text{i}}(A) \to B$ or equivalently, thanks to Lemma 2.25, by a map $\tilde{f} : \mathscr{P}^{\text{i}} \to \mathbf{End}^A_B$. Decomposing $\tilde{f}$ by arity, we see that an ∞-morphism is determined by a series of maps $A^{\otimes n} \to B$, being enough to specify one such map for each generator of $\mathscr{P}^{\text{i}}(n)$ as a Σ_n-module.

We call $f_1 := \tilde{f}(I)\colon A \to B$ the *base map* of the ∞-morphism. Notice that this is only a morphism of cochain complexes and we interpret the other components of f as the homotopies controlling the failure of f_1 to intertwine the $\mathscr{P}$ or $\mathscr{P}_\infty$ structures. If $\tilde{f}$ vanishes in arities ≥ 2, then f_1 is a strict morphism of $\mathscr{P}_\infty$-algebras.

Example 3.6 [LV12, Section 10.2.6]

- A_∞-morphisms:
  Let A and B be A_∞-algebras described by maps μ_i^A and μ_i^B respectively. An ∞-morphism $f : A \rightsquigarrow B$ is given by maps

$$f_n : A[1]^{\otimes n} \to B[1]$$

  for all each $n \geq 1$. The condition that the induced map $\mathbf{Ass}^{\text{¡}}(A) \to \mathbf{Ass}^{\text{¡}}(B)$ preserves the differential is, for all $n \geq 1$:

$$\sum_{0 \leq i+j \leq n} f_{i+j+1} \circ (1^{\otimes i} \otimes \mu_{n-i-j}^A \otimes 1^{\otimes j}) = \sum_{r \geq 1} \sum_{i_1+\cdots+i_r=n} \mu_r^B \circ (f_{i_1} \otimes \cdots \otimes f_{i_r})$$

- L_∞-morphisms:
  Let A and B be L_∞-algebras described by maps ℓ_i^A and ℓ_i^B respectively (with $\ell_1^A = d_A$ and $\ell_1^B = d_B$). An ∞-morphism $f : A \rightsquigarrow B$ is given by maps $f_n : \mathrm{Sym}^n(A[1]) \to B[1]$ for all each $n \geq 1$. The condition that the induced map $\mathbf{Lie}^{\text{¡}}(A) \to \mathbf{Lie}^{\text{¡}}(B)$ preserves the differential is a symmetrization of the A_∞ situation and for all $n \geq 1$:

$$\sum_{i+j=n} \sum_{\sigma \in \mathrm{Sh}(i,j)} \epsilon(\sigma)(f_{j+1} \circ_1 \ell_i^A)^\sigma = \sum_{r \geq 1} \sum_{i_1+\cdots+i_r=n} \sum_{\sigma \in \mathrm{Sh}(i_1,\cdots,i_r)} \epsilon(\sigma)\ell_r^B \circ (f_{i_1} \otimes \cdots \otimes f_{i_r})^\sigma$$

Definition 3.7 We say that f is an ∞-*(quasi)-isomorphism* if the induced map $f_1 : A \to B$ is a (quasi)-isomorphism. An ∞-*isotopy* between two $\mathscr{P}_\infty$-structures on A given by $\alpha, \beta : \mathscr{P}_\infty \to \mathbf{End}_A$ is an ∞-morphism $f : \mathscr{P}^{\text{¡}} \to \mathbf{End}_A^A = \mathbf{End}_A$ between α and β such that $f_1 := \mathrm{id}_A$.

Remark 3.8 A morphism $f : A \to B$ of complexes between two $\mathscr{P}_\infty$-algebras is said to extend to a ∞-morphism of $\mathscr{P}_\infty$-algebra if there exists an ∞-morphism $f_\infty : A \rightsquigarrow B$ such that $f = (f_\infty)_1$.

3.1.2 Homotopy Transfer Theorem

As a general heuristic, there is a trade-off between how small a $\mathscr{P}$-algebra is and how simple its structure is, while preserving the quasi-isomorphism type. For example, a quasi-free resolution of an algebra A (such as $\Omega \mathbf{B} A$) is simple in the sense that it is there are no relations, but free objects tend to be quite big.

Trying to go in the other direction, the smallest possible $\mathscr{P}$-algebra that has any chance of being quasi-isomorphic to A is its own homology. In case A is not formal, the homotopy transfer theorem tells us that in that case we can insist on working with the cochain complex $H(A)$ but the price to pay is to have to endow $H(A)$ with a *different* $\mathscr{P}_\infty$-algebra structure on it.

It might look odd to consider a $\mathscr{P}_\infty$-structure on a complex with trivial differential, but this just means that while there exist higher homotopies, the $\mathscr{P}$-algebra equations are satisfied on the nose. In that situation, the $\mathscr{P}_\infty$ structure on $H(A)$ will recover the canonical $\mathscr{P}$-algebra structure.

Given a quasi-isomorphism $A \to B$ of cochain complexes, we would like to know when we can transfer a $\mathscr{P}$-algebra structure (or even $\mathscr{P}_\infty$-structure) from A to B (or vice versa). We start by describing the conditions on our quasi-isomorphism that permits such a transfer. This procedure gives a way to produce $\mathscr{P}_\infty$-structures on cohomology, transferring from the deformation retract of cochain complexes on cohomology (see Proposition 2.44).

Definition 3.9 We say that (B, d_B) is a *homotopy retract* of (A, d_A) if there are maps of cochain complexes:

$$h \circlearrowright (A, d_A) \underset{i}{\overset{p}{\rightleftarrows}} (B, d_B)$$

such that i is a quasi-isomorphism and $\mathrm{Id}_A - ip = d_A h + h d_A$. This pair is a *deformation retract* if moreover $pi = \mathrm{id}$, and a *strong deformation retract* if we have the side conditions $ph = hi = h^2 = 0$.

To go from a deformation retract to a strong one, it suffices to change the homotopy h to $h'' = -h' d_A h'$, where $h' = (d_A h + h d_A) h (d_A h + h d_A)$.

Theorem 3.10 (Homotopy Transfer Theorem, [LV12, Theorem 10.3.1]) *Let $\mathscr{P}$ be a Koszul operad and let (B, d_B) a homotopy retract of (A, d_A). Then, any $\mathscr{P}_\infty$-algebra structure on A can be transferred to a $\mathscr{P}_\infty$-algebra structure on B such that i extends to a ∞-quasi-isomorphism i_∞. Moreover, the $\mathscr{P}_\infty$ structure on A and i_∞ can be explicitly described (see Remarks 3.12 and 3.13).*

The proof is based on the Rosetta Stone (Proposition 2.40) so that finding a $\mathscr{P}_\infty$-structure on B boils down to finding a map $\mathscr{P}^{\text{¡}} \to \mathbf{BEnd}_B$, which can be reduced to finding a map, $\Psi : \mathbf{BEnd}_A \to \mathbf{BEnd}_B$ that will make the following diagram commute:

$$\begin{array}{ccc} \mathscr{P}^{\text{¡}} & \longrightarrow & \mathbf{BEnd}_A \\ & \searrow & \downarrow \Psi \\ & & \mathbf{BEnd}_B \end{array}$$

This is the goal of the following Lemma.

Lemma 3.11 ([LV12, Section 10.3.2]) *We can construct a map of cooperads Ψ : $\mathbf{BEnd}_A \to \mathbf{BEnd}_B$ that extends (by cofreeness) the map sending $m \in \mathbf{End}_A(n)$ to $p \circ m \circ i^{\otimes n} \in \mathbf{End}_B(n)$.*

Remark 3.12 We will denote by sm the element of a cochain complex $V[-1]$ corresponding to $m \in V$ but in degree 1 higher. To understand the map obtained in Lemma 3.11, recall from Definition 2.39 that $\mathbf{BEnd}_A$ is the cofree cooperad on $\mathbf{End}_A[-1]$ together with the bar differential. The idea is to define a map that will send a tree in $\mathbf{BEnd}_A$ to an element in $\mathbf{End}_B$ by adding i to each leaf of the tree, p to the root, and h to each internal edge:

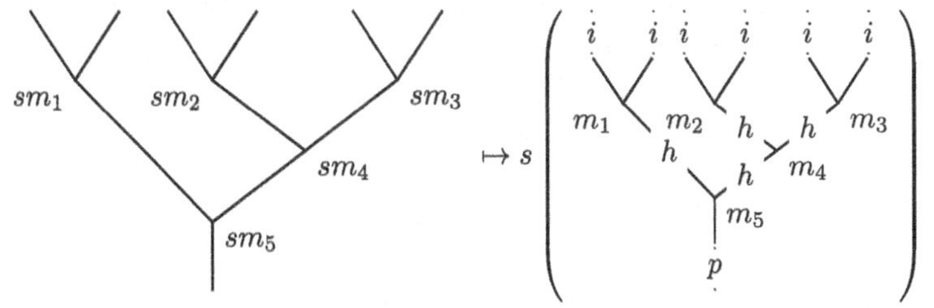

Composing all the maps in the tree on the right hand side gives an element in $\mathbf{End}_B$ of same degree as the left hand side, and the map $\mathbf{BEnd}_A$ to $\mathbf{End}_B$ naturally extends to the map of Lemma 3.11.

Remark 3.13 The ∞-morphism i_∞ is constructed by defining a map $\mathscr{P}^{¡} \to \mathbf{End}_A^B$ as:

$$i_\infty : \overline{\mathscr{P}}^{¡} \xrightarrow{\Delta} \mathfrak{T}^{co}\overline{\mathscr{P}}^{¡} \xrightarrow{\mathfrak{T}^{co} s\mu_A} \mathfrak{T}^{co}(\mathbf{End}_A[-1]) \xrightarrow{\tilde{\Psi}} \mathbf{End}_A^B$$

$$i_\infty : I \mapsto i \in \mathrm{Hom}_k(B, A) \subset \mathbf{End}_A^B$$

The morphism $\tilde{\Psi} : \mathfrak{T}^{co}(\mathbf{End}_A[-1]) \to \mathbf{End}_A^B$ is defined just as Ψ, except for the root which is not labeled by p:

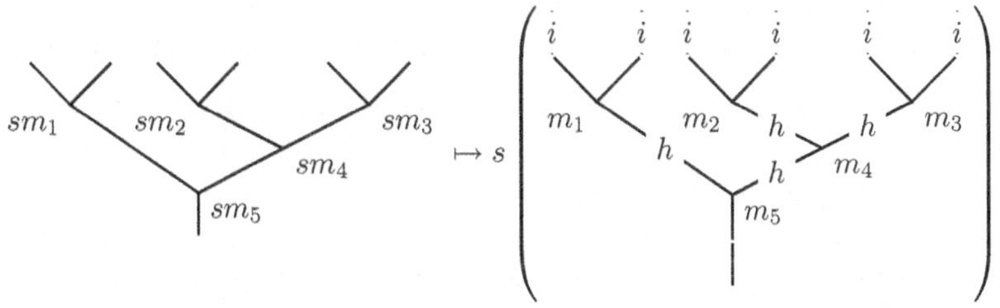

Focusing on the case where B is the chain complex given by the homology of A, Proposition 2.44 guarantees the existence of a deformation retract.

Proposition 3.14 ([LV12, Proposition 10.3.9]) *There exists a strong deformation retract from A to its homology:*

$$h \circlearrowright (A, d_A) \underset{i}{\overset{p}{\rightleftarrows}} (H(A), 0)$$

In this case, p can also be extended to an ∞-quasi-isomorphism of $\mathscr{P}_\infty$-algebras:

$$p_\infty : A \rightsquigarrow H(A)$$

In particular, the homology of a $\mathscr{P}$-algebra always carries a transferred $\mathscr{P}_\infty$-structure. This structure depends on the choice of deformation retract, and therefore it is only unique up to ∞-isomorphism.

Example 3.15 For any topological space X, the cup product $\cup$ endows the singular cochain complex $C^*_{\mathrm{Sing}}(X)$ with an associative algebra structure. Choosing a deformation retract to the cohomology, we transfer an A_∞ structure to $H^*(X)$. For $n = 3$ the product μ_3 recovers essentially the Massey products of X [Mas58], which are generalized to operadic algebras in [Mur23]. Classically, the Massey product of classes $x, y, z \in H^*(X)$ is only defined if $xy = 0$ or $yz = 0$ and the result takes value in a quotient. The disappearance of this ambiguity in our case comes from the choice of deformation retract. More generally, the higher ($\mu_{n\geq 4}$) products obtained from the homotopy transfer theorem are also related to the so-called higher Massey products, but the relation is a bit more subtle [FCMF23].

The homotopy transfer theorem enables a proof of Theorem 3.1 without appealing to any model categorical arguments. More generally, the same arguments show that all inclusions we can write between the four categories given by strict/homotopy $\mathscr{P}$-algebras with strict/∞ morphisms induce equivalences at the level of homotopy categories, appealing to the property known as *rectification*: Every $\mathscr{P}_\infty$-algebra A is equivalent to a strict $\mathscr{P}$-algebra, namely $\Omega_\kappa \mathbf{B}_\kappa A$.

It can be shown (and we invite the reader to do so) that Theorem 3.1 is a formal consequence of Theorem 2.71.

Sketch of Proof of Theorem 3.1 The proof of this result rests on two key points. Given an ∞-quasi-isomorphism of $\mathscr{P}$-algebras $f: X \rightsquigarrow Y$:

(1) There is a zig-zag of quasi-isomorphisms connecting X to Y,
(2) There is $g: Y \rightsquigarrow X$ such that the two composites are the identity in homology.

For the first point, notice that by the bar-cobar adjunction $f: \mathbf{B}X \to \mathbf{B}Y$ is equivalent to a map $\Omega\mathbf{B}X \to Y$. One can therefore consider the short zig-zag $X \xleftarrow{\sim} \Omega\mathbf{B}X \xrightarrow{\sim} Y$.

For the second point, one first consider the case where $d_X = 0 = d_Y$, i.e., we need to know that ∞-isomorphisms of $\mathscr{P}_\infty$ are indeed invertible. This can be shown by an argument similar to the invertibility of formal power series (see [LV12, Theorem 10.4.1]).

In case of non-trivial differentials, by the homotopy transfer theorem we can reduce the problem to the homology. More precisely, the composite of ∞-quasi-isomorphisms

$$H(X) \underset{i_\infty}{\overset{\sim}{\rightsquigarrow}} X \underset{f}{\overset{\sim}{\rightsquigarrow}} Y \underset{p_\infty}{\overset{\sim}{\rightsquigarrow}} H(Y) \tag{5}$$

is an ∞-isomorphism, therefore we set g to be

$$Y \underset{p_\infty}{\overset{\sim}{\rightsquigarrow}} H(Y) \overset{\sim}{\rightsquigarrow} H(X) \underset{i_\infty}{\overset{\sim}{\rightsquigarrow}} X,$$

where the middle map is the inverse of the composition given by (5). □

For a general study of the operadic homotopy transfer theorem in the spirit of the homological perturbation lemma see [Ber14]. This framework gives an extension of i, p and h.

3.2 Deformation of Algebraic Structures

This section is devoted to the study of deformations of algebraic structures. Given $A \in \mathrm{Mod}_k$, the set of $\mathscr{P}_\infty$-structures on A is by definition given by $\mathrm{Tw}\left(\mathscr{P}^{\mathrm{i}}, \mathbf{End}_A\right)$, the set of Maurer–Cartan elements of the convolution Lie algebra (Definitions 3.29 and 2.38). In Sect. 3.2.1, we define a set, groupoid and ∞-groupoid of Maurer–Cartan elements of a Lie algebra, commonly called the Deligne (∞-)groupoid. This is a priori unclear what these objects have to do with deformations, but in Sect. 3.2.3, we define a set, groupoid and ∞-groupoid of $\mathfrak{R}$-deformations of a $\mathscr{P}_\infty$-structure on A with $\mathfrak{R}$ a local (Artinian) commutative ring (for example $\mathfrak{R} = k[\epsilon]$, the ring of dual numbers). Theorem 3.33 then tells us that the deformation ∞-groupoid is homotopy equivalent to a Deligne ∞-groupoid of a certain twisted Lie algebra.

3.2.1 Maurer–Cartan Space and Gauge Group

The set of Maurer–Cartan elements (Definition 2.37) is not only good to define twisting morphism but is known to describe the set of deformations controlled by a Lie algebra $\mathfrak{g}$. We are interested in such deformations up to equivalences.

Such equivalences usually coincide with the general notion of *gauge* equivalences: Just as a finite dimensional classical (i.e. $\mathfrak{g} = \mathfrak{g}^0$) Lie algebra integrates to a Lie group via an exponential map, a (dg) Lie algebra $\mathfrak{g}$ formally integrates to the so-called; gaugegroup $\exp(\mathfrak{g}^0)$. The gauge group will act naturally on $\mathrm{MC}(\mathfrak{g})$.

To talk about exponentials and deal with convergence issues, we will need to use complete filtered Lie algebras. Taking complete filtered Lie algebras will ensure that all the ensuing infinite sums will converge.

Definition 3.16 We say that $\mathfrak{g}$ is a *complete positively filtered Lie algebras* or just a *complete Lie algebra* if it is equipped with a complete Hausdorff descending filtration:

$$\mathfrak{g} := F^1\mathfrak{g} \supset F^2\mathfrak{g} \supset \cdots$$

Being complete Hausdorff here means that there is a equivalence $\mathfrak{g} \to \lim_n \mathfrak{g}/_{F^n\mathfrak{g}}$. To any filtered Lie algebra, one can associate a complete Lie algebra, $\hat{\mathfrak{g}} = \lim_n \mathfrak{g}/_{F^n\mathfrak{g}}$, called its *completion.*

We also assume that the filtration is compatible with the differential and the Lie bracket:

$$d(F^n\mathfrak{g}) \subset F^n\mathfrak{g} \qquad \left[F^p\mathfrak{g}, F^q\mathfrak{g}\right] \subset F^{p+q}\mathfrak{g}$$

We will denote by $\mathfrak{g}_0$ the set of degree 0 elements.

For now, let us assume that the Lie structure on $\mathfrak{g}$ is obtained from a unital associative product $[x, y] = x \star y - y \star x$. Then, assuming convergence, we can write down the usual formula for the exponential:

$$\begin{aligned} \exp : \mathfrak{g}_0 &\longrightarrow \mathfrak{g}_0 \\ X &\mapsto e^X = \sum_k \frac{X^{\star k}}{k!} \end{aligned}$$

The image of this map forms a group with respect to $\star$ and unit 1, which we would like to call the *gauge group* G. Taking the formal logarithm, we can equivalently define the gauge group as a group structure on the underlying set of $\mathfrak{g}_0$:

$$G = (\mathfrak{g}_0, X \cdot_G Y := \log(e^X \star e^Y), 0).$$

However, this product does not make sense for a general Lie algebra. To address this issue, we can value this exponential in the *complete* universal enveloping algebra of $\widehat{\mathfrak{g}_0}$:

$$\begin{aligned} \exp : \mathfrak{g}_0 &\longrightarrow \widehat{\mathfrak{U}}\mathfrak{g}_0 \\ X &\mapsto e^X = \sum_k \frac{X^{\star k}}{k!} \end{aligned}$$

In fact, the tentative formula that we wrote before in the associative case $\log(e^X \star e^Y)$ turns out to be expressible purely in terms of the Lie bracket of $\mathfrak{g}_0$ in a universal way. This formula is called the BCH (Baker–Campbell–Hausdorff) formula, and its initial terms are:

$$\mathrm{BCH}(X, Y) = X + Y + \frac{1}{2}[X, Y] + \frac{1}{12}([X, [X, Y]] + [Y, [Y, X]]) + \ldots.$$

Proposition 3.17 *There exist a universal BCH formula, such that:*

$$e^X e^Y = e^{\mathrm{BCH}(X,Y)}$$

In particular, the exponential map induces a bijection between $\mathfrak{g}_0$ and its image:

$$\exp : \mathfrak{g}_0 \to G \subset \widehat{\mathfrak{U}}\mathfrak{g}_0$$

We call the group $G := \exp(\mathfrak{g}_0) \cong (\mathfrak{g}_0, \mathrm{BCH}, 0)$, the *gauge group* of $\mathfrak{g}$.

Proof We follow an approach close to [BF11, Section 4.1]. To show a universal BCH formula, it is enough to show that such a formula exists on the free complete Lie algebra on two generators $\widehat{\mathbf{Lie}}(X, Y)$. Note that its universal enveloping algebra is the free complete associative algebra on X, Y, i.e. the algebra of non-commutative formal power series $\widehat{\mathbf{Ass}}(X, Y) = k\widehat{\langle X, Y \rangle}$.

It is easy to check that the exponential map is one to one into the set of formal power series with constant term equal to 1 and indeed an inverse is given by the formal logarithm.

$$\exp : \widehat{k\langle X,Y\rangle} \rightleftarrows 1 + \widehat{k\langle X,Y\rangle}_{\geq 1} : \log .$$

We claim that $v \in \widehat{k\langle X, Y\rangle}$ is primitive if and only if $\exp v$ is grouplike (see Sect. 2.2.1). One implication is a direct verification and for the other one can observe that

$$\Delta(\exp v) = \exp v \otimes \exp v \Rightarrow \exp(\Delta v) = \exp(1 \otimes v + v \otimes 1),$$

where the exponentials on the right hand side are being taken in the algebra $\widehat{k\langle X, Y\rangle}\hat{\otimes}\widehat{k\langle X, Y\rangle}$. The claim follows from the injectivity of the exponential.

By the complete version of the Poincaré–Birkoff–Witt theorem, the set of primitive elements of $\widehat{k\langle X, Y\rangle}$ is precisely the Lie algebra $\widehat{\mathbf{Lie}}(X, Y)$ and the group-like elements form a group with respect to the product and unit of the universal enveloping algebra.

The BCH formula is therefore obtained by taking the exponentials of $X, Y \in \widehat{\mathbf{Lie}}(X, Y)$, multiplying them into a group-like element $e^X \cdot e^Y$ and taking the inverse of the exponential.

□

Proposition 3.18 *Given a Lie algebra* $\mathfrak{g}$ *and* $\phi \in \mathrm{MC}(\mathfrak{g})$*, we can twist* $\mathfrak{g}$ *by* ϕ *and define:*

$$\mathfrak{g}^{\phi} = \left(\mathfrak{g}, [-, -]_{\mathfrak{g}}, d_{\phi} = d + [\phi, -]_{\mathfrak{g}}\right)$$

This is also a Lie algebra and a Maurer–Cartan elements in $\mathfrak{g}^{\phi}$ *is exactly an element* $\psi \in \mathfrak{g}$ *such that* $\phi + \psi$ *is a Maurer–Cartan element of* $\mathfrak{g}$*. In other words, we can think of them as the "deformations" of the Maurer–Cartan element* ϕ*.*

We now wish to describe an action of the gauge group on the Maurer–Cartan set of a complete Lie algebra $\mathfrak{g}$. As a motivation, if we assume for a bit $\mathfrak{g}_1$ and $\mathfrak{g}_2$ to be finite dimensional, the solutions of the Maurer–Cartan equation $\mathrm{MC}(\mathfrak{g})$, form an affine variety (an intersection of quadrics) and therefore its tangent space at a Maurer–Cartan element x is $T_x\mathrm{MC}(\mathfrak{g}) = \{\alpha \in \mathfrak{g}_1 | d_x\alpha = 0\}$. It follows that any element $\lambda \in \mathfrak{g}_0$ induces a tangent vector $d_x\lambda \in T_x\mathrm{MC}(\mathfrak{g})$. Varying x, we get that λ induces a vector field on $\mathrm{MC}(\mathfrak{g})$. Two Maurer–Cartan elements connected by the flow of such λ and said to be *gauge equivalent*.

Definition 3.19 Given a complete Lie algebra, there is an action of the gauge group, $\exp(\mathfrak{g}) = (\mathfrak{g}_0, \mathrm{BCH})$, on the set of Maurer–Cartan elements $\mathrm{MC}(\mathfrak{g})$, defined by:

$$e^{\lambda} \cdot x = x + \frac{e^{\mathrm{ad}_{\lambda}} - \mathrm{id}}{\mathrm{ad}_{\lambda}}(d_x\lambda) := x - \sum_{k=0}^{n} \left(\frac{[\lambda, -]^n}{(n+1)!}\right)(d_x\lambda)$$

where $d_x := [x, -] + d$ is the differential on $\mathfrak{g}$ twisted by the Maurer–Cartan element x. The associated action groupoid on the set of Maurer–Cartan elements is called the *Deligne groupoid*, denoted $\underline{\mathrm{Del}}(\mathfrak{g})$.

To relate the gauge action formula with the discussion above, we point out that one can consider the *differential trick*, which gets rid of the differential of a dg Lie algebra by making it internal. We consider the graded Lie algebra $\mathfrak{g}^+ = \mathfrak{g} \oplus k\delta$, where δ is a degree 1 nilpotent element such that for $x \in \mathfrak{g}$, $[\delta, x] = d_{\mathfrak{g}}x$. It it easy to see that the Maurer–Cartan elements of $\mathfrak{g}$ correspond to the set of degree 1 square-zero elements in g^+ of the form $\delta + x$.

The gauge action formula from Definition 3.19 follows from applying the exponential of the operator ad_λ to $\delta + x$. Indeed, since ad_λ, its exponential is a morphism of graded Lie algebras and therefore preserves square-zero elements, which justifies that $e^\lambda \cdot x$ is still a Maurer–Cartan element. For more details see [DSV24, Theorem 1.53].

When dealing with homotopy theory, we often need an ∞-groupoid to obtain a full "space" with higher homotopies. It turns out that there is an ∞-groupoid version of the Deligne groupoid.

Definition 3.20 (Deligne ∞-Groupoid, [DR15, Section 2.2]) Consider the simplicial commutative k-algebra $\Omega[\Delta^\bullet]$ (sometimes written $\Omega_\bullet$) defined by:

$$\Omega[\Delta^n] := {k[t_0, \cdots, t_n, dt_0, \cdots, dt_n]}\Big/\left(\sum t_i = 1,\ \sum dt_i = 0\right)$$

with the differential sending t_i to dt_i.

For any Lie algebra $\mathfrak{g}$, $\mathfrak{g} \otimes \Omega[\Delta^\bullet]$ is a simplicial Lie algebra where for all $n \in \mathbb{N}$, $\mathfrak{g} \otimes \Omega[\Delta^n]$ is the Lie algebra with Lie bracket:

$$[v \otimes \alpha, w \otimes \beta] = [v, w]_{\mathfrak{g}} \otimes \alpha.\beta$$

Then the *Deligne ∞-groupoid* is the simplicial set defined by:

$$\mathbf{Del}(\mathfrak{g}) := \mathrm{MC}(\mathfrak{g}\hat{\otimes}\Omega[\Delta^\bullet])$$

with $\mathfrak{g}\hat{\otimes}\Omega[\Delta^\bullet] := \lim_n \left({\mathfrak{g}}/{F_n\mathfrak{g}}\hat{\otimes}\Omega[\Delta^\bullet]\right)$. This definition from [DR15] enables an extension of Theorem 3.22 to Theorem 3.24. This simplicial set is an ∞-groupoid (i.e. a Kan complex) thanks to [DR17, Proposition 4.1].

Proposition 3.21 ([DR15, Lemma B.2]) *There is a bijection between the connected components of the Deligne groupoid and those of the ∞-groupoid of Maurer–Cartan elements:*

$$\mathrm{MC}(\mathfrak{g})/\!\sim\ := \pi_0(\underline{\mathrm{Del}}(\mathfrak{g})) \simeq \pi_0\left(\mathbf{Del}(\mathfrak{g})\right)$$

The equivalence relation $\sim$ is given by the gauge equivalences obtained from the action of the gauge group.

Theorem 3.22 (Goldman–Milson, [DR15]) *Given $f : \mathfrak{g} \to \mathfrak{h}$, a filtered quasi-isomorphism (see 3.23) of filtered complete Lie algebras, we have a bijection:*

$$\mathrm{MC}(\mathfrak{g})/\!\sim\ \to \mathrm{MC}(\mathfrak{g})/\!\sim$$

From the deformation theoretical perspective, this result tells us that two Lie algebras that are quasi-isomorphic encode equivalent deformation problems. In practice, one might be interested in showing that a Lie algebra is formal in order to have the "simplest possible description" of the deformation problem at hand.

Remark 3.23 The hypothesis that $f\colon \mathfrak{g} \to \mathfrak{h}$ is a *filtered* quasi-isomorphism, meaning that f induces a quasi-isomorphism on the associated graded, is a stronger requirement than just a quasi-isomorphism.

As an example, let $\mathfrak{g}$ be the two dimensional Lie algebra spanned by an element x in degree 1, and an element $[x, x]$ in degree 2, such that $dx = -\frac{1}{2}[x, x]$. This Lie algebra is quasi-isomorphic to 0, but not filtered quasi-isomorphic to 0 for any possible complete filtration. Indeed, $\mathfrak{g}$ has two Maurer–Cartan elements 0 and x which are not equivalent, since the gauge group is trivial.

One can suspect that the Goldman–Milson Theorem should generalise to the L_∞ setting, since ∞-morphisms are equivalent to zig-zags of strict morphisms and L_∞-algebras are rectifiable to strict Lie algebras. Indeed, given a complete L_∞-algebra $\mathfrak{g}$, the corresponding Maurer–Cartan equation is:

$$\mathrm{MC}(\mathfrak{g}) = \{x \in \mathfrak{g}_1 \mid dx + \frac{1}{2!}\ell_2(x, x) + \frac{1}{3!}\ell_3(x, x, x) + \cdots = 0\}.$$

This is functorial with respect to any ∞-morphism $f\colon \mathfrak{g} \rightsquigarrow \mathfrak{h}$ with components $(f_n)_{n\geq 1}$. We get a map:

$$\begin{aligned} f\colon \mathrm{MC}(\mathfrak{g}) &\to \mathrm{MC}(\mathfrak{h}) \\ x &\mapsto \sum_{n\geq 1} \frac{1}{n!} f_n(x, \ldots, x). \end{aligned}$$

The Goldman–Milson Theorem indeed generalises to this setting and furthermore, thanks to Proposition 3.21, we can interpret the Goldman–Milson theorem as the π_0 case of an equivalence of the Deligne ∞-groupoid:

Theorem 3.24 ([DR15, Theorem 1.1]) *Let $f\colon L \to L'$ be an ∞-morphism between complete positively filtered L_∞-algebras compatible with the filtration. If the linear term $f_1 : L \to L'$ is a filtered quasi-isomorphism, then it induces a weak homotopy equivalence:*

$$\mathbf{Del}(L) \to \mathbf{Del}(L')$$

3.2.2 A Digression on Variations of the Deligne Groupoid

Given a complete Lie algebra $\mathfrak{g}$, it follows from $\Omega[\Delta^0] = k$ that the 0-simplices of the Deligne groupoid $\mathbf{Del}(\mathfrak{g})$ are precisely the Maurer–Cartan elements of $\mathfrak{g}$. In light of Proposition 3.21, it would be natural to expect the 1-simplices of $\mathbf{Del}(\mathfrak{g})$ to correspond to the gauge equivalences. This is however not true: the set of 1-simplices is too big (even though it encodes the same gauge equivalence relation). In [Get09], Getzler solved this problem by considering the normalised cellular complex of the n-simplex $C_n := C^*(\Delta^n)$, which has the advantage of being much smaller than $\Omega[\Delta^n]$ and the disadvantage of not being a commutative algebra.

Proposition 3.25 (Dupont Contraction) *There exists a simplicial deformation retract:*

$$h_\bullet \circlearrowright \Omega[\Delta^\bullet] \underset{i_\bullet}{\overset{p_\bullet}{\rightleftarrows}} C_\bullet.$$

The word "simplicial" in the statement means that $i_\bullet$, $p_\bullet$ and $h_\bullet$ commute with the respective simplicial structures.

Theorem 3.26 ([Get09]) *Given a complete Lie algebra* $\mathfrak{g}$*, the simplicial set*

$$\gamma_\bullet(\mathfrak{g}) := \mathrm{MC}(\mathfrak{g}\hat{\otimes}\Omega[\Delta^\bullet]) \cap \ker h_\bullet$$

is isomorphic to the nerve of the Deligne groupoid of $\mathfrak{g}$*. In particular it is a* ∞*-groupoid and the inclusion* $\gamma_\bullet(\mathfrak{g}) \to \mathbf{Del}(\mathfrak{g})$ *is a homotopy equivalence. We call it the* Getzler ∞-groupoid.

Despite its apparent simplicity, the condition $h_\bullet(x) = 0$ hides the difficult combinatorics of the Dupont contraction, which make some categorical properties of $\gamma_\bullet$ more difficult to show.

Let us briefly describe the operadic approach that led Robert-Nicoud and Vallette [RV20] to give a more conceptual presentation of Getzler's ∞-groupoid.

The usual construction of the Deligne ∞-groupoid involves taking a simplicial framing of a Lie algebra $\mathfrak{g}$ by associating to it the simplicial Lie algebra $\mathfrak{g} \otimes \Omega[\Delta^\bullet]$. Notice that its Maurer–Cartan elements can be identified with a mapping space. Let $k\epsilon$ be the 1-dimensional $\mathbf{Lie}^{\text{i}}$ coalgebra, with $\deg \epsilon = 1$ and $\Delta\epsilon = \frac{1}{2}\epsilon \otimes \epsilon$. Then if $\kappa : \mathbf{Lie}^{\text{i}} \to \mathbf{Lie}$ is the canonical Koszul morphism, we have

$$\mathbf{Del}(\mathfrak{g}) = \mathrm{Hom}_{\mathbf{Alg_{Lie}}}(\Omega_\kappa(k\epsilon), \mathfrak{g} \otimes \Omega[\Delta^\bullet]).$$

Another way to cook up a simplicial set out of a Lie algebra $\mathfrak{g}$ is to consider all maps from a cosimplicial Lie algebra into $\mathfrak{g}$. The cosimplicial Lie algebra we will consider is denoted $\mathfrak{mc}^\bullet$ is in fact an L_∞-algebra that is Koszul dual to the C_∞-algebra structure on $C_\bullet$ obtained by applying the homotopy transfer theorem (on every simplicial degree) to the Dupont contraction (see [RV20, Section 2.2]).

Theorem 3.27 ([RV20, Theorem 2.17]) *There is an isomorphism*

$$\gamma_\bullet(\mathfrak{g}) \cong \mathrm{Hom}_{L_\infty}(\mathfrak{mc}^\bullet, \mathfrak{g}).$$

This description automatically gives some functoriality properties of $\gamma_\bullet$ relative to some refined version of ∞-morphisms of Lie algebras and immediately produces the left adjoint $\mathfrak{L}$ of $\gamma_\bullet : \mathbf{Alg}_{\mathscr{L}_\infty} \to$ sSets.

As a beautiful application, let us point out that one of the key steps in [Get09] is showing that $\gamma_\bullet(\mathfrak{g})$ is an ∞-groupoid. Using the explicit formula for $\mathfrak{L}$ one can easily show that given any horn $\Lambda^n_k \to \gamma_\bullet(\mathfrak{g})$, the set of corresponding horn fillers is in bijection with the degree $-n$ elements $\mathfrak{g}_{-n}$. It follows that there always exist horn fillers since $\mathfrak{g}_{-n}$ is a vector space and is therefore non-empty! Furthermore, a canonical horn filler exists: $0 \in \mathfrak{g}_{-n}$. Such objects are called *algebraic Kan complexes*.

Proposition 3.28 (**[RV20, Section 5.2]**) *Let $x, y \in \exp g$ be elements of the gauge group of g. They define a horn $\Lambda^2_1 \to \gamma_\bullet(\mathfrak{g})$. The horn filler correponding to 0 produces the BCH formula:*

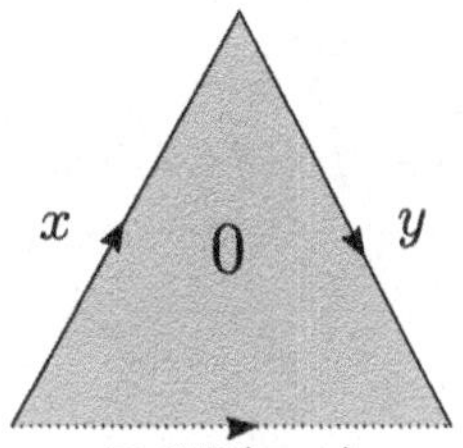

3.2.3 Deformation Complex

In this section we define a groupoid of deformations of algebraic structures obtained by considering algebraic structures on the tensor product with a local commutative algebra such that the quotient by the maximal ideal recovers the initial structure we want to deform. For example, if $\mathscr{O}_X$ is some algebra of functions, a classical question is to understand its deformations as an associative algebra along the algebra of formal power series $k[[t]]$. This is given by an associative product $\star$ on $\mathscr{O}_X \otimes_k k[[t]] = \mathscr{O}_X[[t]]$ such that for all $a, b \in \mathscr{O}_X$, $a.b = a \star b \mod (t)$.

First let us describe the space of $\mathscr{P}$-algebra structures on a given cochain complex A. Such a structure naturally defines a map $\mu : \mathscr{P}_\infty \to \mathbf{End}_A$. We know from Proposition 2.40 that this is equivalent to the datum of a twisting morphism in the convolution pre-Lie algebra.

Definition 3.29 Given a Koszul operad $\mathscr{P}$, we define the *convolution Lie algebra* of the cochain complex A (Definition 2.36) as:

$$\mathfrak{g}_{\mathscr{P}_\infty, A} := \mathrm{Tot}(\mathrm{Conv}(\mathscr{P}^{\text{!`}}, \mathbf{End}_A)) := \prod_{n \geq 0} \underline{\mathrm{Hom}}_{\Sigma_n}\left(\mathscr{P}^{\text{!`}}(n), \mathrm{End}_A(n)\right)$$

together with the Lie structure induced by the natural pre-Lie product of Definition 2.36. As such, we have an equivalence between $\mathscr{P}_\infty$-structures and Maurer–Cartan elements in

$\mathfrak{g}_{\mathscr{P}_\infty, A}$:

$$\{\mathscr{P}_\infty - \text{structures on } A\} \simeq \mathrm{Tw}(\mathscr{P}^{\text{i}}, \mathbf{End}_A) := \mathrm{MC}(\mathfrak{g}_{\mathscr{P}_\infty, A})$$

Notice that the convolution Lie algebra is naturally complete filtered with filtration given by the arity.

Definition 3.30 Let A be a $\mathscr{P}_\infty$-algebra, represented by $\phi \in \mathrm{MC}(\mathfrak{g}_{\mathscr{P}_\infty, A})$. We define the *deformation complex of* A (or deformation complex at ϕ) to be the twisted Lie algebra $\mathfrak{g}^{\phi}_{\mathscr{P}_\infty, A}$ (using the notation of Proposition 3.18).

Remark 3.31 The construction of the twist of a Lie algebra $\mathfrak{g}^{\phi}$ is functorial in the pair $(\mathfrak{g}, \phi)$. More precisely, for a morphism:

$$f : (\mathfrak{g}, \phi) \to (\mathfrak{h}, \phi')$$

given by a morphism of Lie algebras $f : \mathfrak{g} \to \mathfrak{h}$ such that $f(\phi) = \phi'$, then we get a morphism of Lie algebras:

$$f : \mathfrak{g}^{\phi} \to \mathfrak{h}^{\phi'}$$

To see this, we only need to show that f exchanges d_ϕ and $d_{\phi'}$:

$$\begin{aligned} f \circ d_\phi &= f \circ (d + [\phi, -]_{\mathfrak{g}}) \\ &= f \circ d + f\left([\phi, -]_{\mathfrak{g}}\right) \\ &= d \circ f + [f(\phi), f(-)]_{\mathfrak{h}} \\ &= d_{\phi'} \circ f. \end{aligned}$$

It therefore follows from Remark 2.67, that deformation complexes are functorial with respect to morphisms of operads.

We will now explain how this deformation complex does indeed encode deformations of the $\mathscr{P}$-algebra structure determined by ϕ. To do so we fix a classical (non-dg) local commutative algebra $\mathfrak{R}$ with maximal ideal $\mathfrak{M}$ that will parameterize the type of deformation we are interested in. For example we can consider $\mathfrak{R}$ a local Artinian algebra (Definition 4.1), the algebra of dual numbers $\mathfrak{R} = k[\epsilon]$ (we take the geometers' convention that $\epsilon^2 = 0$) for "first-order" deformations (more generally any $\mathfrak{R}$ with $\mathfrak{M}^2 = 0$ will encode "first-order" deformations) or the pro-Artinian algebra (a cofiltered limit of Artinian algebras) $\mathfrak{R} = k[[t]]$ to encode formal deformations.

Definition 3.32 We define the set of deformations of a $\mathscr{P}$-algebraic structure on A, $\phi : \mathscr{P} \to \mathbf{End}_A$, as the set of $\mathfrak{R}$-linear $\mathscr{P}_\infty \otimes \mathfrak{R}$-algebra structure on $A \otimes \mathfrak{R}$ that coincide with ϕ modulo $\mathfrak{M}$. This can be described as:

$$\mathrm{Def}_\phi(\mathfrak{R}) := \left\{\Phi \in \mathrm{MC}\left(\mathfrak{g}^{\phi}_{\mathscr{P}_\infty, A} \otimes \mathfrak{R}\right),\ \Phi \cong \phi \bmod \mathfrak{M}\right\},$$

see [LV12, Lemma 12.2.5] for details. The functor sending $\mathfrak{R}$ to $\mathrm{Def}_\phi(\mathfrak{R})$ is what we would like to think as a prototypical example of a *deformation functor*. A more thorough study of deformation functors is done in Sect. 4.2. We also refer to [Nit06] for a more classical approach to deformation functors.

Following [DSV16], an equivalence between two such deformations is given by an $\mathfrak{R}$-linear ∞-isomorphism $f : (A \otimes \mathfrak{R}, \Phi) \to (A \otimes \mathfrak{R}, \Psi)$ such that $f_1 \cong \mathrm{id} \bmod \mathfrak{M}$.

Together with this notion of equivalence, $\mathrm{Def}_\phi(\mathfrak{R})$ forms a groupoid. We will denote this groupoid by $\underline{\mathrm{Def}}_\phi(\mathfrak{R})$ and its quotient by:

$$Def_\phi(\mathfrak{R}) := {}^{\mathrm{Def}_\phi(\mathfrak{R})}\!/_{\sim} = \pi_0\left(\underline{\mathrm{Def}}_\phi(\mathfrak{R})\right)$$

Theorem 3.33 ([LV12, Theorem 12.2.10]) *For $\mathfrak{R}$ an Artinian local algebra (Definition 4.1), there is a equivalence between the deformation groupoid and the Deligne groupoid (Definition 3.19):*

$$\underline{\mathrm{Def}}_\phi(\mathfrak{R}) \simeq \underline{\mathrm{Del}}\left(\mathfrak{g}^{\phi}_{\mathscr{P}_\infty, A} \otimes \mathfrak{M}\right)$$

Remark 3.34 For the pro-Artinian algebra $\mathfrak{R} = k[\![t]\!] = \lim \left({}^{k[t]}\!/_{(t^n)}\right)$, we could *define* the value of $\underline{\mathrm{Def}}$ and $\underline{\mathrm{Del}}$ as the limits:

$$\underline{\mathrm{Def}}_\phi(k[\![t]\!]) := \lim\, \underline{\mathrm{Def}}_\phi\left({}^{k[t]}\!/_{(t^n)}\right) \qquad \underline{\mathrm{Del}}(k[\![t]\!]) := \lim\, \underline{\mathrm{Del}}\left({}^{k[t]}\!/_{(t^n)}\right)$$

so that the equivalence also extends to pro-Artinian algebras.

Remark 3.35 For $\mathfrak{R} = k[\epsilon]$, the algebra of dual numbers, we have $\mathfrak{M} = k$ and we find that $\mathrm{Def}_\phi(k[\epsilon]) = Z^1(\mathfrak{g}^{\phi}_{\mathscr{P}_\infty, A})$ and $Def_\phi(k[\epsilon]) = H^1(\mathfrak{g}^{\phi}_{\mathscr{P}_\infty, A})$. In that sense, the complex underlying the Lie algebra $\mathfrak{g}^{\phi}_{\mathscr{P}_\infty, A}$ encodes the first order infinitesimal deformations of ϕ.

If one has a first order infinitesimal deformation of ϕ, of the form $\phi + t\alpha_1$ living over $\mathfrak{R} = k[t]/(t^2)$ and wishes to extend it to a second order deformation of the form $\phi + t\alpha_1 + t^2\alpha_2$ living over $k[t]/(t^3)$, simplifying the Maurer–Cartan equation one sees that the condition is that $2[\phi, \alpha_2] + [\alpha_1, \alpha_1] = 0$. In other words, the condition is that the class of $[\alpha_1, \alpha_1]$ vanishes in $H^2(\mathfrak{g}^{\phi}_{\mathscr{P}_\infty, A})$.

This line of reasoning concludes that the obstruction to extending an infinitesimal deformation to a formal deformation is controlled by $H^2(\mathfrak{g}^{\phi}_{\mathscr{P}_\infty, A})$ [LV12, Theorem 12.2.14].

As we will see in Proposition 4.18, the deformation complex is closely related to the operadic tangent complex which classifies higher deformations including higher homology as explained in Remarks 4.11 and 4.42.

Example 3.36

- Let us compute the deformation complex of an associative[10] algebra A.

$$
\begin{aligned}
\mathfrak{g}_{A_\infty,A} &= \prod_{n\geq 0} \underline{\mathrm{Hom}}_{\Sigma_n}\left(\mathbf{Ass}^{\text{¡}}(n), \mathbf{End}_A(n)\right)\\
&= \prod_{n\geq 1} \underline{\mathrm{Hom}}_{\Sigma_n}\left(k[\Sigma_n], \mathbf{End}_A(n)\right)[n-1]\\
&= \prod_{n\geq 1} \underline{\mathrm{Hom}}_{k}(A^{\otimes n}, A)[n-1]
\end{aligned}
$$

When twisting $\mathfrak{g}_{A_\infty,A}$ with a Maurer–Cartan element (i.e. an associative multiplication), the differential on $\mathfrak{g}^m_{A_\infty,A}$ is given by $[m,-]$. If we explicitly write out the differential of an element f living in the $n=2$ component, we obtain

$$
df(a_1,a_2,a_3) = a_1 f(a_2,a_3) \pm f(a_1a_2,a_3) \pm f(a_1,a_2a_3) \pm f(a_1,a_2)a_3.
$$

This coincides with the Hochschild cochain complex (shifted by 1) of the algebra (A,m) with the Gerstenhaber bracket, which was mentioned in Example 1.1.

From Remark 3.35, if we take m an associative multiplication, the infinitesimal deformations are classified by the first Hochschild cohomology group, and the obstruction to extending to higher formal deformations in controlled by the second Hochschild cohomology group.

- The deformation complex for a Lie[11] algebra A is computed as follows:

$$
\begin{aligned}
\mathfrak{g}_{L_\infty,A} &= \prod_{n\geq 0} \underline{\mathrm{Hom}}_{\Sigma_n}\left(\mathbf{Lie}^{\text{¡}}(n), \mathbf{End}_A(n)\right)\\
&= \prod_{n\geq 1} \underline{\mathrm{Hom}}_{\Sigma_n}\left(k, \mathbf{End}_A(n)\right)[n-1]\\
&= \prod_{n\geq 1} \underline{\mathrm{Hom}}_{k}\left(\left(\mathrm{Sym}^n A[-1]\right)[n], A\right)[n-1]\\
&= \prod_{n\geq 1} \underline{\mathrm{Hom}}_{k}\left(\mathrm{Sym}^n(A[-1]), A[-1]\right)
\end{aligned}
$$

This coincides with the Chevalley–Eilenberg algebra valued in A with trivial Lie algebra structure on A. Given a Lie structure on A (corresponding to a Maurer–Cartan element ϕ in $\mathfrak{g}_{L_\infty,A}$), we obtain that $\mathfrak{g}^\phi_{L_\infty,A}$ is the Chevalley–Eilenberg algebra associated to A (with the structure associated to ϕ) valued in A.

[10] Recall from Example 3.4 that $\mathbf{Ass}^{\text{¡}}(n)$ is isomorphic to $k[\Sigma_n]$ in arity n and degree $1-n$.

[11] Recall from Example 3.4 that $\mathbf{Lie}^{\text{¡}}(n)$ is 1-dimensional concentrated in degree $1-n$.

Example 3.37 (Deformation Quantization of Poisson Manifolds) Given a Poisson manifold $(M, \{-,-\})$, a (formal deformation) *quantization* of M is a so-called *star-product*, i.e., an $\mathbb{R}[[\hbar]]$-linear associative product $-\star-$ on formal power series of functions $C^{\infty}(M)[[\hbar]]$, such that:

(i) The star-product is associative.
(ii) The star-product is a deformation of the original product: $f \star g = f \cdot g + O(\hbar)$, for $f, g \in C^{\infty}(M)$.
(iii) The star-product quantizes the Poisson bracket: $f \star g - g \star f = \{f, g\}\hbar + O(\hbar^2)$.

Given a star-product on M, one can check that

$$\{f, g\}_{\star} = \lim_{\hbar \to 0} \frac{f \star g - g \star f}{\hbar}$$

defines a Poisson bracket on M. If our original Poisson bracket is obtained from a star-product in this form, we say that $\star$ *quantizes* $(M, \{-,-\})$. The natural questions being raised are:

(1) Can every Poisson manifold be quantized?
(2) If so, can one provide explicit formulas?
(3) Are quantizations unique?

In his seminal paper [Kon03], Kontsevich used the machinery described above to give a complete solution to the previous questions: (1) yes, (2) yes up to computing some integrals and (3) essentially yes.

The first observation is that the data of a star-product is encoded by the (Hochschild) deformation complex of $A = \mathscr{C}^{\infty}(M)$ seen as an associative algebra. Ignoring condition (iii), a star-product is a Maurer–Cartan element of $\mathfrak{g}_{A_\infty, \mathscr{C}^{\infty}(M)} \hat{\otimes} \mathbb{R}[[\hbar]]$. More precisely, requiring the components of the star-product to be multidifferential operators leads us to consider the Lie subalgebra D_{poly}, where

$$D^{n-1}_{\text{poly}}(M) = \left\{ D \colon \mathscr{C}^{\infty}(M)^{\otimes n} \to \mathscr{C}^{\infty}(M) \,\middle|\, D \overset{\text{loc.}}{=} \sum f \frac{\partial}{\partial x_{I_1}} \otimes \cdots \otimes \frac{\partial}{\partial x_{I_n}} \right\}.$$

On the other hand, the datum of a Poisson bracket is equivalently given by a bivector field $\pi \in \Gamma(M, T_X \wedge T_X)$, such that $[\pi, \pi] = 0$ with respect to the Schouten–Nijenhuis bracket. One can therefore interpret the Poisson structure as a Maurer–Cartan element in the Lie algebra (with $d = 0$) of polyvector fields $T_{\text{poly}}(M)$, where $T^{-1}_{\text{poly}}(M) = \mathscr{C}^{\infty}(M)$ and $T^{n-1}_{\text{poly}}(M) = \Gamma(M, T_X^{\wedge n})$.

Theorem 3.38 (Kontsevich Formality, [Kon03]) *For every manifold M, there exists an ∞-quasi-isomorphism*

$$\mathscr{U} \colon T_{\text{poly}}(M) \rightsquigarrow D_{\text{poly}}(M).$$

The first component of the ∞-morphism is the Hochschild–Kostant–Rosenberg (HKR) map, which is obtained by de-symmetrization. The higher components are parametrized by certain kinds of graphs, with coefficients given by integrals over configuration spaces of points on the upper half-plane.

It now follows from the Goldman–Milson Theorem 3.22 that for every Poisson bivector π, its image under $\mathscr{U}$ is a star-product quantizing it. Furthermore, any such quantization is unique up to gauge equivalence.

Functoriality of the Deformation Complex

The construction of the deformation complex is functorial in the following sense: Suppose $f : \mathscr{Q} \to \mathscr{P}$ is a map of Koszul operads, induced by a map from the generators of $\mathscr{Q}$ to the generators of $\mathscr{P}$. Then, there is an associated map of cooperads $f^{¡} : \mathscr{Q}^{¡} \to \mathscr{P}^{¡}$, as well as the Koszul dual map $f^! : \mathscr{P}^! \to \mathscr{Q}^!$. Then, $(f^{¡}, f)$ form a map between the canonical twisting morphisms as in Remark 2.67. The map $f^{¡}$ induces a Lie algebra morphism at the level of the respective deformation complexes:

$$\mathfrak{g}_{\mathscr{P}_\infty, A} \to \mathfrak{g}_{\mathscr{Q}_\infty, A}$$

Remark 3.39 The morphisms of operads:

$$\mathbf{Lie} \to \mathbf{Ass} \to \mathbf{Com}$$

which are the same as:

$$\mathbf{Com}^! \to \mathbf{Ass}^! \to \mathbf{Lie}^!$$

are dual (up to some shifts) to:

$$\mathbf{Lie}^{¡} \to \mathbf{Ass}^{¡} \to \mathbf{Com}^{¡}$$

This induces morphisms between the deformation complexes:

$$\mathfrak{g}_{\mathbf{Com}_\infty, A} \to \mathfrak{g}_{\mathbf{Ass}_\infty, A} \to \mathfrak{g}_{\mathbf{Lie}_\infty, A}$$

If A is a commutative algebra, applying the Maurer–Cartan functor to the first map recovers the usual inclusion from the Harrison complex into the Hochschild complex. Similarly, if A is an associative algebra, applying the Maurer–Cartan functor to the second map yields the classical projection of the Hochschild complex into the Chevalley–Eilenberg complex of the Lie algebra structure on A.

3.3 A Deformation Approach to $\mathscr{P}_\infty$-Algebras

In this section we will show how the operadic and deformation tools introduced previously apply to address a problem that is not (or at least does not seem) deformation theoretical in nature.

A bit more concretely, assume $\mathscr{P}$ is a Koszul operad and let A and B be two $\mathscr{P}$-algebras. We consider the following questions:

Question 3.40

- Weak equivalence: Do we have a zig-zag of quasi-isomorphisms of $\mathscr{P}$-algebras from A to B? We will denote this equivalence relation by $A \sim_{\mathscr{P}} B$.
- Formality: Is $A \sim_{\mathscr{P}} H(A)$?

Given a map of operads $f : \mathscr{Q} \to \mathscr{P}$ and a $\mathscr{P}$ algebra A, to avoid excessive notation we will refer to $f^*(A)$ as "A seen as a $\mathscr{Q}$-algebra" (see Proposition 2.29). We will be particularly interested in seeing a commutative algebra as just an associative algebra. This abuse of notation justifies the $\mathscr{P}$ in $\sim_{\mathscr{P}}$, which would otherwise be redundant.

Remark 3.41 Let us consider the unit operad $I = I(1) = k$. Every I-algebra is given by some $A \in \mathrm{Mod}_k$ and is formal since k a field. In particular, when forgetting the $\mathscr{P}$-structures (via the unit $I \to \mathscr{P}$), every $\mathscr{P}$-algebra is equivalent to its homology as a cochain complex, but not as a $\mathscr{P}$-algebra a priori.

Recall that Theorem 3.1 implies in particular that the weak-equivalence problem can be fully reformulated in terms of ∞-morphisms.

Proposition 3.42 *$A \sim_{\mathscr{P}} B$ if and only if there is a $\mathscr{P}_\infty$-quasi-isomorphism $A \rightsquigarrow B$.*

This allows us to simplify the question of whether A and B are weakly equivalent $\mathscr{P}$-algebras: First, supposing $H(A) \cong H(B)$ (otherwise the result is trivial), the homotopy transfer theorem allows us to intepret the $\mathscr{P}$-algebra structure of both A and B as a $\mathscr{P}_\infty$ structure in the same graded vector space $H := H(A) \cong H(B)$, which we denote α and β respectively. Secondly, Proposition 3.42 implies that the existence of such a weak equivalence is equivalent to an ∞-quasi-isomorphism:

$$f : (H, \alpha) \rightsquigarrow (H, \beta)$$

notice that since H has null differential, an ∞-quasi-isomorphism $H \rightsquigarrow H$ is just an ∞-isomorphism. We can in fact suppose that $f_1 : H \to H$ is id_H and therefore get an ∞-isotopy (Definition 3.7).

We can therefore reformulate Questions 3.40 as follows:

Reformulation 3.43

- Weak equivalence: Given two $\mathscr{P}_\infty$-algebra structures α and β on H, are they ∞-isotopic?
- Formality: Is the transferred $\mathscr{P}_\infty$ structure on $H(A)$ ∞-isotopic to a strict one?

The following result gives as a deformation theoretical interpretation of the reformulation above.

Theorem 3.44 ([DSV16, Theorem 3]) *The action of the gauge group of $\mathfrak{g}_{\mathscr{P}_\infty,A}$ on $\mathrm{MC}(\mathfrak{g}_{\mathscr{P}_\infty,A})$ corresponds to the action of ∞-isotopies $f : A \rightsquigarrow A$ on the space of $\mathscr{P}_\infty$ structures on A. In other words we have:*

$$\underline{\mathrm{Del}}(\mathfrak{g}_{\mathscr{P}_\infty,A}) \cong (\mathscr{P}_\infty\text{- algebra structures on } A, \infty\text{- isotopies})$$

To finish the section let us mention briefly what kind of tools go into the proof of this result. The proof of this theorem makes use of the fact that the Lie bracket on the deformation complex $\mathfrak{g}_{\mathscr{P}_\infty,A}$ arises as the anti-symmetrisation of a pre-Lie product $\star$, as defined in Definition 2.36.

In [DSV16], Dotsenko–Shadrin–Vallette develop a method of pre-Lie exponentials, making use of the fact that the expression $\sum_k \frac{X^{\star k}}{k!} := \sum_k \frac{(((X\star X)\dots)\star X}{k!}$ makes sense for complete pre-Lie algebras, in order to produce explicit formulas for the gauge group product and action. The explicit formula for the exponential shows that the image consists indeed of the ∞-isotopies. Together with the explicit computation of [DSV16, Lemma 2], one shows that the gauge group structure (the BCH formula) agrees with the composition of ∞-isotopies, noted $\circledcirc$. Using an argument of uniqueness of solutions of formal differential equations, one can show that the gauge group acts indeed by ∞-isotopies, via the formula $\lambda \cdot \alpha = (e^\lambda \star \alpha) \circledcirc e^{-\lambda}$, see [DSV16, Proposition 5].

3.3.1 Commutative and Associative Algebras

Let us start this section with an example arising from rational homotopy theory [Qui69, Sul77], which is at first unrelated to any ∞-structures.

Example 3.45 In rational homotopy theory à la Sullivan, we consider the functor Ω : $\mathbf{Top} \to \mathbf{Alg}^{\mathbb{Q}/}_{\mathbf{Com}}$ of rational differential forms. One might naively wonder whether we could consider instead the simpler functor of rational singular cochains $C^*(-,\mathbb{Q})$ landing in $\mathbf{Alg}^{\mathbb{Q}/}_{\mathbf{Ass}}$. Despite being quasi-isomorphic, it is unclear whether this functor remembers the homotopy type of the space. This is because Corollary 10.10 in [FHT01] only gives a zig-zag of natural quasi-isomorphisms between the singular rational cochains $C^*(-,\mathbb{Q})$ functor and $F \circ \Omega$ where F is the forgetful functor $\mathbf{Alg}^{\mathbb{Q}/}_{\mathbf{Com}} \to \mathbf{Alg}^{\mathbb{Q}/}_{\mathbf{Ass}}$. It raises the following question: given $X, Y \in \mathbf{Top}$ such that $C^*(X,\mathbb{Q}) \sim_{\mathbf{Alg}^{\mathbb{Q}/}_{\mathbf{Ass}}} C^*(Y,\mathbb{Q})$, are they rationally equivalent, i.e. do we have $\Omega(X) \sim_{\mathbf{Alg}^{\mathbb{Q}/}_{\mathbf{Ass}}} \Omega(Y)$? Thanks to Corollary 10.10 in [FHT01], this reduces to proving the following implication:

$$\Omega(X) \sim_{\mathbf{Alg}^{\mathbb{Q}/}_{\mathbf{Ass}}} \Omega(Y) \Rightarrow \Omega(X) \sim_{\mathbf{Alg}^{\mathbb{Q}/}_{\mathbf{Com}}} \Omega(Y)$$

This problem is actually algebraic. Motivated by this example, we consider the following question:

Question 3.46 If A, B are commutative algebras over a field of characteristic zero, do we have:

$$A \sim_{\mathbf{Ass}} B \Rightarrow A \sim_{\mathbf{Com}} B$$

In other words, if there is a zig-zag of quasi-isomorphisms passing by associative algebras, must there be one passing by commutative algebras? A first attempt to show this directly could be to abelianize the associative zig-zag. Unfortunately, abelianization does not commute with taking homology, as we saw in Warning 2.48. Nevertheless, the answer to the question is yes.

Theorem 3.47 ([CPRW19, Theorem (A) and (B)]) *Two (non-)unital commutative algebras A and B are quasi-isomorphic if and only if they are quasi-isomorphic as (non-)unital associative algebras.*

At first this question might look too simple to have a difficult answer. The answer being yes, one might wonder whether this is a formal consequence of some simple property of the (fully faithful) forgetful functor F from commutative to associative algebras, which is also induced from the map of operads $\mathbf{Ass} \to \mathbf{Com}$. This does not seem to be the case: While it is indeed true that F induces a faithful functor at the level of the homotopy categories (this not trivial, see [CPRNW22]), $h(F)$ is not full as can be seen in the following example.

Example 3.48 Consider the algebra $A = k[x, y]$ and a commutative algebra S. Then, in the homotopy category, we have:

$$\mathrm{Hom}_{h\mathbf{Alg}_{\mathbf{Com}}}(k[x, y], S) \sim H^0(S) \oplus H^0(S)$$

$$\mathrm{Hom}_{h\mathbf{Alg}_{\mathbf{Ass}}}(k[x, y], S) \sim H^0(S) \oplus H^0(S) \oplus H^{-1}(S)$$

This comes from the fact that $k[x, y]$ is not cofibrant as an associative algebra. It needs to be replaced by the quasi-free associative algebra generated by x, y and ϵ of degree -1 such that $d\epsilon = xy - yx$.

Making use of Reformulation 3.43 and denoting $H := H(A) \cong H(B)$, we can also reformulate this question as follows.

Reformulation 3.49 If α and β are two C_∞-structure on a graded vector space H that are A_∞ isotopic, are they C_∞-isotopic?

To answer that question, using Theorem 3.44, we need to ask whether an associative gauge transformation λ between α and β in $\underline{\mathrm{Del}}\left(\mathfrak{g}_{A_\infty,H}\right)$ (corresponding to the two A_∞ structures on H) can be lifted to a commutative gauge $\tilde{\lambda}$ in $\underline{\mathrm{Del}}\left(\mathfrak{g}_{C_\infty,H}\right)$ such that the image of $\tilde{\lambda}$ through the map between the Deligne groupoids induced by the map $\mathfrak{g}_{C_\infty,H} \to \mathfrak{g}_{A_\infty,H}$ from Example 3.39 is λ. If $\mathfrak{g}_{C_\infty,H} \to \mathfrak{g}_{A_\infty,H}$ were a quasi-isomorphism, then our conclusion would follow from the Goldmann–Milson Theorem 3.22. In our case, the map is injective and it even remains injective in homology upon twisting by a C_∞ structure, but this does not generally suffice to obtain an injective map at the level of gauge equivalence classes of Maurer–Cartan elements. The following theorem is the appropriate "injectivity" refinement of the Goldmann–Milson Theorem.

Theorem 3.50 (Theorem 1.7 in [CPRW19]) *Let $i : \mathfrak{h} \hookrightarrow \mathfrak{g}$ a map of filtered complete differential graded Lie algebra. Suppose that i has a retract r as $\mathfrak{h}$-modules (i.e. $r[i(h), x] = [h, r(x)]$). Then the following map is injective:*

$$\mathrm{MC}(\mathfrak{h})/_{\sim} \hookrightarrow \mathrm{MC}(\mathfrak{g})/_{\sim}$$

Idea of Proof of Theorem 3.47 Admitting Theorem 3.50, we need to construct such a retract of $i : \mathfrak{g}_{C_\infty,H} \to \mathfrak{g}_{A_\infty,H}$. If there existed a retract of $f : \mathbf{Lie} = \mathbf{Com}^! \to \mathbf{Ass} = \mathbf{Ass}^!$ as operads, then it would induce a retract of i as Lie algebras. The much weaker existence of a retract of f as symmetric sequences, essentially only tells us about the injectivity of i.

It turns out that to obtain a retract as $\mathfrak{g}_{C_\infty,H}$-modules, it suffices to construct retract of $\mathbf{Lie} \to \mathbf{Ass}$ as infinitesimal $\mathbf{Lie}$-bimodules, which is a type of structure whose strength sits between the two.

To obtain such a retract, we can use a suitable variant of the Poincaré–Birkhoff–Witt theorem 2.33 to express $\mathbf{Ass} \cong \mathrm{Sym}(\mathbf{Lie})$ and then projecting onto the summand $\mathrm{Sym}^1(\mathbf{Lie}) = \mathbf{Lie}$. The reader can find the details in [CPRW19, Section 2]. □

We point out that in Theorem 3.50, the gauge equivalence $\tilde{\lambda} \in \exp(\mathfrak{h})$ one constructs is not just $r(\lambda)$. It is rather obtained by an iterative procedure which is not fully explicit. Therefore, while the answer to Question 3.46 is yes, even given explicit deformation retracts of A and B into H, there is no known constructive way of exhibiting such a zig-zag.

3.3.2 Lie Algebras and Their Enveloping Algebras

Let us consider a seemingly unrelated question to the previous section:

Question 3.51 Can we recover a Lie algebra from its universal enveloping algebra?

Recalling the digression in Sect. 2.2.1, the immediate answer should be yes: $\mathfrak{g}$ is isomorphic (as a Lie algebra) to the primitive elements of $\mathfrak{U}\mathfrak{g}$. However, this makes use of the Hopf algebra structure. Interpreting $\mathfrak{U}$ as a functor into associative algebras, we can rephrase the previous question more similarly to Question 3.46

Question 3.52 If $\mathfrak{g}$ and $\mathfrak{h}$ are Lie algebras, do we have:

$$\mathfrak{U}\mathfrak{g} \sim_{\mathbf{Ass}} \mathfrak{U}\mathfrak{h} \Rightarrow \mathfrak{g} \sim_{\mathbf{Lie}} \mathfrak{h}.$$

Let us start by pointing out two differences between Questions 3.46 and 3.52. The first one is that the converse of Question 3.46 is obviously true: the same zig-zag exhibiting the equivalence between two commutative algebras exhibits their equivalence as associative algebras. The same argument holds for Question 3.52, as soon as we know the much less trivial fact that universal enveloping algebras preserve quasi-isomorphisms. Indeed, by the

PBW Theorem 2.33, as a cochain complex $\mathfrak{U}\mathfrak{g} \cong \mathrm{Sym}\,\mathfrak{g}$ and the symmetric algebra functor preserves quasi-isomorphisms. Notice that both of these facts require char $k = 0$.

The second difference is that in the classical (i.e. non-dg) setting, Question 3.52 is non-trivial. In fact, in characteristic 0 the following is still an open question:

Question 3.53 (The Isomorphism Problem for Enveloping Algebras) Let $\mathfrak{g}$ and $\mathfrak{h}$ be Lie algebras in vector spaces, such that the associative algebras $\mathfrak{U}\mathfrak{g}$ and $\mathfrak{U}\mathfrak{h}$ are isomorphic. Are $\mathfrak{g}$ and $\mathfrak{h}$ necessarily isomorphic?

Question 3.46 concerns the restriction functor associated to the operad morphism $f\colon \mathbf{Ass} \to \mathbf{Com}$, while Question 3.52 concerns the induction functor associated to the Koszul dual morphism $f^{!}\colon \mathbf{Lie} \to \mathbf{Ass}$. In a sense that we will make precise, Koszul duality intertwines these two functors in a way that makes the two questions almost equivalent. The word almost is present because Koszul duality, as presented in Sect. 2.5 is only an equivalence between $\mathscr{P}$-algebras and $\mathscr{P}^{\text{i}}$-coalgebras and some nilpotence assumptions are required in order to have an equivalence between $\mathscr{P}$-algebras and $\mathscr{P}^{!}$-algebras (see Warning 2.70).

Theorem 3.54 *Let $\mathfrak{g}$ and $\mathfrak{h}$ be Lie algebras in non-negative degrees that are either nilpotent or concentrated in strictly positive degrees. Then we have:*

$$\mathfrak{U}\mathfrak{g} \sim_{\mathbf{Ass}} \mathfrak{U}\mathfrak{h} \Rightarrow \mathfrak{g} \sim_{\mathbf{Lie}} \mathfrak{h}$$

Proof We consider the bar-cobar adjunction associated to the twisting morphism $\alpha : \mathscr{P}^{\text{i}} \dashrightarrow \mathscr{P}$ (Definition 2.64) for $\mathscr{P} = \mathbf{Lie}$ and $\mathscr{P} = \mathbf{Ass}$:

$$\Omega_\alpha : \mathbf{coAlg}^{\mathrm{conil}}_{\mathscr{P}^{\text{i}}} \rightleftarrows \mathbf{Alg}_{\mathscr{P}} : \mathbf{B}_\alpha$$

Recall that we have a morphism of twisting morphisms (Definition 2.38)

$$\begin{array}{ccc} \mathbf{Com}^\vee\{-1\} \simeq \mathbf{Lie}^{\text{i}} & \overset{\alpha}{\dashrightarrow} & \mathbf{Lie} \\ \downarrow & & \downarrow \\ \mathbf{Ass}^\vee\{-1\} \simeq \mathbf{Ass}^{\text{i}} & \overset{\beta}{\dashrightarrow} & \mathbf{Ass} \end{array}$$

the induces the following diagram of adjunctions thanks to the naturality of the bar-cobar construction (see Remark 2.67):

$$\begin{array}{ccc} \mathbf{coAlg}_{\mathbf{Lie}^{\text{i}}} & \underset{\mathbf{B}_\alpha}{\overset{\Omega_\alpha}{\rightleftarrows}} & \mathbf{Alg}_{\mathbf{Lie}} \\ \uparrow\downarrow F & & \uparrow\downarrow \mathfrak{U} \\ \mathbf{coAlg}_{\mathbf{Ass}^{\text{i}}} & \underset{\mathbf{B}_\beta}{\overset{\Omega_\beta}{\rightleftarrows}} & \mathbf{Alg}_{\mathbf{Ass}} \end{array}$$

The diagram of left (respectively right) adjoint commute and therefore $\mathfrak{U}\circ\Omega_\alpha = \Omega_\beta \circ F$, where F the is the forgetful functor.

Let us now suppose $\mathfrak{U}\mathfrak{g} \sim_{\mathbf{Ass}} \mathfrak{U}\mathfrak{h}$. The rest of the proof decomposes into the following cochain of implications explained below:

(1) $\mathfrak{U}\Omega_\alpha \mathbf{B}_\alpha \mathfrak{g} \sim_{\mathbf{Ass}} \mathfrak{U}\Omega_\alpha \mathbf{B}_\alpha \mathfrak{h}$.
(2) $\Omega_\beta \mathbf{B}_\alpha \mathfrak{g} \sim_{\mathbf{Ass}} \Omega_\beta \mathbf{B}_\alpha \mathfrak{h}$.
(3) $\mathbf{B}_\alpha \mathfrak{g} \sim_{\mathbf{Ass}^{\text{i}}} \mathbf{B}_\alpha \mathfrak{h}$.
(4) $\mathbf{B}_\alpha \mathfrak{g} \sim_{\mathbf{Lie}^{\text{i}}} \mathbf{B}_\alpha \mathfrak{h}$.
(5) $\mathfrak{g} \sim_{\mathbf{Lie}} \mathfrak{h}$.

The first point follows from the PBW theorem 2.33, which implies that

$$\mathfrak{U}\Omega_\alpha \mathbf{B}_\alpha \mathfrak{g} \xrightarrow{\sim} \mathfrak{U}\mathfrak{g}$$

is a resolution. The second point follows from the equality $\mathfrak{U} \circ \Omega_\alpha = \Omega_\beta \circ F$ established before.

The third point follows from applying $\mathbf{B}_\beta$ and the fact that id $\to \mathbf{B}_\beta \Omega_\beta$ is a natural quasi-isomorphism since **Lie** is Koszul and using Proposition 2.68. Notice that we are also using that bar constructions preserve quasi-isomorphisms (see Proposition 2.68), since we are applying $\mathbf{B}_\beta$ to the whole zig-zag.

We find ourselves in a situation where we have two (shifted) cocommutative algebras that are quasi-isomorphic only as (shifted) coassociative algebras. The same methods used to prove Theorem 3.47 can be applied in the coalgebra setting to show that such coalgebras must also be quasi-isomorphic as cocommutative algebras [CPRW19, Theorem 4.27], showing point 3.3.2.

Finally, to deduce point 3.3.2 we would like to proceed as in point 3.3.2 and apply Ω_α, but unfortunately cobar constructions do not preserve quasi-isomorphisms in general.

What one can instead do is to apply the *complete* cobar construction $\Omega_\alpha^\wedge$, which does preserve quasi-isomorphisms. The resulting Lie algebras are no longer quasi-isomorphic to $\mathfrak{g}$ and $\mathfrak{h}$, but rather to their *homotopy completions*. The hypotheses of the theorem are such that the homotopy completions of $\mathfrak{g}$ and $\mathfrak{h}$ are equivalent to $\mathfrak{g}$ and $\mathfrak{h}$, thus concluding the proof. □

4 Formal Moduli Problems and Koszul Duality

This section aims at explaining recent results on "operadic deformation theory" and extend the classical philosophy that the space of deformations is classified by a Lie algebra.

We start with the Sect. 4.1.1 where we define deformations from a geometric point of view and recall the notion of infinitesimal deformations, and their relationship with the tangent complex (Sect. 4.1.2). In Sect. 4.1.3 we will look at deformations of algebras and compare with what was studied in Sect. 3.2.

Then Sect. 4.2 sets up a general context to describe deformation theories (following [Lur11]). We will see a generalization of the notion of infinitesimal objects using the formalism of small objects and morphisms. This leads to the definition of *formal moduli problem* (Definition 4.31) which gives an axiomatic framework for what a deformation functor[12] is in a given deformation context. Formal moduli problems, assemble into an ∞-category noted **FMP**.

With the right definitions in place, this allows us to formalize the classical slogan that deformation problems correspond to Lie algebras into a rigorous equivalence of ∞-categories.

Theorem 3.36 ([Lur11, Pri10]) *There is an equivalence of ∞-categories:*

$$\mathrm{MC}\colon \mathbf{Alg}_{\mathrm{Lie}} \underset{}{\overset{\sim}{\rightleftarrows}} \mathbf{FMP}\colon \mathbb{T}[-1]$$

This equivalence is roughly given by the Deligne ∞-groupoid in one direction and by the shifted tangent space in the other direction.

As we will see, this is in fact an instance of the Koszul duality between the operads **Com** (deformations are parametrized by commutative algebras) and **Lie**. In fact, we can consider deformations parameterized by more general algebraic structures.

Indeed, in Sect. 4.3 we describe the categories of formal moduli problems over a class of deformation contexts for $\mathscr{P}$-algebras. We finish by providing the tools that go into understanding and showing the following generalization of Theorem 3.36:

Theorem 3.64 ([CCN22]) *For $\mathscr{P}$ a Koszul binary quadratic operad concentrated in non-positive degrees, there is an equivalence of ∞-categories:*

$$\mathbf{FMP}_{\mathscr{P}} \simeq \mathbf{Alg}_{\mathscr{P}^!}.$$

4.1 Classical Deformation Theory

4.1.1 Deformation Theory of Varieties

Let us start by reviewing classical deformation theory. We refer the readers to [Har09], as well as the classical books [Ill71, Ill72] First of all, we recall that classically, the space along which we deform is often chosen to be the spectrum of a local Artinian algebra. Such algebras are infinitesimal extensions of a point and therefore will encode infinitesimal deformations.

Definition 4.1 The category Art of *classical local Artinian k-algebras* with residue field k is the category of local non-graded commutative k-algebras A, with residue field k, such that the maximal ideal satisfies $\mathfrak{M}_A^n = 0$ for $n \gg 0$.

The adjective "classical" in the definition above is non-standard. The reason for this is to distinguish it from the differential graded version of Artinian algebras which will be central starting from Sect. 4.2.

---

[12] Essentially, a deformation functor is a functor that sends a "small" deformation parameter (e.g. $k[\epsilon]$) to the "space of all deformations" along this parameter space.

Remark 4.2 Geometrically, Artinian algebras contain only one closed point (given by the augmentation $A \to k$), and they correspond to infinitesimal neighborhoods of this point. For example, the algebras $k[x]/(x^n)$ are classical local Artinian algebras.

The following definition is given for k-schemes but the reader can keep in mind the example of proper smooth $\mathbb{C}$-varieties.

Definition 4.3 (Deformation of Algebraic Schemes) Let X be an algebraic scheme (locally finitely generated k-scheme). Let (S, s) be a pointed scheme. Then, an (S, s) deformation is given by a scheme $\widetilde{X}$ together with a flat surjective morphism $\pi : \widetilde{X} \to S$ such that the following square is a pullback:

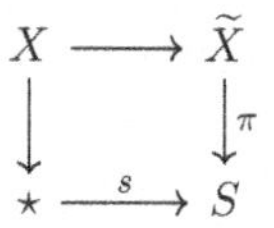

Deformations along Spec $(k[\epsilon])$, where $\epsilon^2 = 0$, are called *first-order* deformations. More generally, we say that a deformation is *infinitesimal* if $S = \mathrm{Spec}(A)$ with A a local Artinian k-algebra with residue field k.

Two deformations $\widetilde{X}_1$ and $\widetilde{X}_2$ are said to be isomorphic if there is an isomorphism of schemes $\phi : \widetilde{X}_1 \to \widetilde{X}_2$ making the following diagram commute:

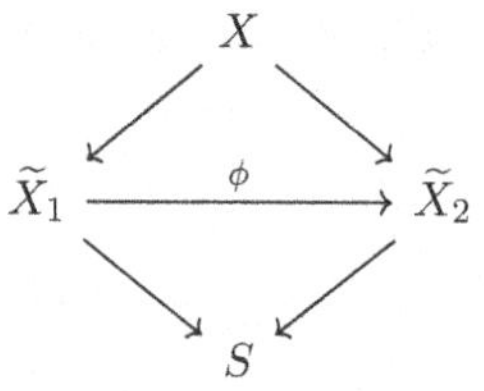

Observe that this implies that ϕ induces the identity on the pullback, X.

Remark 4.4

- If $X = \mathrm{Spec}(B)$ and $S = \mathrm{Spec}(A)$ are affine and A is a classical local Artinian algebra, then it is shown in [Ser07, Section 1.2.2] that any deformation $\widetilde{X}$ is also affine. We can rephrase the notion of deformation of an affine space by saying that a deformation of B along A is the data of a commutative algebra B' together with a flat map $A \to B'$ and an isomorphism $B' \otimes_A k \xrightarrow{\sim} B$.

  As before, an equivalence of two such deformations is given by an isomorphism $\phi : B'_1 \to B'_2$ of A-algebras making the following diagram commute:

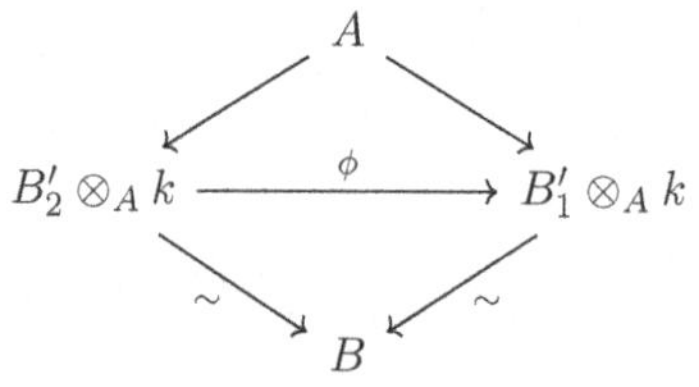

- Smooth algebras are rigid. If B is a smooth (non-graded) algebra, then Corollary 4.8 in [Har09] shows that if A is a classical local Artinian algebra, then any deformation of B along A is trivial, that is, any deformation B' is equivalent to $B \otimes_k A$.

Let us now mention two classical results on deformations of complex manifolds that will motivate the relationship between first-order deformations, the tangent space and a set of Maurer–Cartan elements. These relations will then be explored in detail and generalized later on.

Theorem 4.5 (Kodaira–Spencer, [KS58]) *First order deformations (for $A = \mathbb{C}[\epsilon]/\epsilon^2$) of a complex analytic manifold are controlled by the cohomology of X with values on the tangent bundle $H^*(X, T_X)$, namely:*

$$H^1(X, T_X) \overset{1:1}{\longleftrightarrow} \text{deformations along Spec } (\mathbb{C}[\epsilon])$$

$$H^2(X, T_X) \overset{1:1}{\longleftrightarrow} \text{Obstructions to extending the deformation to higher order.}$$

This statement is a geometric version of [LV12, Theorem 12.2.14] and Remark 3.35. We will see the relationship between algebraic and geometric deformations in Sect. 4.1.3. In this setting, the Lie algebra controlling infinitesimal deformations of the complex manifold X is precisely the Dolbeault complex $\Omega^{0,*}(T_X)$, with degree and differential coming from the differential forms, and inheriting a bracket from the Lie bracket of vector fields.

Theorem 4.6 ([Man05, Theorem V.55]) *Given a classical local Artinian algebra A with maximal ideal $\mathfrak{M}_A$ we have an isomorphism of groupoids:*

$$\text{Aut} \circlearrowright \{\text{Deformations of } X \text{ along } A\} \leftrightarrow \text{MC}\left(\Omega^{0,*}(T_X) \otimes \mathfrak{M}_A\right) \circlearrowleft \text{Gauge}$$

4.1.2 Cotangent Complex and Higher Deformations

As we have seen in Proposition 3.21, the right hand side of Theorem 4.6 is part of the full Deligne ∞-groupoid. The construction of **Del** involves tensoring the Lie algebra $\Omega^{0,*}(T_X) \otimes \mathfrak{M}_A$, living in non-negative degrees with differential forms, also living in non-negative degrees. Since the Maurer–Cartan equation lives in degree 2, even for the Deligne ∞-groupoid we only see the degrees 0, 1 and 2 of $\Omega^{0,*}(T_X)$ playing a part. It is natural to wonder what is the role of the higher degrees of $\Omega^{0,*}(T_X)$, or more generally, of any Lie algebra controlling a deformation problem. In short, these higher degrees control *derived deformations*.

This section will explore the relationship between infinitesimal deformation, square zero extensions, (higher) infinitesimal deformations, higher **Ext** groups and the cotangent complex.

Definition 4.7 A *square zero extension* of $B \in \mathbf{Alg}_{\mathbf{Com}}$ is a commutative algebra $B' \in \mathbf{Alg}_{\mathbf{Com}}$ together with a surjection $\pi : B' \to B$ that fits in an exact sequence in $\mathbf{Alg}_{\mathbf{Com}}$:

$$0 \longrightarrow I \longrightarrow B' \stackrel{\pi}{\longrightarrow} B \longrightarrow 0$$

such that $I^2 = 0$. This is an algebraic version of the extension of a scheme described in [Ser07, Section 1.1.3]. Moreover, if $B \in \mathbf{Alg}_{\mathbf{Com}}^{A/}$, i.e. B comes with a map $A \to B$ for some fixed $A \in \mathbf{Alg}_{\mathbf{Com}}$, then the same definition makes sense with all objects under A.

Given any B-module M, we can construct the *trivial square zero extension* $B \oplus M$ given by the direct sum as a B-module, and with the product defined by $(b, m) \times (b', m') = (bb', bm' + b'm)$. This square zero extension corresponds to the split exact sequence:

$$0 \longrightarrow M \longrightarrow B \oplus M \stackrel{\pi}{\longrightarrow} B \longrightarrow 0$$

A *morphism between two square zero extensions* B'_1 and B'_2 is a morphism $\phi : B'_1 \to B'_2$ that commutes with the projections on B.

Example 4.8 Any iterated square zero extension of k by a finite dimensional vector space is classical local Artinian.

Conversely, any classical local Artinian algebra can be obtained as an iterated square zero extension of the trivial algebra k. Indeed, if A is such that $\mathfrak{M}_A^n = 0$ and $\mathfrak{M}_A^{n-1} \neq 0$, then:

$$0 \to \mathfrak{M}_A^{n-1} \to A \to A/_{\mathfrak{M}_A^{n-1}} \to 0$$

exhibits A as a square zero extension of an algebra of lower nilpotence.

As a particular example, the quotient $k[x]/_{(x^n)} \to k[x]/_{(x^{n-1})}$ is a square zero extension.

Remark 4.9 Let us explain how trivial square zero extensions are functorially related to derivations.

- Given $B \in \mathbf{Alg}_{\mathbf{Com}}^{A/}$, the trivial square zero extension construction defines a functor:

$$B \oplus - : \begin{array}{c} \mathrm{Mod}_B \longrightarrow \mathbf{Alg}_{\mathbf{Com}}^{A//B} \\ M \longmapsto B \oplus M \end{array}$$

- A-linear sections of the projection $B \oplus M \to B$ are precisely A-linear M-valued derivations[13] on B:

$$\mathrm{Hom}_{\mathbf{Alg}_{\mathbf{Com}}^{A//B}}(B, B \oplus M) \cong \mathrm{Der}_A(B, M)$$

[13] $\mathrm{Der}_A(B, M)$ denotes the B-module of A-linear M-valued derivations concentrated in degree 0 and $\mathbb{R}\mathrm{Der}_A(B, M)$ its derived functor.

- The square zero extension functor $B \oplus -$ has a left adjoint:

$$L : \mathbf{Alg}_{\mathbf{Com}}^{A//B} \to \mathrm{Mod}_B$$

  Moreover this adjunction is Quillen for the standard model structures on $\mathbf{Alg}_{\mathbf{Com}}^{A//B}$ and Mod_B.

Definition 4.10 Given $B \in \mathbf{Alg}_{\mathbf{Com}}^{A/}$, the *module of Kähler differentials* is defined as the B-modules $\Omega^1_{B/A}$ representing A-linear derivations:

$$\mathrm{Hom}_B\left(\Omega^1_{B/A}, M\right) \cong \mathrm{Der}_A(B, M)$$

Remark 4.9 ensures that:

$$\Omega^1_{B/A} = L(\mathrm{id} : B \to B)$$

A direct construction of $\Omega^1_{B/A}$ is obtained by constructing the free B-module on formal generators $d_{\mathrm{dR}}b$ for all $b \in B$, subject to the relations $d_{\mathrm{dR}}a = 0$ for $a \in A$, $d_{\Omega^1}d_{\mathrm{dR}}x = d_{\mathrm{dR}}d_Bx$ and $d_{\mathrm{dR}}(xy) = (d_{\mathrm{dR}}x)y + xd_{\mathrm{dR}}y$.

The derived version is called the *cotangent complex* $\mathbb{L}_{B/A}$ representing the derived module of derivations:

$$\underline{\mathrm{Hom}}_B\left(\mathbb{L}_{B/A}, M\right) \simeq \mathbb{R}\mathrm{Der}_A(B, M) := \mathbb{R}\underline{\mathrm{Hom}}_{\mathbf{Alg}_{\mathbf{Com}}^{A//B}}(B, B \oplus M).$$

The B-linear dual of $\mathbb{L}_{B/A}$ is called the *tangent complex* and we have:

$$\mathbb{T}_{B/A} := \underline{\mathrm{Hom}}_B\left(\mathbb{L}_{B/A}, B\right) \simeq \mathbb{R}\mathrm{Der}_A(B, B)$$

Remark 4.11 Here are a few basic properties of the cotangent complex:

- Using the derived functor of the adjunction in Remark 4.9, we obtain:

$$\mathbb{L}_{B/A} := \mathbb{L}(\mathrm{Id} : B \to B) := \Omega^1_{Q(B)/A} \otimes_{Q(B)} B,$$

  where Q is a cofibrant replacement functor in $\mathbf{Alg}_{\mathbf{Com}}^{A//B}$ and $\mathbb{L}$ denotes the left derived functor of L.
- Given $f : A \to B$ and $g : B \to C$, we have a homotopy fiber sequence:

$$\mathbb{L}_{B/A} \otimes_B C \longrightarrow \mathbb{L}_{C/A} \longrightarrow \mathbb{L}_{C/B}$$

  In particular for $A = k$ we write $\mathbb{L}_{B/k} = \mathbb{L}_B$ and we get:

$$\mathbb{L}_B \otimes_B C \longrightarrow \mathbb{L}_C \longrightarrow \mathbb{L}_{C/B}$$

- The isomorphism classes of square zero extensions of B along I are in one to one correspondence with $\mathbf{Ext}^1_A(\mathbb{L}_B, I)$. In particular, for $I = B$ and $A = B$, we get $\mathbf{Ext}^1_B(\mathbb{L}_B, B) = \mathbf{Ext}^1_B(B, \mathbb{T}_B)$. Moreover, the trivial square zero extension of B by M described in Remark 4.9 represents the class of $0 \in \mathbf{Ext}^1(\mathbb{L}_B, M)$.
- Following [Vez10], the cotangent complex enables us to consider deformations in higher derived directions. Such higher deformations are encoded by higher **Ext** groups, and those are controlled by derived mappings from higher infinitesimal disks into the k-scheme X we want to deform.

  More precisely, for any R, a commutative k-algebra, a R-point in X, $x : \mathrm{Spec}(R) \to X$, and any $M \in \mathrm{Mod}_R$. Then we define the derived i-th order infinitesimal disk over R by:

$$\mathbf{D}_{R,i}(M) := \mathrm{Spec}\,(R \oplus M[i])$$

  More generally, for $\mathscr{M}$ a quasi-coherent sheaf on X, we define:

$$\mathbf{D}_{X,i}(\mathscr{M}) := \mathrm{Spec}_X\,(\mathscr{O}_X \oplus \mathscr{M}[i])$$

  Then Proposition 2.1 in [Vez10] tells us that this is related to the **Ext**[14] functor via the equivalences:

$$\mathbf{Ext}^i_R\left(\mathbb{L}_{X,x}, M\right) \simeq \mathbb{R}\underline{\mathrm{Hom}}_\star(\mathbf{D}_{R,i}(M), (X, x))$$

$$\mathbf{Ext}^i_{\mathscr{O}_X}(\mathbb{L}_X, \mathscr{M}) \simeq \mathbb{R}\underline{\mathrm{Hom}}_X(\mathbf{D}_{X,i}(\mathscr{M}), X)$$

  Above, the first Hom is required to be pointed and the second Hom is required to restrict to the identity on X. Extensions of the Kodaira–Spencer morphism and obstruction are also discussed in [Vez10, Section 3].

We can now show how *first-order* derivations of smooth schemes are related to the tangent cohomology.

Proposition 4.12 *Let X be a smooth algebraic variety, then first-order deformations up to equivalence are in bijection with $H^1(X, T_X)$.*

Proof First, notice that a first-order deformation of a smooth algebra $\mathrm{Spec}(B)$ is trivial as explained in Remark 4.4, and given by $B' := B \otimes k[\epsilon]$, with $\epsilon^2 = 0$. It is not difficult to see that B' coincides with the trivial square zero extension of B by B itself. In other words $B' = B \oplus B$. To simplify the notations, we will write $B' = B[\epsilon]$ and the product is given by $(b + b'\epsilon)(a + a'\epsilon) = ba + (ba' + b'a)\epsilon$ as expected when having $\epsilon^2 = 0$.

[14] $\mathbf{Ext}_A(M, N)$ is the derived functor of $\underline{\mathrm{Hom}}_A(-, -)$ evaluated at M, N.

Now a first order deformation $\widetilde{X}$ can be covered by smooth affine opens. Take $\{\mathrm{Spec}(B_i')\}$ such a smooth open chart. Then pulling back this covering along $X \to \widetilde{X}$ gives a covering of X. Then, we have the following pullback:

$$\begin{array}{ccc} \mathrm{Spec}(B_i) & \longrightarrow & \mathrm{Spec}(B_i') \\ \downarrow & & \downarrow \\ X & \longrightarrow & \widetilde{X} \end{array}$$

Since $\mathrm{Spec}(B_i)$ is an open of X which is smooth, then $\mathrm{Spec}(B)$ is also smooth, and by rigidity (see Remark 4.4) we have $B_i' = B_i[\epsilon]$. $\widetilde{X}$ is therefore only characterized by the gluing data of the $B_i[\epsilon]$, lifting the gluing data of X. In other, if B_i and B_j intersect, we need a lifting of the isomorphism $\phi_{ij} : B_{i,j} \to B_{j,i}$[15] characterizing the gluing on X:

$$\begin{array}{ccc} B_{i,j} & \xrightarrow{\phi_{ij}} & B_{j,i} \\ \downarrow & & \downarrow \\ B_{i,j}[\epsilon] & \overset{\phi_{ij}^{\epsilon}}{\dashrightarrow} & B_{j,i}[\epsilon] \\ \downarrow & & \downarrow \\ B_{i,j} & \xrightarrow{\phi_{ij}} & B_{j,i} \end{array}$$

(with curved arrows labelled id from the top row to the bottom row on each side)

Since ϕ is an isomorphism, giving a filling is equivalent to taking a section of $B_{j,i}[\epsilon] \to B_{j,i}$. We can also show that such a lift is necessarily an isomorphism.

Since such sections are equivalent to derivations of $B_{j,i}$, a deformation is given by a family of derivations on pairwise intersections of X.

Moreover the maps, ϕ_{ij}, satisfy a cocycle condition on triple intersections that lift to a cocycle condition on the family of derivations. Such a family is then exactly an element in the first cohomology class of T_X. □

4.1.3 Deformation Theory of Algebras

We want to describe deformations from the algebraic point of view. To do so, we start by describing deformations of commutative algebras similarly to the affine geometric situation of Remark 4.4. This definition extends naturally from commutative algebras to general $\mathscr{P}$-algebras and we will compare this notions of deformation that mimics geometric deformations, with deformations of $\mathscr{P}$-algebraic structures as defined in Sects. 3.2.1 and 3.2.3. From now on we will assume that $\mathscr{P}$ is Koszul, without differential.

[15] B_{ij} denotes the restriction of B_i to the intersection of B_i with B_j. In other words, $B_{i,j}$ is a localisation of B_i by some ideal realising this intersection.

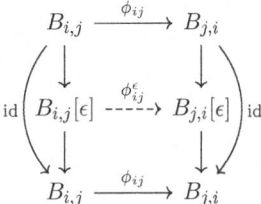

We consider the canonical model structure on groupoids in which weak equivalences are equivalences of categories and fibrations are isofibrations.[16] This model structure is the natural model structure restricting the model structure on the category of small categories, or equivalently the transferred model structure from **sSet** via the nerve adjunction (see [Hol08, Theorem 2.1 and Corollary 2.3]).

Definition 4.13 The *groupoid of geometric deformations* of a $\mathscr{P}$-algebra B along a local Artinian algebra A (Proposition 4.27) is defined as the homotopy pullback of groupoids:[17]

$$\begin{array}{ccc} \widetilde{\underline{\mathrm{Def}}}_B^{\mathscr{P}}(A) & \longrightarrow & \star \\ \downarrow & & \downarrow B \\ \mathbf{Alg}^{\mathrm{iso}}_{\mathscr{P}\otimes_k A}(\mathrm{Mod}_A) & \longrightarrow & \mathbf{Alg}^{\mathrm{iso}}_{\mathscr{P}} \end{array}$$

where the lower functor sends B' to $B' \otimes_A k$. For groupoids all objects are fibrant so we only need to have one fibration. To do that, we need to replace the map B by an equivalent fibration given by the following factorization:

$$\star \xrightarrow{\text{triv.cof}} (\mathbf{Alg}^{\mathrm{iso}}_{\mathscr{P}})^{/B} \xrightarrow{\text{fib.}} \mathbf{Alg}^{\mathrm{iso}}_{\mathscr{P}}$$

Computing the strict pullback with this replacement, we obtain that $\widetilde{\underline{\mathrm{Def}}}_B^{\mathscr{P}}(A)$ is equivalent to the groupoid whose objects are elements $B' \in \mathbf{Alg}_{\mathscr{P}}(\mathrm{Mod}_A)$ together with an isomorphism $B' \otimes_A k \xrightarrow{\sim} B$. This corresponds to the affine version of the set of geometric A-deformations as described in Remark 4.4.

The morphisms in $\widetilde{\underline{\mathrm{Def}}}_B^{\mathscr{P}}(A)$ correspond to the notion of equivalence described in Remark 4.4. In other words, two deformations B_1' and B_2' are *equivalent* if there exists an isomorphism $\phi : B_1' \to B_2'$ in $\mathbf{Alg}_{\mathscr{P}\otimes_k A}(\mathrm{Mod}_A)$ such that the following diagram commutes:

$$\begin{array}{ccc} B_1' \otimes_A k & \xrightarrow{\phi\otimes_A k} & B_2' \otimes_A k \\ & \searrow \quad \swarrow & \\ & B & \end{array}$$

In that situation, the $\mathscr{P}$-algebra structure on B is given by a map $\mathscr{P} \to \mathbf{End}_B$. Recall that the set of deformations of a $\mathscr{P}$-algebra structure $\phi : \mathscr{P} \to \mathbf{End}_B$ along A, given by Definition 3.32, is the set:

$$\mathrm{Def}_\phi(A) := \left\{ \Phi \in \mathrm{MC}\left(\mathfrak{g}^{\phi}_{\mathscr{P}_\infty, B} \otimes A\right),\ \Phi \cong \phi \bmod \mathfrak{M} \right\}$$

[16] An isofibration is a functor $p : \mathscr{C} \to \mathscr{D}$ such that for any object e of $\mathscr{C}$ and any isomorphism $f : p(e) \to b$, there exists an isomorphism $\tilde{f} : e \to e'$ such that $p(\tilde{f}) = f$.

[17] The superscript ^{iso} denotes the subcategory given by the maximal subgroupoid.

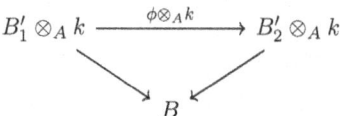

It turns out that these *algebraic* deformations coincide with the more *geometric* deformations (of affine):

Proposition 4.14 *Given B a $\mathscr{P}$-algebra with $\phi : \mathscr{P} \to \mathbf{End}_B$ and A a classical local Artinian k-algebra, there is a map of groupoids:*

$$\underline{\mathrm{Def}}_\phi(A) \hookrightarrow \widetilde{\underline{\mathrm{Def}}}_B^{\mathscr{P}_\infty}(A)$$

Proof An element in $\underline{\mathrm{Def}}_\phi(A)$ is a $\mathscr{P}_\infty$-structure on $B \otimes_k A$ such that modulo $\mathfrak{M}_A$, it recovers the $\mathscr{P}$-structure on B. In other words, the natural map:

$$B \otimes_k A \otimes_A k \to B$$

is an isomorphism on $\mathscr{P}$-algebras. Therefore there is an inclusion of the set of objects.

Furthermore, the definition of morphisms in $\underline{\mathrm{Def}}_\phi(A)$ (Definition 3.32) induces a morphism in $\widetilde{\underline{\mathrm{Def}}}_B^{\mathscr{P}_\infty}(A)$ defining the faithfull functor we seek. □

We will now extend the previous construction to the full ∞-groupoids of deformations. This ∞-groupoid will both extend the previous ones to higher homotopies of deformations and will take into account the homotopy theory of $\mathscr{P}$-algebras.

Definition 4.15 ([Hin04, Definition 2.1]) Let A be an Artinian commutative algebra concentrated in non-positive degree (Proposition 4.27) and $\mathscr{P}$ a Koszul operad. We define $\mathscr{W}^c(\mathscr{P}, A)$ to be the simplicial category whose objects are cofibrant $\mathscr{P} \otimes_k A$-algebras in Mod_A and whose n-morphisms from B to C are given by quasi-isomorphisms $B \to \Omega(\Delta^n) \otimes C$. Then the *higher deformation groupoid* $\mathbf{Def}_A(B)$ of $B \in \mathbf{Alg}_{\mathscr{P}}$ is defined as the homotopy pullback:

$$\begin{array}{ccc}
\mathbf{Def}_A(B) & \longrightarrow & \star \\
\downarrow & & \downarrow{\scriptstyle \tilde{B}} \\
N(\mathscr{W}^c(\mathscr{P}, A)) & \longrightarrow & N(\mathscr{W}^c(\mathscr{P}, k))
\end{array}$$

Where $\tilde{B} \to B$ is a cofibrant resolution of B and N denotes the simplicial nerve functor.

Remark 4.16 In [Hin04], it is shown thanks to Theorem 2.1.2, that for $\mathscr{P}$ and B non-positively graded, the deformation ∞-groupoid $\mathbf{Def}_A(B)$ is naturally equivalent to the higher Deligne groupoid $\mathbf{Del}\left(\mathbb{T}_B^{\mathscr{P}} \otimes_k A\right)$ of the $\mathscr{P}$-operadic tangent Lie algebra (see Definition 4.17).

The operadic tangent and cotangent complexes are defined in a very similar way to what was explained in Sect. 4.1.2 for $\mathscr{P} = \mathbf{Com}$.

Definition 4.17 (Operadic Tangent Complex) Let A be a $\mathscr{P}$-algebra. A *derivation* of A valued in A is a map $D \colon A \to A$ such that for all $p \in \mathscr{P}(n)$,

$$D(p(b_1, \dots, b_n)) = \sum_{i=1}^{n} p(b_1, \dots, Db_i, \dots, b_n).$$

The k-module of all derivations $\mathrm{Der}(A, A)$ is a Lie algebra with bracket given by the commutator of derivations.

The derived derivations of A, which we denote $\mathbb{T}_A^{\mathscr{P}}$, is called the *tangent complex* of A.

As the name indicates, the derived derivations can be obtained as a derived functor. Indeed, there is an operadic notion of module of Kähler differentials, $\Omega_A^{\mathscr{P}}$ and the *operadic cotangent complex* $\mathbb{L}_A^{\mathscr{P}}$, see [LV12, Sections 12.3.8–12.3.10]. One could equivalently define $\mathbb{T}_A^{\mathscr{P}} = \underline{\mathrm{Hom}}_A\left(\mathbb{L}_A^{\mathscr{P}}, A\right)$.

To complete this story, we would like to have a relationship between $\mathbb{T}_A^{\mathscr{P}}$ and the deformation complex, to have a relationship between the higher deformation groupoid of Definition 4.15 and the higher Deligne groupoid (Definition 3.20).

To obtain a model of $\mathbb{T}_A^{\mathscr{P}}$ (which is only defined up to quasi-isomorphism) we choose the cofibrant resolution of A given by the bar-cobar resolution. We have then

$$\mathbb{T}_A^{\mathscr{P}} \simeq \mathrm{Der}_k(\Omega_\kappa \mathbf{B}_\kappa A, \Omega_\kappa \mathbf{B}_\kappa A) = \mathrm{Hom}_k(\mathbf{B}_\kappa A, \Omega_\kappa \mathbf{B}_\kappa A) \xrightarrow{p} \mathrm{Hom}_k\,(\mathbf{B}_\kappa A, A)\,.$$

The map p is induced by the bar-cobar resolution and is therefore a quasi-isomorphism of complexes. In fact, $\mathrm{Hom}_k\,(\mathbf{B}_\kappa A, A)$ can be identified with coderivations on $\mathbf{B}_\kappa A$ and therefore has a Lie algebra structure making p a quasi-isomorphism of Lie algebras. Notice that ignoring differentials $\mathbf{B}_\kappa A = \mathscr{P}^{\text{i}} \circ A$ and therefore as graded vector spaces we have

$$\underline{\mathrm{Hom}}_k\,(\mathbf{B}_\kappa A, A) \cong \prod_{n\geq 1} \underline{\mathrm{Hom}}_k\left(\mathscr{P}^{\text{i}}(n), \mathrm{End}_A(n)\right)$$

We recognize here the underlying graded vector space of the deformation complex, up to a caveat: In the deformation complex one should remove the counit by taking $\overline{\mathscr{P}^{\text{i}}}$ instead, see Remark 2.41.

Keeping track of the differential and the bracket, we see that they correspond precisely to the differential and bracket of the deformation complex. Taking this model for $\mathbb{T}_A^{\mathscr{P}}$, the projection $\mathscr{P}^{\text{i}} \to \overline{\mathscr{P}^{\text{i}}}$ yields a well defined inclusion of Lie algebras:

$$\mathfrak{g}^{\phi}_{\mathscr{P}_\infty, A} \to \mathbb{T}_A^{\mathscr{P}}$$

This map is the Lie algebra version of the map of groupoids of Proposition 4.14. In short, we have shown the following Proposition:

Proposition 4.18 *Let $\phi : \mathscr{P} \to \mathbf{End}_A$ be a $\mathscr{P}$-algebra structure on $A \in \mathrm{Mod}_k$. There is a homotopy fiber sequence of Lie algebras:*

$$\mathfrak{g}^{\phi}_{\mathscr{P}_\infty, A} \to \mathbb{T}_A^{\mathscr{P}} \to \mathrm{Hom}(A, A)$$

Remark 4.19 Heuristically, Proposition 4.18 expresses that the "difference" between the two deformation problems is the deformation of the underlying k-module, which is encoded by $\mathrm{Hom}(A, A)$.

Notice that the Lie algebra structure on $\mathrm{Hom}(A, A)$ in fact arises from an associative algebra structure. From the deformation perspective, this means that the corresponding deformation problem is also richer. This will correspond to an *associative* (as opposed to commutative, not Lie) formal moduli problem and its formalisation will be the topic of the next sections.

Remark 4.20 (Global Deformations) So far, we only talked about deformations of "affine objects", that is $\mathscr{P}$-algebras. However, we could be interested in deforming not only $\mathscr{P}$-algebras, but a sheaf of $\mathscr{P}$-algebras in the spirit of the first part of Remark 4.4.

In [Hin05], it is shown that the deformations of a sheaf of $\mathscr{P}$-algebras $\mathscr{A}$ are controlled, under some conditions on $\mathscr{P}$ and $\mathscr{A}$ given in [Hin05, Assumption 3.5.1], by the higher Deligne groupoid of the derived global sections of the presheaf of local tangent Lie algebras, which by definition gives the tangent cohomology.

4.2 *Formal Moduli Problems*

This section is about axiomatizing the notion of deformation theory. After seeing how deformations from an algebraic and from a geometric point of view are related, we expect them to have very similar properties which we will encode in the notion of formal moduli problems.

The idea is that deformations can be understood as a functor sending an "Artinian" object (which we think of as a parameter space) to a space of deformations along this object. For example such a deformation functor assigns for each choice of Artinian local algebra A a set, groupoid or ∞-groupoid of deformations of X along A. As the category of local Artinian algebras and the notion of "infinitesimal" object will play a more important role from here, we will start, following [Lur11], by defining a general framework in which we can speak of deformations and Artinian objects.

In this section, $\mathscr{P}$ will be an augmented Koszul operad.

4.2.1 Artinian and Small Algebras in a Deformation Context

The general idea behind Artinian algebras is to have a class of algebras considered "small" so that deforming along those algebras amounts to consider *infinitesimal deformations*, looking at the *formal neighborhood* of what we want to deform. We will always assume that our ambient category $\mathscr{A}$, in which we will define the notion of Artinian object, has a terminal object. This leads to the general framework of deformation context:

Definition 4.21 ([Lur11, Definition 1.1.3]) A *deformation context* is a pair $(\mathscr{A}, \{E_\alpha\}_\alpha)$ where $\mathscr{A}$ is a presentable[18] ∞-category and $\{E_\alpha\}_\alpha$ is a set of objects in $\mathbf{Stab}(\mathscr{A})$, the stabilization[19] of $\mathscr{A}$.

Intuitively, $\{E_\alpha\}_\alpha$ corresponds to the first-order objects we want to deform along. To compare with the classical picture, we take $\mathscr{A}$ to be the category of commutative k-algebras. Since previously we took $k[\epsilon]$ as the algebra representing first-order deformations, we now take it as the underlying object of the spectrum object $E = (\cdots, k[\epsilon_2], k[\epsilon_1], k[\epsilon_0], \cdots)$, where $k[\epsilon_i]$ is the 2-dimensional augmented commutative algebra with $\deg \epsilon_i = -i$ and $\epsilon_i^2 = 0$. We will see in Sect. 4.3.1 that deformations along $k[\epsilon_i]$ are both related to the higher tangent complex and control obstructions to lifting deformations.

In all cases we consider, the family $\{E_\alpha\}_\alpha$ contains only a single object. Here are some other examples of deformation contexts.

Remark 4.22

- Given a commutative algebra A, the stabilization of A-augmented commutative A-algebra, $\mathbf{Stab}\left(\mathbf{Alg}_{\mathbf{Com}}^{A//A}\right)$ can be identified with Mod_A using the square zero extension functor, see [dRBM04, Theorem 3.7]

$$\begin{aligned} \mathrm{Mod}_A &\xrightarrow{\sim} \mathbf{Stab}\left(\mathbf{Alg}_{\mathbf{Com}}^{A//A}\right) \\ M &\longmapsto (A \oplus M[n])_{n\in\mathbb{Z}} \end{aligned}$$

- To understand $\mathbf{Stab}\left(\mathbf{Alg}_{\mathbf{Com}}\right)$ first observe that there is an equivalence:

$$\mathbf{Alg}_{\mathbf{Com}} \rightleftarrows \mathbf{Alg}_{\mathbf{Com}}^{k//k} \tag{6}$$

sending a non-unital commutative algebra B to its unitalization $k \oplus B$ and an augmented k-algebra A to the fiber of the augmentation $A \to k$. This equivalence induces an equivalence:

$$\mathbf{Stab}\left(\mathbf{Alg}_{\mathbf{Com}}\right) \rightleftarrows \mathbf{Stab}\left(\mathbf{Alg}_{\mathbf{Com}}^{k//k}\right)$$

As such $\mathbf{Stab}\left(\mathbf{Alg}_{\mathbf{Com}}\right)$ is also equivalent to Mod_k and under this equivalence the spectrum object corresponding to $M \in \mathrm{Mod}_k$ is $(M[n])_{n\in\mathbb{Z}}$ with the trivial commutative product on each $M[n]$.

[18] $\mathscr{A}$ is presentable if it is generated by a small set of "small objects" under homotopy colimits.

[19] Recall that objects in $\mathbf{Stab}(\mathscr{A})$ correspond to spectrum objects in $\mathscr{A}_*$, i.e., a sequence $E = (\cdots, a_2, a_1, a_0, a_{-1}, \cdots)$ of pointed objects of $\mathscr{A}$, such that a_i is equivalent to the loop space Ωa_{i+1}. For $n \geq 0$, we denote $\Omega^{\infty-n} E = a_n$.

- Similarly $\mathbf{Stab}(\mathbf{Alg}_{\mathscr{P}})$ will be given by the category of $\mathscr{P}$-modules. A $\mathscr{P}$-module M corresponds the spectrum object $(M[n])_{n\in\mathbb{Z}}$ where $M[n]$ is viewed with the trivial $\mathscr{P}$-algebra structure.[20]

Example 4.23

- For $\mathscr{A} = \mathrm{Mod}_A$ (with $A \in \mathbf{Alg_{Com}}$), we can define the deformation context $(\mathrm{Mod}_A, \{E\})$ with only one spectrum object $E = (A[n])_{n\in\mathbb{Z}}$ [CG21, Example 2.2].
- Given A a commutative algebra, we consider $\mathscr{A} = \mathbf{Alg}_{\mathbf{Com}}^{A//A}$ and we can define the following deformation context:

$$\left(\mathbf{Alg}_{\mathbf{Com}}^{A//A}, \{E = (A \oplus A[n])_{n\in\mathbb{Z}}\}\right)$$

  where $A \oplus A[n]$ is the square zero extension of A by $A[n]$ [CG21, Remark 2.5]. When $A = k$, we will call this deformation context $\mathscr{A}_{\mathbf{Com}}^{\mathrm{aug}}$ for short.
- For $\mathscr{A} = \mathbf{Alg_{Com}}$ we consider the following deformation context:

$$\left(\mathbf{Alg_{Com}}, \{(k[n])_{n\in\mathbb{Z}}\}\right)$$

  where $k[n]$ is the trivial commutative algebra on $k[n]$. This deformation context can also be seen as obtained from a transfer of deformation contexts along the adjunction of Eq. (6) (see [CG21, Lemma 2.6]).
- For $\mathscr{A} = \mathbf{Alg}_{\mathscr{P}}$ we consider the deformation context:

$$\left(\mathbf{Alg}_{\mathscr{P}}, (k[n])_{n\in\mathbb{Z}}\right)$$

  We will denote it by $\mathscr{A}_{\mathscr{P}}$ for short.

From a deformation context, we can define the notion of "small" objects and morphisms using $\Omega^{\infty-n}E_\alpha$ as building blocks.

Definition 4.24 ([Lur11, Definition 1.1.14]) Given a deformation context $(\mathscr{A}, \{E_\alpha\}_{\alpha\in T})$ we say that:

- A morphism $f : A \to A'$ is *elementary* if it is given by a pullback of the form:

$$\begin{array}{ccc} A & \longrightarrow & \star \\ \downarrow f & & \downarrow \\ A' & \longrightarrow & \Omega^{\infty-n}(E_\alpha) \end{array}$$

  for some $\alpha \in T$ and $n \geq 1$.
- A morphism $f : A \to A'$ is *small* if it is a finite composition of elementary morphisms.

[20] Since we assume that $\mathscr{P}$ is augmented, Remark 2.31 gives us a functor $\mathrm{Mod}_k \to \mathbf{Alg}_{\mathscr{P}}$ giving the trivial $\mathscr{P}$-algebra structure to a module.

- An object $A \in \mathscr{A}$ is *Artinian* if the morphism $\mathscr{A} \to \star$ is small.[21] We denote by $\mathbf{Art}_{\mathscr{A}}$ the full sub-category of $\mathscr{A}$ given by small objects.

Example 4.25 Going back to the example of (augmented) commutative algebras, we have $\Omega^{\infty-n}E = k[\epsilon_n]$ so that in order to compute the homotopy pullback of $\star \to \Omega^{\infty-n}E$, one can replace the point $\star = k$ by the algebra $k \oplus k\epsilon_n \oplus k\epsilon_{n-1}$ (notice that $n \geq 1$), with the only non-zero product being the unit and with differential sending ϵ_n to ϵ_{n-1}.

The strict pullback exhibits therefore A as a square zero extension of A' along $k\epsilon_{n-1}$. In particular, an Artinian commutative algebra concentrated in degree zero is a classical Artinian local algebra in the sense of Definition 4.1, thanks to Example 4.8.

Conversely, suppose A is Artinian. Then for any square zero extension by the algebra $k[\epsilon_n]$:

$$k[\epsilon_n] \longrightarrow B \longrightarrow A,$$

B is also Artinian. Indeed up to pulling back to a quasi-free resolution of A, the surjection $B \to A$ splits as graded algebras and $B \simeq A \oplus k\epsilon_n$ with differential $d(a, v) = (da, \chi(a))$. We can then check that B can be realised as the homotopy pullback square:

$$\begin{array}{ccc} B & \longrightarrow & k \\ \downarrow & & \downarrow \\ A & \xrightarrow{\chi} & k[\epsilon_{n+1}] \end{array}$$

where χ is obtained as a (shift of) the part of the differential on $A \oplus k[n]$ going from A to $k[n]$. In particular, we have $B \simeq A \times^h_{k[n+1]} 0$ and therefore B is Artinian.

We will now specialize these definitions to the deformation contexts we are interested in, namely $\mathscr{A}^{\mathrm{aug}}_{\mathbf{Com}}$ and $\mathscr{A}_{\mathscr{P}}$.

Proposition 4.26 (Artinian Commutative Augmented k-Algebras) *The category $\mathbf{Art}_{\mathscr{A}^{\mathrm{aug}}_{\mathbf{Com}}}$ is the smallest full sub-category of $\mathbf{Alg}^{k//k}_{\mathbf{Com}}$ such that:*

- *$k \oplus k[n] \in \mathbf{Art}_{\mathscr{A}^{\mathrm{aug}}_{\mathbf{Com}}}$ for all $n \geq 0$*
- *For any $A \in \mathbf{Art}_{\mathscr{A}^{\mathrm{aug}}_{\mathbf{Com}}}$ and any map $A \to k \oplus k[n]$ with $n \geq 1$, the homotopy pullback $A \times_{k\oplus k[n]} k$ is also Artinian:*

$$\begin{array}{ccc} A \times_{k\oplus k[n]} k & \longrightarrow & k \\ \downarrow & & \downarrow \\ A & \longrightarrow & k \oplus k[n] \end{array}$$

This is a variation of Proposition 4.28.

[21] Artinian objects can be though as "small" objects. In fact they are called "small" in [Lur11].

We can also describe Artinian algebras in more classical terms:

Proposition 4.27 ([Lur11, Proposition 1.1.11]) *$A \in \mathbf{Alg}_{\mathbf{Com}}^{k//k}$ is Artinian if and only if:*

- *$H^i(A) = 0$ for all $i > 0$.*
- *$H^*(A)$ is of total finite dimension.*
- *Artinian condition:*

$$\ker\left(H^0(A) \to k\right) \text{ is nilpotent.}$$

In particular $H^0(A) \in$ Art and there is an embedding of strict categories Art $\to$ $\mathbf{Art}_{\mathscr{A}_{\mathbf{Com}}^{\mathrm{aug}}}$. Indeed, as we have seen in Example 4.8, classical local Artinian algebras are created by successive square-zero extensions.

With Definition 4.24 and the deformation context on $\mathbf{Alg}_{\mathscr{P}}$ of Example 4.23 we have now a suitable notion of Artinian algebra over an operad.

Proposition 4.28 *The ∞-category $\mathbf{Art}_{\mathscr{P}}$ of Artinian $\mathscr{P}$-algebras is the smallest full subcategory of the ∞-category $\mathbf{Alg}_{\mathscr{P}}$ such that:*

- *$k[n] \in \mathbf{Art}_{\mathscr{P}}$ for all $n \geq 0$*
- *For any $A \in \mathbf{Art}_{\mathscr{P}}$ and any map in $\mathbf{Alg}_{\mathscr{P}}$, $A \to k[n]$ with $n \geq 1$, the following homotopy pullback $A \times_{k[n]} 0$ is also Artinian:*

$$\begin{array}{ccc} A \times_{k[n]} 0 & \longrightarrow & 0 \\ \downarrow & & \downarrow \\ A & \longrightarrow & k[n] \end{array}$$

This is corresponds to [CCN22, Definition 2.2].

Proof By definition, we have that $k[n]$ is Artinian since the map $k[n] \to 0$ is elementary, given by the homotopy pullback:

$$\begin{array}{ccc} k[n] & \longrightarrow & 0 \\ \downarrow & & \downarrow \\ 0 & \longrightarrow & k[n+1] \end{array}$$

Moreover, the map $A \times_{k[n]} 0 \to A$ is elementary, therefore if A is Artinian, then the composition $A \times_{k[n]} 0 \to A \to 0$ is Artinian as well. Therefore $\mathbf{Art}_{\mathscr{P}}$ satisfies the properties given in the proposition. It is the smallest such category because any Artinian object can be obtained from some $k[n]$ in finitely many pullbacks along some $0 \to k[n]$ for $n \geq 1$. □

Similarly we can rephrase the Artinian $\mathscr{P}$-algebra property as a simpler condition.

Lemma 4.29 ([CCN22, Lemma 2.8]) *Suppose the operad $\mathscr{P}$ is non-positively graded. A $\mathscr{P}$-algebra $A \in \mathbf{Alg}_{\mathscr{P}}$ is Artinian if:*

- $H^i(A) = 0$ *for* $i > 0$.
- $H^*(A)$ *has total finite dimension.*
- $H^i(A)$ *is nilpotent with respect to the* $H^0(\mathscr{P})$ *algebra* $H^0(A)$ *i.e. for all* $a_1, \cdots, a_n \in H^0(A)$, $P \in H^0(\mathscr{P})(n+1)$ *the map:*

$$\mathscr{P}(a_1, \cdots, a_n, -) : H^i(A) \to H^i(A)$$

is such that the k-th iteration of this map is 0 *for some* $k \in \mathbb{N}$.

Remark 4.30 If $A \in \mathbf{Art}_{\mathscr{P}}$, then for any square zero extension by the trivial $\mathscr{P}$-algebra $k[n]$:

$$k[n] \to B \to A$$

B is also in $\mathbf{Art}_{\mathscr{P}}$. The argument is precisely the same as in Remark 4.25, see also [CCN22, Example 2.3].

4.2.2 Formal Moduli Problems

In Sects. 4.2.1 and 3.2, we defined, given a local Artinian algebra A, different notions of sets, groupoids or even ∞-groupoids of deformations of a object along A.

These assignments that send A to such spaces of deformations define functors which will be the prototypical examples of what we will call "formal moduli problems". In other words, given an object X (for example a $\mathscr{P}$-algebra, a complex manifold, etc...), we will be interested in studying the functor that sends an Artinian object A to the *space* of all deformations of X along A. To be precise about what we mean by "space", we take $\mathscr{S}$ to be the ∞-category of ∞-groupoids.

Definition 4.31 (Formal Moduli Problem, [Lur11, Definition 1.1.14]) Given a deformation context $(\mathscr{A}, \{E_\alpha\}_{\alpha \in T})$, a functor of ∞-categories $F : \mathbf{Art}_{\mathscr{A}} \to \mathscr{S}$ is called a *formal moduli problem* if it satisfies the following conditions:

- Deformations along the point (terminal object) are trivial:

$$F(\star) \simeq \star$$

- Any homotopy pullback in $\mathbf{Art}_{\mathscr{A}}$ along a small morphism $\phi : A \to B$ is sent to a homotopy pullback in $\mathscr{S}$:

$$F\left(\begin{array}{ccc} A' & \longrightarrow & A \\ \downarrow & & \downarrow{\scriptstyle \phi} \\ B' & \longrightarrow & B \end{array}\right) = \begin{array}{ccc} F(A') & \longrightarrow & F(A) \\ \downarrow & & \downarrow{\scriptstyle F(\phi)} \\ F(B') & \longrightarrow & F(B) \end{array}$$

The ∞-category of such functors will be denoted $\mathbf{FMP}(\mathscr{A}, \{E_\alpha\}_{\alpha\in T})$ or $\mathbf{FMP}_{\mathscr{A}}$ for short if the choice of the collection of spectrum objects E_α is clear.

Example 4.32 A formal moduli problem on the deformation context $\mathscr{A}_{\mathscr{P}}$ is a functor $F : \mathbf{Art}_{\mathscr{P}} \to \mathscr{S}$ such that $F(\star) \simeq \star$ and sends the following homotopy pullbacks of Artinian objects to homotopy pullbacks (for all $n \geq 0$):

$$F\left(\begin{array}{ccc} A & \longrightarrow & 0 \\ \downarrow & & \downarrow \\ B & \longrightarrow & k[n+1] \end{array}\right) = \begin{array}{ccc} F(A) & \longrightarrow & \star \\ \downarrow & & \downarrow \\ F(B) & \longrightarrow & F(k[n+1]) \end{array}$$

This is the class of examples we will be interested in. We will denote it by $\mathbf{FMP}_{\mathscr{P}}$. The category $\mathbf{FMP}_{\mathbf{Com}}$ will be denoted **FMP**.

Remark 4.33 Given a square zero extension of an Artinian object along $k[n]$, we have seen in Remark 4.30 that it is equivalent to a homotopy pullback of the small morphism $0 \to k[n]$. In particular, we have a homotopy pullback diagram:

$$\begin{array}{ccc} F(B) & \longrightarrow & \star \\ \downarrow & & \downarrow{\scriptstyle F(0)} \\ F(A) & \longrightarrow & F(k[n+1]), \end{array}$$

where (up to taking a cofibrant resolution) the graded vector space underlying B can be taken to be $A \oplus k[n]$.

Therefore an A-deformation, given by a point $\star \to F(A)$, lifts to a B-deformation $\star \to F(B)$ if and only if the composition $\star \to F(A) \to F(k[n+1])$ is homotopic to 0. In other words, the image of the point in $F(k[n+1])$ is the obstruction to a lift and if the obstruction vanishes, the choice of a lift corresponds to the choice of a 1-simplex of $F(k[n+1])$.

We will see in Sect. 4.3.1 that the collection of $F(k[n+1])$ defines the notion of tangent complex which controls the obstructions to lifting deformations and indeed a deformation lifts if it induces an exact element in the tangent complex.

Example 4.34 For any $B \in \mathbf{Alg}_{\mathbf{Com}}^{k//k}$ the following functor is a formal moduli problem:

$$\begin{array}{rccc} \mathrm{Spf}(B): & \mathbf{Art}_{\mathscr{A}^{\mathrm{aug}}_{\mathbf{Com}}} & \longrightarrow & \mathscr{S} \\ & A & \longmapsto & \mathrm{Map}\,(B, A) \end{array}$$

Moreover we get a functor $\mathbf{Alg}_{\mathbf{Com}} \to \mathbf{FMP}$ sending B to $\mathrm{Spf}(B)$.

Spf(B) is called the *formal spectrum* of B. More generally, we can define the formal spectrum in a similar way for $B \in \mathbf{Alg}_{\mathscr{P}}$ as:

$$\begin{aligned}\mathrm{Spf}(B): \quad \mathbf{Art}_{\mathscr{P}} &\longrightarrow \mathscr{S} \\ A &\longmapsto \mathrm{Map}\,(B, A)\end{aligned}$$

Proposition 4.35 *Following [Lur11, Lemma 1.1.20], a morphism $f : A \to B$ between Artinian commutative algebras, $A, B \in \mathbf{Art}_{\mathscr{A}^{\mathrm{aug}}_{\mathbf{Com}}}$ is small if and only if it induces a surjection of commutative rings $H^0(A) \to H^0(B)$. Therefore the second condition of Definition 4.31 can be rephrased as follows:*

Any homotopy pullback in $\mathbf{Art}_{\mathscr{A}^{\mathrm{aug}}_{\mathbf{Com}}}$ such that $H^0(A) \to H^0(B)$ or $H^0(B') \to H^0(B)$ is surjective is sent to a pullback in $\mathscr{S}$:

$$F\left(\begin{array}{ccc} A' & \longrightarrow & A \\ \downarrow & & \downarrow \phi \\ B' & \longrightarrow & B \end{array}\right) = \begin{array}{ccc} F(A') & \longrightarrow & F(A) \\ \downarrow & & \downarrow F(\phi) \\ F(B') & \longrightarrow & F(B) \end{array}$$

It implies in particular the classical Schlessinger condition (see [Sch68a]).

As we have mentioned multiple times, it is an old philosophy that deformation problems are classified by Lie algebras. This classical idea is formalized by the following theorem:

Theorem 4.36 ([Lur11, Pri10]) *There is an equivalence of ∞-categories:*

$$\mathbf{MC}\colon \mathbf{Alg}_{\mathbf{Lie}} \overset{\sim}{\rightleftarrows} \mathbf{FMP}\colon \mathbb{T}[-1]$$

Warning 4.37 For $\mathfrak{g}$ a Lie algebra, the one should think heuristically of **MC** as the functor assigning to an Artinian algebra A the space $\mathbf{Del}(\mathfrak{g} \otimes A)$, that is, the space of Maurer–Cartan elements. Unfortunately, $\mathbf{Del}(\mathfrak{g} \otimes -)$ does not generally define a formal moduli problem, essentially for the same reasons the Goldmann–Milson theorem 3.22 does not preserve all weak-equivalences, but only the filtered ones. On finite dimensional A, the formula $\mathbf{Del}(\mathfrak{g} \otimes A)$ gives the correct result on objects. The functor **MC** is in some sense more general in order to circumvent this problem.

In the next sections we will explore the techniques used to prove this result, as well as explain its generalisation to operadic algebras.

4.3 Operadic Formal Moduli Problems and Koszul Duality

The goal of this section is to make sense of a generalization of Theorem 3.36. The main insight is that the equivalence between (commutative) formal moduli problems and Lie algebras arises from the Koszul duality between the operads **Com** and **Lie**. With this in mind, this Theorem generalizes to Theorem 3.64, stating the existence of an equivalence of ∞-categories:

$$\mathbf{Alg}_{\mathscr{P}} \simeq \mathbf{FMP}_{\mathscr{P}^!}$$

for $\mathscr{P}$ a Koszul binary quadratic operad concentrated in non-positive degrees.

The key observation is that if the generators $E = \mathscr{P}(2)$ are finite dimensional, there is a map of operads $\mathbf{Lie} \to \mathscr{P} \otimes \mathscr{P}^!$, defined by sending the generator ℓ_2 of **Lie** to $\sum_{e \in E} e \otimes e^\vee$, where the sum runs over a basis of E, and $e^\vee$ represents a dual basis. In this case, if $\mathfrak{g}$ is now a $\mathscr{P}$-algebra and A is an Artinian $\mathscr{P}^!$-algebra, $\mathfrak{g} \otimes A$ is a Lie algebra and, with the same caveats of Warning 4.37, the formula $\mathbf{Del}(\mathfrak{g} \otimes A)$ is the correct one.

We will start in Sect. 4.3.1 by explaining how the inverse of **MC** is related to the tangent complex. Then Sects. 4.3.2 and 4.3.3 are devoted to the study of Koszul duality in the context of deformation theory. Then the main result and its functoriality are discussed in Sects. 4.3.4 and 4.3.5.

4.3.1 Tangent Complex of a Formal Moduli Problem

The goal of this section is to explain that the algebra controlling a given formal moduli problem is the "tangent" of this formal moduli problem. In particular this explains why first-order deformations and obstructions to lifting those deformations are controlled by the tangent complex.

To motivate the following definition, we will start by explaining the example of the formal spectrum (Example 4.34) of an augmented commutative algebra B. We have that $\mathrm{Spf}(B)(A) = \mathrm{Map}(B, A)$. In the case of first order deformations, i.e. $A = k[\epsilon]$, with $\epsilon^2 = 0$, this mapping space corresponds to the tangent complex of $\mathrm{Spec}(B)$ at the point[22] corresponding to the augmentation of B, $x : \star \to \mathrm{Spec}(B)$. This is a consequence of the following equivalences:

$$\begin{aligned}\mathrm{Map}_{\mathbf{Alg}_{\mathbf{Com}}^{k//k}}(B, k[\epsilon]) &\simeq \mathrm{Map}_{\mathbf{Alg}_{\mathbf{Com}}^{k//B}}(B, B \oplus x_* k[n]) \\ &\simeq \mathrm{Map}_{\mathrm{Mod}_B}(\mathbb{L}_B, x_* k[n])\end{aligned}$$

[22] Given a map $B \to A$ with A local Artinian, the composition $B \to A \to k$ defines a unique point which is deformed within $\mathrm{Spec}(B)$. The tangent complex at a point x is defined as the k-linear derivations of B at x, with $\mathrm{Der}_k^x(B, M) := \mathrm{Hom}_{\mathbf{Alg}_{\mathbf{Com}}^{k//k}}(B, k \oplus M)$ with B viewed over k via x (see [TV08, Definition 1.4.1.2]).

$$\simeq \mathrm{Map}_{\mathrm{Mod}_k}\left(x^*\mathbb{L}_B, k[n]\right)$$
$$\simeq \left|\mathbb{T}_{B,x}[n]\right|$$

The goal will be to generalize this by applying the functor not only to $k[\epsilon]$ but to all "first-order elements" of a given deformation context. Such "first-order" objects correspond to the elements in $\{\Omega^{\infty-n}(E_\alpha)\}_{\alpha\in T, n\in\mathbb{N}}$ (for example the square zero extensions $k \oplus k[n]$ for $n \in \mathbb{N}$ in the case of augmented commutative algebras). Therefore, we will define the tangent functor of a general formal moduli problem by evaluating it at first-order elements.

Definition 4.38 (Tangent Functor, [Lur11, Section 1.2]) Given a deformation context $(\mathscr{A}, \{E_\alpha\}_{\alpha\in T})$, (see Definition 4.21), we define the *tangent space* of a formal moduli problem $F : \mathbf{Art}_{\mathscr{A}} \to \mathscr{S}$ on this deformation context at $\alpha \in T$ to be $T_\alpha F := F(\Omega^\infty E_\alpha) \in \mathscr{S}$.

To remember higher deformations, it is not enough to evaluate F at $\Omega^\infty E_\alpha$ but we should also evaluate it at all $\Omega^{\infty-n} E_\alpha$ for $n \geq 0$. We define *the tangent complex* of F at α to be the spectrum object $F(E_\alpha) \in \mathbf{Sp} := \mathbf{Stab(sSet)}$, which we denote[23] $\mathbb{T}_F$ in case there is only a single $\alpha \in T$.

This spectrum object verifies $\Omega^{\infty-n} F(E_\alpha) \simeq F(\Omega^{\infty-n} E_\alpha)$ for all $n \geq 0$ (see [Lur11, Remark 1.2.7]). Notice that the tangent complex $\mathbb{T}_F$ is not actually a chain complex but rather only a spectrum object a priori. These two categories are related by the composition:

$$\mathrm{Mod}_k \overset{\text{Forget}}{\longrightarrow} \mathrm{Mod}_{\mathbb{Z}} \overset{DK}{\simeq} \mathbf{Stab(sAb)} \overset{\text{Forget}}{\longrightarrow} \mathbf{Stab(sSet)} = \mathbf{Sp} \tag{7}$$

In the setting of operadic formal moduli problems, the following proposition justifies the terminology.

Proposition 4.39 ([CCN22, Lemma 2.15]) *A formal moduli problem $F \in \mathbf{FMP}_{\mathscr{P}}$ has a unique pre-image in* Mod_k *under the map* (7)*, which we also denote by $\mathbb{T}_F$. Moreover for any $n \geq 0$, we have an equivalence:*

$$\mathrm{Map}_{\mathrm{Mod}_k}(k[-n], \mathbb{T}_F) \simeq F(k[n])$$

As a motivation for part of the k-module structure, notice that the ground field k acts naturally on $k[n]$ by multiplication, which implies that it acts on $\mathbb{T}_F$ by functoriality.

In the case of the formal spectrum of unital commutative rings, we saw that $\mathrm{Spf}(B)(k[\epsilon]) \simeq |\mathbb{T}_{B,x}|$ (and similarly when shifting ϵ). Now the associated spectrum object to $\mathbb{T}_{B,x}$ (given by the composition (7)) coincides with the spectrum object given by $\mathrm{Spf}(B)(k[\epsilon_n])$ so that $\mathbb{T}_{B,x}$ is the representative of $\mathbb{T}_{\mathrm{Spf}(B)}$ given by Proposition 4.39.

[23] This notation is by analogy to the tangent complex but is a priori only a (collection of) spectrum objects. However, we will see that $\mathbb{T}_F$ can be "represented" by objects with more structure (see the discussion following Propositions 4.39 and 4.53).

It turns out that this also extends to the formal spectrum of a $\mathscr{P}$-algebra A. Indeed, the tangent functor is given by the collection for each n of the spaces:

$$\mathrm{Spf}(A)(k[n]) \simeq \mathrm{Map}_{\mathbf{Alg}_{\mathscr{P}}}(A, k[n])$$

where $k[n]$ is endowed with the trivial $\mathscr{P}$-algebra structure (induced by the augmentation $\mathscr{P} \to I$). Taking the trivial algebra is a right adjoint functor (see Remark 2.31) and we get:

$$\mathrm{Spf}(A)(k[n]) \simeq \mathrm{Map}_{\mathbf{Alg}_{\mathscr{P}}}(A, k[n]) \simeq \mathrm{Map}_{\mathrm{Mod}_k}(I \circ_{\mathscr{P}} \tilde{A}, k[n])$$

where $\tilde{A}$ is a cofibrant resolution of A. In particular, we can take the cobar-bar resolution, $\tilde{A} := \Omega_\kappa \mathbf{B}_\kappa A$. Note that $I \circ_{\mathscr{P}} \tilde{A}$ is the *space of derived indecomposables*, which is the derived quotient by all products in A. We have the following observations:

- There is an equivalence:

$$I \circ_{\mathscr{P}} \Omega_\kappa \mathbf{B}_\kappa A \simeq \left(I \circ_{\mathscr{P}} \mathscr{P} \circ_I \mathbf{B}_\kappa A, d_\Omega + d_{\mathbf{B}_\kappa A}\right)$$

- We have:

$$I \circ_{\mathscr{P}} \mathscr{P} \circ_I \mathbf{B}_\kappa A \simeq \mathbf{B}_\kappa A$$

- Since d_Ω has positive weight on the free algebra $\mathscr{P} \circ_I \mathbf{B}_\kappa A$ and the functor $I \circ_{\mathscr{P}}$ kills all the positive weight parts, the differential induced on $I \circ_{\mathscr{P}} \mathscr{P} \circ_I \mathbf{B}_\kappa A$ is exactly $d_{\mathbf{B}_\kappa A}$ and we have an equivalence:

$$I \circ_{\mathscr{P}} \Omega_\kappa \mathbf{B}_\kappa A \simeq \mathbf{B}_\kappa A$$

Therefore we get:

$$\mathrm{Spf}(A)(k[n]) \simeq \mathrm{Map}_{\mathrm{Mod}_k}(\mathbf{B}_\kappa A, k[n]) \simeq \left|\mathbf{Der}^{\mathscr{P}}(A, k[n])\right| \simeq \left|\mathbb{T}^{\mathscr{P}}_{A,x}[n]\right|$$

Proposition 4.40 *For the deformation context $\mathscr{A}^{\mathrm{aug}}_{\mathbf{Com}}$, the tangent complex of a formal moduli problem F is given by the Ω-spectrum determined*[24] *by $(F(k \oplus k[n]))_{n\geq 0}$.*

For the deformation context $\mathscr{A}_{\mathscr{P}}$, the tangent functor of a formal moduli problem F is given by the Ω-spectrum defined by $(F(k[n]))_{n\geq 0}$.

Proof Essentially, the properties of F ensure that

$$F(k \oplus k[n]) \to \Omega F(k \oplus k[n+1])$$

---

[24] An Ω-spectrum object is completely determined by a collection $(x_i)_{i\in\mathbb{N}}$, *indexed by* $\mathbb{N}$, such that there are weak equivalences $x_i \to \Omega x_{i+1}$. We recover the full spectrum object by applying Ω successively to the x_i.

for $n \geq 0$ are equivalences making $F(k \oplus k[n])$ an Ω-spectrum. See the discussion preceding [CCN22, Definition 2.14]. □

Remark 4.41 Following Remark 4.33, this implies that the obstruction to lifting deformations along square zero extensions is controlled by the cohomology of $\mathbb{T}_F$.

Remark 4.42 Let $B \in \mathbf{Alg}_{\mathbf{Com}}^{k//k}$ and let $x : \star \to \mathrm{Spec}(B)$ be the point corresponding to the augmentation. Then we have:

$$\mathbb{T}_x \mathrm{Spf}(B) := \left(\mathrm{Map}_{\mathbf{dSch}_{/\star}}\left(\mathbf{D}_{k,n}(k), \mathrm{Spec}(B)\right)\right)_{n \geq 0}$$

and we recover the higher deformations of $\mathrm{Spec}(B)$ as discussed in 4.11 (see also [Vez10]). In Mod_k this corresponds to the tangent complex at the point x, $\mathbb{T}_{B,x}$.

Remark 4.43 There is a classical version of a deformation functor of a variety X taking values in sets, $F : \mathrm{Art} \to \mathbf{Set}$, given by deformations of X over the artinian algebra modulo isomorphism and similar to our more general version the natural notion of the tangent of this functor is given by $F(k[\epsilon])$. Schlessinger proves [Sch68b] (see also [Ser07, Theorem 2.4.1]) that given X a smooth proper variety over k, the tangent of the deformation functor Def_X is given by the k-vector space $H^1(X, T_X)$.

4.3.2 Koszul Duality Context

Going back again to the setting of a general deformation context $\mathscr{A}$, we are looking to identify an ∞-category of algebraic objects $\mathscr{B}$ which is equivalent to $\mathbf{FMP}_{\mathscr{A}}$.

The formal spectrum Spf from Example 4.34 is a functorial way to construct formal moduli problems out of objects of $\mathscr{A}$. Assuming the existence of the desired equivalence, we could construct a functor:

$$\mathfrak{D} \colon \mathscr{A}^{\mathrm{op}} \xrightarrow{\mathrm{Spf}} \mathbf{FMP}_{\mathscr{A}} \xrightarrow{\sim} \mathscr{B}$$

This functor should be interpreted as a "weak duality" functor, which is not an equivalence, but at least its restriction to Artinian objects should behave as an equivalence into a subcategory of $\mathscr{B}$ of "good" objects. Of course, we are interested in the case where $\mathscr{B}$ are $\mathscr{P}$-algebras and $\mathscr{A}$ are $\mathscr{P}^!$-algebras. In that case the reader can think of $\mathfrak{D}$ as the dual of the bar construction, see Warning 2.70.

In this section, we introduce the notion of Koszul duality context as the appropriate axiomatic framework that enables us to obtain the desired equivalence, as shown in Theorem 4.52.

Definition 4.44 (Dual Deformation Context, [CG21, Definition 2.11]) A pair $(\mathscr{B}, \{F_\alpha\}_{\alpha \in T})$ is called a *dual deformation context* if $\mathscr{B}$ is a presentable ∞-category and $F_\alpha \in \mathbf{Stab}(\mathscr{B}^{\mathrm{op}})$.

We say that an object (resp. morphism) of $\mathscr{B}$ is *good* if it is Artinian when considered in $(\mathscr{B}^{\mathrm{op}}, \{F_\alpha\}_{\alpha\in T})$.[25] We denote by $\mathscr{B}^{\mathrm{gd}}$ the full sub-category of good objects of $\mathscr{B}$.

Example 4.45

- If $A \in \mathbf{Alg}_{\mathbf{Com}}$, then $\left(\mathrm{Mod}_A, (A[n])_{n\in\mathbb{Z}}\right)$ is a deformation context and since taking the opposite category exchanges the suspension and desuspension functor, we get the dual deformation context $\left(\mathrm{Mod}_A, (A[-n])_{n\in\mathbb{Z}}\right)$.

  If A is bounded and concentrated in non-positive degrees, [CG21, Lemma 2.16] tells us that an element $M \in \mathrm{Mod}_A$ in this dual deformation context is good if it is perfect and cohomologically concentrated in positive degrees (see also [CG21, Remark 2.17]).
- Dual deformation contexts can be transferred along left adjoints (see [CG21, Section 2.2]). Using this on the free-forget adjunction:

$$\mathrm{Mod}_k \rightleftarrows \mathbf{Alg}_{\mathscr{P}}$$

  we get a dual deformation context:

$$\left(\mathbf{Alg}_{\mathscr{P}}, \{(\mathscr{P} \circ k[-n])_{n\in\mathbb{Z}}\}\right)$$

Definition 4.46 ([CG21, Definition 2.18]) A *weak Koszul duality context* is the data of:

- A deformation context $(\mathscr{A}, \{E_\alpha\}_{\alpha\in T})$
- A dual deformation context $(\mathscr{B}, \{F_\alpha\}_{\alpha\in T})$
- An adjunction:

$$\mathfrak{D} : \mathscr{A} \rightleftarrows \mathscr{B}^{\mathrm{op}} : \mathfrak{D}'$$

such that for all $n \geq 0$, there is an equivalence $\Omega^{\infty-n} E_\alpha \simeq \mathfrak{D}'\left(\Omega^{\infty-n} F_\alpha\right)$.

It is called a *Koszul Duality context* if the following holds:

(1) For every object $B \in \mathscr{B}^{\mathrm{gd}}$, the counit morphism $\mathfrak{D}\mathfrak{D}'B \to B$ is an equivalence.
(2) For each α, the functor:

$$\Theta_\alpha : \mathscr{B} \to \mathbf{Sp}$$

sending $B \in \mathscr{B}$ to the spectrum object:

$$\left(\mathbf{Hom}_{\mathscr{B}}(\Omega^{\infty-n} E_\alpha, B)\right)_{n\in\mathbb{Z}} \in \mathbf{Sp}$$

under the functor is conservative and preserves sifted colimits.

[25] Here $(\mathscr{B}^{\mathrm{op}}, \{F_\alpha\}_{\alpha\in T})$ is generally not a deformation context since $\mathscr{B}^{\mathrm{op}}$ will in general not be presentable. However the definition of Artinian object still makes sense in $\mathscr{B}^{\mathrm{op}}$.

Remark 4.47 A weak Koszul duality context together with condition (1) gives us a weak deformation theory according to the terminology of Lurie, [Lur11, Definition 1.3.1]. A Koszul duality context is an example of a *deformation theory* according to [Lur11, Definition 1.3.9].

Remark 4.48 The name "Koszul duality context" comes from the fact that Koszul duality for operads $(\mathscr{C}, \mathscr{P})$ induces a weak Koszul duality context between $\mathscr{P}$-algebras and $\mathscr{C}^{\vee}$-algebras (see Proposition 4.58). Under the assumptions of Theorem 4.60, this is even a Koszul duality context.

Example 4.49 Let $A \in \mathbf{Alg}_{\mathbf{Com}}$. We have a Koszul duality context given by the dualization:

$$(-)^{\vee} : \left(\mathrm{Mod}_A, (A[n])_{n\in\mathbb{Z}}\right) \rightleftarrows \left(\mathrm{Mod}_A^{\mathrm{op}}, (A[-n])_{n\in\mathbb{Z}}\right) : (-)^{\vee}$$

Proposition 4.50 ([CG21, Proposition 2.22]) *Given a Koszul duality context (using the same notation as in Definition 4.46), we have the following:*

- $\mathfrak{D}\left(\Omega^{\infty-n} E_\alpha\right) \simeq \Omega^{\infty-n} F_\alpha$ *for all* $n \geq 0$.
- *For every Artinian object* $A \in \mathbf{Art}_{\mathscr{A}}$*, the unit map* $A \to \mathfrak{D}'\mathfrak{D}A$ *is an equivalence.*
- *The adjunction* $\mathfrak{D} \dashv \mathfrak{D}'$ *induces an equivalence:*

$$\mathfrak{D} : \mathbf{Art}_{\mathscr{A}} \rightleftarrows \left(\mathscr{B}^{\mathrm{gd}}\right)^{\mathrm{op}} : \mathfrak{D}'$$

- *If* $M \in \mathbf{Art}_{\mathscr{A}}$ *and* $f : A \to B$ *is a small morphism in* $\mathscr{A}$ *then* $\mathfrak{D}$ *sends the pullback diagram:*

$$\begin{array}{ccc} P & \longrightarrow & A \\ \downarrow & & \downarrow f \\ M & \longrightarrow & B \end{array}$$

to a pullback diagram.

Corollary 4.51 *Given a Koszul duality context, we can construct the following functor:*

$$\psi : \mathscr{B} \longrightarrow \mathbf{Fun}\left(\mathscr{B}^{\mathrm{op}}, \mathscr{S}\right) \xrightarrow{\circ\mathfrak{D}} \mathbf{Fun}\left(\mathscr{A}, \mathscr{S}\right)$$

Then ψ *factors through* $\mathbf{FMP}(\mathscr{A}, \{E_\alpha\})$ *and we get a map:*

$$\Psi : \mathscr{B} \to \mathbf{FMP}(\mathscr{A}, \{E_\alpha\})$$

It turns out that the definition of Koszul duality context ensures that the map $\Psi : \mathscr{B} \to \mathbf{FMP}(\mathscr{A}, \{E_\alpha\})$ is in fact an equivalence.

Theorem 4.52 ([CG21, Theorem 2.33] or [Lur11, Theorem 1.3.12]) *Given a Koszul duality context as above, the functor Ψ is an equivalence:*

$$\Psi : \mathscr{B} \to \mathbf{FMP}(\mathscr{A}, \{E_\alpha\})$$

Proposition 4.53 ([CG21, Proposition 2.36]) *We have the following commutative diagram:*

$$\begin{array}{ccc} \mathscr{B} & \xrightarrow{\quad \Psi \quad} & \mathbf{FMP}\,(\mathscr{A}, \{E_\alpha\}) \\ & {\scriptstyle \Theta} \searrow \quad \swarrow {\scriptstyle T} & \\ & \mathbf{Sp} & \end{array}$$

where Θ is the functor described in Definition 4.46 and T is the tangent complex of F (Definition 4.38).

Remark 4.54 This diagram shows that TF is equivalent to $\Theta(\mathbb{T}_F)$ with $\mathbb{T}_F \in \mathscr{B}$ so this tangent functor has more structure that being just a spectrum object. When we will work with the Koszul Duality context for operads (Definition 4.59), with Spf(B), we get that $\mathbb{T}_F$ is exactly $\mathbb{T}^{\mathscr{P}}_{B,x}$ the $\mathscr{P}$-tangent complex together with a $\mathscr{P}^!$ structure.

Definition 4.55 ([CG21, Definition 2.23]) Given two (weak) Koszul duality contexts:

$$\mathfrak{D}_1 : (\mathscr{A}_1, \{E^1_\alpha\}) \rightleftarrows (\mathscr{B}_1^{\mathrm{op}}, \{F^1_\alpha\}) : \mathfrak{D}'_1$$

$$\mathfrak{D}_2 : (\mathscr{A}_2, \{E^2_\alpha\}) \rightleftarrows (\mathscr{B}_2^{\mathrm{op}}, \{F^2_\alpha\}) : \mathfrak{D}'_2$$

A *morphism of (weak) Koszul duality context* is given by a diagram of adjunctions:

$$\begin{array}{ccc} \mathscr{A}_1 & \underset{\mathfrak{D}'_1}{\overset{\mathfrak{D}_1}{\rightleftarrows}} & \mathscr{B}_1^{\mathrm{op}} \\ {\scriptstyle R_1}\uparrow\downarrow{\scriptstyle L_1} & & {\scriptstyle R_2}\uparrow\downarrow{\scriptstyle L_2} \\ \mathscr{A}_2 & \underset{\mathfrak{D}'_1}{\overset{\mathfrak{D}_2}{\rightleftarrows}} & \mathscr{B}_2^{\mathrm{op}} \end{array}$$

such that the diagram of right adjoint commute:

$$\begin{array}{ccc} \mathscr{A}_1 & \underset{\mathfrak{D}'_1}{\longleftarrow} & \mathscr{B}_1^{\mathrm{op}} \\ {\scriptstyle R_1}\uparrow & & \uparrow{\scriptstyle R_2} \\ \mathscr{A}_2 & \underset{\mathfrak{D}'_1}{\longleftarrow} & \mathscr{B}_2^{\mathrm{op}} \end{array}$$

and we have equivalences $R_2\left(\Omega^{\infty-n} F^2_\alpha\right) \simeq F^1_\alpha$ for all $\alpha \in T$ and $n \geq 0$.

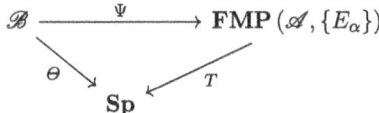

Proposition 4.56 ([CG21, Proposition 2.39]) *Given a morphism of Koszul duality contexts (following the notation of Definition 4.55), we get a commutative diagram:*

$$\begin{array}{ccccc} \mathscr{B}_1 & \xrightarrow{\Psi_1} & \mathbf{FMP}\left(\mathscr{A}_1, \{E^1_\alpha\}\right) & & \\ & & & \overset{T}{\searrow} & \\ \Big\downarrow{\scriptstyle L_2} & & \Big\downarrow{\scriptstyle R_1} & & \mathbf{Sp} \\ & & & \underset{T}{\nearrow} & \\ \mathscr{B}_2 & \xrightarrow{\Psi_2} & \mathbf{FMP}\left(\mathscr{A}_2, \{E^2_\alpha\}\right) & & \end{array}$$

4.3.3 Koszul Duality Context from Koszul Duality

In this section we will study examples of (weak) Koszul duality contexts that arise from operadic Koszul duality. The typical instance of such Koszul duality context can be exemplified by the following proposition which can be extracted from [Lur11, Theorem 2.3.1], see [CG21, Proposition 2.30].

Proposition 4.57 *We have a Koszul duality context:*

$$\mathfrak{D} : \left(\mathbf{Alg}^{\mathrm{aug}}_{\mathbf{Com}}, (k \oplus k[n])\right) \rightleftarrows \left(\mathbf{Alg}^{\mathrm{op}}_{\mathbf{Lie}}, (\mathbf{Lie}(k[-n-1]))_{n \in \mathbb{Z}}\right) : \mathrm{CE}^{\bullet}$$

where the right adjoint is the (∞-categorical) Chevalley–Eilenberg cochain complex, see Example 2.69.

While [CG21] and [Lur11] do not construct left adjoint $\mathfrak{D}$ explicitly (they deduce it from the Adjoint Functor Theorem), one can check that the functor $\mathfrak{D}$ coincides with the so-called Harrison complex, constructed as the functor (also denoted $\mathfrak{D}$) preceding Example 2.69. These duality functors preserve quasi-isomorphisms and therefore define functors of ∞-categories, but notice that they do not arise from a Quillen adjunction. The proof of this proposition makes use of the fact that we are in the conditions of Warning 2.70.

It is clear that this result has more to do with the Koszul duality between the operads **Com** and **Lie** and with the operads themselves. We will show how construct a (weak) Koszul duality context from operadic Koszul duality, which as we will see is the main tool to prove Theorem 3.64.

Throughout this section, we will work with the deformation context $\mathscr{A}_{\mathscr{P}}$ described in Example 4.23. From now on, we will consider $\mathscr{P}$ a k-augmented k-operad so that $\mathscr{P} = \bar{\mathscr{P}} \oplus I$. Given a twisting morphism $\phi : \mathscr{C} \dashrightarrow \mathscr{P}$, we use the bar–cobar adjunction associated to ϕ to define the following functor:

$$\mathfrak{D}_\phi : \mathbf{Alg}_{\mathscr{P}} \xrightarrow{\mathbf{B}_\phi} \mathbf{coAlg}_{\mathscr{C}} \xrightarrow{(-)^\vee} (\mathbf{Alg}_{\mathscr{C}^\vee})^{\mathrm{op}}$$

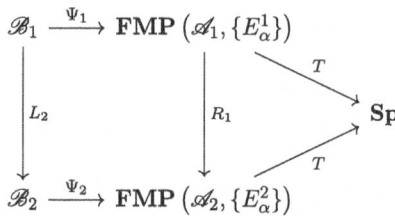

Proposition 4.58 *If ϕ is a weakly Koszul twisting morphism (see Definition 2.55), then we have the following:*

- *For any $A \in \mathbf{Alg}_{\mathscr{P}}$, we have:*

$$\mathfrak{D}_\phi(A) \simeq \mathbb{R}\mathrm{Der}_{\mathscr{P}}(A, k)$$

- *$\mathfrak{D}_\phi$ preserves all colimits and therefore has a right adjoint:*

$$\mathfrak{D}_\phi : \mathbf{Alg}_{\mathscr{P}} \rightleftarrows \mathbf{Alg}^{\mathrm{op}}_{\mathscr{C}^\vee} : \mathfrak{D}'_\phi$$

- *The aforementioned adjunction defines a weak Koszul duality context with $\mathbf{Alg}_{\mathscr{P}}$ seen with the deformation context $\mathscr{A}_{\mathscr{P}}$ and with the dual deformation context $\left(\mathbf{Alg}_{\mathscr{C}^\vee}, \left(\mathscr{C}^\vee(k[-n])\right)_{n\in\mathbb{Z}}\right)$ described in Example 4.45.*

Proof The first two parts can be extracted from [CCN22, Lemma 4.1], while the last claim is a consequence of [CCN22, Corollary 4.7] where we apply the result to $V = k[n] = \Omega^{\infty-n}F$. Since $\mathscr{P}(k[n]) = k[n]$ (the free $\mathscr{P}$-algebra structure on $k[n]$ is the trivial one), we obtain that $\mathfrak{D}_\phi(k[n]) = \mathfrak{D}_\phi(\mathscr{P}(k[n])) = \mathrm{triv}(k[n]) = k[n]$ which is the criterion for the adjunction to be a weak Koszul duality context. □

Definition 4.59 Given $\mathscr{P}$ an augmented operad, we denote $\mathfrak{D}(\mathscr{P}) := (\mathbf{B}\mathscr{P})^\vee$ and $\pi : \mathbf{B}\mathscr{P} \dashrightarrow \mathscr{P}$ the universal twisting morphism (Definition 2.55). Then the associated weak Koszul duality context is denoted by:

$$\mathfrak{D} : \mathbf{Alg}_{\mathscr{P}} \rightleftarrows \mathbf{Alg}^{\mathrm{op}}_{\mathfrak{D}(\mathscr{P})} : \mathfrak{D}'$$

We would like to know when this weak Koszul duality context is a Koszul duality context so that under those conditions, formal moduli problems over $\mathscr{P}$ will be equivalent to $\mathbf{Alg}_{\mathfrak{D}(\mathscr{P})}$. The necessary conditions are given by the following theorem:

Theorem 4.60 *Let $\mathscr{P}$ be an augmented operad concentrated in non-positive degrees which is* splendid *(see Remark 4.61). Then the following holds:*

- *For any $A \in \mathbf{Art}_{\mathscr{P}}$, the unit map $A \to \mathfrak{D}'\mathfrak{D}(A)$ is an equivalence.*
- *$\mathfrak{D}(k[n])$ is freely generated by $k[-n]$ for all $n \geq 0$, that is:*

$$\mathfrak{D}(k[n]) \simeq \mathfrak{D}(\mathscr{P})(k[-n])$$

- *$\mathfrak{D}$ sends a pullback square in $\mathbf{Art}_{\mathscr{P}}$:*

$$\begin{array}{ccc} A & \longrightarrow & 0 \\ \downarrow & & \downarrow \\ A' & \longrightarrow & k[n] \end{array}$$

to a pullback square in $\mathbf{Alg}_{\mathfrak{D}(\mathscr{P})}$.

- *In general (without the assumptions of the theorem) the functor:*

$$\mathbf{Alg}_{\mathfrak{D}(\mathscr{P})} \xrightarrow{\mathbf{MC}} \mathbf{FMP}_{\mathscr{P}} \xrightarrow{T} \mathbf{Sp}$$

is conservative and preserves sifted colimits.

Then, we have a Koszul duality context:

$$\mathfrak{D} : \mathbf{Alg}_{\mathscr{P}} \rightleftarrows \mathbf{Alg}^{\mathrm{op}}_{\mathfrak{D}(\mathscr{P})} : \mathfrak{D}'$$

for the deformation context $\left(\mathbf{Alg}_{\mathscr{P}}, \{(k[n])_{n\in\mathbb{Z}}\}\right)$ *of Example 4.23 and dual deformation context* $\left(\mathbf{Alg}_{\mathfrak{D}(\mathscr{P})}, \{(\mathfrak{D}(\mathscr{P})(k[-n]))_{n\in\mathbb{Z}}\}\right)$ *of Example 4.45.*

Proof The first four items of the Theorem are proven in [CCN22, Theorem 5.1]. The last one is proven at the end of the proof of [CCN22, Theorem 4.18]. For the last part, we will show that these axioms imply the axioms of a Koszul duality context (Definition 4.46).

Using that the unit of the adjunction is an equivalence on Artinian algebras (similarly to Warning 2.70), and that $\mathfrak{D}(k[n]) \simeq \mathfrak{D}(\mathscr{P})(k[-n])$ we get an equivalence:

$$k[n] \simeq \mathfrak{D}'\mathfrak{D}(k[n]) \simeq \mathfrak{D}'(\mathfrak{D}(\mathscr{P})(k[-n]))$$

This shows that the adjunction is a weak Koszul duality context.

More generally, we can show that there is an equivalence:

$$\mathfrak{D} : \mathbf{Art}_{\mathscr{P}} \rightleftarrows \left(\mathbf{Alg}^{\mathrm{gd}}_{\mathfrak{D}(\mathscr{P})}\right)^{\mathrm{op}} : \mathfrak{D}'$$

Therefore the counit restricted to good object is also an equivalence. Finally, the following functor is conservative and preserves sifted colimits:

$$\mathbf{Alg}_{\mathfrak{D}(\mathscr{P})} \xrightarrow{\mathbf{MC}} \mathbf{FMP}_{\mathscr{P}} \xrightarrow{T} \mathbf{Sp}$$

but thanks to Proposition 4.53, this functor is exactly the functor Θ from Definition 4.46 that we want to show is conservative and preserves sifted colimits. □

Remark 4.61 The condition being *splendid* is a technical condition detailed [CCN22, Definition 3.37], which imposes that a certain derived relative composite product

$$\mathscr{P}(1) \circ^h_{\mathscr{P}_{\geq 1}} \mathscr{P}(1)$$

has cohomology living only in more and more negative degrees.

The main example to keep in mind is that any (non-necessarily binary) Koszul quadratic operad generated in non-positive degrees and in bounded arity is splendid.

In particular, **Com**, **Ass** and **Lie** are splendid and satisfy all the conditions of Theorem 4.60.

4.3.4 Main Result

Putting together [CCN22, Theorem 4.18] and Theorem 4.60, we obtain the following result:

Theorem 4.62 *Suppose that $\mathscr{P}$ is an augmented operad that is connective and splendid, then we have an equivalence:*

$$\Psi^{\mathscr{P}} : \mathbf{Alg}_{\mathfrak{D}(\mathscr{P})} \to \mathbf{FMP}_{\mathscr{P}}$$

that sends $\mathfrak{g} \in \mathbf{Alg}_{\mathfrak{D}(\mathscr{P})}$ to $\mathrm{Map}_{\mathfrak{D}(\mathscr{P})}(\mathfrak{D}(-), \mathfrak{g})$. Moreover the inverse functor sends a formal moduli problem F to its tangent complex $\mathbb{T}_F$ endowed with some $\mathfrak{D}(\mathscr{P})$-algebra structure, with $\mathfrak{D}(\mathscr{P}) = (\mathbf{B}\mathscr{P})^\vee \simeq \mathscr{P}^!\{-1\}$ (see Remark 4.54).

Remark 4.63 The functor $\Psi^{\mathscr{P}}$ of Theorem 4.62 is, for $\mathscr{P} = \Omega\mathscr{C}$ and other mild conditions, given by some space of Maurer–Cartan elements.

Let us sketch why this is the case, see [CCN22, Theorem 7.18] for more details. Under these conditions, given any $\mathfrak{g} \in \mathbf{Alg}_{\mathscr{C}^\vee}$ and $A \in \mathbf{Art}_{\mathscr{P}}$, we want to understand the space:

$$\mathrm{Map}_{\mathbf{Alg}_{\mathscr{C}^\vee}}(\mathfrak{D}(A), \mathfrak{g})$$

With some finiteness assumptions on A, we have a natural isomorphism $\Omega_{\alpha^\dagger}(A^\vee) \to \mathfrak{D}_\alpha(A)$, obtained from the twisting morphisms $\alpha : \mathscr{C} \dashrightarrow \mathscr{P}$ and $\alpha^\dagger : \mathbf{B}(\mathscr{C}^\vee) \dashrightarrow \mathscr{C}^\vee$.

There is a Rosetta Stone for algebras similar to the case of operads 2.40, where we can identify maps from a cobar construction with Maurer–Cartan elements of a certain L_∞-algebra [Wie19, Theorem 7.1]:

$$\mathrm{Map}_{\mathbf{Alg}_{\mathscr{C}^\vee}}\left(\Omega_{\alpha^\dagger}(A^\vee), \mathfrak{g}\right) \simeq \mathrm{MC}\left(\underline{\mathrm{Hom}}_{k\text{-Mod}}\left(A^\vee, \mathfrak{g} \otimes \Omega(\Delta^\bullet)\right)\right)$$

Since A is Artinian, as complexes, the L_∞-algebra above is isomorphic to $A \otimes_k \mathfrak{g} \otimes \Omega(\Delta^\bullet)$. Expanding on the discussion at the beginning of Sect. 4.3, there is a (shifted) L_∞-algebra structure on $\mathscr{P} \otimes \mathscr{C}^\vee$-algebras given by the twisting morphism (see [CCN22, Lemma 7.15]) and these two L_∞-structures on $A \otimes_k \mathfrak{g} \otimes \Omega(\Delta^\bullet)$ coincide. We conclude that

$$\mathrm{Map}_{\mathbf{Alg}_{\mathscr{C}^\vee}}\left(\Omega_{\alpha^\dagger}(A^\vee), \mathfrak{g}\right) \simeq \mathrm{MC}\left(A \otimes_k \mathfrak{g} \otimes \Omega(\Delta^\bullet)\right) \simeq \mathbf{Del}\left(A \otimes_k \mathfrak{g}\right)$$

When $\mathscr{P}$ is a Koszul operad there is an equivalence between $\mathfrak{D}(\mathscr{P})$ and $\mathscr{P}^!\{-1\}$. This implies the result:

Theorem 4.64 ([CCN22]) *For $\mathscr{P}$ a Koszul binary quadratic operad concentrated in non-positive degrees, there is an equivalence of ∞-categories:*

$$\mathbf{FMP}_{\mathscr{P}} \simeq \mathbf{Alg}_{\mathscr{P}^!}.$$

Proof We have weak equivalences of operads $\Omega\mathscr{P}^{\text{i}} \to \mathscr{P}$ and $\mathfrak{D}(\mathscr{P}) \to \mathscr{P}^{!}\{-1\}$.
$\mathscr{P}$ satisfies the condition of Theorem 4.62, and we get a zig-zag of equivalences:

$$\mathbf{FMP}_{\mathscr{P}} \xrightarrow{\mathbb{T}} \mathbf{Alg}_{\mathfrak{D}(\mathscr{P})} \xleftarrow{\simeq} \mathbf{Alg}_{\mathscr{P}^{!}\{-1\}} \xrightarrow{V \mapsto V[-1]} \mathbf{Alg}_{\mathscr{P}^{!}}$$

□

Example 4.65

- For $\mathscr{P} = \mathbf{Com}$ we get back Theorem 3.36:

$$\mathbf{FMP} \xrightarrow{\sim} \mathbf{Alg}_{\mathbf{Lie}}$$

- For $\mathscr{P} = \mathbf{Ass}$, we get the result of Lurie, in [Lur11]:

$$\mathbf{FMP}_{\mathbf{Ass}} \xrightarrow{\sim} \mathbf{Alg}_{\mathbf{Ass}}$$

- For $\mathscr{P} = \mathbf{Lie}$ we obtain a new equivalence:

$$\mathbf{FMP}_{\mathbf{Lie}} \xrightarrow{\sim} \mathbf{Alg}_{\mathbf{Com}}$$

- For the permutative operad we get:

$$\mathbf{FMP}_{\mathbf{Perm}} \xrightarrow{\sim} \mathbf{Alg}_{\mathbf{preLie}}$$

The reader can find many examples treated in detail in Section 3 of [CCN22].

4.3.5 Naturality of the Main Result

A final important point to make concerns the naturality Theorem 3.64 with respect to the operads involved. The main example of deformation problems studied in Sect. 3 were encoded by the deformation complex $\mathfrak{g}^{\phi}_{P_\infty,A}$. If we consider deformations of the trivial algebra structure on A, i.e. when $\phi = 0$, the deformation complex is in fact a pre-Lie algebra. The associated permutative formal moduli problem yields by restriction a commutative moduli problem and the naturality of Theorem 3.64 asserts that the two commutative formal moduli problems are the same.

Indeed, the following diagram of restrictions commutes:

$$\begin{array}{ccc}
\mathbf{FMP}_{\mathbf{Com}} & \xleftarrow[\sim]{} & \mathbf{Alg}_{\mathbf{Lie}} \\
\uparrow & & \uparrow \\
\mathbf{FMP}_{\mathbf{Perm}} & \xleftarrow[\sim]{} & \mathbf{Alg}_{\mathbf{preLie}} \\
\uparrow & & \uparrow \\
\mathbf{FMP}_{\mathbf{Ass}} & \xleftarrow[\sim]{} & \mathbf{Alg}_{\mathbf{Ass}} \\
\uparrow & & \uparrow \\
\mathbf{FMP}_{\mathbf{Lie}} & \xleftarrow[\sim]{} & \mathbf{Alg}_{\mathbf{Com}}
\end{array}$$

More generally, following Theorem 4.56, we can formulate a notion of naturality in the Koszul duality contexts obtained from universal twisting morphisms. We recall that Koszul twisting morphism form a category (Definition 2.55).

Proposition 4.66 ([CCN22, Proposition 6.5]) *The construction in Sect. 4.3.3 defines a natural transformation:*[26]

$$\mathbf{Kos} \;\underset{\mathbf{Alg}^{\mathrm{dual}}}{\overset{\mathbf{Alg}}{\Downarrow \mathfrak{D}}}\; \mathrm{Cat}_{\infty}^{L}$$

such that $\mathfrak{D}(\alpha) = \mathfrak{D}_{\alpha} : \mathbf{Alg}_{\mathscr{P}} \to \mathbf{Alg}_{\mathscr{C}}^{\mathrm{op}}$

If we restrict to using the universal twisting morphism as in Definition 4.59, we get the natural transformation [CCN22, Corollary 6.7]:

$$\mathbf{Op}_{k}^{\mathrm{aug}} \;\underset{\mathbf{Alg}^{\mathrm{dual}}}{\overset{\mathbf{Alg}}{\Downarrow \mathfrak{D}}}\; \mathrm{Cat}_{\infty}^{L}$$

[26] $\mathrm{Cat}_{\infty}^{L}$ denotes the ∞-category of ∞-categories with left adjoint ∞-functor as morphisms.

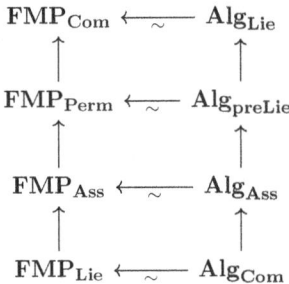

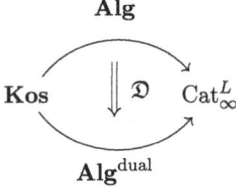

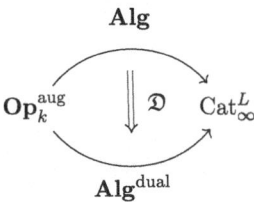

Note that the functor $\mathfrak{D}$ gives us weak Koszul duality context (and even Koszul duality context for nice enough operads). Moreover, Proposition 4.56 gives us some naturality with respect to morphisms of Koszul duality context. In the end, we obtain the following result:

Proposition 4.67 ([CCN22, Proposition 6.10]) *Let* $\mathbf{Op}_k^+$ *denote the category of splendid connective k-operads. Then we have a natural equivalence:*[27]

$$\mathbf{Op}_k^+ \underset{\mathbf{FMP}}{\overset{\mathbf{Alg}_{\mathfrak{D}}}{\rightrightarrows}} \mathrm{Pr}^R, \qquad \Psi : \mathbf{Alg}_{\mathfrak{D}} \Rightarrow \mathbf{FMP}$$

that sends $f : \mathscr{P} \to \mathscr{Q}$ *to the commutative square of right adjoint:*

$$\begin{array}{ccc} \mathbf{Alg}_{\mathfrak{D}(\mathscr{P})} & \longrightarrow & \mathbf{FMP}_{\mathscr{P}} \\ \big\downarrow{\scriptstyle \mathfrak{D}(f)^*} & & \big\downarrow{\scriptstyle (f^*)^*} \\ \mathbf{Alg}_{\mathfrak{D}(\mathscr{Q})} & \longrightarrow & \mathbf{FMP}_{\mathscr{Q}} \end{array}$$

4.3.6 Some Vague Comments on Positive Characteristic

The whole text we worked over a field of characteristic zero with good reason. Many of the homological results do not generally hold in positive characteristic, for instance the bar-cobar "resolution" can be constructed but is not a quasi-isomorphism and the Maurer–Cartan equation and gauge groups cannot even be defined.

On the other hand, whenever the concerned operads are non-symmetric or at least the symmetric groups act freely (Σ-free) there is some hope to retain some homotopy invariance. For instance, the Maurer–Cartan equation for A_∞-algebras (which are indeed equivalent to associative algebras) reads

$$0 = d\mu + m \star m + m \star m \star m + \ldots$$

which makes sense in every characteristic and can therefore be studied with operadic methods, see [dKW18].

One approach to operadic deformation theory in positive characteristic involves therefore considering Σ-free resolutions of operads. A particularly important example is the case of the commutative operad, whose Σ-free resolutions are called E_∞-operads. Notice that since the commutative operad is the unit for the tensor product of operads (if we are careful with assumptions on arities ≤ 1), given such an E_∞-operad $\mathscr{E}$, $\mathscr{P} \otimes \mathscr{E}$ is

[27] Pr^R denote the ∞-category of presentable ∞-categories with morphisms given by right adjoint functors.

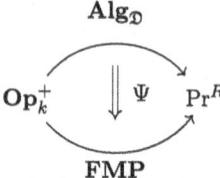

a Σ-free resolution of $\mathscr{P}$. It is therefore important to have access to small models $\mathscr{E}$, but there is no similar Koszul duality for algebraic operads able to produce minimal models in positive characteristic. The best known such model is the Barratt–Eccles operads [BF04], but others exist [BF04, MM20, BCN21].

One of the goals of Dehling–Vallette [DV21] is to study the homotopy theory of operads in positive characteristic by including Σ-action on the definition of the operad of operads, as in Sect. 2.6.1.

An analogue of the Lurie–Pridham Theorem 3.36 is also available in positive characteristic: Depending on whether the formal moduli problems are parameterized by E_∞-rings or simplicial commutative rings, Brantner–Mathew [BM19] have shown that these are equivalent to the ∞-categories of spectral or derived partition Lie algebras, respectively.

Acknowledgments This survey was based on a mini-course the first author gave on the Workshop on Higher Structures and Operadic Calculus organized at the CRM in Barcelona. We would like to thank all participants and organizers and Bruno Vallette in particular. We are deeply grateful to Vladimir Dotsenko and Joost Nuiten for a very thorough review of the paper. We would also like to thank Pelle Steffens for helpful discussions during the writing of this review and Miguel Barata and José Moreno-Fernandez for comments on the first ArXiV version.

This project has received funding from the grant ANR-20-CE40-0016 HighAGT and from the European Research Council (ERC) under the European Union's Horizon 2020 research and innovation programme (grant agreement No 768679).

The first author was supported by ANR-20-CE40-0016 HighAGT. The second author received funding from the European Research Council (ERC) under the European Union's Horizon 2020 research and innovation programme (grant agreement No 768679)

References

[AST76] M. Artin, C.S. Seshadri, and A. Tannenbaum. *Lectures on Deformations of Singularities*. Lectures on mathematics and physics. Tata Institute of Fundamental Research, 1976.

[BCN21] Lukas Brantner, Ricardo Campos, and Joost Nuiten. Pd operads and explicit partition lie algebras. *arXiv preprint arXiv:2104.03870*, 2021.

[Ber14] Alexander Berglund. Homological perturbation theory for algebras over operads. *Algebr. Geom. Topol.*, 14(5):2511–2548, 2014.

[BF04] Clemens Berger and Benoit Fresse. Combinatorial operad actions on cochains. *Math. Proc. Cambridge Philos. Soc.*, 137(1):135–174, 2004.

[BF11] A. Bonfiglioli and R. Fulci. *Topics in Noncommutative Algebra: The Theorem of Campbell, Baker, Hausdorff and Dynkin*. Lecture Notes in Mathematics. Springer Berlin Heidelberg, 2011.

[BM03] Clemens Berger and Ieke Moerdijk. Axiomatic homotopy theory for operads. *Comment. Math. Helv.*, 78(4):805–831, 2003.

[BM19] Lukas Brantner and Akhil Mathew. Deformation theory and partition lie algebras. *arXiv preprint arXiv:1904.07352*, 2019.

[CCN22] Damien Calaque, Ricardo Campos, and Joost Nuiten. Moduli problems for operadic algebras. *Journal of the London Mathematical Society*, 106(4):3450–3544, 2022.

[CG21] Damien Calaque and Julien Grivaux. Formal moduli problems and formal derived stacks. In *Derived algebraic geometry*, pages 85–145. Paris: Société Mathématique de France (SMF), 2021.

[CL01] Frédéric Chapoton and Muriel Livernet. Pre-Lie algebras and the rooted trees operad. *International Mathematics Research Notices*, 2001(8):395–408, 01 2001.

[CPRNW22] Ricardo Campos, Dan Petersen, Daniel Robert-Nicoud, and Felix Wierstra. Commutative homotopical algebra embeds into non-commutative homotopical algebra. *arXiv preprint arXiv:2211.02387*, 2022.

[CPRW19] Ricardo Campos, Dan Petersen, Daniel Robert-Nicoud, and Felix Wierstra. Lie, associative and commutative quasi-isomorphism. *arXiv e-prints*, page arXiv:1904.03585, April 2019.

[CT20] Ricardo Campos and Pedro Tamaroff. Differential forms on smooth operadic algebras. *arXiv e-prints*, page arXiv:2010.08815, October 2020.

[DCH16] Gabriel C. Drummond-Cole and Joseph Hirsh. Model structures for coalgebras. *Proc. Amer. Math. Soc.*, 144(4):1467–1481, 2016.

[dKW18] Niek de Kleijn and Felix Wierstra. Lie theory for complete curved a_∞-algebras. *arXiv preprint arXiv:1809.07743*, 2018.

[DR15] Vasily A. Dolgushev and Christopher L. Rogers. A version of the Goldman-Millson theorem for filtered L_∞-algebras. *J. Algebra*, 430:260–302, 2015.

[DR17] Vasily A. Dolgushev and Christopher L. Rogers. On an enhancement of the category of shifted L_∞-algebras. *Appl. Categ. Structures*, 25(4):489–503, 2017.

[dRBM04] María del Rosario Basterra and Michael A. Mandell. Homology and cohomology of E_∞ ring spectra. *Mathematische Zeitschrift*, 249:903–944, 2004.

[Dri14] Vladimir Drinfeld. A letter from Kharkov to Moscow. *EMS Surv. Math. Sci.*, 1(2):241–248, 2014.

[DSV16] Vladimir Dotsenko, Sergey Shadrin, and Bruno Vallette. Pre-Lie deformation theory. *Mosc. Math. J.*, 16(3):505–543, 2016.

[DSV24] Vladimir Dotsenko, Sergey Shadrin, and Bruno Vallette. *Maurer-Cartan methods in deformation theory. The twisting procedure*, volume 488 of *Lond. Math. Soc. Lect. Note Ser.* Cambridge: Cambridge University Press, 2024.

[DV21] Malte Dehling and Bruno Vallette. Symmetric homotopy theory for operads. *Algebr. Geom. Topol.*, 21(4):1595–1660, 2021.

[FCMF23] Oisín Flynn-Connolly and José M. Moreno-Fernández. Higher order massey products for algebras over algebraic operads. 2023.

[FHT01] Yves Félix, Stephen Halperin, and Jean-Claude Thomas. *Rational homotopy theory*, volume 205 of *Graduate Texts in Mathematics*. Springer-Verlag, New York, 2001.

[FN57] Alfred Frölicher and Albert Nijenhuis. A theorem on stability of complex structures. *Proc. Nat. Acad. Sci. U.S.A.*, 43:239–241, 1957.

[Get09] Ezra Getzler. Lie theory for nilpotent L_∞-algebras. *Ann. of Math. (2)*, 170(1):271–301, 2009.

[GK94] Victor Ginzburg and Mikhail Kapranov. Koszul duality for operads. *Duke Math. J.*, 76(1):203–272, 1994.

[GM88] William M. Goldman and John J. Millson. The deformation theory of representations of fundamental groups of compact Kähler manifolds. *Inst. Hautes Études Sci. Publ. Math.*, (67):43–96, 1988.

[Har09] R. Hartshorne. *Deformation Theory*. Graduate Texts in Mathematics. Springer New York, 2009.

[Hin97] Vladimir Hinich. Homological algebra of homotopy algebras. *Comm. Algebra*, 25(10):3291–3323, 1997.

[Hin01] Vladimir Hinich. DG coalgebras as formal stacks. *J. Pure Appl. Algebra*, 162(2–3):209–250, 2001.

[Hin04] Vladimir Hinich. Deformations of homotopy algebras. *Comm. Algebra*, 32(2):473–494, 2004.

[Hin05] Vladimir Hinich. Deformations of sheaves of algebras. *Adv. Math.*, 195(1):102–164, 2005.

[Hol08] Sharon Hollander. A homotopy theory for stacks. *Israel J. Math.*, 163:93–124, 2008.

[Hov07] M. Hovey. *Model Categories*. Mathematical surveys and monographs. American Mathematical Society, 2007.

[Ill71] L. Illusie. *Complexe cotangent et déformations*. Number vol. 1 in Complexe cotangent et déformations. Springer-Verlag, 1971.

[Ill72] L. Illusie. *Complexe cotangent et deformations II.* Lecture notes in mathematics. Springer, 1972.

[Kon03] Maxim Kontsevich. Deformation quantization of Poisson manifolds. *Lett. Math. Phys.*, 66(3):157–216, 2003.

[KS58] K. Kodaira and D. C. Spencer. On deformations of complex analytic structures. I, II. *Ann. of Math. (2)*, 67:328–466, 1958.

[KS60] K. Kodaira and D. C. Spencer. On deformations of complex analytic structures. III. Stability theorems for complex structures. *Ann. of Math. (2)*, 71:43–76, 1960.

[Lor15] Fosco Loregian. Coend calculus. *arXiv e-prints*, page arXiv:1501.02503, January 2015.

[Lur09] Jacob Lurie. *Higher topos theory*, volume 170 of *Annals of Mathematics Studies*. Princeton University Press, Princeton, NJ, 2009.

[Lur11] Jacob Lurie. Derived algebraic geometry x: Formal moduli problems. 2011. available at: https://www.math.ias.edu/~lurie/papers/DAG-X.pdf.

[LV12] Jean-Louis Loday and Bruno Vallette. *Algebraic operads*, volume 346 of *Grundlehren der Mathematischen Wissenschaften [Fundamental Principles of Mathematical Sciences]*. Springer, Heidelberg, 2012.

[Man02] Marco Manetti. Extended deformation functors. *Int. Math. Res. Not.*, (14):719–756, 2002.

[Man05] Marco Manetti. Lectures on deformations of complex manifolds. *arXiv Mathematics e-prints*, page math/0507286, July 2005.

[Mas58] W. S. Massey. Some higher order cohomology operations. In *Symposium internacional de topología algebraica International symposium on algebraic topology*, pages 145–154. Universidad Nacional Autónoma de México and UNESCO, Mexico City, 1958.

[MM20] Anibal M. Medina-Mardones. A finitely presented E_∞-prop I: algebraic context. *High. Struct.*, 4(2):1–21, 2020.

[Mur23] Fernando Muro. Massey products for algebras over operads. *Communications in Algebra*, 51(8):3298–3313, 2023.

[Nit06] Nitin Nitsure. Notes on Deformation Theory. page 45, November 2006. Lecture.

[Pri70] Stewart B. Priddy. Koszul resolutions. *Transactions of the American Mathematical Society*, 152(1):39–60, 1970.

[Pri10] J. P. Pridham. Unifying derived deformation theories. *Adv. Math.*, 224(3):772–826, 2010.

[Qui67] Daniel G. Quillen. *Homotopical algebra.* Springer, 1967.

[Qui69] Daniel Quillen. Rational homotopy theory. *Annals of Mathematics*, 90(2):205–295, 1969.

[RV20] Daniel Robert-Nicoud and Bruno Vallette. Higher Lie theory. *arXiv e-prints*, page arXiv:2010.10485, October 2020.

[Sch68a] Michael Schlessinger. Functors of Artin rings. *Trans. Amer. Math. Soc.*, 130:208–222, 1968.

[Sch68b] Michael Schlessinger. Functors of artin rings. *Transactions of the American Mathematical Society*, 130(2):208–222, 1968.

[Ser07] E. Sernesi. *Deformations of Algebraic Schemes.* Grundlehren der mathematischen Wissenschaften. Springer Berlin Heidelberg, 2007.

[Sul77] Dennis Sullivan. Infinitesimal computations in topology. *Publications Mathématiques de l'Institut des Hautes Études Scientifiques*, 47(1):269–331, Dec 1977.

[TV08] Bertrand Toën and Gabriele Vezzosi. Homotopical algebraic geometry. II. Geometric stacks and applications. *Mem. Amer. Math. Soc.*, 193(902):x+224, 2008.

[Val20] Bruno Vallette. Homotopy theory of homotopy algebras. *Ann. Inst. Fourier (Grenoble)*, 70(2):683–738, 2020.

[vdL04] Pepijn van der Laan. Coloured koszul duality and strongly homotopy operads, 2004.

[Vez10] Gabriele Vezzosi. A note on the cotangent complex in derived algebraic geometry. *arXiv e-prints*, page arXiv:1008.0601, August 2010.

[War21] Benjamin C Ward. Massey Products for Graph Homology. *International Mathematics Research Notices*, 2022(11):8086–8161, 01 2021.

[Wie19] F. Wierstra. Algebraic Hopf invariants and rational models for mapping spaces. *Journal of Homotopy and Related Structures*, 14(3):719–747, 2019. arXiv:1612.07762.

Index of Notations

Operads

- $\mathbf{Seq}_\Sigma$: Category of symmetric sequences (Definition 2.1).
- $\otimes_{\mathrm{Day}}$: Day convolution (Definition 2.5).
- $\circ$: Monoidal product on $\mathbf{Seq}_\Sigma$ defining the composition of symmetric sequences (Definition 2.8).
- $\hat{\circ}$: Complete composition of symmetric sequences (Proposition 2.15).
- $\bar{\circ}$: Composition of symmetric sequences for cooperads (Eq. 4).
- $\mathbf{Op}$, $\mathbf{coOp}$, $\mathbf{coOp}^{\mathrm{conil}}$: Category of operads (Definition 2.9), cooperads (Definition 2.17) and conilpotent cooperads (Definition 2.19).
- $\mathbf{Op}_k^+$: Full sub-category of $\mathbf{Op}$ given by splendid connective k-operads
- $\mathbf{cOp}$: Category of colored operads (Sect. 2.6.1).
- $\circ_i$: Partial composition of an operad (Definition 2.10).
- Δ_i: Partial cocomposition of a cooperad.
- $\mathfrak{T}$, $\mathfrak{T}^{\mathrm{co}}$: Free and cofree operad functor (see Constructions 2.11 and 2.20 respectively).
- $\mathscr{P}(E, R)$: Operad generated by $E \in \mathbf{Seq}_\Sigma$ with relations generated by $R \subset \mathfrak{T}E$ (Definition 2.10).
- $\mathscr{C}(E, R)$: Cooperad cogenerated by $E \in \mathbf{Seq}_\Sigma$ with corelations generated by $R \subset \mathfrak{T}^{\mathrm{co}}E$ (Construction 2.20).
- I: Unit (co)operad concentrated in arity 1 and with $I(1) = k$ (Example 2.14).
- **Ass**, **Com**, **Lie**: Associative, commutative and Lie operads (Example 2.14).
- **Perm** and **preLie**: Permutative and pre-Lie operads (Definition 2.35).
- $\mathbf{End}_V$, $\mathbf{coEnd}_V$: Endomorphism and coendomorphism operads (Example 2.14).
- $\mathbf{coEnd}_N^M$: Symmetric sequence of map from M to $N^{\otimes \bullet}$ (Definition 2.24).
- $\mathscr{O}$: Colored operad encoding symmetric operads (Sect. 2.6.1).
- $\mathscr{P}^\vee$, $\mathscr{C}^\vee$: $\mathscr{P}^\vee$ is the arity-wise dual of the operad $\mathscr{P}$ and $\mathscr{C}^\vee$ is the operad given by the arity-wise dual of the cooperad $\mathscr{C}$ (Construction 2.20).
- $\mathbf{End}_N^M$: Symmetric sequence of map from $M^{\otimes \bullet}$ to N (Definition 2.24).
- $\mathrm{Conv}(\mathscr{C}, \mathscr{P})$: Convolution Operad (Definition 2.34).
- $\mathbf{B} \dashv \Omega$: Bar-cobar adjunction (Definition 2.39).
- $\mathscr{Q}\{-n\}$: $\mathscr{Q} \otimes \mathbf{End}_{k[n]}$.
- $\mathscr{P}^!$: Koszul Dual Operad (Definition 2.60).
- $\mathscr{P}^{¡}$: Koszul dual cooperad (Definition 2.60).
- $\mathscr{P}_\infty$: Minimal cofibrant resolution of $\mathscr{P}$, $\mathscr{P}_\infty := \Omega\mathscr{P}^{¡}$ (Convention 2.63).
- **Tw**: Category of all twisting morphisms (Definition 2.38).
- **Koszul**: Full sub-category of **Tw** given by Koszul morphisms (Definition 2.55).
- A_∞, L_∞, C_∞: Notation for $\mathbf{Ass}_\infty$, $\mathbf{Lie}_\infty$ and $\mathbf{Com}_\infty$ (Convention 2.63).

Other

- k: Field of characteristic 0.
- Σ_n: Group of permutations of n-elements.
- $\mathrm{Der}_A(B, M)$, $\underline{\mathrm{Der}}_A(B, M)$ and $\mathbb{R}\mathrm{Der}_A(B, M)$: B-Module, differential graded module and derived module of M-valued, A-linear derivations on B respectively.

- $k[[t]]$: Algebra of formal power series in t over k.
- $\sim$: Denote an equivalence relation. In general in a model category, $A \sim B$ if there is a zig-zag of weak-equivalences between them.
- $\mathfrak{M}_A$: Maximal ideal of a local ring A.
- $B \oplus -$: Square zero extension functor sending a B-module to the trivial square zero extension $B \oplus M$ (Definition 4.7).
- $\Omega^1_{A/B}$: A-module of Kähler differential on A relative to B (Definition 4.10).
- $\mathbb{L}_{A/B}$: Cotangent complex of A relative to B (Definition 4.10).

Categories

- Mod_k: Category of cochain complexes over k.
- Mod_R: Category of cochain complexes over a commutative ring R.
- $\underline{\mathrm{Hom}}$: Internal Hom for categories enriched over themselves.
- $\mathscr{C}^{A/}$, $\mathscr{C}^{/A}$ and $\mathscr{C}^{A//B}$: Category of objects under A, objects over A and objects both over B and under A respectively.
- $\mathrm{Fun}(\mathscr{C}, \mathscr{D})$: Category (or ∞-category) of functors from $\mathscr{C}$ to $\mathscr{D}$.
- $\mathscr{C}^{\mathrm{iso}}$: Given a category $\mathscr{C}$, $\mathscr{C}^{\mathrm{iso}}$ denotes the sub-category of $\mathscr{C}$ with same objects and morphism being the isomorphism in $\mathscr{C}$.
- $\mathbb{N}^{\sim}$: Category with objects the sets $\{1, \ldots, n\}$ for all $n \in \mathbb{N}$ and morphisms given by the permutations of these sets.
- $\mathbb{N}_S$: Category with objects being k-tuples of elements on S for all $k \in \mathbb{N}$ and morphisms given by all permutations of these tuples, $(a_1, \cdots, a_k) \to (a_{\sigma_1}, \cdots, a_{\sigma_k})$.
- $\int_{c\in\mathscr{C}}$ and $\int^{c\in\mathscr{C}}$: End and coend functors sending a functor $F : \mathscr{C}^{\mathrm{op}} \times \mathscr{C} \to \mathscr{D}$ to and object $\int^{c\in F} F(c, c) \in \mathscr{D}$ (see [Lor15]).
- $\mathrm{Mon}(\mathscr{C})$: Categorie of monoid in a monoidal category $\mathscr{C}$.
- $C - \mathrm{Mod}$: Categorie of C-module for $C \in \mathrm{Mon}(\mathscr{C})$.
- N: Denotes all the nerve functors, including simplicial nerves and nerve of a category or groupoid.
- $\mathscr{S}$: ∞-categories of spaces.
- **sSet**: Category of simplicial sets.
- $|-|$: Geometric realisation functor $|-| : \mathbf{sSet} \to \mathbf{Top}$.
- $h\mathscr{C}$: Homotopy category of a model or ∞-category (Sect. 2.4.1).
- $\mathbf{Stab}(\mathscr{C})$: Stable ∞-category associated to an ∞-category $\mathscr{C}$.
- $\Omega^{\infty-n}$: Functor from $\mathbf{Stab}(\mathscr{C})$ to $\mathscr{C}$ sending a spectrum object $(A_k)_{k\in\mathbb{Z}}$ to A_n.
- **Sp**: Category of spectra, $\mathbf{Sp} = \mathbf{Stab}(\mathbf{Top})$.
- $\underline{\mathrm{Def}}_\phi(\mathfrak{R})$: Groupoid of deformation of an algebraic structure $\phi : \mathscr{P} \to \mathbf{End}_A$ along the local algebra $\mathfrak{R}$.

Deformation Theory

- MC: Functor sending a Lie algebra $\mathfrak{g}$ to its set of Maurer–Cartan elements, $\mathrm{MC}(\mathfrak{g})$ (Definition 2.37).
- $k[\epsilon]$: k-algebra of dual numbers.

$$k[\epsilon] := {}^{k[x]}\!\big/_{(x^2)}$$

- $\Omega[\Delta^\bullet]$: Simplicial commutative k-algebra of differential forms on simplices (Definition 3.20).
- $\mathbf{Del}(\mathfrak{g})$: Deligne ∞-groupoid associated to a Lie algebra $\mathfrak{g}$ (Definition 3.20).
- $\underline{\mathrm{Del}}(\mathfrak{g})$: Deligne groupoid associated to a Lie algebra $\mathfrak{g}$ (Definition 3.19).
- $\widetilde{\mathrm{Def}}_B^{\mathscr{P}}(A)$: Groupoid of geometric deformation of a $\mathscr{P}$-algebra structure on B along A (Definition 4.13).
- $\mathbf{Def}_B^{\mathscr{P}}(A)$: ∞-groupoid of geometric deformation of a $\mathscr{P}$-algebra structure on B along A (Definition 4.15).
- $\mathscr{A}^{\mathrm{aug}}_{\mathbf{Com}}$: Deformation context for augmented commutative k-algebras (Example 4.23).
- $\mathscr{A}_{\mathscr{P}}$: Augmentation context for $\mathscr{P}$-algebras (Example 4.23).
- Art, $\mathbf{Art}_{\mathscr{A}^{\mathrm{aug}}_{\mathbf{Com}}}$ and $\mathbf{Art}_{\mathscr{P}}$: Category of classical commutative local Artinian algebra, differential graded commutative local Artinian algebra and Artinian algebra in $\mathbf{Art}_{\mathscr{P}}$ (Definition 4.1 and 4.24).
- **FMP**, $\mathbf{FMP}_{\mathscr{A}}$ and $\mathbf{FMP}_{\mathscr{P}}$: Category of formal moduli problem associated to commutative deformations, the deformation context $\mathscr{A}$ and the deformation context $\mathscr{A}_{\mathscr{P}}$ respectively (Definition 4.31 and Example 4.32).
- Spf: Formal spectrum functor (Example 4.34).

Algebraic Structures

- $\mathbf{Alg}_{\mathbf{Com}\otimes_k A}(\mathrm{Mod}_A)$: Category of A-linear commutative algebra in Mod_A.
- **CE**: Chevalley-Eilenberg functor.
- Art: Category of local Artinian k-algebras (Definition 4.1).
- $\mathbf{Alg}_{\mathscr{P}}$, $\mathbf{coAlg}_{\mathscr{P}}$ and $\mathbf{coAlg}_{\mathscr{P}}^{\mathrm{conil}}$: Category of all $\mathscr{P}$-algebras, $\mathscr{P}$-coalgebras and conilpotent $\mathscr{P}$-coalgebras respectively (Definition 2.21).
- $\mathbf{Alg}_{\mathscr{C}}$, $\mathbf{coAlg}_{\mathscr{C}}$ and $\mathbf{coAlg}_{\mathscr{C}}^{\mathrm{conil}}$: Category of all $\mathscr{C}$-algebras, $\mathscr{C}$-coalgebras and conilpotent $\mathscr{C}$-coalgebras respectively (Definition 2.22).
- $\mathscr{P}(-)$ and $\mathscr{C}(-)$: Free $\mathscr{P}$-algebra functor and cofree conilpotent $\mathscr{C}$-coalgebra functor respectively (Proposition 2.27).
- $\mathrm{Conv}(\mathscr{C}, \mathscr{P})$: Convolution operad (Definition 2.34).
- $\mathrm{Tot}(\mathscr{P})$: Totalisation pre-Lie algebra (Definition 2.36).
- $(-)^\sharp$: Forgetful functor forgetting the $\mathscr{P}$ (or $\mathscr{C}$) (co)algebra structure.
- $f_! \dashv f^*$: Free-forget adjunction associated to a map of operad $f : \mathscr{P} \to \mathscr{Q}$ (Proposition 2.29). In particular $f_!$ is the free $\mathscr{Q}$-algebra functor relative to $\mathscr{P}$.
- $\mathfrak{U}$: Universal enveloping algebra functor (Sect. 2.2.1).
- Sym: Symmetric algebra functor.
- $\Omega_\alpha \dashv \mathbf{B}_\alpha$: Algebraic cobar–bar adjunction associated to a twisting morphism α (Definition 2.64).
- $\mathfrak{D}$: Koszul dual algebra (Definition 4.59).
- $f : A \rightsquigarrow B$: ∞-morphism from A to B (Definition 3.5).
- $F^\bullet\mathfrak{g}$: Filtration on a Lie algebra $\mathfrak{g}$ (Definition 3.16).
- $\widehat{-}$: Completion functor for filtered algebras (Definition 3.16).
- BCH: The BCH product satisfying $e^X e^Y = e^{\mathrm{BCH}(X,Y)}$ (Proposition 3.17).
- $\mathfrak{g}_{\mathscr{P}_\infty, A}$: Convolution Lie Algebra of the cochain complex A (Definition 3.29).
- $\mathfrak{g}^\phi$: Twisted Lie algebra by a Maurer–Cartan element ϕ of $\mathfrak{g}$ (Proposition 3.18).
- $\mathfrak{g}^\phi_{\mathscr{P}_\infty, A}$: Deformation complex of A with $\mathscr{P}_\infty$ structure $\phi \in \mathrm{MC}(\mathfrak{g}_{\mathscr{P}_\infty, A})$ (Definition 3.30).

Weight Structures and Formality

Coline Emprin and Geoffroy Horel

Contents

C. Emprin
Département de mathématiques et applications, École normale supérieure, Paris, France
e-mail: coline.emprin@ens.psl.eu

G. Horel (✉)
Université Sorbonne Paris Nord, Laboratoire de Géométrie, Analyse et Applications, CNRS, UMR 7539, Villetaneuse, France
e-mail: horel@math.univ-paris13.fr

B. Vallette (ed.), *Higher Structures and Operadic Calculus*, Advanced Courses in Mathematics - CRM Barcelona, https://doi.org/10.1007/978-3-031-77779-0_3

1 Introduction

The Notion of Formality Let A be a chain complex equipped with an algebraic structure (e.g. an associative algebra, a commutative algebra, an operad, etc.). The homology of this complex inherits the same type of structure. In general, this induced structure does not retain all the homotopical information contained in A. For example, it is well-known that the homology of a differential graded algebra can have additional non-trivial *Massey products* that witness homotopical information about the algebra. In some cases all these Massey products vanish, namely if A is homotopy equivalent to an algebra whose differential is identically zero. If so, the algebra is said to be *formal*. The idea of formality originated in rational homotopy theory. In this context, a topological space X is formal if its Sullivan's algebra of polynomial forms $\Omega^*_{PL}(X)$ is connected to its cohomology $H^*(X; \mathbb{Q})$ by a string of quasi-isomorphisms of commutative differential graded algebras.

The Case of Compact Kähler Manifolds A central formality result was proved by Deligne, Griffiths, Morgan, and Sullivan in [DGMS75]. Using Hodge theory, they showed that any compact Kähler manifold is formal, see Sect. 3. However, as explained in their introduction, their intuition came from the Weil's conjectures and the following observation : formality of an algebra can be viewed as a multiplicative splitting of the canonical filtration. Indeed, if that is the case, the algebra is quasi-isomorphic to the associated graded of the canonical filtration, which is exactly the cohomology algebra. The abstract statement that emerges from this intuition is the following one.

Theorem *If a dg-algebra A^* admits a multiplicative "weight decomposition"*

$$A^* = \oplus_{i \in \mathbb{Z}} A_i^*$$

with the property that $H^i(A)$ is concentrated in weight i, then A is formal.

A functorial version of this theorem appears as Proposition 4.3, whose proof is basic. The strength of this result comes from combining it with deep results from algebraic geometry producing the desired weight decompositions.

Mixed Hodge Structures The first type of such decompositions studied in this survey comes from mixed Hodge theory. By work of Deligne, the cohomology of complex algebraic varieties carries a canonical mixed Hodge structure. It turns out that this structure can be lifted at the chains level. Moreover, a result of Deligne shows that mixed Hodge structure are functorialy split. From these two facts, we can obtain a functorial weight decomposition as in the Theorem above for many varieties. We refer the reader to Sect. 6 for a survey of this result based on joint work of the second author and Joana Cirici, see [CH20b].

Galois Group Actions One can try to use similar techniques in order to prove formality results over $\mathbb{F}_p$ instead of $\mathbb{Q}$. In that case, mixed Hodge theory does not make sense anymore. However, one can use Frobenius actions on étale cohomology. In the context a smooth projective variety over a finite field, the canonical filtration can be split by eigenvalues of a Frobenius action. This idea is explained in Sect. 7 based on a joint paper of the second author with Joana Cirici, see [CH22]. In this context, one does not obtain full $\mathbb{Z}$-graded weight decompositions as in the Theorem above but merely $\mathbb{Z}/(h)$-graded decompositions for some integer h.

Gauge Formality By using the operadic calculus, one can have another approach to formality which boils down to a deformation problem. This leads to the notion of *gauge formality* which is presented in Sect. 8. We present a joint work of Gabriel Drummond-Cole and the second author (see [DCH21]) which revisit the results of the previous sections, using the approach of gauge formality. This idea of gauge formality gives rise to an obstruction theory to formality due initially to Kaledin [Kal07] in the context of associative algebras and pushed further by Melani-Rubio and the first author, see [MR19, Emp24]. One significant consequence of this theory are formality descent results, see Theorem 8.19.

Notations and Conventions

- Let R be a commutative ground ring.
- We generically write $\otimes$ for the tensor product over a commutative ground ring that should always be clear from context.
- If A is a chain complex and $x \in A$ is a homogeneous element, we denote by $|x|$ its homological degree.
- The abbreviation "dg" stands for the words "differential graded".
- We use the notations of [LV12] for operads.

2 The Notion of Formality

In this section, we define the formality of various algebraic structures and discuss its origins in rational homology theory.

2.1 *Formality of Algebraic Structures*

Let $\mathscr{P}$ be an operad in the category of R-modules. Let (A, ϕ) be a dg $\mathscr{P}$-algebra, i.e. a chain complex A over R endowed with an operad morphism

$$\varphi : \mathscr{P} \longrightarrow \mathrm{End}_A$$

between $\mathscr{P}$ and the endomorphism operad associated to A. The induced map $\varphi_* : \mathscr{P} \to \mathrm{End}_{H(A)}$ turns the homology into a $\mathscr{P}$-algebra. We will refer to it as the *induced structure* in homology.

Definition 2.1 (Formality of dg $\mathscr{P}$-Algebras) The dg $\mathscr{P}$-algebra (A, φ) is *formal* if there exists a zig-zag of dg $\mathscr{P}$-algebra quasi-isomorphisms

$$(A, \varphi) \xleftarrow{\sim} \cdot \xrightarrow{\sim} \cdots \xleftarrow{\sim} \cdot \xrightarrow{\sim} (H(A), \varphi_*)$$

relating it to its induced structure in homology.

Remark 2.2 The number of quasi-isomorphisms involved in a formality zig-zag is arbitrary. However, it can be reduced to a length two zig-zag in many cases, e.g. if R is a field, under additional assumptions on $\mathscr{P}$ in the positive characteristic case. Under these assumptions, the category of dg $\mathscr{P}$-algebras is equipped with a transferred model category structure, see [Hin97, Theorem 4.1.1]. Then, any zig-zag of quasi-isomorphisms induces an isomorphism in the associated homotopy category and can be represented by an actual weak equivalence between a cofibrant replacement of the source and a fibrant one of the target. Any object being fibrant in this context, this leads to a zig-zag

$$(A, \varphi) \xleftarrow{\sim} Q(A, \varphi) \xrightarrow{\sim} (H(A), \varphi_*) ,$$

where $Q(A, \varphi)$ denotes a cofibrant replacement of (A, φ).

Definition 2.3 (Lax Symmetric Monoidal Functor) A *lax monoidal functor*

$$(F, \kappa, \eta) : (\mathscr{C}, \otimes, \mathbf{1}) \longrightarrow (\mathscr{D}, \otimes, \mathbf{1}')$$

is a functor $F : \mathscr{C} \to \mathscr{D}$ between monoidal categories together with maps

$$\kappa_{X,Y} : F(X) \otimes F(Y) \longrightarrow F(X \otimes Y)$$

that are natural in the objects X and Y of $\mathscr{C}$, and a morphism of $\mathscr{D}$,

$$\eta : \mathbf{1}' \longrightarrow F(\mathbf{1})$$

that are compatible with the constraints of associativity and unit. The functor F is said to be *lax symmetric monoidal*, if κ is compatible with the commutativity constraint. A lax monoidal functor is called *strong* if κ and η are isomorphisms. We refer the reader to [EGNO15, Chapter 2] for more details.

Remark 2.4 In the sequel, strong monoidal functors will not play an important role and we shall often use "symmetric monoidal functor" to refer to a lax symmetric monoidal functor.

Example 2.5 Let $(\mathscr{A}, \otimes, \mathbf{1})$ be an abelian symmetric monoidal category with infinite direct sums. The homology functor $H : \mathrm{Ch}_*(\mathscr{A}) \to \mathrm{Ch}_*(\mathscr{A})$ is lax symmetric monoidal, via the usual Künneth morphism. If $\mathscr{A}$ is the category of vector spaces over a field, then this functor is strong symmetric monoidal.

Definition 2.6 (Formality of Symmetric Monoidal Functors) Let $(\mathscr{C}, \otimes, \mathbf{1})$ a symmetric monoidal category. A symmetric monoidal functor $F : \mathscr{C} \to \mathrm{Ch}_*(R)$ is *formal* if it is weakly equivalent to $H \circ F$, i.e. if there exists a zig-zag of natural transformations of symmetric monoidal functors

$$F \xleftarrow{\Phi_1} F_1 \longrightarrow \cdots \longleftarrow F_n \xrightarrow{\Phi_n} H \circ F$$

such that $\Phi_i(X)$ is a quasi-isomorphisms for every object X of $\mathscr{C}$.

If we allow operads to be *colored*, the formality of symmetric monoidal functors appears as a particular case of Definition 2.1 (see Proposition 2.12 below). A colored operad is an operad in which each input or output comes with a color chosen in a given set. A composition is possible whenever the colors of the corresponding input and output involved match. We now give a precise definition. Let $\langle n\rangle$ denote the finite set $\{0, 1, \ldots, n\}$.

Definition 2.7 (Set Colored Operads) Let I be a set of colors. Fix $(\mathscr{C}, \otimes, \mathbf{1})$ a symmetric monoidal category. An *I-colored operad* in $\mathscr{C}$ is a set

$$\{\mathscr{P}(n, i)\}_{i:\langle n\rangle \to I}$$

of objects of $\mathscr{C}$ indexed by all maps $i : \langle n\rangle \to i$, for $n \geqslant 0$, together with

- composition maps for all $l \leqslant n$,

$$\circ_l : \mathscr{P}(n, i) \otimes \mathscr{P}(m, j) \to \mathscr{P}(m + n - 1, i \circ_l j)$$

  where $i(l) = j(0)$ and $i \circ_l j : \langle m + n - 1\rangle \to I$ is defined by

$$i \circ_l j(k) = \begin{cases} i(k) & \text{if } k < l \\ j(k - l + 1) & \text{if } l \leqslant k < l + m \\ i(k - m) & \text{if } l + m \leqslant k\,; \end{cases}$$

- a right $\mathbb{S}_n$-action on

$$\bigoplus_{i:\langle n\rangle \to I} \mathscr{P}(n, i)\,;$$

- an identity $\mathrm{id}_\alpha \in \mathscr{P}(1, c_\alpha)$, for each $\alpha \in I$, where $c_\alpha : \langle 1\rangle \to I$ is the constant map with value α. These identities act as units with respect to any well defined composition.

These data satisfy the compatibility relations for $\circ_l$-operations of an operad (associativity, equivariance, etc.) whenever these make sense.

Example 2.8 (The Endomorphism Colored Operad) Let $A = \{A_\alpha\}_{\alpha \in I}$ be a family of chain complexes. The associated endomorphism I-colored operad End_A is defined for all $n \geqslant 0$, and all $i : \langle n\rangle \to I$, by

$$\mathrm{End}_A(n, i) = \mathrm{Hom}\left(A_{i(1)} \otimes \cdots \otimes A_{i(n)}, A_{i(0)}\right) ,$$

where the composition products (resp. the $\mathbb{S}_n$-actions) are induced by substitution (resp. permutation) of the tensor factors.

Definition 2.9 (Algebras Over a Set Colored Operad) Let $\mathscr{P}$ be a I-colored operad. A dg $\mathscr{P}$-algebra is a family $A = \{A_\alpha\}_{\alpha \in I}$ of chain complexes endowed with a morphism of I-colored operads

$$\mathscr{P} \longrightarrow \mathrm{End}_A .$$

Remark 2.10 In order to encode symmetries, one can also consider groupoid colored operads where colors are chosen in a given groupoid $\mathbb{V}$ instead of a set I. The associated Koszul duality theory was developed by Ward in [War21]. We refer the reader to [RiL22, Section 5] for more details.

Example 2.11 There exists an $\mathbb{N}$-colored operad $\mathscr{O}$ such that $\mathscr{O}$-algebras are exactly non-symmetric operads, see [VdL03, Section 4]. Similarly, there exists an $\mathbb{N}$-colored operad encoding symmetric operads (see [CH20a, Definition 5.1.5])

Proposition 2.12 *Let $(\mathscr{C}, \otimes, 1)$ be a symmetric monoidal category. There is an associated $Ob(\mathscr{C})$−colored operad defined for all $i : \langle n \rangle \to Ob(\mathscr{C})$ by*

$$\mathscr{Q}(n, i) := \mathrm{Hom}_{\mathscr{C}}(i(1) \otimes \cdots \otimes i(n), i(0)) .$$

A dg $\mathscr{Q}$-algebra over this operad is the same data as a lax symmetric monoidal functor

$$F : \mathscr{C} \to \mathrm{Ch}_*(R) .$$

Proof Let $\{A_\alpha\}_{\alpha \in Ob(\mathscr{C})}$ be a family of chain complexes and let

$$\varphi : \mathscr{Q} \to \mathrm{End}_A$$

be a dg $\mathscr{Q}$-algebra structure. Setting $F(\alpha) := A_\alpha$ for all $\alpha \in Ob(\mathscr{C})$, we obtain a symmetric monoidal functor $\mathscr{C} \to \mathrm{Ch}_*(R)$ such that

$$\kappa_{\alpha,\beta} := \varphi(2, i)(\mathrm{id}_{\alpha \otimes \beta})$$

where $i : \langle 2 \rangle \to Ob(\mathscr{C})$ is defined by $i(0) = \alpha \otimes \beta$, $i(1) = \alpha$ and $i(2) = \beta$. Conversely, out of a symmetric monoidal functor $F : \mathscr{C} \to \mathrm{Ch}_*(R)$ one defines a dg $\mathscr{Q}$-algebra (A, φ) with

$$A := \{F(c)\}_{c \in Ob(\mathscr{C})} \quad \text{and} \quad \varphi(n, i) := F \circ \kappa$$

for all $i : \langle n \rangle \to Ob(\mathscr{C})$, where κ denotes the successive compositions giving

$$F(i(1)) \otimes \cdots \otimes F(i(n)) \to F(i(1) \otimes \cdots i(n)) .$$

□

The two following propositions are direct consequences of the definitions.

Proposition 2.13 *Let $\mathscr{P}$ be an operad in sets. If $F : \mathscr{C} \to \mathrm{Ch}_*(R)$ is a formal symmetric monoidal functor and if A is a $\mathscr{P}$-algebra in $\mathscr{C}$ (resp. operad), then $F(A)$ is a formal $\mathscr{P}$-algebra (resp. operad).*

Proof A lax monoidal functor sends $\mathscr{P}$-algebras to $\mathscr{P}$-algebras. A natural transformation between lax monoidal functors sends $\mathscr{P}$-algebras to morphisms of $\mathscr{P}$-algebras. If we evaluate the zig-zag connecting F to $H \circ F$ on A, we obtain a formality zig-zag for A. □

Proposition 2.14 *Let $U : \mathscr{B} \to \mathrm{Ch}_*(R)$ be a formal symmetric monoidal functor. For every symmetric monoidal functor $F : \mathscr{C} \to \mathscr{B}$, the composition*

$$U \circ F : \mathscr{C} \to \mathrm{Ch}_*(R)$$

is a formal symmetric monoidal functor.

Remark 2.15 (An Application of the Formality of an Operad) Over a characteristic zero field $\mathbf{k}$, there is a Quillen equivalence between algebras over a formal dg operad $\mathscr{P}$ and algebras encoded by its homology $H(\mathscr{P})$, see [Hin97, Theorem 4.7.4]. This leads to Quillen equivalences between

- A_∞-algebras and associative algebras;
- C_∞-algebras and commutative algebras;
- L_∞-algebras and Lie algebras;
- Gerstenhaber algebras and algebras over the dg-operad $C_*(\mathscr{D}_2; \mathbf{k})$, since the little disks operad $\mathscr{D}_2$ is formal and its homology is the Gerstenhaber operad, see Example 4.2.

Kontsevich formality theorem can be improved using the last Quillen equivalence. Given a smooth manifold M, it asserts that the Gerstenhaber algebra of polyvector fields $\Gamma(\Lambda TM)$ is weakly equivalent to the $C_*(\mathscr{D}_2)$-algebra of Hochschild cochains on the algebras of smooth functions on M. This version of Kontsevich formality is due to Tamarkin and generalizes the classical formulation in terms of L_∞-algebras. The fact that the Hochschild cochain complex carries an action of $C_*(\mathscr{D}_2)$ is a highly non-trivial theorem called Deligne's conjecture and was initially proved by Tamarkin, see [Tam98, Hin03].

2.2 *Origins in Rational Homotopy Theory*

The idea of formality originated in the field of rational homotopy theory. For an overview of rational homotopy theory, we refer the reader to [BS24]. Very briefly, the rational homotopy category $\mathrm{Ho(Top)}_{\mathbb{Q}}$ is obtained from the category of topological spaces by inverting maps that induce isomorphisms on homology with rational coefficients. This is a localization of the usual homotopy category in which one only inverts weak homotopy equivalences. The set of morphisms in this category are much more computable than in the usual homotopy category and still capture interesting invariants of homotopy types. For instance, the rational homotopy groups of a simply connected topological space X can be computed as maps from a sphere to X in the rational homotopy category:

$$\pi_n(X) \otimes_{\mathbb{Z}} \mathbb{Q} \cong [S^n, X]_{\mathrm{Ho}(\mathrm{Top})_{\mathbb{Q}}}.$$

The most fundamental theorem in the field of rational homotopy theory is due to Sullivan. It relies on the construction of a functorial commutative differential graded algebra (CDGA), the Sullivan algebra of polynomial forms

$$X \mapsto \Omega^*_{PL}(X)\,,$$

that faithfully reflects the rational homotopy type of X under mild hypotheses. The cohomology of this algebra is isomorphic to the cohomology of X.

We shall now explain this construction with more details. We start with the construction of the Sullivan algebra of polynomial forms. For every chain complex V, we denote by $S(V)$ the associated *symmetric algebra* defined by

$$S(V) := \bigoplus_{r \geqslant 0} \left(V^{\otimes r}\right)_{\mathbb{S}_r}\,,$$

and equipped with the only differential extending the differential of V and compatible with the Leibniz rule.

Definition 2.16 For $n \geq 0$, let K_n^* be the cochain complex over $\mathbb{Q}$ defined by

$$K_n^* := \bigoplus_{i=0}^{n} \mathbb{Q}t_i \to \bigoplus_{i=0}^{n} \mathbb{Q}dt_i \to 0 \to \dots,$$

with the differential being obvious from the notation. The CDGA of polynomial forms on Δ^n is defined as

$$\Omega^*_{PL}\left(\Delta^n\right) := \frac{S(K_n^*)}{\langle \sum t_i - 1\rangle}.$$

This induces a simplicial object in the category of CDGAs. This definition extends to a functor of polynomial forms

$$\begin{aligned} \Omega^*_{PL} : \mathrm{sSet} &\longrightarrow \mathrm{CDGA}^{op} \\ X &\longmapsto \mathrm{Hom}_{\mathrm{sSet}}\left(X, \Omega^*_{PL}\left(\Delta^\bullet\right)\right) \end{aligned}$$

This functor is the left adjoint in an adjunction

$$\Omega^*_{PL} : \mathrm{sSet} \leftrightarrows \mathrm{CDGA}^{op} : \langle - \rangle$$

where the right adjoint is then simply given by the formula

$$\langle A\rangle_n := \mathrm{Hom}_{\mathrm{CDGA}}(A, \Omega^*_{PL}(\Delta^n)) .$$

Given a topological space X, one defines

$$\Omega^*_{PL}(X) := \Omega^*_{PL}(S_\bullet(X))$$

where $S_\bullet(X) := \mathrm{Hom}_{\mathrm{Top}}(\Delta^\bullet, X)$ is the singular simplicial set associated to X. The fundamental theorems of rational homotopy theory states that the induced functor from the rational homotopy category of simplicial sets

$$\Omega^*_{PL} : \mathrm{Ho}(\mathrm{sSet})_{\mathbb{Q}} \to \mathrm{Ho}(\mathrm{CDGA})^{\mathrm{op}}$$

is fully faithful when restricted to nilpotent simplicial of finite type.

Definition 2.17 A topological space X is formal, if $\Omega^*_{PL}(X)$ is related to $H^*(X;\mathbb{Q})$ by a zig-zag of quasi-isomorphisms of CDGAs.

It follows that, if X is formal, one can reconstruct the rational homotopy type of X simply from the datum of the cohomology algebra of X. In particular, if X is simply connected one gets the following formula for homotopy groups

$$\pi_i(X)\otimes\mathbb{Q}\cong\pi_i\langle Q(H^*(X,\mathbb{Q}))\rangle$$

where $Q(H^*(X,\mathbb{Q}))$ denotes a cofibrant replacement of $H^*(X;\mathbb{Q})$ in the model category of CDGAs. It follows from this discussion that the determination of rational homotopy groups of a formal space becomes a purely algebraic computation.

Example 2.18

(1) The sphere $\mathbb{S}^n$ is formal, for all $n \geqslant 1$. Its cohomology is given by

$$H^*\left(\mathbb{S}^n;\mathbb{Q}\right)\cong\mathbb{Q}[x]/x^2 ,$$

where x is such that $|x| = n$. Let us introduce the following CDGA

$$\mathscr{M}_n = \begin{cases} \mathbb{Q}[u], & |u| = n \qquad d = 0 \quad \text{if } n \text{ is odd} \\ \mathbb{Q}[u, v], & \begin{matrix} |u| = n & du = 0 \\ |v| = 2n-1 & dv = u^2 \end{matrix} \quad \text{if } n \text{ is even} \end{cases}$$

There is a zig-zag of quasi-isomorphisms

$$\Omega^*_{PL}(\mathbb{S}^n) \xleftarrow[\sim]{f} \mathcal{M}_n \xrightarrow[\sim]{g} H^*(\mathbb{S}^n;\mathbb{Q})\,,$$

where f and g are defined as follows. Let us set $g(u) = x$ for all n and $g(v) = 0$ for n even. Since the cohomology of $\Omega^*_{PL}(\mathbb{S}^n)$ is given by $H^*(\mathbb{S}^n;\mathbb{Q})$, there exist a cocycle $\tilde{u} \in \Omega^n_{PL}(\mathbb{S}^n)$ such that $[\tilde{u}] = x$ for all n and $\tilde{v} \in \Omega^{2n-1}_{PL}(\mathbb{S}^n)$ such that $d\tilde{v} = \tilde{u}^2$ in the case where n even. One defines a quasi-isomorphism f by setting $f(u) = \tilde{u}$ and $f(v) = \tilde{v}$. Although the integral homotopy groups of spheres are still mostly unknown, the previous result provides a method to efficiently calculate their rational homotopy groups. If n is odd, this leads to

$$\pi_*(S^n)\otimes\mathbb{Q} \cong \begin{cases} \mathbb{Q} \text{ if } * = n \\ 0 \text{ otherwise} \end{cases},$$

and if n is even, we have

$$\pi_*(S^n)\otimes\mathbb{Q} \cong \begin{cases} \mathbb{Q} \text{ if } * = n, 2n-1 \\ 0 \text{ otherwise} \end{cases}.$$

(2) Complex projective space $\mathbb{CP}^n$, for all $n \geqslant 0$ are formal. The proof is very similar to the proof for even spheres.
(3) Lie groups are formal. The proof is similar to the one of spheres and generalizes to any simply connected spaces X whose rational cohomology is free as a graded algebra.

3 The Example of Compact Kähler Manifolds

There is a long tradition of using Hodge theory as a tool for proving formality results. This section focuses on the first result in this direction: the formality of compact Kähler manifolds established by P. Deligne, P. Griffiths, J. Morgan and D. Sullivan in [DGMS75]. This result can be stated as formality of some functors, through two theorems that are very much related: a contravariant version and a covariant one.

3.1 The Contravariant Version

Recall that a compact Kähler manifolds is a manifold with three mutually compatible structures: a complex structure, a Riemannian structure, and a symplectic structure. One important source of examples is given by smooth projective complex varieties. Those are Kähler by pulling back the Kähler structure of the complex projective space in which they embed. Complex projective spaces have a Kähler structure given by the Fubini-Study metric, see [Voi02, Section 3.3.2]. Let Käh be the category of compact Kähler manifolds.

Theorem 3.1 ([DGMS75, Main Theorem]) *The functor of differential forms*

$$\mathscr{E}^* : \text{Käh}^{op} \to \text{Ch}^*(\mathbb{R})$$

is a formal symmetric monoidal functor.

Proof Let M be a compact Kähler manifold. Since M is a complex manifold, its tangent bundle is equipped with an endomorphism $J : TM \to TM$ satisfying $J^2 = -id$. By dualizing J, it induces an endomorphism of the cotangent vector bundle and therefore an automorphism of the de Rham complex $\mathscr{E}^*(M)$. Thus, this complex is equipped with its usual differential denoted d, but also with another operator called d^c defined by $d^c = -JdJ$. These operators make $\mathscr{E}^*(X)$ into a bicomplex that moreover satisfies a lemma, called dd^c-lemma at the heart of the proof.

Lemma 3.2 ([DGMS75, dd^c-Lemma]) *If x is a differential form such that $dx = 0$ and $x = d^c y$, then $x = dd^c(z)$ for some z.*

Let $\mathscr{E}^*(M)$ be the real de Rham complex of M, ${}^c\mathscr{E}^*(M)$ be the subcomplex of d^c-closed forms, and $H^*_{d^c}(M)$ be the quotient complex ${}^c\mathscr{E}^*(M)/d^c(\mathscr{E}^*(M))$. Then we have a diagram of the form

$$(\mathscr{E}^*(M), d) \xleftarrow{i} ({}^c\mathscr{E}^*(M), d) \xrightarrow{\pi} (H^*_{d^c}(M), d)$$

where i corresponds to the inclusion of the subcomplex and π to the quotient. Let us first prove that i induces an isomorphism in cohomology. For all

$$[x] \in H^*(\mathscr{E}^*(M), d) ,$$

the form $d^c x$ satisfies the hypothesis of the dd^c-lemma. Thus, there exists an element y such that $d^c x = dd^c y$. Setting $z = x + dy$, we get $d^c(z) = 0$ and i induces a surjection in cohomology. Let $y \in {}^c\mathscr{E}^*(M)$ be a closed form which is exact in $\mathscr{E}^*(M)$. Then, we have $d^c y = 0 = dy$ and $y = dz$. Thus, there exists w such that $y = dd^c w$ and y is necessarily trivial. This shows that i_* is injective and thus, an isomorphism. Let us now prove that π induces an isomorphism in cohomology. For all

$$[y] \in H^*({}^c\mathscr{E}^*(M), d) ,$$

the element y is d^c-closed. Thus dy satisfies the hypothesis of the dd^c-lemma. There exists z such that $dy = dd^c z$. Setting $x = y + d^c z$, then $dx = 0$ and $[x] = [y]$. Thus, π is surjective. Finally, let y be such that $[y] = 0$ in $(H^*_{d^c}(M), d)$. Then $y = d^c(w)$ and by the dd^c-lemma, there exists z such that $y = dd^c z$ and π is injective.
Note that the differential induced by d on $H^*_{d^c}(M)$ is 0. Indeed, if $d^c y = 0$, then by the dd^c-lemma, there exists w such that $dy = dd^c w$ and $dy \in \text{Im}(d^c)$. We deduce that $[dy] = 0$ in $H^*_{d^c}(M)$. Thus, there exists an isomorphism

$$H^*_{d^c}(M) \cong H^*(\mathscr{E}^*(M)) .$$

Furthermore, since d^c satisfies the Leibniz rule, ${}^c\mathscr{E}^*$ inherits a monoidal symmetric functor structure from that of $\mathscr{E}^*$. Finally, since the morphisms i and π are natural and compatible with the structure of monoidal symmetric functors, we conclude that $\mathscr{E}^*$ is formal. □

The functor of de Rham forms $\mathscr{E}^*(-)$ on smooth differentiable manifolds is naturally quasi-isomorphic to the functor $\Omega^*_{PL}(-) \otimes_{\mathbb{Q}} \mathbb{R}$, see [GM13, Corollary 9.9]. This leads to the following corollary.

Corollary 3.3 ([DGMS75, Corollary 1]) *Let M a compact Kähler manifold. The real homotopy type of M, that is the homotopy type of the real CDGA*

$$\Omega^*_{PL}(M) \otimes_{\mathbb{Q}} \mathbb{R}\,,$$

is determined by the real cohomology algebra of M. In particular, if M is pointed and simply connected, there is an isomorphism

$$\pi_*(M) \otimes_{\mathbb{Z}} \mathbb{R} \cong \pi_*\langle Q(H^*(M, \mathbb{R}))\rangle$$

where $\langle - \rangle$ is the real version of Sullivan realization functor and Q denotes a cofibrant replacement in the model category of CDGAs.

In light of Definition 2.17, it is very natural to wonder to what extent formality depends on the coefficient ring. A formality result with coefficients in a certain field $\mathbb{K}$ will also be satisfied for any extension $\mathbb{L}$ of this field. Formality descent gives a partial converse to this result.

Theorem 3.4 (Formality Descent) *Let $\mathbb{L} \subset \mathbb{K}$ be two characteristic zero fields. Let A be a* CDGA *over $\mathbb{L}$ with finite type cohomology. The algebra A is formal if and only if $A \otimes_{\mathbb{L}} \mathbb{K}$ is formal.*

Proof This theorem is proved in [HS79, Corollary 6.9]. See also Theorem 8.19 in these notes for a somewhat different point of view. □

Remark 3.5 Two CDGAs over $\mathbb{L}$ may become quasi-isomorphic after extending the scalars to $\mathbb{K}$ without being quasi-isomorphic over $\mathbb{K}$. As a simple example, one can take the following two commutative algebras over $\mathbb{R}$,

$$A = \mathbb{C} \quad \text{and} \quad B = \mathbb{R} \times \mathbb{R}\,.$$

These two algebras are not isomorphic though we have isomorphisms

$$A \otimes_{\mathbb{R}} \mathbb{C} \cong \mathbb{C} \times \mathbb{C} \cong B \otimes_{\mathbb{R}} \mathbb{C}.$$

Using Theorem 3.4 and Corollary 3.3, one deduces the following result.

Corollary 3.6 *Let M a compact Kähler manifold. The rational homotopy type of M is determined by its cohomology as commutative graded algebras. In particular, if M is* *simply connected, there is an isomorphism*

$$\pi_*(M) \otimes_{\mathbb{Z}} \mathbb{Q} \cong \pi_*\langle Q(H^*(M, \mathbb{Q}))\rangle \ .$$

Remark 3.7 One issue with this corollary is that we have lost functoriality in the process of descending formality. For instance, if G is a discrete group acting on a compact Kähler manifold, it is not at all obvious that the isomorphism

$$\pi_*(M) \otimes_{\mathbb{Z}} \mathbb{Q} \cong \pi_*\langle Q(H^*(M, \mathbb{Q}))\rangle$$

is an isomorphism of representations of G. It is however be an isomorphism of G-representations after extending the scalars to $\mathbb{R}$ thanks to the functoriality of the Deligne–Griffiths–Morgan–Sullivan Theorem. As we shall see later in these notes, functoriality does hold over $\mathbb{Q}$ but requires a different argument (see Theorem 6.22).

3.2 The Covariant Version

In the paper [GSNPR05], the authors elaborate on the method of [DGMS75] and prove that operads (as well as cyclic operads, modular operads, etc.) internal to the category of compact Kähler manifolds are formal. They establish the following covariant version to Theorem 3.1.

Theorem 3.8 ([GSNPR05, Corollary 3.2.3]) *The functor of singular chains*

$$C_*(-; \mathbb{R}) : \text{Käh} \to \text{Ch}_*(\mathbb{R})$$

is a formal lax symmetric monoidal functor.

Remark 3.9 The lax symmetric monoidal structure on the functor $C_*(-; \mathbb{R})$ comes from the shuffle product

$$\delta_{X,Y} : C_*(X; \mathbb{R}) \otimes C_*(Y; \mathbb{R}) \to C_*(X \times Y; \mathbb{R})$$

and the obvious unit morphism, see [EML53, Theorem 5.2].

Let M be a differentiable manifold. Consider $\mathscr{E}^*(M)$, the complex of differential forms. We can make it into a locally convex topological vector space by giving it the topology of compact convergence for all derivatives of forms. We can then consider its topological dual, denoted $\mathscr{E}'_*(M)$ equipped with the strong topology. An element of $\mathscr{E}'_*(M)$ is called a de Rham current with compact support. We thus have a covariant functor

$$\mathscr{E}'_* : \text{Dif} \to \text{Ch}_*(\mathbb{R}) \ ,$$

where Dif denotes the category of differentiable manifolds and smooth maps. This functor has a symmetric monoidal structure inherited from the wedge product of differential forms.

More precisely, given $S \in \mathscr{E}_*(M)$ and $T \in \mathscr{E}_*(N)$, we define $\kappa(S \otimes T) \in \mathscr{E}_*(M \times N)$ by the formula

$$\langle \kappa(S \otimes T), \pi_M^*(\omega) \wedge \pi_N^*(\nu) \rangle = \langle S, \omega \rangle \cdot \langle T, \nu \rangle$$

valid for all $\omega \in \mathscr{E}^*(M)$ and $\nu \in \mathscr{E}^*(N)$, where π_M and π_N denote the projections on M and N respectively. It turns out that this formula is sufficient for defining a current as the map

$$\begin{aligned} \mathscr{E}^*(M) \otimes \mathscr{E}^*(N) &\longrightarrow \mathscr{E}^*(M \times N) \\ (\omega, \nu) &\longmapsto \pi_M^*(\omega) \wedge \pi_N^*(\nu) \end{aligned}$$

has dense image.

Theorem 3.10 ([GSNPR05, Proposition 2.4.1]) *The two functors*

$$C_*, \mathscr{E}_* : \mathrm{Dif} \to \mathrm{Ch}_*(\mathrm{R})$$

are weakly equivalent symmetric monoidal functors.

Proof The proof is dual to de Rham's theorem giving a quasi-isomorphism between the de Rham complex $\mathscr{E}^*$ and the functor of singular cochains. For every differentiable manifold M, let $C_*^\infty(M; \mathbb{Z})$ be the sub-complex of singular chains generated by the $\mathscr{C}^\infty$-maps. The shuffle product of $\mathscr{C}^\infty$-singular chains is also a $\mathscr{C}^\infty$-singular chain. We obtain a symmetric monoidal functor

$$C_*^\infty : \mathrm{Dif} \to \mathrm{Ch}_*(\mathrm{R})\,.$$

Let M be a differentiable manifold and let $c : \Delta^p \to M$ be a $\mathscr{C}^\infty$-singular simplex. By Stokes theorem, integration along c induces a morphism

$$C_*^\infty(M) \to \mathscr{E}_*(M), \quad c \mapsto \int_c$$

defined by $\int_c \omega := \int_{\Delta^p} c^*(\omega)$, see [Bre93, Chapter V. 5]. De Rham theorem's implies that integration induces an isomorphism in homology. Thus

$$\int : C_*^\infty \to \mathscr{E}_*$$

is a weak equivalence between functors, see [Bre93, Chapter V. Theorem 9.1]. To conclude that this is a weak monoidal equivalence, one needs to verify that integration is compatible with the monoidal structure. We refer the reader to [GSNPR05, Proposition 2.4.1] for more details about this. Furthermore, the natural inclusion of $\mathscr{C}^\infty$-singular chains in the singular chains defines a symmetric monoidal natural transformation $C_*^\infty \to C_*$, since the structures of symmetric monoidal functors are both defined with the shuffle product. Finally, the inclusion induces an isomorphism in homology, see for example [Bre93,

Page 291]. By composing the two weak equivalences, we get that C_* and $\mathscr{E}'_*$ are weakly equivalent symmetric monoidal functors. □

Remark 3.11 There is a small mistake in [GSNPR05]. Instead of $\mathscr{E}'_*$, the authors use the functor of de Rham currents $\mathscr{D}'_*$ which is the topological dual of the functor of de Rham differential forms with compact support. The issue is that this functor is not quasi-isomorphic to singular chains but instead computes Borel-Moore homology of the manifold. However, the restriction of the two functors $\mathscr{E}'_*$ and $\mathscr{D}'_*$ to compact manifolds are isomorphic and [GSNPR05] only applies the previous to compact manifolds.

Since C_* and $\mathscr{E}'_*$ are two weakly equivalent symmetric monoidal functors, Theorem 3.8 is a consequence of the following.

Theorem 3.12 *The functor of currents* $\mathscr{E}'_* : \text{Käh} \to \text{Ch}_*(\text{R})$ *is a formal symmetric monoidal functor.*

Proof The proof is essentially the same than the one of Theorem 3.1. One can prove that the Kähler identities between the operators $d, d^c, \Delta, \ldots$ of the de Rham complex of differential forms are also satisfied by the corresponding dual operators on the de Rham complex of currents. Thus, we have a similar version of the dd^c-lemma that holds for the complex $\mathscr{E}'_*$. Let M be a compact Kähler manifold. Let ${}^c\mathscr{E}'_*(M)$ be the subcomplex of $\mathscr{E}'_*(M)$ defined by the d^c-closed currents and let us denote

$$H_*^{d^c}(M) := {}^c\mathscr{E}'_*(M)/d^c(\mathscr{E}'_*(M)) .$$

As in the proof of Theorem 3.1, we obtain a diagram

$$(\mathscr{E}'_*(M), d) \xleftarrow{i} ({}^c\mathscr{E}'_*(M), d) \xrightarrow{\pi} (H_*^{d^c}(M), d) ,$$

where i corresponds to the inclusion and π to the quotient. Both maps are weak equivalences and the differential induced by d on the quotient is zero (using dd^c-lemma). Since the morphisms i and π are morphisms of lax monoidal functors, we conclude that $\mathscr{E}'_*$ is formal. □

Corollary 3.13 *If* $\mathscr{O}$ *is an operad in* Käh *then* $C_*(\mathscr{O}, \mathbb{R})$ *is formal.*

Proof This follows directly from Theorem 3.8 and Proposition 2.13. □

Example 3.14 (Moduli Spaces of Stable Algebraic Curves of Genus 0) Let

$$\mathscr{O} = \{\overline{\mathscr{M}}_{0,l}\}_l$$

be the (cyclic) operad of moduli spaces of stable algebraic curves of genus 0, defined as follows. We denote by $\mathscr{M}_{0,l}$ the moduli space of ℓ-tuples, $(x_1, \ldots, x_\ell)$, of distinct points of the complex projective line $\mathbb{CP}^1$ modulo projective automorphisms; that is the transformations of the form

$$\mathbb{CP}^1 \to \mathbb{CP}^1, \quad [\xi_1, \xi_2] \mapsto [a\xi_1 + b\xi_2, c\xi_1 + d\xi_3]$$

for $a, b, c, d \in \mathbb{C}$ such that $ad - bc \neq 0$. The space $\overline{\mathcal{M}}_{0,l}$, originally defined in [DM69], corresponds to a certain compactification of $\mathcal{M}_{0,l}$ and is the moduli space of stable algebraic curves of genus 0 with ℓ-marked points. Briefly, a stable curve is a curve that can have nodal singularities but that has a finite group of automorphisms (this imposes that each component of the curve has at least 3 points that are either nodal points or marked points). The operations of gluing two stable curves along marked points turns the collection of spaces $\overline{\mathcal{M}}_{0,l}$ into a cyclic operad $\mathcal{O}$ in the category of smooth projective varieties, so in particular in the category of compact Kähler manifolds. Using the previous theorem, one can conclude that $C_*(\mathcal{O}, \mathbb{R})$ is formal.

Remark 3.15 The above discussion generalizes to the modular operad obtained by taking moduli spaces of stable curves of all genera. The only issue is that the higher genus moduli spaces are not compact Kähler manifolds anymore. However, they are Deligne-Mumford stacks and the technology that we have just explained extends to this context. In [GSNPR05], the authors are able to prove that this modular operad is formal.

4 Purity Implies Formality

The aim of this section is to give an equivalent characterization of formality of symmetric monoidal functors in terms of weight decompositions. We are going to prove that if a certain condition called *purity* is satisfied, then formality is guaranteed.

4.1 An Equivalent Definition of Formality

Let $\mathbf{k}$ be a field. We denote by grVect the category of graded vector spaces over $\mathbf{k}$. This abelian category inherits a symmetric monoidal structure from that of $\mathbf{k}$-vector spaces (the symmetry isomorphism does not involve any sign).

Every chain complex in grVect has two gradings. One comes from the underlying category grVect and is called "weight". We will denote it with a superscript. The other one corresponds to the homological grading. We will keep the term "degree" for this grading denoted with a subscript. The category $\mathrm{Ch}_*(\mathrm{grVect})$ also inherits a symmetric monoidal structure given by

$$(C \otimes D)^p_m = \bigoplus_{n,q \in \mathbb{Z}} C^q_n \otimes D^{p-q}_{m-n}$$

The symmetry isomorphism involves the usual Koszul sign for the homological degree but not for the weight.

Definition 4.1 An object V of $\mathrm{Ch}_*(\mathrm{grVect})$ has *pure homology* if

$$H_n(V)^p = 0 \quad \text{for all } p \neq n.$$

Denote by $\mathrm{Ch}_*(\mathrm{grVect})^{pure}$ the full subcategory of $\mathrm{Ch}_*(\mathrm{grVect})$ consisting of chain complexes with pure homology.

The following proposition is straightforward exercise of linear algebra. A proof can be found in [CH20b, Proposition 2.7]. Associated to Proposition 2.14, it is the basis for many formality results that we will establish.

Proposition 4.2 *The forgetful functor defined by forgetting the weight*

$$U : \mathrm{Ch}_*(grVect)^{pure} \to \mathrm{Ch}_*(\boldsymbol{k})$$

is formal as lax symmetric monoidal functor.

From this, one can derive the following formality criterion.

Proposition 4.3 *A symmetric monoidal functor $F : \mathscr{C} \to \mathrm{Ch}_*(\boldsymbol{k})$ is formal if and only if it is weakly equivalent to a functor $\tilde{F}$ that admits a factorisation,*

$$\begin{array}{ccc} & & \mathrm{Ch}_*(grVect)^{pure} \\ & \overset{G}{\nearrow} & \big\downarrow U \\ \mathscr{C} & \xrightarrow[\tilde{F}]{} & \mathrm{Ch}_*(\mathbf{k}) \end{array}$$

Proof If F is formal, it is weakly equivalent to the functor $H \circ F$ which splits as a direct sum $\oplus_n H_n \circ F$. This splitting satisfies the desired condition. Conversely, if $\tilde{F}$ admits such a factorization, then $\tilde{F}$ is formal using the previous proposition and Proposition 2.14. □

The following Proposition gives a method for producing pure weight gradings.

Proposition 4.4 *A symmetric monoidal functor $F : \mathscr{C} \to \mathrm{Ch}_*(\boldsymbol{k})$ is formal if it is weakly equivalent to a functor $\tilde{F}$ such that*

(1) *$\tilde{F}$ is object-wise and degree-wise finite dimensional.*
(2) *$\tilde{F}$ has an endomorphism σ which acts as multiplication by λ^n on $H_n(\tilde{F}(c))$ for all $n \geqslant 0$ and $c \in Ob(\mathscr{C})$, where $\lambda \in \boldsymbol{k}$ is a unit of infinite order.*

Proof Since the property of being formal is stable under weak equivalences, it is sufficient to check that $\tilde{F}$ is formal. This functor admits a sub-complex given by the direct sum

$$i : \bigoplus_{n \in \mathbb{Z}} \tilde{F}^n \hookrightarrow \tilde{F}$$

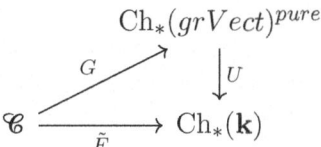

where each $\tilde{F}^n$, for $n \geqslant 0$, is the corresponding generalized eigenspace for the eigenvalue λ^n. Note that this is a sub-complex since the differential has to preserve generalized eigenspaces. The inclusion i is a lax symmetric monoidal functor and a quasi-isomorphism since the other generalized eigenspaces will not contribute to homology because of condition (2). Then, the sub-functor

$$G := \bigoplus_{n\in\mathbb{Z}} \tilde{F}^n$$

has a canonical weight grading by construction, and admits a factorisation as in Proposition 4.3. This implies that G and hence $\tilde{F}$ and F are formal. □

4.2 The Formality of the Little Disks Operad

The original proof of formality of the little disks operad $\mathscr{D}_2$ is due to Kontsevich [Kon99] and Tamarkin [Tam98] independently. In this section, we present another proof relying on Proposition 4.4 which is due to Petersen [Pet14].

Let **PaB** be the operad parenthesized braids. This is the operad in groupoids such that $\mathbf{PaB}(n)$ are parenthesized permutations of $\{1, \ldots n\}$ and morphisms are braids on n strands maintaining the same label at the start and at the end of each strand. We refer the reader to [CRiL24, Section 2.3] in this volume for more details. We also consider its $\mathbb{Q}$-pro-unipotent completion, denoted $\widehat{\mathbf{PaB}}_{\mathbb{Q}}$. The $\mathbb{Q}$-pro-algebraic Grothendieck-Teichmüller group is defined as

$$\widehat{GT}_{\mathbb{Q}} := \mathrm{Aut}^{+}_{\mathrm{OpGrpd}}\left(\widehat{\mathbf{PaB}}_{\mathbb{Q}}\right) .$$

where Aut^+ denotes the group of automorphisms in the category of operads in groupoids that induce the identity map on the objects of the groupoid in each arity. The operad **PaB** is weakly equivalent to the operad of fundamental groupoids of $\mathscr{D}_2$. Since $\mathscr{D}_2(n)$ is a $K(\pi, 1)$-space, we obtain a weak equivalence

$$\mathscr{D}_2 \simeq \mathrm{B}(\mathbf{PaB})$$

where B denotes the classifying space functor. It turns out that the configuration spaces are also rational $K(\pi, 1)$, see [PY99]. This implies that the weak equivalence above induces an equivalence

$$(\mathscr{D}_2)_{\mathbb{Q}} \simeq \mathrm{B}\widehat{\mathbf{PaB}}_{\mathbb{Q}} .$$

From this, we obtain a quasi-isomorphism of differential graded operads

$$C_*(\mathscr{D}_2; \mathbb{Q}) \cong C_*\left(\mathrm{B}\widehat{\mathbf{PaB}}_{\mathbb{Q}}, \mathbb{Q}\right) .$$

The natural action of $\widehat{GT}_{\mathbb{Q}}$ on $\widehat{\mathbf{PaB}}_{\mathbb{Q}}$ extends to an action on the right-hand side. Recall that the homology of the little two-disks operad is the Gerstenhaber operad. The induced action of $\widehat{GT}_{\mathbb{Q}}$ on $H_*(\mathscr{D}_2)$ is given

$$\begin{array}{ccc} \widehat{GT}_{\mathbb{Q}} \times H_n(\mathscr{D}_2) & \longrightarrow & H_n(\mathscr{D}_2) \\ (\sigma, a) & \longmapsto & \sigma \cdot a = \chi(\sigma)^n a \end{array}$$

where $\chi : \widehat{GT}_{\mathbb{Q}} \to \mathbb{Q}^\times$ is the cyclotomic character. The map χ is surjective by Drinfeld [Dri90, Section 5]. Let $\alpha \in \mathbb{Q}^\times$ of infinite order. By surjectivity, there is a lift

$$\tilde{\alpha} \in \widehat{GT}_{\mathbb{Q}}$$

such that $\chi(\tilde{\alpha}) = \alpha$. This lift induces an endomorphism of

$$C_*\left(\mathrm{B}\widehat{\mathbf{PaB}}_{\mathbb{Q}}, \mathbb{Q}\right)$$

which acts by multiplication by α^n on the homology group of degree n for all $n \in \mathbb{N}$. The dg operad at stake is not finite dimensional in each degree and arity. Let us consider

$$M \xrightarrow{\sim} C_*\left(\mathrm{B}\widehat{\mathbf{PaB}}_{\mathbb{Q}}, \mathbb{Q}\right)$$

to be a minimal model as chain operad. This can easily be shown to be finite dimensional so it satisfies the condition (1) of the previous proposition. This minimal model inherits an endomorphism $\tilde{\alpha}$ satisfying the condition (2) of the proposition 4.4 and this leads to the desired formality result.

Remark 4.5 A similar argument was used by Boavida de Brito and the second author in order to prove formality of the higher dimensional little disks operads in [BdBH21]. The original formality argument is also due to Kontsevich with a detailed proof by Lambrechts-Volić (see [Kon99, LV14]).

Remark 4.6 We expand a little bit about a subtle point of this section. The fact that configuration spaces of points in $\mathbb{R}^2$ are rational $K(\pi, 1)$-spaces is non-trivial. It can happen that the rationalization of a $K(\pi, 1)$ space has non-zero higher homotopy groups. As an example, we can consider the classifying space of the infinite linear group of the integers $BGL_\infty(\mathbb{Z})$. This is of course a $K(\pi, 1)$-space but, the canonical map to Quillen's plus construction

$$BGL_\infty(\mathbb{Z}) \to BGL_\infty(\mathbb{Z})^+$$

induces an isomorphism in homology with coefficient in $\mathbb{Z}$ and hence also with coefficients in $\mathbb{Q}$. It follows that the two spaces have the same rational homotopy type. But the higher homotopy groups of the rationalization of $BGL_\infty(\mathbb{Z})^+$ are, by definition, the rationalization of the higher K-groups of $\mathbb{Z}$. Those are known to be non-trivial in degree $4k + 1$ for each $k > 0$, by work of Borel, see [Bor74, Proposition 12.2].

5 Interlude: Infinity Categories

The Sects. 6 and 7 uses the formality criteria of Proposition 4.3 in order to deduce formality. To do so, it will be convenient to use the extra flexibility provided by working with ∞-categories. The aim of this section is to introduce some concepts related to this subject.

5.1 *Classical Infinity Categories*

Definition 5.1 (Nerve of a Category) The *nerve* of a given category $\mathscr{C}$ is the simplicial set $\mathrm{N}(\mathscr{C})$ defined by

$$\mathrm{N}(\mathscr{C})_n = \mathrm{Fun}([n], \mathscr{C}) .$$

This definition extends to a fully faithful functor $\mathrm{N} : \mathrm{Cat} \to \mathrm{sSet}$.

Example 5.2 The nerve of the category $[n]$ is the standard n-simplex, i.e.

$$\mathrm{N}([n]) \cong \Delta^n ,$$

for all $n \geqslant 0$. Recall that the k-th horn $\Lambda^n_k \subset \partial\Delta^n$ for $0 \leqslant k \leqslant n$, is obtained from $\partial\Delta^n$ by removing the k-th face $\partial_k\Delta^n$

Definition 5.3 (Infinity Categories) A simplicial set $\mathscr{C}$ is an ∞-*category* if every inner horn

$$\Lambda^n_k \to \mathscr{C} ,$$

for $0 < k < n$ can be extended to an n-simplex $\Delta^n \to \mathscr{C}$.

Example 5.4 The nerve $\mathrm{N}(\mathscr{C})$ of a category $\mathscr{C}$ is an infinity category. Through its nerve, any category can be seen as an ∞-category. From this example, we see that the 0-simplices of an ∞-category should be thought of as objects and the 1-simplices as morphisms. In addition, every extension of an inner horn $\Lambda^2_1 \to \mathrm{N}(\mathscr{C})$ to a two simplex $\Delta^2 \to \mathrm{N}(\mathscr{C})$ corresponds to the composition of morphisms. The higher dimensional inner horn extensions witness higher coherence in the associativity of compositions.

Definition 5.5 (Functors Between Infinity Categories) A *functor* between two ∞-categories $\mathscr{C}$ and $\mathscr{D}$ is a morphism of simplicial sets $\mathscr{C} \to \mathscr{D}$. We denote by $\mathrm{Fun}(\mathscr{C}, \mathscr{D})$ the simplicial set of functors from $\mathscr{C}$ to $\mathscr{D}$. This is an ∞-category and we shall refer to it as the ∞-category of functors from $\mathscr{C}$ to $\mathscr{D}$.

Definition 5.6 (Homotopy Category) Let $\mathscr{C}$ an ∞-category. Let $f, g : x \to y$ be two morphisms in $\mathscr{C}$. They are said *homotopic* if there is a 2-simplex $\sigma : \Delta^2 \to \mathscr{C}$ with boundary $\partial\sigma$ corresponding to (g, f, id_x). The homotopy relation is an equivalence relation on the set of 1-simplices with faces x and y. Furthermore, there is an ordinary category $\mathrm{Ho}(\mathscr{C})$, the *homotopy category* of $\mathscr{C}$, with the same objects as $\mathscr{C}$ and morphisms the homotopy classes of morphisms in $\mathscr{C}$. We say that a 1-simplex of an ∞-category is an *equivalence* if it induces an isomorphism in the homotopy category.

Remark 5.7 This homotopy category $\mathrm{Ho}(\mathscr{C})$ does not capture all the homotopical information contained in $\mathscr{C}$. This can be enriched by constructing a *mapping simplicial set* (usually called mapping space) denoted $\mathrm{map}_{\mathscr{C}}(x, y)$ for each pair of objects x and y. These mapping spaces can be composed in a coherent homotopy associative way. The homotopy category of $\mathscr{C}$ is then simply the category obtained by applying π_0 to each of the mapping spaces.

Definition 5.8 A functor $f : \mathscr{C} \to \mathscr{D}$ between ∞-categories is called an *equivalence* if it induces an equivalence of categories

$$\mathrm{Ho}(\mathscr{C}) \to \mathrm{Ho}(\mathscr{D})$$

and weak equivalences of simplicial sets

$$\mathrm{map}_{\mathscr{C}}(x, y) \to \mathrm{map}_{\mathscr{D}}(f(x), f(y))$$

for any choice of x and y two objects of $\mathscr{C}$.

Definition 5.9 ([Lur17, Definition 1.3.4.1]) Let $\mathscr{C}$ and $\mathscr{D}$ be two ∞-categories and W be a collection of morphisms in $\mathscr{C}$. We say that a morphism $f : \mathscr{C} \to \mathscr{D}$ *exhibit* $\mathscr{D}$ *as the localization of* $\mathscr{C}$ *with respects to* W if, for every ∞-category $\mathscr{E}$, the composition with f induces a fully faithful embedding

$$\mathrm{Fun}(\mathscr{D}, \mathscr{E}) \to \mathrm{Fun}(\mathscr{C}, \mathscr{E}),$$

whose essential image is the collection of functors $F : \mathscr{C} \to \mathscr{E}$ which carry each morphism in W to an equivalence in $\mathscr{E}$. In this case, the ∞-category $\mathscr{D}$ is determined uniquely up to equivalence by $\mathscr{C}$ and W, and is denoted by

$$\mathscr{C}[W^{-1}]\,.$$

If $\mathscr{C}$ is an ordinary category, we denote this localization by $N_W(\mathscr{C})$.

Our main tool for constructing ∞-categories will be the following theorem.

Theorem 5.10 ([Hin16]) *Let $\mathscr{C}$ be an ∞-category and W be a collection of morphism in $\mathscr{C}$. The ∞-category $\mathscr{C}[W^{-1}]$ exists. In particular, when $\mathscr{C}$ is an ordinary category, the homotopy category* $\mathrm{Ho}(N_W(\mathscr{C}))$ *is the one-categorical localization of $\mathscr{C}$ with respect to the maps of W.*

5.2 Symmetric Monoidal ∞-Categories

Let $\mathbf{Fin}_*$ be the category of based finite sets. For every object I of $\mathbf{Fin}_*$, we may choose a pointed bijection

$$\alpha : I \cong \langle n \rangle := \{*, 1, \cdots, n\},$$

where $n + 1$ is the cardinality of I. Using these isomorphisms, we may identify $\mathbf{Fin}_*$ with its full subcategory spanned by the objects $\langle n \rangle$ and we shall do so implicitly. For every pair of integers $1 \leqslant i \leqslant n$, we let $\rho^i : \langle n \rangle \to \langle 1 \rangle$ denote the morphism given by $\rho^i(j) = 1$ if $i = j$ and $*$ otherwise.

Definition 5.11 A *symmetric monoidal* ∞*-category* is a coCartesian fibration of simplicial sets $p : \mathscr{C}^{\otimes} \to \mathrm{N}(\mathbf{Fin}_*)$ satisfying the following Segal condition. Let $\mathscr{C}^{\otimes}_{\langle n \rangle}$ denote the fiber of p over the object $\langle n \rangle$. For each $n \geqslant 0$, the maps

$$\{\rho^i : \langle n \rangle \to \langle 1 \rangle\}_{1 \leqslant i \leqslant n}$$

induce functors $\rho^i_! : \mathscr{C}^{\otimes}_{\langle n \rangle} \to \mathscr{C}^{\otimes}_{\langle 1 \rangle}$ assembling into a map

$$\mathscr{C}^{\otimes}_{\langle n \rangle} \to (\mathscr{C}^{\otimes}_{\langle 1 \rangle})^n.$$

The Segal condition is requiring that this map is an equivalence of ∞-categories.

Definition 5.12 A *strong symmetric monoidal functor* between ∞-categories is a simply a morphism of coCartesian fibrations. A *lax symmetric monoidal functor* is a map that is only required to preserve certain cocartesian morphisms (the so-called "inert morphisms").

Our two main examples are the following.

Example 5.13 Let $(\mathscr{C}, \otimes, \mathbf{1})$ be a symmetric monoidal category.

(1) In [Lur17, Construction 2.0.0.1] Lurie constructs a symmetric monoidal ∞-category

$$\mathrm{N}(\mathscr{C})^{\otimes} \to \mathrm{N}(\mathbf{Fin}_*)$$

whose fiber $\mathrm{N}(\mathscr{C})_{\langle 1 \rangle}$ is identified with $\mathrm{N}(\mathscr{C})$. This way, ordinary symmetric monoidal categories can be seen as symmetric monoidal ∞-categories.

(2) Suppose that $\mathscr{C}$ is equipped with a collection of weak equivalences W. Assume that for all $X \in \mathscr{C}$ all $f : A \to B \in W$, the induced map $X \otimes A \to X \otimes B$ is in W, then $\mathrm{N}_W\mathscr{C}$ inherits a structure of symmetric monoidal ∞-category (see [Hin15, Proposition 3.2.2]) that we shall denote by $N_W(\mathscr{C})^{\otimes}$.

Definition 5.14 Let $(\mathscr{A}, \otimes, \mathbf{1})$ be an abelian symmetric monoidal category with infinite direct sums. We denote by $\mathbf{Ch}_*(\mathscr{A})$ the ∞-category obtained from $\mathrm{Ch}_*(\mathscr{A})$ by inverting the quasi-isomorphisms :

$$\mathbf{Ch}_*(\mathscr{A}) := \mathrm{N}_W \mathrm{Ch}_*(\mathscr{A}),$$

where W is the class of all quasi-isomorphisms.

Assume now that the tensor product of $\mathscr{A}$ is exact in each variables. Let $X \in \mathrm{Ch}_*(\mathscr{A})$ and $f : A \to B$ a quasi-isomorphism. Then the induced map $X \otimes A \to X \otimes B$ is also a quasi-isomorphism. Example 5.13 implies that $\mathbf{Ch}_*(\mathscr{A})$ inherits a structure of symmetric monoidal ∞-category that we shall denote $\mathbf{Ch}_*(\mathscr{A})^{\otimes}$.

Definition 5.15 Let $\mathscr{C}$ be a symmetric monoidal category and

$$F : \mathrm{N}(\mathscr{C}) \to \mathbf{Ch}_*(\mathscr{A})$$

a lax symmetric monoidal functor in the ∞-categorical sense. We say that F is a *formal symmetric monoidal ∞-functor* if F and $H \circ F$ are equivalent in the ∞-category of symmetric monoidal functors from $\mathrm{N}(\mathscr{C})$ to $\mathbf{Ch}_*(\mathscr{A})$.

Remark 5.16 Clearly, a formal symmetric monoidal functor $\mathscr{C} \to \mathrm{Ch}_*(\mathscr{A})$ induces a formal symmetric monoidal ∞-functor $\mathrm{N}(\mathscr{C}) \to \mathbf{Ch}_*(\mathscr{A})$. The following theorem due to Hinich gives a partial converse.

Theorem 5.17 ([Hin15]) *Let $\mathscr{C}$ be a small symmetric monoidal category and let $\boldsymbol{k}$ be a characteristic zero field. If two symmetric monoidal functors*

$$F, G : \mathscr{C} \to \mathrm{Ch}_*(\boldsymbol{k})$$

are equivalent as symmetric monoidal ∞-functors $\mathrm{N}(\mathscr{C}) \to \mathbf{Ch}_*(\boldsymbol{k})$, they are weakly equivalent as symmetric monoidal functors.

Corollary 5.18 *Let $\boldsymbol{k}$ be a characteristic zero field. Let $\mathscr{C}$ be a small symmetric monoidal category. Let $F : \mathscr{C} \to \mathrm{Ch}_*(\boldsymbol{k})$ be a symmetric monoidal functor. If F is formal as a symmetric monoidal ∞-functor $\mathrm{N}(\mathscr{C}) \to \mathbf{Ch}_*(\boldsymbol{k})$, then F is formal as a symmetric monoidal functor.*

6 Mixed Hodge Structures

In paper [CH20b], Cirici and the second author use mixed Hodge theory to produce decompositions for the singular chains functor and dually for Sullivan's functor in order to deduce formality. The purpose of this section is to explain these results. We denote by $\mathrm{Var}_{\mathbb{C}}$ the category of complex schemes that are reduced, separated and of finite type. We will use the word variety for an object of this category.

6.1 The Definition of Mixed Hodge Structures

We start by recalling what mixed Hodge structures are as well as their properties.

Definition 6.1 (Pure Hodge Structures) A *pure Hodge structure* over $\mathbb{Q}$ of weight n is a finite dimensional $\mathbb{Q}$-vector space V together with a decomposition of its complexification into a finite direct sum of complex subspaces

$$V \otimes_{\mathbb{Q}} \mathbb{C} = \bigoplus_{p=-\infty}^{+\infty} U_{p,n-p}, \quad \text{such that } \overline{U}_{p,n-p} = U_{n-p,p} .$$

Remark 6.2 An equivalent definition is obtained by replacing the direct sum decomposition of $V \otimes_{\mathbb{Q}} \mathbb{C}$ with the *Hodge filtration*, a finite decreasing filtration of $H := V \otimes_{\mathbb{Q}} \mathbb{C}$ by complex subspaces $F^p H$ for $p \in \mathbb{Z}$, subject to the condition

$$F^p H \oplus \overline{F^{n+1-p} H} = H$$

for all $p \in \mathbb{Z}$. The relation between these two descriptions is given by

$$U_{p,q} = F^p H \cap \overline{F^q H} \quad \text{and} \quad F^p H = \bigoplus_{i \geqslant p} U_{i,n-i} .$$

Definition 6.3 (Mixed Hodge Structures) A *mixed Hodge structure* on a finite dimensional $\mathbb{Q}$-vector space V is given by

(1) a finite increasing filtration W of V, called the weight filtration;
(2) a decreasing filtration F on $H := V \otimes_{\mathbb{Q}} \mathbb{C}$, called the Hodge filtration;

such that for all $m \geqslant 0$, each $\mathbb{Q}$-vector space

$$Gr_W^m V := W_m V / W_{m-1} V$$

is a pure Hodge structure with respect to the filtration induced by F on

$$Gr_W^m V \otimes_{\mathbb{Q}} \mathbb{C} .$$

Morphisms of mixed Hodge structures are given by morphisms $f : V \to V'$ of $\mathbb{Q}$-vector spaces that are compatible with filtrations. We denote by $\mathrm{MHS}_{\mathbb{Q}}$ the category of mixed Hodge structures over $\mathbb{Q}$.

Remark 6.4 Mixed Hodge structures form an abelian category by Deligne [Del71, Theorem 2.3.5]. Deligne shows that the morphisms are necessarily strictly compatible with both filtrations. The kernels and cokernels in this category coincide with the usual kernels and cokernels in the category of vector spaces, with the induced filtrations. Moreover, there is a symmetric monoidal structure on the category of mixed Hodge structures given by the usual tensor product of underlying vector spaces equipped with the induced filtrations.

The theory of mixed Hodge structures on the cohomology of algebraic varieties was introduced by Deligne in 1970s (see [Del70, Del71, Del74]).

Theorem 6.5 (Deligne) *Let X be an algebraic variety over $\mathbb{C}$ and $n \geqslant 0$.*

(1) *The cohomology group $H^n(X; \mathbb{Q})$ carries a canonical mixed Hodge structure.*
(2) *This structure is functorial and compatible with the Künneth isomorphism.*
(3) *If X is smooth and proper, then the mixed Hodge structure of $H^n(X; \mathbb{Q})$ is pure of weight n.*

Ideas of the Proof First, the cohomology with complex coefficients is given by the hypercohomology of the holomorphic de Rham complex :

$$H^n(X; \mathbb{C}) = \mathbb{H}^n(X; \Omega^*) ,$$

see [Gro66]. In the smooth and proper case, the Hodge filtration is then the filtration induced by the so-called stupid filtration on this complex of sheaves,

$$F^p H^n(X; \mathbb{C}) = \mathrm{im}\left(\mathbb{H}^n\left(X; \Omega^{*\geqslant p}\right) \to \mathbb{H}^n\left(X; \Omega^*\right)\right) .$$

It can then be shown using Hodge theory that this filtration is indeed a pure Hodge structure of weight n. In the smooth case, we can proceed as follows for the weight filtration. Using Nagata's compactification theorem and Hironaka's theorem on resolution of singularities, the variety X can be embedded in a smooth complete variety $\overline{X}$ so that the complement $\overline{X} - X$ is a normal crossing divisor. This means that the inclusion $X \subseteq \overline{X}$ is locally a union of coordinate hyperplanes in $\mathbb{C}^n$. Considering the Leray-Serre spectral sequence associated to the inclusion $X \subset \overline{X}$, we obtain a spectral sequence which will converge to the cohomology of X. The E^1-page of this spectral sequence looks as follows

$$E_1^{-s,t} = H^{t-2s}\left(D^{(s)}, \mathbb{Q}\right)$$

where $D^{(s)}$ is the disjoint union of all s-fold intersections of components of the divisor. The groups $H^{t-2s}(D^{(s)}; \mathbb{Q})$ correspond to the cohomology of smooth and proper varieties so they carry pure Hodge structures. As with any spectral sequence, we get a filtration on the target whose associated graded is the page E_∞ of the spectral sequence. In this particular case, the filtration on $H^n(X; \mathbb{Q})$ is, up to a shift, the weight filtration. Note that the associated graded of this filtration will be sub-quotients of cohomology groups of smooth and proper varieties so they carry pure Hodge structures. The functoriality of this construction is not obvious since the compactifications are not functorial. However, it turns out that the E^2 page of the Leray-Serre spectral sequence is natural (although it is not the case for E^1). □

Let FVect_K be the category of filtered vector spaces over a certain field K. The weight filtration induces a functor

$$W : \mathrm{MHS}_{\mathbb{Q}} \longrightarrow \mathrm{FVect}_{\mathbb{Q}} .$$

Theorem 6.6 ([CH20b, Lemma 4.4]) *The weight filtration of* $\mathrm{MHS}_{\mathbb{Q}}$ *naturally splits over* $\mathbb{Q}$ *as a strong symmetric monoidal functor, i.e. the diagram*

$$\begin{array}{ccc} & & \mathrm{grVect}_{\mathbb{Q}} \\ & \overset{G}{\nearrow} & \downarrow T \\ \mathrm{MHS}_{\mathbb{Q}} & \underset{W}{\longrightarrow} & \mathrm{FVect}_{\mathbb{Q}} \end{array}$$

where $G(V)^p = Gr_p^W(V)$ *and* T *is the totalization functor*

$$W^p(T(V)) = \oplus_{i \leqslant p} V^i \ ,$$

commutes up to a natural isomorphism.

Proof This theorem was proved over the complex numbers by Deligne [Del71, 1.2.11]. He constructs a functor $G_{\mathbb{C}} : \mathrm{MHS}_{\mathbb{Q}} \to \mathrm{grVect}_{\mathbb{C}}$ defined by

$$G_{\mathbb{C}}(V)^p = Gr_p^W\left(V \otimes_{\mathbb{Q}} \mathbb{C}\right)$$

that makes the following diagram commute

$$\begin{array}{ccc} & & \mathrm{grVect}_{\mathbb{C}} \\ & \overset{G_{\mathbb{C}}}{\nearrow} & \downarrow T \\ \mathrm{MHS}_{\mathbb{Q}} & \underset{W}{\longrightarrow} & \mathrm{FVect}_{\mathbb{C}} \end{array}$$

where T denotes the totalization functor. As proved in [CH20b], this result can be descended to $\mathbb{Q}$ using the fact that the set of such splittings form a torsor over a pro-unipotent algebraic group. Since this torsor has a $\mathbb{C}$-point (given by Deligne's splitting) we obtain that it has a $\mathbb{Q}$-point. □

Remark 6.7 It is important to emphasize that the construction of this splitting over $\mathbb{Q}$ is through obstruction theory and does not give a formula for the splitting (contrary to Deligne's formula over $\mathbb{C}$).

Remark 6.8 The reasoning used in the above is strongly reminiscent of the proof of existence of Drinfeld's'd associators over $\mathbb{Q}$. In that case one can explicitly construct associators over $\mathbb{C}$ using the Knizhnik–Zamolodchikov equation. Since the set of associators is a torsor over a pro-unipotent group scheme, there must exists associators over $\mathbb{Q}$ (see [Dri90]).

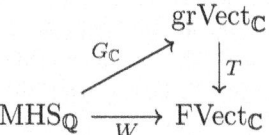

6.2 Purity

Let $\mathscr{A}$ be a symmetric monoidal abelian category. Denote by $\mathrm{gr}\mathscr{A}$ the category of graded objects of $\mathscr{A}$ which inherits a symmetric monoidal structure as before. Let α be a rational number. We denote by

$$\mathrm{Ch}_*(gr\mathscr{A})^{\alpha\text{-}pure}$$

the full subcategory of $\mathrm{Ch}_*(gr\mathscr{A})$ spanned by those graded complexes $V = \bigoplus V_n^p$ with *α-pure homology*, i.e. such that

$$H_n(V)^p = 0 \quad \text{for all } p \neq \alpha n.$$

Proposition 4.2 can be generalized as follows.

Proposition 6.9 ([CH20b, Proposition 2.7.]) *Let α be a non-zero rational number. The forgetful functor defined by forgetting the degree*

$$\mathrm{Ch}_*(gr\mathscr{A})^{\alpha\text{-}pure} \longrightarrow \mathrm{Ch}_*(\mathscr{A})$$

is a formal as lax symmetric monoidal functor.

Definition 6.10 (α-Pure Variety) A smooth algebraic variety X over $\mathbb{C}$ is called *α-pure* if $H^k(X; \mathbb{Q})$ is a pure Hodge structure of weight αk if $\alpha k \in \mathbb{Z}$ and 0 otherwise.

Example 6.11

(1) Smooth proper algebraic varieties over $\mathbb{C}$ are 1-pure by Theorem 6.5.

(2) The open variety $\mathbb{C}^* := \mathbb{C}\backslash\{0\}$ is 2-pure. Indeed, let us consider the standard covering of $\mathbb{CP}^1$ given by two copies of $\mathbb{C}$ whose intersection is $\mathbb{C}^*$. The Mayer-Vietoris long exact sequence gives an isomorphism,

$$H^{n-1}\left(\mathbb{C}^*; \mathbb{Q}\right) \cong H^n\left(\mathbb{CP}^1; \mathbb{Q}\right) \cong \begin{cases} \mathbb{Q} \text{ if } n = 2 \\ 0 \ \text{ otherwise} \end{cases}$$

for all $n \geqslant 2$. In this situation, all the morphisms involved in the long exact sequence are morphisms of mixed Hodge structures. In particular, the isomorphisms above are isomorphisms of mixed Hodge structures. Since $\mathbb{CP}^1$ is proper and thus 1-pure by the first example, $H^2\left(\mathbb{CP}^1; \mathbb{Q}\right)$ is a pure Hodge structure of weight 2, and therefore, so is $H^1\left(\mathbb{C}^*; \mathbb{Q}\right)$. Since $H^0\left(\mathbb{C}^*; \mathbb{Q}\right)$ is always a pure Hodge structure of weight 0 and the higher cohomology groups are trivial, the variety $\mathbb{C}^*$ is 2-pure.

(3) The variety $\mathbb{C}^d\backslash\{0\}$ is $2d/(2d-1)$-pure. This space is homotopy equivalent to $\mathbb{S}^{2d-1}$ and thus for all $n \geqslant 0$,

$$H^n(\mathbb{C}^d\backslash\{0\}; \mathbb{Q}) \cong \begin{cases} \mathbb{Q} \text{ if } n = 0 \text{ or } n = 2d-1 \\ 0 \ \text{ otherwise.} \end{cases}$$

The cohomology group is degree 0 is of weight 0. It remains to compute the weight in degree $2d-1$. Consider the following fibration,

$$\mathbb{C}^* \longrightarrow \mathbb{C}^d\backslash\{0\} \longrightarrow \mathbb{CP}^{d-1} .$$

The cohomology of the middle term can be computed using the Leray-Serre spectral sequence that will again be compatible with the mixed Hodge structures. We get

$$H^*(\mathbb{CP}^{d-1};\mathbb{Q}) \otimes H^*(\mathbb{C}^*;\mathbb{Q}) \implies H^*(\mathbb{C}^d\backslash\{0\};\mathbb{Q}) .$$

Moreover, we know that

$$H^n(\mathbb{CP}^{d-1};\mathbb{Q}) \cong \begin{cases} \mathbb{Q} \text{ if } n \text{ is even and } 0 \leqslant n \leqslant 2(d-1) \\ 0 \ \text{ otherwise} . \end{cases}$$

Computing the cohomology of $\mathbb{C}^d\backslash\{0\}$ through the spectral sequence and by counting weight, we can show that $H^{2d-1}(\mathbb{C}^d\backslash\{0\};\mathbb{Q})$ is of weight $2d$ which implies that the variety $\mathbb{C}^d\backslash\{0\}$ is $2d/(2d-1)$-pure.

To give another example, which generalizes example (3) above, we introduce the following definition.

Definition 6.12 (Good Arrangements) Let V be a finite dimensional $\mathbb{C}$-vector space. We say that a finite set $\{H_i\}_{i\in I}$ of subspaces of V is a *good arrangement of codimension d subspaces* if for any $J \subset I$, the intersection $\cap_{i\in J} H_i$ has codimension a multiple of d.

Proposition 6.13 *Let $\{H_1, \ldots, H_k\}$ be a good arrangement of codimension d subspaces of $\mathbb{C}^n$. The algebraic variety $\mathbb{C}^n - \cup_i H_i$ is $2d/(2d-1)$-pure.*

Proof See [CH20b, Proposition 8.6]. Note that the definition of good arrangement of codimension d subspaces used in [CH20b] is not quite the same as the one above but it is easy to check that the proof of [CH20b, Proposition 8.6] applies with the above definition. □

Remark 6.14 Proposition 6.13 holds for any hyperplane arrangement, since Definition 6.12 is automatically satisfied in codimension 1. In that case, Proposition 6.13 reduces to the main result of Kim's paper [Kim94]. In the higher codimension case, the above proposition follows from the more general result proved in [DGM00, Example 1.14].

Example 6.15 Consider the space of ordered n-configuration points in $\mathbb{C}^d$ denoted $\mathrm{Conf}_n(\mathbb{C}^d)$. Let (i, j) be an unordered pair of distinct elements in $\{1, \cdots, n\}$, and consider the diagonal

$$\Delta_{i,j} = \{(x_1, \ldots, x_n) \in (\mathbb{C}^d)^n, x_i = x_j\} .$$

The collection $\{\Delta_{i,j}\}_{(i,j)}$ of codimension d subspaces of $\left(\mathbb{C}^d\right)^n$ is easily seen to be a good arrangement and the complement

$$(\mathbb{C}^d)^n - \cup_{(i,j)}\Delta_{i,j}$$

is exactly $\mathrm{Conf}_n\left(\mathbb{C}^d\right)$. Using the previous proposition, one can conclude that this space is $2d/(2d-1)$-pure.

6.3 Formality of the Singular Chains Functor

Let $\mathrm{Var}_{\mathbb{C}}^{\alpha\text{-}pure}$ denotes the full subcategory of $\mathrm{Var}_{\mathbb{C}}$ whose objects are varieties that are α-pure.

Theorem 6.16 ([CH20b, Theorem 7.3]) *Let α be a non-zero rational number. The singular chains functor*

$$C_*(-;\mathbb{Q}) : \mathrm{Var}_{\mathbb{C}}^{\alpha\text{-}pure} \to \mathrm{Ch}_*(\mathbb{Q})$$

is formal as lax symmetric monoidal functor.

Ideas of the Proof By Corollary 5.18, it suffices to prove that this functor is formal as an ∞-lax symmetric monoidal functor. The main ingredient of the proof is that there exists a formal functor

$$\mathscr{D}_*(-)_{\mathbb{Q}} : \mathrm{N}(\mathrm{Var}_{\mathbb{C}})^{\times} \to \mathbf{Ch}_*(\mathbb{Q})^{\otimes}$$

which is weakly equivalent to $C_*(-;\mathbb{Q})$ in the category of strong symmetric monoidal ∞-functors. The construction of this functor involves the notion of mixed Hodge complexes introduced by Deligne in [Del74].

Definition 6.17 (Mixed Hodge Complex) A *mixed Hodge complex* over $\mathbb{Q}$ is the data of

- a filtered chain complex $(K_{\mathbb{Q}}, W)$ over $\mathbb{Q}$;
- a bifiltered chain complex $(K_{\mathbb{C}}, W, F)$ over $\mathbb{C}$;
- a finite string of filtered quasi-isomorphisms of filtered complexes of $\mathbb{C}$-vector spaces

$$(K_{\mathbb{Q}}, W) \otimes \mathbb{C} \xrightarrow{\alpha_1} (K_1, W) \xleftarrow{\alpha_2} \cdots \xrightarrow{\alpha_{l-1}} (K_{l-1}, W) \xrightarrow{\alpha_l} (K_{\mathbb{C}}, W).$$

We call l the *length* of the mixed Hodge complex. The following axioms must furthermore be satisfied:

(MH_0) The homology $H_*(K_{\mathbb{Q}})$ is bounded and finite-dimensional.
(MH_1) The differential of $Gr_W^p K_{\mathbb{C}}$ is strictly compatible with F.
(MH_2) The filtration on $H_n\left(Gr_W^p K_{\mathbb{C}}\right)$ induced by F makes $H_n\left(Gr_W^p K_{\mathbb{Q}}\right)$ into a pure Hodge structure of weight $p+n$.

Morphisms of mixed Hodge complexes are given by levelwise bifiltered morphisms of complexes making the corresponding diagrams commute. We denote by $\mathrm{MHC}_{\mathbb{Q}}$ the category of mixed Hodge complexes of a certain fixed length, omitted in the notation. We can view it as a symmetric monoidal category, with the filtered variant of the Künneth formula, since the tensor product of mixed Hodge complexes is again a mixed Hodge complex.

Denote by $\mathbf{MHC}_{\mathbb{Q}}$ the ∞-category obtained by inverting weak equivalences of mixed Hodge complexes. It can be equipped with a structure of symmetric monoidal ∞-category (see [Dre15]). Beilinson gave an equivalence of categories between the derived category of mixed Hodge structures and the homotopy category of shifted version of mixed Hodge complexes. This equivalence can be lifted at the level of symmetric monoidal ∞-categories under the form of the following theorem originally due to Drew (see [Dre15]).

Theorem 6.18 ([CH20b, Theorem 5.4.]) *There exists an equivalence of symmetric monoidal ∞-categories $\boldsymbol{Ch}_*(\mathrm{MHS}_{\mathbb{Q}})^{\otimes} \to \boldsymbol{MHC}_{\mathbb{Q}}^{\otimes}$.*

The functor $\mathscr{D}_*(-)_{\mathbb{Q}}$ is then obtained as the pre-composition of the forgetful functor $\mathbf{MHC}_{\mathbb{Q}}^{\otimes} \to \mathbf{Ch}_*(\mathbb{Q})^{\otimes}$ by a symmetric monoidal functor

$$\mathscr{D}_* : \mathrm{N}(\mathrm{Var}_{\mathbb{C}})^{\times} \to \mathbf{MHC}_{\mathbb{Q}}^{\otimes} .$$

In the case of smooth varieties, it suffices to take a functorial mixed Hodge complex model for the cochains as constructed for instance in [NA87] and dualize it, see [CH20b, Section 6.1]. Once one has constructed this functor for smooth varieties, it can be extended to more general varieties by standard descent arguments, see [CH20b, Theorem 6.7]. Finally, one can show that the functor $\mathscr{D}_*(-)_{\mathbb{Q}}$ is formal as follows. Let $\mathscr{E}$ be a strong monoidal inverse of the equivalence of Theorem 6.18. Consider the composite

$$\mathscr{E} \circ \mathscr{D}_* : \mathrm{N}(\mathrm{Var}_{\mathbb{C}})^{\times} \to \mathbf{Ch}_*(\mathrm{MHS}_{\mathbb{Q}})^{\otimes}.$$

Using Theorem 6.18, the functor $\mathscr{D}_*(-)_{\mathbb{Q}}$ is weakly equivalent to $U \circ \mathscr{E} \circ \mathscr{D}_*$ where U is the forgetful functor

$$U : \mathrm{MHS}_{\mathbb{Q}} \to \mathrm{Vect}_{\mathbb{Q}} .$$

The restriction of $\mathscr{E} \circ \mathscr{D}_*$ to $\mathrm{Var}_{\mathbb{C}}^{\alpha\text{-}pure}$ lands in $\mathbf{Ch}_*(\mathrm{MHS}_{\mathbb{Q}})^{\alpha\text{-}pure}$, the full subcategory of $\mathbf{Ch}_*(\mathrm{MHS}_{\mathbb{Q}})$ spanned by chain complexes whose homology is α-pure. However, Theorem 6.6 and Theorem 6.9 imply that the restriction of the functor

$$U : \mathbf{Ch}_*(\mathrm{MHS}_{\mathbb{Q}})^{\otimes} \to \mathbf{Ch}_*(\mathbb{Q})^{\otimes}$$

to $\mathbf{Ch}_*(\mathrm{MHS}_{\mathbb{Q}})^{\alpha\text{-}pure}$ is formal. Proposition 2.14 thus implies that $U \circ \mathscr{E} \circ \mathscr{D}_*$ is formal, so does $\mathscr{D}_*(-)_{\mathbb{Q}}$. □

Example 6.19 (Noncommutative Little Disks Operad) There is a topological model of the noncommutative little disks operad and non-commutative framed little disks operad introduced in [DSV15]. The operads at stake are two non-symmetric topological operads

$$\mathcal{A}s_{\mathbb{S}^1} \quad \text{and} \quad \mathcal{A}s_{\mathbb{S}^1} \rtimes \mathbb{S}^1$$

that are given in each arity by a product of copies of $\mathbb{C}^*$. Using Example 6.11 and Künneth formula, one shows that such a product is 2-pure. Theorem 6.16 implies that the operads $C_*(\mathcal{A}s_{\mathbb{S}^1}; \mathbb{Q})$ and $C_*(\mathcal{A}s_{\mathbb{S}^1} \rtimes \mathbb{S}^1; \mathbb{Q})$ are formal.

Example 6.20 (Cazanave's Monoid) We denote by F_d the algebraic variety of degree d algebraic maps from $\mathbb{CP}^1$ to itself that send the point ∞ to 1. An element in this variety can be seen as a pair (f, g) of degree d monic polynomials without any common roots. One can show that the weight filtration on $H^*(F_d; \mathbb{Q})$ is 2-pure, see [CH20b, Proposition 7.6.]. In [Caz12, Proposition 3.1.], Cazanave shows that the variety $\sqcup_d F_d$ has the structure of a graded monoid which is algebraic and compatible in a homotopical sense with the loop space structure on $\text{Map}_*(\mathbb{S}^2, \mathbb{S}^2)$. This implies that the graded monoid in chain complexes $\bigoplus_d C_*(F_d; \mathbb{Q})$ is formal.

Example 6.21 (Vaintrob) There is an operad in log schemes whose complex points recover the (framed) little disks operad, see [Vai19]. Log schemes have a Hodge structure on their cohomology and we can lift it at the level of chains. Using this, we can deduce the formality of the chain complex associated to (framed) little disks operad. This approach is developed in [Vai21].

6.4 *Formality of Sullivan's Polynomial Forms Functor*

The following is inspired by Cirici and Horel [CH20b, Section 8.]. Recall the functor of polynomial forms

$$\Omega^*_{PL} : \text{sSet} \to \text{CDGA}^{op}$$

introduced in Sect. 2.2. Using this functor, one can obtain a contravariant version of Theorem 6.16 as follows.

Theorem 6.22 ([CH20b, Theorem 8.1.]) *Let α be a non-zero rational number. Sullivan's polynomial forms functor*

$$\Omega^*_{PL} : (\text{Var}_{\mathbb{C}})^{op} \to \text{Ch}_*(\mathbb{Q})$$

is formal as lax symmetric monoidal functor when restricted to varieties whose weight filtration in cohomology is α-pure.

Proof The proof is almost the same as the proof of Theorem 6.16, using a contravariant version of the functor $\mathcal{D}_*$. □

Example 6.23 Theorem 6.22 and Proposition 6.13 imply that complements of good codimension d arrangements are formal over $\mathbb{Q}$ (in the sense of Definition 2.17). The

Deligne-Griffiths-Morgan-Sullivan theorem on formality of compact Kähler manifolds is strongly related to the case $\alpha = 1$ of the previous theorem. Indeed compact Kähler manifolds include smooth and projective algebraic varieties to which the above theorem applies.

6.5 Formality of Hopf Cooperads

Given a topological group G, the graded vector space $H^*(G; \mathbb{Q})$ is a Hopf algebra in which the multiplication comes from the diagonal of G and the comultiplication comes from the multiplication of G. With this structure on cohomology, one may be interested in the formality of $C^*(G; \mathbb{Q})$ as a Hopf algebra. One has to deal with the issue that the multiplication in the Hopf algebra structure at the cochains level is not strictly commutative. On the other hand, if we consider $\Omega^*_{PL}(G)$, the multiplication is strictly commutative but the comultiplication is only coassociative up to homotopy. A similar problem also arises with operads in spaces, the cohomology of an operad in spaces is a Hopf cooperad (see definition below) but the formality as a Hopf cooperad is not so easy to define since the structure is not strict at the cochain level. The purpose of this subsection is to give a framework for studying this question.

Definition 6.24 (Hopf Cooperad) A *Hopf cooperad* over a field $\mathbf{k}$ is an operad in the symmetric monoidal category

$$(\mathrm{CDGA}_{\mathbf{k}}^{op}, \otimes)\,.$$

If we unravel this definition, a Hopf cooperad is a collection of CDGAs indexed by the positive integers $\{A(n)\}_{n\in\mathbb{N}}$ together with

(1) a map $A(1) \to \mathbf{k}$,
(2) a symmetric group action of $\mathbb{S}_n$ on $A(n)$ for each n,
(3) maps of CDGAs

$$\circ_i : A(m+n-1) \to A(m) \otimes A(n)$$

defined for each integer $i \in \{1, \cdots, m\}$.

satisfying the dual axioms of those of an operad.

We can make sense of this definition for more general algebraic structures. To do so, we will introduce the language of algebraic theories.

Definition 6.25 An *algebraic theory* is a small category T with finite products. For $\mathscr{C}$ a category with finite products, a T-*algebra* in $\mathscr{C}$ is a finite product preserving functor $T \to \mathscr{C}$. The category of T-algebras is the category whose objects are T-algebras and whose morphisms are natural transformation of functors.

Example 6.26

(1) Let FFGrp be the full sub-category of Grp spanned by free groups on a finite set of generators. Then FFGrp^{op} is an algebraic theory. It is an instructive exercise to check that the category of T-algebras is equivalent to the category of groups. One side of this equivalence is given by the functor

$$\mathrm{Grp} \to \mathrm{Alg}_T$$

sending G to the functor $F \mapsto \mathrm{Hom}(F, G)$.

(2) Similarly, there exist algebraic theories for which the T-algebras are monoids, abelian groups, rings, operads, cyclic operads, modular operads etc. They formally look very similar to the previous example. One simply takes the opposite of the category of free objects on finitely many generators.

Definition 6.27 (Hopf T-Coalgebras) Let T be an algebraic theory and $\mathbf{k}$ be a field. Then the *category of dg Hopf T-coalgebras* over $\mathbf{k}$ is the opposite of the category of finite product preserving functors from T to the category CDGA^{op}.

Remark 6.28 Since the cartesian product in CDGA^{op} is the coproduct in CDGA and is simply given by the tensor product, it is clear that the above definition generalizes Definition 6.24.

Definition 6.29 (Weak Hopf T-Coalgebras) Let T be an algebraic theory and $\mathscr{C}$ a category with products and with a notion of weak equivalences (e.g. a model category). A *weak T-algebra* in $\mathscr{C}$ is a functor $F : T \to \mathscr{C}$ such that for each pair (s, t) of objects of T, the canonical map

$$F(t \times s) \to F(t) \times F(s)$$

is a weak equivalence. In particular, if $\mathscr{C}$ is the category CDGA^{op}, we call these objects *weak dg Hopf T-coalgebras.*

Remark 6.30 There are rigidification results due to Bernard Badzioch and Julie Bergner (see [Bad02, Ber06]) that imply that for algebraic theory T, we have an equivalence of homotopy categories

$$\mathrm{Ho}(\text{weak } T\text{-algebras in sSet}) \cong \mathrm{Ho}(T\text{-algebras in sSet}).$$

This is for example true for group, monoids, operads, cyclic operads.

Example 6.31 If $X : T \to \mathrm{sSet}$ is a T-algebra (or even a weak T-algebra), then $\Omega^*_{PL}(X)$ is a weak T-algebra in CDGA^{op}.

Theorem 6.32 ([CH20b, Theorem 8.18]) *Let T be an algebraic theory and let*

$$X : T \to \mathrm{Var}_{\mathbb{C}}$$

be a T-algebra such that for all $t \in T$, the weight filtration on the cohomology of $X(t)$ is α-pure, for $\alpha \in \mathbb{Q}^{\times}$. The weak dg Hopf T-coalgebra $\Omega^*_{PL}(X)$ is formal.

Proof The result is an immediate consequence of Theorem 6.22 since being a weak T-coalgebra is a property of a functor $T^{op} \to$ CDGA that is invariant under quasi-isomorphism. □

Remark 6.33 The fact that $\Omega^*_{PL}(X)$ is formal as a weak dg Hopf T-coalgebra implies that the rational homotopy type of X is determined by $H^*(X; \mathbb{Q})$ as a T-algebra in graded commutative algebras. Indeed, if we apply the derived Sullivan spatial realization functor

$$\langle - \rangle : \mathrm{CDGA}^{op} \to \mathrm{Top},$$

to a weak dg Hopf T-coalgebra, we are going to obtain a weak T-algebra in rational spaces. If X is a T-algebra in spaces, we get a rational model for X in the sense that the map

$$X \longrightarrow \langle \Omega^*_{PL}(X) \rangle$$

is a rational weak equivalence of weak T-algebras whose target is objectwise rational. Thanks to the rigidification results that hold in the case of spaces, see Remark 6.30, the weak T-algebra $\langle \Omega^*_{PL}(X) \rangle$ can be strictified to a strict T-algebra. If X is formal, one also get a rational model for X through

$$\langle H^*(X; \mathbb{Q}) \rangle \ .$$

Remark 6.34 Let us cite some related work that predates [CH20b]. First, Morgan constructed in [Mor78] an explicit small Sullivan model of smooth algebraic varieties which is equipped with a mixed Hodge structure. This model was used by Dupont in order to show that 1 and 2-pure smooth algebraic varieties are formal, see [Dup16]. An alternative argument in the 2-pure case due to Beilinson is explained in [DH18, Proposition 3.4]. Similar ideas are used by Chataur and Cirici in [CC17] in order to prove the formality of some singular projective algebraic varieties.

7 Galois Group Actions

So far, we only considered formality problems with coefficients in a field of characteristic zero. By exploiting Galois group actions on étale cohomology rather than mixed Hodge structures, one can derive formality results with torsion coefficient. The formality results obtained in this case is however only up to a certain degree, which depends on the cardinality of the field of coefficients. In this section, we expose these results, based on [CH22].

Let $\mathscr{A}$ be a symmetric monoidal abelian category.

Definition 7.1 Let N be an integer. A morphism of chain complexes

$$f : A \longrightarrow B \in \mathrm{Ch}_*(\mathscr{A})$$

is called a *N-quasi-isomorphism* if the induced morphism in homology $H_i(f) : H_i(A) \to H_i(B)$ is an isomorphism for all $i \leqslant N$.

Definition 7.2 Let $(\mathscr{C}, \otimes, \mathbf{1})$ be a symmetric monoidal category and let

$$F : \mathscr{C} \to \mathrm{Ch}_*(\mathscr{A})$$

be a lax symmetric monoidal functor. The functor F is said to be a *N-formal lax symmetric monoidal* functor if there is a string of natural transformations of lax symmetric monoidal functors

$$F \xleftarrow{\Phi_1} F_1 \longrightarrow \cdots \longleftarrow F_n \xrightarrow{\Phi_n} H \circ F$$

such that for all X in $\mathscr{C}$, the morphisms $\Phi_i(X)$ are N-quasi-isomorphisms.

Remark 7.3 We can also extend this definition to a notion of N-formal lax symmetric monoidal ∞-functor for functors $\mathrm{N}(\mathscr{C}) \to \mathbf{Ch}_*(\mathscr{A})$ as in Definition 5.15.

7.1 Some Words on Étale Cohomology

Our main tool to prove formality results with torsion coefficients uses a Galois group action on étale cohomology. This section recalls some basic notions around this topic.

Étale cohomology is a particular example of sheaf cohomology. Recall that if X is a Hausdorff, paracompact and locally contractible topological space (this is the case if X is a smooth manifold), the singular cohomology of X can be computed as the sheaf cohomology with values in the constant sheaf. Let A be an abelian group and denote $\underline{A}$ the constant sheaf. Then there is an isomorphism between singular cohomology and sheaf cohomology:

$$H^*_{sing}(X; A) \cong H^*_{sheaf}(X; \underline{A}) .$$

In fact, this statement can be lifted at the level of cochains. If R is a commutative ring, the dg algebra of singular cochains $C^*(X; R)$ is quasi-isomorphic to the sheaf-cohomology complex $\mathrm{R}\Gamma(X; \underline{R})$. This quasi-isomorphism can be chosen to be compatible with the E_∞-algebra structures on both sides, see [Pet22] or [CC22]. If X is a scheme over some base field K, an étale cover of X is a set

$$\{p_i : U_i \to X\}$$

of étale morphisms locally of finite type which are jointly surjective. The notion of étale morphism can be viewed as an algebro-geometric analogue of the notion of local homeomorphism in topology. Then, étale cohomology of X with coefficients in an abelian group A can be defined as the derived global sections of the constant sheaf with value A

on the étale site of X,

$$H^*_{et}(X; A) := H^*\left(\mathrm{R}\Gamma\left(X_{et}; \underline{A}\right)\right).$$

For smooth schemes over $\mathbb{C}$, étale cohomology is related to the classical cohomology thanks to the following theorem.

Theorem 7.4 (Artin) *If X is a smooth scheme over $\mathbb{C}$ and A is a finite abelian group, then there is an isomorphism*

$$H^*_{et}(X; \underline{A}) \cong H^*_{sheaf}(X_{\mathrm{an}}; \underline{A})$$

where X_{an} denotes the complex manifold underlying X.

Let p be a prime number and K be a p-adic field. We denote by k the residue field of K which is a finite field of characteristic p. Let $\iota : \overline{K} \hookrightarrow \mathbb{C}$ be an embedding. We denote by Sch_K the category of schemes over K that are separated and of finite type. Let X be a smooth scheme over K. We denote by X_{an} the complex analytic space underlying $X \times_K \mathbb{C}$. Then we can relate the étale cohomology of $X \times_K \overline{K}$ with the one of $X \times_K \mathbb{C}$ using the embedding ι. We obtain a zig-zag of maps

$$H^*_{et}(X \times_K \overline{K}; A) \xleftarrow{u} H^*_{et}(X \times_K \mathbb{C}; A) \xrightarrow{\cong} H^*_{sheaf}(X_{\mathrm{an}}, \underline{A}),$$

for A any finite abelian group. Following a standard theorem of étale cohomology known as *smooth base change theorem* (see [SGA]), the map u is also an isomorphism. By functoriality of étale cohomology, the group

$$H^*_{et}(X \times_K \overline{K}; A)$$

has an action of the absolute Galois group $\mathrm{Gal}(\overline{K}/K)$. It can be shown that all these isomorphisms lift as E_∞-quasi-isomorphisms at the cochain level (see [Shi23]) and similarly that the Galois action on the left side of this zig-zag lifts to the cochain level. The group $\mathrm{Gal}(\overline{k}/k)$ is isomorphic to the profinite completion of the integers, denoted $\hat{\mathbb{Z}}$, generated by the Frobenius $x \mapsto x^q$, with $q = |k| = p^n$. We make once and for all a choice of a lift φ of the Frobenius in $\mathrm{Gal}(\overline{K}/K)$. The upshot of all this discussion is that, given a finite ring A, we have a zig-zag

$$C^*_{et}(X \times_K \overline{K}, A) \longleftarrow C^*_{et}(X \times_K \mathbb{C}; A) \longrightarrow C^*_{sing}(X_{\mathrm{an}}; A).$$

in which both maps are quasi-isomorphisms of E_∞-algebras and in which the left-hand side is equipped with an automorphism φ. Moreover, this data is functorial in the input X. This discussion can be extended from finite coefficients to ℓ-adic coefficients via the following definition.

Definition 7.5 Let X be a scheme over some base field K. Define

$$H^*_{et}(X; \mathbb{Z}_\ell) := \lim_n H^*_{et}(X; \mathbb{Z}/\ell^n)$$

and

$$H_{et}^*(X; \mathbb{Q}_\ell) := H_{et}^*(X; \mathbb{Z}_\ell) \otimes_{\mathbb{Z}_\ell} \mathbb{Q}_\ell.$$

From the fact that smooth schemes have finite type cohomology, one can also show that for X a smooth scheme over $\mathbb{C}$, there is an isomorphism

$$H_{et}^*(X; \mathbb{Z}_\ell) = H_{sheaf}^*(X_{\mathrm{an}}; \mathbb{Z}_\ell)$$

and in fact all the previous discussion can be applied to the case of $\mathbb{Z}_\ell$ and $\mathbb{Q}_\ell$-coefficients as well.

7.2 Formality Using Étale Cohomology

As in the previous section, K denotes a p-adic field and $k = \mathbb{F}_q$ its residue field. Denote by h the order of q in $\mathbb{F}_\ell^\times$. For ℓ some prime number which is prime to q.

Definition 7.6 (q-Tate Modules) Let V be a finite dimensional $\mathbb{F}_\ell$-vector space and φ an automorphism of V. We say that the pair (V, φ) is a *q-Tate module* if the eigenvalues of φ in $\overline{\mathbb{F}_\ell}$ are powers of q. Let $n \in \mathbb{N}$. A q-Tate module is said to be *pure of weight n* if the only eigenvalue of φ is q^n.

Remark 7.7 It should be noted that the weight of a pure Tate module is only well-defined modulo h. Observe also that the weights have been divided by 2 compared to the Mixed Hodge case.

Remark 7.8 It can be shown that the category of Tate modules denoted TMod is a symmetric monoidal abelian category. The kernels and cokernels are simply the kernels and cokernels of the underlying $\mathbb{F}_\ell$-vector spaces equipped with the induced action of the endomorphism φ.

Definition 7.9 Let α be a rational number satisfying $0 < \alpha < h$. Let $X \in \mathrm{Sch}_K$. We say that the étale cohomology of X is *α-pure* if the following conditions are satisfied

(1) If $\alpha n \notin \mathbb{Z}$, then $H_{et}^n(X \times_K \overline{K}; \mathbb{F}_\ell) = 0$.
(2) If $\alpha n \in \mathbb{Z}$, then $q^{\alpha n}$ is the only eigenvalue of the Frobenius acting on

$$H_{et}^n(X \times_K \overline{K}; \mathbb{F}_\ell)\,.$$

Example 7.10 Let $X = \mathbb{P}^n$. Then the étale cohomology of X is $1/2$-pure.

Definition 7.11 A colored operad $\mathscr{P}$ is *admissible* if the category

$$\mathrm{Alg}_{\mathscr{P}}(\mathrm{Ch}_*(K))$$

admits a model structure transferred along the forgetful functor

$$\mathrm{Alg}_{\mathscr{P}}(\mathrm{Ch}_*(K)) \to \mathrm{Ch}_*(K)^{Ob(\mathscr{P})}.$$

We say that $\mathscr{P}$ is Σ-*cofibrant* if for all $n \geqslant 1$, and by all surjections $i : [n] \to I$, the symmetric group $\mathbb{S}_n$ acts freely on $\mathscr{P}(n, i)$. A colored operad $\mathscr{P}$ in sets is called *homotopically sound* if it is admissible and Σ-cofibrant.

Theorem 7.12 ([CH22, Theorem 6.5]) *Let $\mathscr{P}$ be a homotopically sound operad and let X be a $\mathscr{P}$-algebra in* Sch_K *such that for each color $c = i(k)$ of $\mathscr{P}$,*

$$H^*_{et}(X(c) \times_K \overline{K}; \mathbb{F}_\ell)$$

is α-pure. The dg $\mathscr{P}$-algebra $C_*(X_{\mathrm{an}}; \mathbb{F}_\ell)$ is $\lfloor (h-1)/\alpha \rfloor$-formal.

Let us explain where the homotopically sound hypothesis comes from. Recall that if $\mathbf{k}$ is a field of characteristic zero and if F is formal (resp. N-formal) as lax symmetric monoidal ∞-functor $\mathrm{N}(\mathscr{C}) \to \mathbf{Ch}_*(\mathbf{k})$, then F is formal (resp. N-formal) as a lax symmetric monoidal functor (as in the Corollary 5.18). However for a field which is not of characteristic zero, this corollary fails. This comes from the fact that the homotopy theory of lax monoidal functors is in general not equivalent to the homotopy theory of lax monoidal ∞-functors. In order to circumvent this difficulty, we will restrict to homotopically sound operads for which we have a rigidification result due to Hinich. For $\mathscr{P}$ an operad in sets, we denote by

$$\mathbf{Alg}_{\mathscr{P}}(\mathbf{Ch}_*(\mathbf{k}))$$

the ∞-category of $\mathscr{P}$-algebras in the ∞-category of chain complexes of $\mathbf{k}$-vector spaces. There is an obvious functor

$$\mathrm{Alg}_{\mathscr{P}}(\mathrm{Ch}_*(\mathbf{k})) \to \mathbf{Alg}_{\mathscr{P}}(\mathbf{Ch}_*(\mathbf{k}))$$

which sends quasi-isomorphisms to equivalences. It induces a map

$$\mathrm{N}_W(\mathrm{Alg}_{\mathscr{P}}(\mathrm{Ch}_*(\mathbf{k}))) \to \mathbf{Alg}_{\mathscr{P}}(\mathbf{Ch}_*(\mathbf{k}))$$

Hinich shows that, under the hypothesis that $\mathscr{P}$ is homotopically sound, this functor is an equivalence of ∞-categories. In particular, we obtain the following proposition as a corollary of Hinich's theorem.

Proposition 7.13 *Let $\mathscr{P}$ be a homotopically sound operad in sets. Let A be a $\mathscr{P}$-algebra in* $\mathrm{Ch}_*(\boldsymbol{k})$.

(1) *If A is formal in $\boldsymbol{Alg}_{\mathscr{P}}(\boldsymbol{Ch}_*(\boldsymbol{k}))$, then A is formal in* $\mathrm{Alg}_{\mathscr{P}}(\mathrm{Ch}_*(\boldsymbol{k}))$.
(2) *If A is N-formal in $\boldsymbol{Alg}_{\mathscr{P}}(\boldsymbol{Ch}_{\geqslant 0}(\boldsymbol{k}))$, then A is N-formal in* $\mathrm{Alg}_{\mathscr{P}}(\mathrm{Ch}_{\geqslant 0}(\boldsymbol{k}))$.

Proof The point (1) is an immediate consequence of Hinich's theorem. Point (2) follows from (1) and from the observation that a $\mathscr{P}$-algebra A is N-formal if and only if its truncation $t_{\leqslant N}(A)$ is formal (where $t_{\leqslant N}$ denotes the truncation functor which kills homology in degrees greater than N). □

Let h be a positive integer. If $\mathscr{A}$ is a symmetric monoidal abelian category, we denote by $gr^{(h)}\mathscr{A}$ the category of $\mathbb{Z}/h$-graded objects of $\mathscr{A}$. An object in this category is a collection

$$\{A^a\}_{a\in\mathbb{Z}/h}$$

of objects of $\mathscr{A}$ indexed by the elements of the group $\mathbb{Z}/h$. It is a symmetric monoidal category, with the tensor product given by

$$(A\otimes B)^n := \sum_{a+b\equiv n \pmod h} A^a\otimes B^b.$$

There is a functor Tot : $gr^{(h)}\mathscr{A}\to\mathscr{A}$ given by the formula

$$\{A^a\}_{a\in\mathbb{Z}/h}\longmapsto\bigoplus_{a\in\mathbb{Z}/h} A^a$$

It is straightforward that this functor can be given the structure of a strong symmetric monoidal functor. We have the following version of Theorem 6.6.

Lemma 7.14 ([CH22, Lemma 2.9]) *Let h be the order of q in $(\mathbb{F}_\ell)^\times$. The functor U :* TMod $\to$ $\mathrm{Vect}_{\mathbb{F}_\ell}$ *defined by $(V,\varphi)\mapsto V$ admits a factorization*

$$\begin{array}{ccc} & & gr^{(h)}\mathrm{Vect}_{\mathbb{F}_\ell} \\ & \overset{G}{\nearrow} & \downarrow{\scriptstyle \mathrm{Tot}} \\ \mathrm{TMod} & \xrightarrow[U]{} & \mathrm{Vect}_{\mathbb{F}_\ell} \end{array}$$

into strong symmetric monoidal functors. The functor G is defined by declaring $G(V,\varphi)^n$ to be the generalized eigenspace for the eigenvalue q^n.

To study formality in the torsion case, there is also a version of Theorem 6.9, which deals with chain complexes of $\mathbb{Z}/h$-graded objects. We denote by

$$\mathrm{Ch}_*(gr^{(h)}\mathscr{A})^{\alpha\text{-}pure}\subset\mathrm{Ch}_*(gr^{(h)}\mathscr{A})$$

the full subcategory given by those $\mathbb{Z}/h$-graded complexes $V=\bigoplus V_n^p$ with *α-pure homology* :

$$H_n(V)^p=0 \quad \text{for all } p\neq\alpha n \pmod m.$$

The following proposition is a $\mathbb{Z}/h$-graded version of Proposition 6.9.

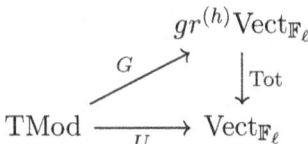

Proposition 7.15 ([CH22, Proposition 5.13.]) *The forgetful functor*

$$U : \mathrm{Ch}_{\geqslant 0}(gr^{(h)}\mathscr{A})^{\alpha\text{-}pure} \to \mathrm{Ch}_{\geqslant 0}(\mathscr{A})$$

is $\lfloor (h-1)/\alpha \rfloor$*-formal as lax symmetric monoidal functor.*

Definition 7.16 (Tate Complex) A *Tate complex* is a pair (C, φ) where C is a chain complex of $\mathbb{F}_\ell$-modules and φ is an endomorphism of C such that the pair $(H_n(C), H_n(\varphi))$ is a Tate module for all n.

We denote by **TComp** the ∞-category of Tate complexes and by $\mathbf{TComp}_{\geqslant 0}^{\alpha\text{-}pure}$ to be the full subcategory of α-pure non-negatively graded Tate complexes.

Proof of Theorem 7.12 It suffices to prove that the forgetful functor

$$\mathbf{TComp}_{\geqslant 0}^{\alpha\text{-}pure} \to \mathbf{Ch}_{\geqslant 0}(\mathbb{F}_\ell)$$

is N-formal as a lax monoidal ∞-functor, with $N = \lfloor (h-1)/\alpha \rfloor$. Indeed, if this is true, this will imply that $C_*(X_{\mathrm{an}}; \mathbb{F}_\ell)$ is N-formal in the ∞-category $\mathbf{Alg}_{\mathscr{P}}(\mathbf{Ch}_*(\mathbf{k}))$ and Proposition 7.13 allows to come back to the standard definition of formality. To do so, we use the same strategy as in the mixed Hodge complexes case. First, one can prove a Beilinson type theorem for the categories of Tate complexes. There is a canonical symmetric monoidal functor $\mathrm{Ch}_*(\mathrm{TMod}) \to \mathrm{TComp}$ that preserves quasi-isomorphisms on both sides. Therefore it induces a symmetric monoidal ∞-functor

$$\mathbf{Ch}_*(\mathrm{TMod}) \to \mathbf{TComp}.$$

One can show that this ∞-functor is in fact an equivalence of symmetric monoidal ∞-categories, see [CH22, Theorem 4.7.]. Using Lemma 7.14, the forgetful functor $\mathbf{TComp} \to \mathbf{Ch}_{\geqslant 0}(\mathbb{F}_\ell)$ factorizes as follows

$$\mathbf{TComp} \to \mathbf{Ch}_*(\mathrm{TMod}) \to \mathbf{Ch}_*\left(gr^{(h)}\mathrm{Vect}_{\mathbb{F}_\ell}\right) \to \mathbf{Ch}_*(\mathbb{F}_\ell).$$

where the first map is an inverse to the equivalence mentioned before. We can restrict all categories to α-pure $\mathbb{Z}/h$-graded objects and then the last functor is N-formal by Proposition 7.15. The result follows from Proposition 2.14. □

We now give some examples of applications of this theorem.

Proposition 7.17 *Let K be any p-adic field. The cohomology $H^*_{et}(\overline{\mathscr{M}_{0,n}} \times_{\mathbb{Z}} \overline{K}, \mathbb{F}_\ell)$ is $1/2$-pure.*

Proof A smooth scheme over K is said to be $1/2$-pure if its étale cohomology is $1/2$-pure. We first make the following claim.

- The set of $1/2$-pure schemes is stable under finite products.
- If $Z \to X$ is a closed embedding of smooth schemes and Z and X are $1/2$-pure, then the blow-up $B_Z(X)$ is also $1/2$-pure.

The first property is an immediate consequence of the Künneth formula in étale cohomology. The second property follows from the blow-up formula which gives an equivariant isomorphism

$$H^*_{et}(B_Z(X); \mathbb{F}_\ell) \cong H^*_{et}(X; \mathbb{F}_\ell) \oplus \left(\bigoplus_{i=1}^{c-1} H^*_{et}(Z; \mathbb{F}_\ell)[-2i] \otimes_{\mathbb{F}_\ell} \mathbb{F}_\ell(-i) \right)$$

where $\mathbb{F}_\ell(-i)$ denotes the Tate module in which the Frobenius acts by multiplication by q^i and c denoted the codimension of Z in X. Now, we can prove the proposition by induction on n. For $n = 3$, this moduli space is a point. For $n = 4$, we have $\overline{\mathscr{M}}_{0,4} \cong \mathbb{P}^1$ and the proposition is a classical computation. Assume that the proposition has been proved for $\{3, 4, \cdots, n\}$. We may use Keel's inductive description of $\overline{\mathscr{M}}_{0,n+1}$ as a sequence of blow-ups starting from $\overline{\mathscr{M}}_{0,n} \times \overline{\mathscr{M}}_{0,4}$ and in which, at each stage, the variety that is blown-up is isomorphic to $\overline{\mathscr{M}}_{0,p+1} \times \overline{\mathscr{M}}_{0,q+1}$ with $p + q = n$, see [Kee92, Section 1]. We conclude by the induction hypothesis and the first claim of the proof. □

Example 7.18 Consider the operad $\mathscr{O} = \{\overline{\mathscr{M}}_{0,*+1}\}$ of moduli spaces of stable algebraic curves of genus 0 of Example 3.14. Let us pick a prime number p that is different from ℓ and such that p has order $\ell - 1$ in $\mathbb{F}_\ell^\times$ (such a prime exists thanks to Dirichlet's theorem on arithmetic progressions). The hypothesis of Theorem 7.12 are satisfied with $\alpha = 1/2$ and $h = \ell - 1$ thanks to Proposition 7.17 above. In this case, the operad $\mathscr{P}$ is the colored operad whose algebras are operads. This operad is homotopically sound. We can therefore conclude that the dg- operad $C_*(\mathscr{O}; \mathbb{F}_\ell)$ is $2(\ell - 2)$ formal.

Example 7.19 Let $\mathscr{D}_2$ denotes the little disks operad. This is not quite an operad in the category of smooth schemes but in any case, one can construct a model $\mathscr{C}$ of $C_*(\mathscr{D}_2; \mathbb{F}_\ell)$ equipped with an action of the profinite Grothendieck-Teichmüller group $\widehat{GT}$. This is very similar to what we explained in Example 4.2. The group $\widehat{GT}$ is equipped with a surjective homomorphism

$$\widehat{GT} \to \widehat{\mathbb{Z}}^\times \cong \prod_p \mathbb{Z}_p^\times$$

where $\widehat{\mathbb{Z}}^\times$ is the group of units in the profinite completion of the ring of integers. We may pick an element φ of $\widehat{GT}$ that lifts $p \in \mathbb{Z}_\ell^\times$ where p is some prime number distinct from ℓ. Then, it can be shown that $\mathscr{C}$ equipped with the automorphism φ is an operad in the category of Tate complexes. Moreover, the homology $H_*(\mathscr{D}_2(k); \mathbb{F}_\ell)$ is 1-pure for each k. Using the method of Theorem 7.12 we deduce that the operad $\mathscr{C}$, and therefore the operad

$$C_*(\mathscr{D}_2(k); \mathbb{F}_\ell) \, ,$$

is $(\ell - 2)$-formal, see [CH22, Theorem 6.7.]. It should be noted that this result is sharp since it can be shown that the operad $C_*(\mathscr{D}_2; \mathbb{F}_\ell)$ is not $(\ell - 1)$-formal.

8 Homotopy Transfer and Formality

Formality aims to measure if the induced structure in homology retains all of the homotopical information contained in a given algebra. Through an operadic approach, one can make this intuition precise and derive another characterization of formality: *gauge formality*. In this section, we present this other approach and formality criteria based on it.

Assumption

- We suppose that every operad or cooperad in $\mathrm{Ch}_*(R)$ is reduced, connected and has an additional weight grading.
- Let $\mathscr{P}$ be an operad in the category of R-modules, concentrated in degree zero. We assume that either R is a $\mathbb{Q}$-algebra or that $\mathscr{P}$ is a non-symmetric operad.
- Let $\mathscr{C}$ be a conilpotent cooperad over R, with coaugmentation coideal $\overline{\mathscr{C}}$ concentrated in strictly positive degree. We assume that we are given the datum of a Koszul morphism $\mathscr{C} \to \mathscr{P}$, i.e. a twisting morphism that induces a quasi-isomorphism

$$\mathscr{P}_\infty := \Omega\mathscr{C} \xrightarrow{\sim} \mathscr{P} .$$

8.1 Gauge Formality

There exist several equivalent characterizations of a $\mathscr{P}_\infty$-algebra structure. We are going to use the one in terms of coderivations, see [LV12, Section 10.1] for more details.

Proposition 8.1 *A $\mathscr{P}_\infty$-algebra structure on a chain complex A is equivalent to a codifferential of $\mathscr{C}(A)$, i.e. a degree -1 square-zero coderivation*

$$M \in \mathrm{Coder}(\mathscr{C}(A))$$

of the cofree conilpotent coalgebra $\mathscr{C}(A)$. An ∞-morphism *between two $\mathscr{P}_\infty$-algebras*

$$F : (A, M) \rightsquigarrow (A', M') .$$

is a map of dg $\mathscr{C}$-coalgebras $(\mathscr{C}(A), M) \to (\mathscr{C}(A'), M')$.

Remark 8.2 A coderivation M is completely determined by its projection on the cogenerators $m : \mathscr{C}(A) \to A$. In other words, there is an isomorphism

$$\mathrm{Coder}(\mathscr{C}(A)) \cong \mathrm{Hom}(\mathscr{C}(A); A) ,$$

see [LV12, Proposition 6.3.8] for more details. These two points of view will be used equally by keeping the uppercase letters for the coderivations and the lowercase letters for the associated projections. Similarly, an ∞-morphism F is completely determined by its projection $f : \mathscr{C}(A) \to A'$.

For every coderivation M, we denote by m_i the restriction to

$$\mathscr{C}_i(A) := \bigoplus_{k\in\mathbb{N}} \mathscr{C}_i(k) \otimes_{\mathbb{S}_k} A^{\otimes k} .$$

Similarly, we denote by f_i the restriction of an ∞-morphism F to $C_i(A)$. Since $\mathscr{C}_0 = I$, the component f_0 is an endomorphism $A \to A$.

Definition 8.3 An ∞-morphism $F : A \rightsquigarrow A'$ is an ∞-*quasi-isomorphism* (resp. ∞-isomorphism) if its first component $f_0 : A \to A'$ is a quasi-isomorphism (resp. isomorphism).

Given a $\mathscr{P}_\infty$-algebra (A, M), there is a natural induced $\mathscr{P}$-algebra structure $(H(A), m_*)$ on the homology. In general, this structure forgets a part of the homotopical information: most structures are not formal. However, when there is a homotopy retraction between A and $H(A)$, there is another way to transfer a given structure to the homology without loss of homotopical information. It is given by the homotopy transfer theorem.

Theorem 8.4 (Homotopy Transfer Theorem) *Let A be a chain complex and let*

$$h \circlearrowright (A,d) \underset{i}{\overset{p}{\rightleftarrows}} (H(A),0)$$

be a homotopy retraction where i is a quasi-isomorphism, $ip - \mathrm{id}_A = d_A h + h d_A$, and $pi = \mathrm{id}_{H(A)}$. Let (A, M) be a $\mathscr{P}_\infty$-algebra structure.

(1) *There exists a transferred $\mathscr{P}_\infty$-algebra structure $(H(A), M^t)$ such that*

$$m_1^t = m_* .$$

(2) *The inclusion i and the projection p extend to mutually quasi-inverse ∞-quasi-isomorphisms, which we will denote by i_∞ and p_∞.*
(3) *The transferred structure is independent of the choice of sections of $H(A)$ on A in up to ∞-isomorphisms.*

Proof We refer the reader to [Ber14] and [LV12, Section 10.3] and references therein. □

Remark 8.5 Of course, the existence of a homotopy retraction of this form is not automatic if the base ring R is not a field. If the base ring is a principal ideal domain it holds if $H(A)$ is degreewise projective.

Definition 8.6 Let $n \in \mathbb{N}^*$ and a dg $\mathscr{P}$-algebra (A, d, m) such that $H(A)$ is a homotopy retract of A. The algebra A is said

- *gauge formal* if there exists an ∞-quasi-isomorphism

$$(H(A), M^t) \overset{\sim}{\rightsquigarrow} (H(A), m_*) .$$

- *gauge n-formal* if there exist a $\mathscr{P}_\infty$-algebra structure $(H(A), R)$ with $r_0 = 0$ and $r_i = 0$ for i such that $2 \leqslant i \leqslant n$ and ∞-quasi-isomorphism

$$(H(A), M^t) \overset{\sim}{\rightsquigarrow} (H(A), R) .$$

Remark 8.7

(1) Over a characteristic zero field, gauge formality is equivalent to formality, see [LV12, Theorem 11.4.9]. Over a general ring (and under the assumption that the operad $\mathscr{P}$ is nonsymmetric) this is also true under mild flatness hypothesis, see [DCH21, Proposition 1.15].
(2) The terminology of gauge formality comes from the equivalence between the existence of ∞-quasi-isomorphisms is this situation and the existence of gauge equivalence in a certain dg Lie algebra, see [Emp24, Section 2] for more details.

8.2 *Automorphism Lifts*

In [DCH21], Drummond-Cole and the second author present another proof of Theorem 6.16 under sightly different assumptions and using gauge formality approach.

Definition 8.8 Let V be a graded R-module. Let α be a unit in R. The *degree twisting* by α, denoted σ_α, is the linear automorphism of V which acts on the degree n homogenous component of V via multiplication by α^n.

Theorem 8.9 ([DCH21, Main Theorem]) *Let A be a chain complex such that $H(A)$ is a homotopy retract. Let (A, M) be a $\mathscr{P}_\infty$-algebra structure. Let α be a unit in R and suppose that the degree twisting σ_α on $H(A)$ admits a chain level lift, i.e. there exists an ∞-quasi-isomorphism v of (A, M) such that $H(v_0) = \sigma_\alpha$.*

- *If $\alpha^k - 1$ is a unit of R for $k \leqslant n$, then (A, M) is gauge n-formal.*
- *If $\alpha^k - 1$ is a unit of R for all k, then (A, M) is gauge formal.*

Remark 8.10 This result generalizes to more general types of algebraic structures (colored operads, properads,...) and to other types of homology automorphism, see [Emp24, Theorem 4.10].

Example 8.11 (Complement of Subspace Arrangements) Let X be a complement of hyperplanes arrangement over $\mathbb{C}$, i.e. a complement of a finite collection of affine hyperplanes in $\mathbb{A}^n_{\mathbb{C}}$ viewed as a scheme over $\mathbb{C}$. Let p and ℓ be two different prime numbers. Suppose that X can be defined over a finite extension K of $\mathbb{Q}_p$, i.e. there exist an embedding $K \to \mathbb{C}$ and a complement of a hyperplane arrangement over K denoted $\mathscr{X}$ such that

$$X \cong \mathscr{X} \times_K \mathbb{C} .$$

We denote by q cardinality of the residue field of the ring of integers of K and h the order of q in the group of units of $\mathbb{F}_\ell$. We can apply Theorem 8.9 to prove that the algebra $C^*_{\text{sing}}(X_{\text{an}}; \mathbb{Z}_\ell)$ is $(h-1)$-formal, where X_{an} denotes the complex analytic space underlying $X_{\mathbb{C}} = X \times_K \mathbb{C}$ and where s is the order of q in $\mathbb{F}_\ell^\times$. Indeed, as in the previous section, there exists a zig-zag of quasi-isomorphisms of dg associative algebras

$$C^*_{\text{sing}}(X_{\text{an}}; \mathbb{Z}_\ell) \xleftarrow{\sim} C^*_{\text{ét}}(\mathcal{X}_{\mathbb{C}}; \mathbb{Z}_\ell) \xrightarrow{\sim} C^*_{\text{ét}}(\mathcal{X}_{\overline{K}}; \mathbb{Z}_\ell) .$$

One can show that the action of a Frobenius lift on $H^n_{et}(\mathcal{X}_{\overline{K}}; \mathbb{Z}_\ell)$ is given by multiplication by q^n (see [Kim94, Theorem 1'] for a proof).

Remark 8.12 The condition of being defined over K is essential here: for each ℓ there exists a complement hyperplane arrangement defined over $\mathbb{C}$ inducing non-trivial Massey products in $H^2(-; \mathbb{F}_\ell)$, see [Mat06].

Remark 8.13 The notion of gauge n-formality is fundamental in this example. In [CH22, Theorem 7.12, (iii)] it is wrongly claimed that under these assumptions, $C^*_{sing}(X_{an}; \mathbb{F}_\ell)$ is $(h-1)$-formal in the sense that there is a zig-zag of morphisms connecting this dg-algebra to its cohomology and that induce isomorphisms in cohomological degree $\leq h-1$. This statement is incorrect as explained in [CH24]. Gauge n-formality seems to be the best way to express the kind of partial formality that we have in this situation. On the other hand, similar results hold for complements of subspace arrangements of higher codimension and in this case the two notions of partial formality can be used.

Example 8.14 (Coformality of Configuration Spaces) For $d \geqslant 3$, the configuration space $X = \text{Conf}_n(\mathbb{R}^d)$ is coformal, i.e. the dg algebra $C_*(\Omega X; \mathbb{Z}_\ell)$ is $(\ell-2)(d-2)$-formal, for a given prime number ℓ, see [DCH21, Theorem 4.16]. This result is proved using an action of the profinite Grothendieck-Teichmüller group on these spaces. Let us mention that the terminology coformal is unusual in this context. Usually a space is called coformal if the Quillen model for its rational homotopy type is formal as a differential graded Lie algebra. This definition does not generalize well when the ring of coefficients is not a field of characteristic zero. However, Saleh has proved that coformality of a based connected topological space X is equivalent to formality of the differential graded algebra of chains on its based loop space $C_*(\Omega X; \mathbb{Q})$ (see [Sal17]). The latter condition makes perfect sense for any ring of coefficients and explains our choice of terminology.

8.3 Kaledin Classes

Given a dg associative algebra over a characteristic zero field, Kaledin constructs in [Kal07] a class in the associated Hochschild cohomology, which vanishes if and only if the algebra is formal. The work of Kaledin was extended by Melani and Rubió in [MR19] for algebras over a binary Koszul operad in characteristic zero and by the first author in [Emp24] for algebras over groupoid colored operad or properad, over a commutative ground ring R.

Let H be a graded R-module and let (H, M) be a $\mathscr{P}_\infty$-algebra structure. Recall that the complex $\mathrm{Coder}(\mathscr{C}(H))$ can be equipped with a complete dg Lie algebra structure, whose filtration is defined for all $p \geqslant 0$ by

$$\mathscr{F}^p\mathrm{Coder}(\mathscr{C}(A)) := \prod_{k \geqslant p} \mathrm{Hom}\left(\mathscr{C}_p(A), A\right) ,$$

and whose differential is given by the operator $d_M := [M, -]$ see e.g. [LV12, Section 6.4]. Let us consider the *prismatic decomposition*

$$\partial_\hbar M = m_2 + 2m_3 + 3m_4 + \cdots$$

This element is a cycle in $\mathfrak{g}^M := (\mathscr{F}^1\mathfrak{g}, d_M)$, see e.g. [Emp24, Proposition 1.14].

Definition 8.15 (Kaledin Class) The Kaledin class of (H, M) is the class

$$K_M := \left[\partial_\hbar M\right] \in H_{-1}\left(\mathfrak{g}^M\right) .$$

The *n-truncation* K_M^n of K_M is the class associated to the cycle

$$m_2 + 2m_3 + \cdots + (n-1)m_n$$

in the cohomology of the complex $\mathscr{F}^1\mathrm{Coder}(\mathscr{C}(A))/\mathscr{F}^{n+1}$.

Lemma 8.16 (Invariance of the Kaledin Classes Under ∞-Quasi-Isomorphism) *An ∞-isomorphism between two $\mathscr{P}_\infty$-algebras $F : (H, M) \rightsquigarrow (H, N)$ induces an isomorphism of dg Lie algebras $F : \mathfrak{g}^M \to \mathfrak{g}^N$ and the following equality holds*

$$K_N = \left[F\left(\partial_\hbar M\right)\right] .$$

The same goes for the n-truncations.

Proof We refer the reader to [Emp24, Lemma 2.38]. □

Proposition 8.17 *Let (H, M) be a $\mathscr{P}_\infty$-algebra. Let $n \geqslant 1$ such that $n!$ is a unit in R. The following propositions are equivalent.*

(1) *The n-truncation K_M^n is zero.*
(2) *There exist a $\mathscr{P}_\infty$-algebra structure (H, N) with $n_0 = 0$ and $n_i = 0$ for i such that $2 \leqslant i \leqslant n$ and ∞-quasi-isomorphism*

$$(H, M) \overset{\sim}{\rightsquigarrow} (H, N) .$$

Proof Let us suppose that R is a characteristic zero field and present the proof of [MR19, Proposition 2.9]. If (2) holds, the Kaledin class K_N^n is zero and so does K_M^n by Lemma 8.16. Let us prove the converse result by induction on n. The case $n = 1$ is clear. Suppose that the result holds for $n-1$ and that $K_M^n = 0$. In particular, we also have $K_M^{n-1} = 0$ and we can assume that

$$m_2 = \cdots = m_{n-1} = 0\,,$$

without loss of generality. Thus, we have

$$K_M^n = [(n-1)m_n] = 0$$

and there exists T in $\mathfrak{g}$ such that

$$d_M(T) \equiv (n-1)m_n \pmod{\mathcal{F}^{n+1}}\,.$$

Considering the coderivation $\tau := \frac{t_{n-1}}{n-1}$, we get $[m_1, \tau] = m_n$. One can define an exponential coderivation E^τ by

$$e^\tau := \mathrm{id} + \tau + \frac{\tau^{\circ 2}}{2} + \cdots + \frac{\tau^{\circ k}}{k!} + \cdots\,.$$

We obtain a $\mathcal{P}_\infty$-algebra (H, N) by considering $N = E^\tau N E^{-\tau}$. By construction, the element E^τ induces an ∞-isomorphism

$$(H, M) \overset{\sim}{\rightsquigarrow} (H, N)\,.$$

By construction, we have $n_i = m_i$ for all $i < n$ and $n_n = m_n - [m_2, \tau] = 0$. We refer the reader to [Emp24, Proposition 2.32] for a proof over a commutative ring R. □

Theorem 8.18 *Let (A, M) be a $\mathcal{P}_\infty$-algebra such that $H(A)$ is a homotopy retract of A. If R is a $\mathbb{Q}$-algebra, the algebra (A, M) is gauge formal if and only if the Kaledin class K_{M^t} of a transferred structure $(H(A), M^t)$ is zero.*

Proof Let us fix $(H(A), M^t)$ a transferred structure. If (A, M) is gauge formal, there exists an ∞-quasi-isomorphism

$$F : (H(A), M^t) \overset{\sim}{\rightsquigarrow} (H(A), m_*)\,,$$

and $K_{M^t} = 0$, by Lemma 8.16. Conversely, suppose that $K_{M^t} = 0$. Let $\iota \geqslant 2$ the smallest integer such that $m_\iota \neq 0$. Taking up the demonstration of the previous proposition, there exists a $\mathcal{P}_\infty$-algebra structure $(H(A), N)$ where $n_1 = m_1, n_2 = \cdots = n_\iota = 0$, and an ∞-isomorphism

$$E^{\tau_\iota} : (H(A), M^t) \overset{\sim}{\rightsquigarrow} (H(A), N)\,,$$

where τ_ι is the projection of a coderivation of weight $\iota - 1$. This procedure can be iterated for any $i \geqslant \iota$. We obtain a series of ∞-isomorphism E^{τ_i}. The composition $F = \cdots \circ$

$E^{\tau_{i+1}} \circ E^{\tau_i}$ is well defined since each τ_i correspond to a coderivation of weight $i - 1$ and leads to the desired ∞-isomorphism. □

As a corollary of this Theorem, we can deduce the following very general result for descent of formality.

Theorem 8.19 (Formality Descent) *Let S be a faithfully flat commutative R-algebra. Let A be a chain complex such that $H(A)$ is an R-module of finite presentation and a homotopy retract of A. Let (A, M) be a $\mathscr{P}_\infty$-algebra. Let us denote*

$$A_S := A \otimes_R S\ .$$

(1) *Let $n \geqslant 1$ be such that $n!$ is a unit in R. The algebra (A, M) is gauge n-formal if and only if $(A_S, M \otimes 1)$ is gauge n-formal.*
(2) *If R is a $\mathbb{Q}$-algebra, the algebra (A, M) is gauge formal if and only if $(A_S, M \otimes 1)$ is gauge formal.*

Proof We refer the reader to [Emp24, Theorem 4.1]. □

Example 8.20 Combining this result with Example 8.11 or Example 8.14, we deduce that the formality result for complement of hyperplane arrangements can be descended to the localized ring $\mathbb{Z}_{(\ell)}$, see [Emp24, Theorem 4.2] and similarly for the coformality result for configuration spaces.

Acknowledgments This survey emerged from a series of three lectures given by the second named author at the workshop Higher Structures and Operadic Calculus at CRM Barcelona in June 2021. We would like to thank all participants and organizers. We are deeply grateful to Joana Cirici and Clément Dupont for a thorough review of these notes.

Both authors were supported ANR-20-CE40-0016 HighAGT.

References

[Bad02] B. Badzioch. Algebraic theories in homotopy theory. *Annals of Mathematics*, 155(3):895–913, 2002. https://doi.org/10.2307/3062135. arXiv : math/0110101.

[BdBH21] P. Boavida de Brito and G. Horel. On the formality of the little disks operad in positive characteristic. *Journal of the London Mathematical Society*, 104(2):634–667, 2021. https://doi.org/10.1112/jlms.12442. arXiv : 1903.09191.

[Ber06] J. E. Bergner. Rigidification of algebras over multi-sorted theories. *Algebraic & Geometric Topology*, 6(4):1925–1955, 2006. https://doi.org/10.2140/agt.2006.6.1925. arXiv : math/0508152.

[Ber14] A. Berglund. Homological perturbation theory for algebras over operads. *Algebraic & Geometric Topology*, 14(5):2511–2548, 2014. https://doi.org/10.2140/agt.2014.14.2511. arXiv : 0909.3485v2.

[Bor74] A. Borel. Stable real cohomology of arithmetic groups. In *Annales scientifiques de l'École Normale Supérieure*, volume 7, pages 235–272, 1974. https://doi.org/10.24033/asens.1269.

[Bre93] G. E. Bredon. *Topology and Geometry*. Number 14 in Graduate Texts in Mathematics. Springer, 1993. https://doi.org/https://doi.org/10.1007/978-1-4757-6848-0.

[BS24] A. Berglund and R. Stoll. Higher structures in rational homotopy theory. *Higher Structure and Operadic Calculus, Advanced Courses in Mathematics - CRM Barcelona, Springer*, 2024. arXiv : 2310.11824.

[Caz12] C. Cazanave. Algebraic homotopy classes of rational functions. *Ann. Sci. Éc. Norm. Supér.*, 45(4):511–534, 2012. https://doi.org/10.24033/asens.2172. arXiv : 0912.2227v2.

[CC17] D. Chataur and J. Cirici. Rational homotopy of complex projective varieties with normal isolated singularities. *Forum Mathematicum*, 29(1):41–57, 2017. https://doi.org/10.1515/forum-2015-0101. arXiv : 1503.05347.

[CC22] D. Chataur and J. Cirici. Sheaves of E-infinity algebras and applications to algebraic varieties and singular spaces. *Trans. Am. Math. Soc.*, 375(2):925–960, 2022. https://doi.org/10.1090/tran/8569. arXiv: 1811.08642.

[CH20a] H. Chu and R. Haugseng. Enriched ∞-operads. *Adv. Math.*, 361:85, 2020. https://doi.org/10.1016/j.aim.2019.106913. arXiv : 1707.08049.

[CH20b] J. Cirici and G. Horel. Mixed Hodge structures and formality of symmetric monoidal functors. *Annales Scientifiques de l'École Normale Supérieure*, 53(4):1071–1104, 2020. https://doi.org/10.24033/asens.2440. arXiv: 1703.06816.

[CH22] J. Cirici and G. Horel. Étale cohomology, purity and formality with torsion coefficients. *Journal of Topology*, 15(4):2270–2297, 2022. https://doi.org/10.1112/topo.12273. arXiv: 1806.03006.

[CH24] J. Cirici and G. Horel. Corrigendum: Étale cohomology, purity and formality with torsion coefficients. *Journal of Topology*, 17(2), 2024. https://doi.org/10.1112/topo.12348.

[CRiL24] D. Calaque and V. Roca i Lucio. Associators from an operadic point of view. *Higher Structure and Operadic Calculus, Advanced Courses in Mathematics - CRM Barcelona, Springer*, 2024. arXiv: 2402.05539.

[DCH21] G. C. Drummond-Cole and G. Horel. Homotopy transfer and formality. *Annales de l'Institut Fourier*, 71(5):2079–2116, 2021. https://doi.org/10.5802/aif.3444. arXiv: 1906.03475v1.

[Del70] P. Deligne. Théorie de Hodge. I. *Actes Congr. Internat. Math.*, 1:425–430, 1970.

[Del71] P. Deligne. Théorie de Hodge. II. *Publications Mathématiques de l'IHÉS*, 40:5–58, 1971. https://doi.org/10.1007/BF02684692.

[Del74] P. Deligne. Théorie de Hodge. III. *Publications Mathématiques de l'IHÉS*, 44:5–77, 1974. https://doi.org/10.1007/BF02685881.

[DGM00] P. Deligne, M. Goresky, and R. MacPherson. L'algèbre de cohomologie du complément, dans un espace affine, d'une famille finie de sous-espaces affines. *Michigan Mathematical Journal*, 48(1):121–136, 2000. https://doi.org/10.1307/MMJ/1030132711.

[DGMS75] P. Deligne, P. Griffiths, J. Morgan, and D. Sullivan. Real homotopy theory of Kähler manifolds. *Invent. Math.*, 29(3):245–274, 1975. https://doi.org/10.1007/BF01389853.

[DH18] C. Dupont and G. Horel. On two chain models for the gravity operad. *Proceedings of the American Mathematical Society*, 146(5):1845–1910, 2018. https://doi.org/10.1090/proc/13874. arXiv: 1702.02479.

[DM69] P. Deligne and D. Mumford. The irreducibility of the space of curves of given genus. *Inst. Hautes Études Sci. Publ. Math.*, 36:75–109, 1969. https://doi.org/10.1007/BF02684599.

[Dre15] B. Drew. Rectification of Deligne's mixed Hodge structures. *arXiv preprint*, 2015. arXiv: 1511.08288v1.

[Dri90] V. G. Drinfeld. On quasitriangular quasi-Hopf algebras and on a group that is closely connected with Gal$(\overline{\mathbb{Q}}/\mathbb{Q})$. *Algebra i Analiz*, 2:149–181, 1990.

[DSV15] V. Dotsenko, S. Shadrin, and B. Vallette. De Rham cohomology and homotopy Frobenius manifolds. *J. Eur. Math. Soc. (JEMS)*, 17(3):535–547, 2015. https://doi.org/10.4171/JEMS/510. arXiv: 1203.5077.

[Dup16] C. Dupont. Purity, formality, and arrangement complements. *International Mathematics Research Notices*, 2016(13):4132–4144, 2016. https://doi.org/10.1093/imrn/rnv260. arXiv: 1505.00717.

[EGNO15] P. Etingof, S. Gelaki, D. Nikshych, and V. Ostrik. *Tensor categories*, volume 205 of *Math. Surv. Monogr.* Providence, RI: AMS, 2015. https://doi.org/10.1090/surv/205.

[EML53] S. Eilenberg and S. Mac Lane. On the groups of $H(\Pi, n)$. I. *Ann. of Math. (2)*, 58:55–106, 1953. https://doi.org/10.2307/1969820.

[Emp24] C. Emprin. Kaledin classes and formality criteria. *arXiv preprint*, 2024. arXiv: 2404.17529.

[GM13] P. Griffiths and J. Morgan. *Rational homotopy theory and differential forms*, volume 16 of *Prog. Math.* New York, NY: Birkhäuser/Springer, 2nd revised and corrected ed. edition, 2013. https://doi.org/10.1007/978-1-4614-8468-4.

[Gro66] A. Grothendieck. On the De Rham cohomology of algebraic varieties. *Publ. Math., Inst. Hautes Étud. Sci.*, 29:95–103, 1966. https://doi.org/10.1007/BF02684807.

[GSNPR05] F. Guillén Santos, V. Navarro, P. Pascual, and A. Roig. Moduli spaces and formal operads. *Duke Mathematical Journal*, 129(2):291–335, 2005. https://doi.org/10.1215/S0012-7094-05-12924-6. arXiv: math/0402098.

[Hin97] V. Hinich. Homological algebra of homotopy algebras. *Comm. Algebra*, 25(10):3291–3323, 1997. https://doi.org/10.1080/00927879708826055. arXiv: q-alg/9702015.

[Hin03] V. Hinich. Tamarkin's proof of Kontsevich formality theorem. *Forum Mathematicum*, 15(4):591–614, 2003. https://doi.org/10.1515/form.2003.032. arXiv: math/0003052.

[Hin15] V. Hinich. Rectification of algebras and modules. *Doc. Math.*, 20:879–926, 2015. https://doi.org/10.4171/dm/508. arXiv: 1311.4130.

[Hin16] V. Hinich. Dwyer-Kan localization revisited. *Homology, Homotopy and Applications*, 18(1):27–48, 2016. https://doi.org/10.4310/HHA.2016.v18.n1.a3. arXiv: 1311.4128.

[HS79] S. Halperin and J. Stasheff. Obstructions to homotopy equivalences. *Advances in mathematics*, 32(3):233–279, 1979. https://doi.org/10.1016/0001-8708(79)90043-4.

[Kal07] D. Kaledin. Some remarks on formality in families. *Mosc. Math. J.*, 7:643–652, 2007. https://doi.org/10.17323/1609-4514-2007-7-4-643-652. arXiv: math/0509699.

[Kee92] S. Keel. Intersection theory of moduli space of stable n-pointed curves of genus zero. *Trans. Amer. Math. Soc.*, 330(2):545–574, 1992. https://doi.org/10.1090/S0002-9947-1992-1034665-0.

[Kim94] M. Kim. Weights in cohomology groups arising from hyperplane arrangements. *Proceedings of the American Mathematical Society.*, 120(3):697–703, 1994. https://doi.org/10.1090/S0002-9939-1994-1179589-0.

[Kon99] M. Kontsevich. Operads and motives in deformation quantization. *Letters in Mathematical Physics*, 48(1):35–72, 1999. https://doi.org/10.1023/A:1007555725247. arXiv: math/9904055.

[Lur17] J. Lurie. *Higher Algebra*. 2017. Can be found at Higher Algebra.

[LV12] J.-L. Loday and B. Vallette. *Algebraic operads*, volume 346 of *Grundlehren der Mathematischen Wissenschaften*. Springer, 2012. https://doi.org/10.1007/978-3-642-30362-3.

[LV14] P. Lambrechts and I. Volić. *Formality of the little N-disks operad*, volume 1079 of *Mem. Am. Math. Soc.* Providence, RI: American Mathematical Society (AMS), 2014. https://doi.org/10.1090/memo/1079. arXiv: abs/0808.0457.

[Mat06] D. Matei. Massey products of complex hypersurface complements. In *Singularity theory and its applications*, pages 205–219. Mathematical Society of Japan, 2006. https://doi.org/10.2969/ASPM/04310205. arXiv: math/0505391.

[Mor78] J. Morgan. The algebraic topology of smooth algebraic varieties. *Publications Mathématiques de l'IHÉS*, 48:137–204, 1978. https://doi.org/10.1007/BF02684316.

[MR19] V. Melani and M. Rubió. Formality criteria for algebras over operads. *Journal of Algebra*, 529:65–88, 2019. https://doi.org/10.1016/j.jalgebra.2019.03.016. arXiv: 1712.09229v2.

[NA87] V. Navarro-Aznar. Sur la théorie de Hodge–Deligne. *Invent. Math.*, 90:11–76, 1987. https://doi.org/10.1007/BF01389031.

[Pet14] D. Petersen. Minimal models, GT-action and formality of the little disk operad. *Selecta Math.*, 20(3):817–822, 2014. https://doi.org/10.1007/s00029-013-0135-5. arXiv: 1303.1448.

[Pet22] D. Petersen. A remark on singular cohomology and sheaf cohomology. *Math. Scand.*, 128(2):229–238, 2022. https://doi.org/10.7146/math.scand.a-132191. arXiv: 2102.06927.

[PY99] S. Papadima and S. Yuzvinsky. On rational $K[\pi, 1]$ spaces and Koszul algebras. *Journal of Pure and Applied Algebra*, 144(2):157–167, 1999. https://doi.org/10.1016/S0022-4049(98)00058-9.

[RiL22] V. Roca i Lucio. Curved operadic calculus. *Bulletin de la Société Mathématique de France*, 152(1):45–147, 2022. arXiv: 2201.07155.

[Sal17] B. Saleh. Noncommutative formality implies commutative and Lie formality. *Algebraic & Geometric Topology*, 17(4):2523–2542, 2017. https://doi.org/10.2140/agt.2017.17.2523. arXiv: 1609.02540v2.

[SGA] SGA4. Théorie des topos et cohomologie étale des schémas. Tome 3. Séminaire de Géométrie Algébrique du Bois-Marie, dirigé par M. Artin, A. Grothendieck, J.-L. Verdier, 1963-1964. *LSLN Springer-Verlag*, (269, 270, 305). https://doi.org/10.1007/BFb0070714.

[Shi23] T. Shin. Prismatic cohomology and p-adic homotopy theory. *J. Homotopy Relat. Struct.*, 18(4):521–541, 2023. https://doi.org/10.1007/s40062-023-00335-0. arXiv : 2107.02256.

[Tam98] D. E. Tamarkin. Another proof of M. Kontsevich formality theorem. 1998. https://doi.org/10.48550/arXiv.math/980302510.48550/arXiv.math/9803025. arXiv : math/9803025.

[Vai19] D. Vaintrob. Moduli of framed formal curves. *arXiv preprint* 1910.11550, 2019.

[Vai21] D. Vaintrob. Formality of little disks and algebraic geometry. *arXiv e-print* 2103.15054, 2021.

[VdL03] P. Van der Laan. Coloured Koszul duality and strongly homotopy operads. *arXiv preprint*, 2003. arXiv : math/0312147.

[Voi02] C. Voisin. *Hodge Theory and Complex Algebraic Geometry I*. Cambridge Studies in Advanced Mathematics. Cambridge University Press, 2002. https://doi.org/10.1017/CBO9780511615344.

[War21] B. Ward. Massey Products for Graph Homology. *International Mathematics Research Notices*, 2022(11):8086–8161, 01 2021. https://doi.org/10.1093/imrn/rnaa346. arXiv: 1903.12055v3.

Associators from an Operadic Point of View

Damien Calaque and Victor Roca i Lucio

Contents

D. Calaque (✉)
IMAG, Université de Montpellier, CNRS, Montpellier, France
e-mail: damien.calaque@umontpellier.fr

V. Roca i Lucio
Université Paris Cité and Sorbonne Université, CNRS, Paris, France
e-mail: rocalucio@imj-prg.fr

B. Vallette (ed.), *Higher Structures and Operadic Calculus*, Advanced Courses in Mathematics - CRM Barcelona, https://doi.org/10.1007/978-3-031-77779-0_4

1 Introduction: Drinfeld Associators

1.1 A Deformation Quantization Problem

One of the original motivations for the introduction of associators by Drinfeld in [Dri91] was a problem of deformation quantization, occuring in quantum group theory. In short terms, it asks for a universal quantization of infinitesimally braided monoidal categories; we are going to formulate the problem precisely.

Let $\mathfrak{g}$ be a Lie algebra over $\Bbbk$, a field of characteristic zero. The category $\mathsf{Rep}(\mathfrak{g})$ of $\mathfrak{g}$-representations is a $\Bbbk$-linear symmetric monoidal category, with monoidal product the tensor product of representations, and monoidal unit the trivial representation $\Bbbk$.

Let now t be a Casimir element, that is, an element in $S^2(\mathfrak{g})^{\mathfrak{g}}$.

Remark 1.1 Notice that $S^2(\mathfrak{g})^{\mathfrak{g}} \subset (\mathfrak{g} \otimes \mathfrak{g})^{\mathfrak{g}} \subset (\mathfrak{U}(\mathfrak{g})^{\otimes 2})^{\mathfrak{g}}$, therefore t acts naturally on any tensor product of two representations of $\mathfrak{g}$. The induced natural transformation from $\otimes$ to itself is called an *infinitesimal braiding*, and turns $\mathsf{Rep}(\mathfrak{g})$ into an *infinitesimally braided monoidal category*. Infinitesimally braided monoidal categories are braided monoidal $(\Bbbk[\hbar]/\hbar^2)$-linear categories that are symmetric monoidal mod $\hbar$; i.e. they are first order braided deformations of symmetric monoidal $\Bbbk$-linear categories. They can be viewed as Gerstenhaber algebras (also known as $\mathbb{P}_2$-algebra) inside a suitable 2-category of $\Bbbk$-linear categories.

Problem 1 Given a pair $(\mathfrak{g}, t)$ as above, find a formal deformation of the symmetric monoidal category $(\mathsf{Rep}(\mathfrak{g}), \otimes, \Bbbk)$ as a braided monoidal category over $\Bbbk[[\hbar]]$ such that:

(1) The braiding σ can be written as $\sigma = R\tau$, where τ is the natural symmetry isomorphism of $\mathsf{Rep}(\mathfrak{g})$ and

$$R = 1 \otimes 1 + \frac{t}{2}\hbar + O(\hbar^2) \quad \text{in} \quad (\mathfrak{U}(\mathfrak{g})^{\otimes 2})^{\mathfrak{g}}[[\hbar]] .$$

(2) The associator Φ can be written as[1]

$$\Phi = 1 \otimes 1 \otimes 1 + O(\hbar) \quad \text{in} \quad (\mathfrak{U}(\mathfrak{g})^{\otimes 3})^{\mathfrak{g}}[[\hbar]] .$$

Remark 1.2 In view of Remark 1.1, Problem 1 asks for the existence of a formal extension to all orders of the first order braided deformation of $\mathsf{Rep}(\mathfrak{g})$ given by t. Note that braided monoidal categories can be viewed as $\mathbb{E}_2$-algebras in the 2-category of categories. Moreover, the Gerstenhaber operad being the homology of the $\mathbb{E}_2$ operad, one can state the problem of quantizing a Gerstenhaber algebra into an $\mathbb{E}_2$-algebra. In other words, Problem 1 amounts to searching for a deformation quantization of the Gerstenhaber algebra $\mathsf{Rep}(\mathfrak{g})$ as an $\mathbb{E}_2$-algebra in a suitable 2-category of $\mathbb{k}$-linear categories.

1.2 Universal Reformulation

Drinfeld noticed in [Dri91] that one could restrict to the case where $R = e^{\frac{\hbar t}{2}}$ and where Φ is given by a universal formula involving only copies of t and iterated Lie brackets. This is analogous to what happens with the Baker–Campbell–Hausdorff formula, that only involves iterated Lie brackets of two elements of a given Lie algebra. In order to reformulate this formal deformation problem in a universal manner, one first introduces the family $(\mathfrak{t}_n)_{n\geq 0}$ of *Drinfeld–Kohno* Lie algebras. In analogy with the Baker–Campbell–Hausdorff universal formula living in the free complete Lie algebra on two generators, a universal formula for Φ shall live in a completion of the universal enveloping algebra of $\mathfrak{t}_3$.

Definition 1.3 (Drinfeld–Kohno Lie Algebras) The n-th *Drinfeld–Kohno* Lie algebra $\mathfrak{t}_n$ is given by the following presentation:

$$\mathfrak{t}_n := \frac{\mathcal{L}ie\{t_{ij} \mid 1 \leq i, j \leq n, i \neq j\}}{\left(t_{ij} = t_{ji}\ ,\ [t_{ij}, t_{ik} + t_{kj}] = 0\ ,\ [t_{ij}, t_{kl}] = 0\right)} .$$

Assigning degree 1 to generators turns $\mathfrak{t}_n$ into a graded Lie algebra.

Note that $\mathfrak{t}_0 = \mathfrak{t}_1 = \{0\}$.

[1] Strictly speaking, the associator should be Φa, where a is the natural associativity isomorphism of $\mathsf{Rep}(\mathfrak{g})$. But thanks to the Mac Lane coherence theorem, we can pretend that a is the identity.

The n-th Drinfeld–Kohno Lie algebra is indeed universal in the following sense:

Lemma 1.4 *Let $\mathfrak{g}$ be a Lie algebra together with a Casimir element*

$$t = \sum_{\alpha \in I} a_\alpha \otimes b_\alpha \in S^2(\mathfrak{g})^{\mathfrak{g}}\,.$$

The assignment

$$\begin{aligned} \varphi_t^n : \mathfrak{t}_n &\longrightarrow (\mathfrak{U}(\mathfrak{g})^{\otimes n})^{\mathfrak{g}} \\ t_{ij} &\longmapsto \sum_{\alpha \in I} a_\alpha^{(i)} b_\alpha^{(j)} \end{aligned}$$

defines a morphism of Lie algebras. Here, $x^{(i)} := 1^{\otimes i-1} \otimes x \otimes 1^{\otimes n-i}$, and $\mathfrak{U}(\mathfrak{g})^{\otimes n}$ is endowed with the Lie bracket given by the commutator of the associative product.

Proof It is enough to check that the images of the generators t_{ij} satisfy the relations defining Drinfeld–Kohno Lie algebra. Since t is symmetric, $\varphi_t^n(t_{ij}) = \varphi_t^n(t_{ji})$. One can check that since t is $\mathfrak{g}$-invariant, we have

$$[\varphi_t^n(t_{ij}), \varphi_t^n(t_{ik}) + \varphi_t^n(t_{kj})] = \sum_{\alpha,\beta \in \mathrm{I}} \left(\left[a_\alpha^{(i)} b_\alpha^{(j)}, a_\beta^{(i)} b_\beta^{(k)} \right] + \left[a_\alpha^{(i)} b_\alpha^{(j)}, a_\beta^{(j)} b_\beta^{(k)} \right] \right) = 0\,.$$

The relation $[\varphi_t^n(t_{ij}), \varphi_t^n(t_{kl})] = 0$ follows from the definition of φ_t^n. □

Since the universal enveloping algebra is left adjoint, we get a universal morphism $\varphi_t^n : \mathfrak{U}(\mathfrak{t}_n) \longrightarrow (\mathfrak{U}(\mathfrak{g})^{\otimes n})^{\mathfrak{g}}$ of associative algebras for any pair $(\mathfrak{g}, t)$. This indicates that a universal reformulation of Problem 1 can be made using the Drinfeld–Kohno Lie algebras.

Remark 1.5 A useful variant of the above, where one sends t_{ij} to

$$\hbar \sum_{\alpha \in \mathrm{I}} a_\alpha^{(i)} b_\alpha^{(j)}$$

gives an algebra morphism $\varphi_{\hbar t}^n : \hat{\mathfrak{U}}(\mathfrak{t}_n) \to (\mathfrak{U}(\mathfrak{g})^{\otimes n})^{\mathfrak{g}}[[\hbar]]$, where $\hat{\mathfrak{U}}(\mathfrak{t}_n)$ is the degree completion with respect to the grading from Definition 1.3 (given by assigning degree 1 to the generators t_{ij}).

The family of Drinfeld–Kohno Lie algebras $(\mathfrak{t}_n)_{n \geq 0}$ comes equipped with an extra structure, often called *insertion-coproduct morphisms.*

Definition-Proposition 1.6 (Insertion-Coproduct Morphisms) For every partially defined map $f : \{1, \ldots, m\} \to \{1, \ldots, n\}$, that one can define as a pointed map $\{*, 1, \ldots, m\} \to \{*, 1, \ldots, m\}$, there is a Lie algebra morphism $(-)^f : \mathfrak{t}_n \longrightarrow \mathfrak{t}_m$

that sends t_{ij} to

$$t_{ij}^{f} := \sum_{\substack{k \in f^{-1}(i) \\ l \in f^{-1}(j)}} t_{kl} \,.$$

Insertion-coproduct morphisms are compatible with composition, making $n \mapsto \mathfrak{t}_n$ a presheaf of graded Lie algebras on (the skeleton of) the category Fin_* of finite pointed sets

Notation *Since a partially defined map $f : \{1, \ldots, m\} \to \{1, \ldots, n\}$ is completely determined by the familly of level sets $(f^{-1}(i))_{1 \le i \le n}$, one sometimes write $(-)^{f^{-1}(1), \ldots, f^{-1}(n)} = (-)^f$. For instance, if $x \in \mathfrak{t}_3$, then $x^{13,2,5} \in \mathfrak{t}_6$ corresponds to x^f for $f : \{1, \ldots, 6\} \to \{1, 2, 3\}$ defined by $f^{-1}(1) = \{1, 3\}$, $f^{-1}(2) = \{2\}$, and $f^{-1}(3) = \{5\}$.*

Observe that $(\mathfrak{U}(\mathfrak{g})^{\otimes n})^{\mathfrak{g}}$ is isomorphic to the algebra $\mathsf{End}(\otimes^n)$ of natural endomorphisms of the n-fold tensor product functor $\otimes^n : \mathsf{Rep}(\mathfrak{g})^n \to \mathsf{Rep}(\mathfrak{g})$.[2] For every partially defined map $f : \{1, \ldots, m\} \to \{1, \ldots, n\}$ there is a functor $f_* : \mathsf{Rep}(\mathfrak{g})^m \to \mathsf{Rep}(\mathfrak{g})^n$ taking $(V_1, \ldots, V_m)$ to $(W_1, \ldots, W_n)$ with $W_i := \otimes_{j \in f^{-1}(i)} V_j$. Precomposing with f_* gives an algebra map $(-)^f : \mathsf{End}(\otimes^n) \to \mathsf{End}(\otimes^m)$, making $n \mapsto \mathsf{End}(\otimes^n) \cong (\mathfrak{U}(\mathfrak{g})^{\otimes n})^{\mathfrak{g}}$ a presheaf of associative algebras on Fin_*.

We let the reader check the following properties:

- The morphisms $(-)^f : (\mathfrak{U}(\mathfrak{g})^{\otimes n})^{\mathfrak{g}} \to (\mathfrak{U}(\mathfrak{g})^{\otimes m})^{\mathfrak{g}}$ can be expressed using coproducts, insertion of 1's, and counits (that one can see as degenerate iterated coproducts), whence the name *insertion-coproduct morphisms*;
- The morphisms $\varphi_t^n : \mathfrak{U}(\mathfrak{t}_n) \longrightarrow (\mathfrak{U}(\mathfrak{g})^{\otimes n})^{\mathfrak{g}}$ from Lemma 1.4 commute with insertion-coproduct morphisms, thus defining a morphism of presheaves of algebras on Fin_*. The same remains true with the variant φ_{ht}^n from Remark 1.5.

Going back to Problem 1, the conditions that $R \in \mathsf{Aut}(\otimes^2)$ and $\Phi \in \mathsf{Aut}(\otimes^3)$ must satisfy[3] to define a braided monoidal structure on $\mathsf{Rep}(\mathfrak{g})$, where the monoidal product functor $\otimes$ and the monoidal unit remain the same, can be rephrased in terms of insertion-coproduct morphisms. We refer to [JS93] for the original conditions that we now state in this "new" way:

(1) Unit condition: $\Phi^{1,\emptyset,2} = \mathrm{id}$, in $\mathsf{End}(\otimes^2)$.
(2) Inverse condition: $\Phi^{-1} = \Phi^{3,2,1}$, in $\mathsf{End}(\otimes^3)$.
(3) Pentagon equation: $\Phi^{2,3,4}\Phi^{1,23,4}\Phi^{1,2,3} = \Phi^{1,2,34}\Phi^{12,3,4}$, in $\mathsf{End}(\otimes^4)$.
(4) Hexagon equations: $\Phi^{2,3,1}R^{1,23}\Phi^{1,2,3} = R^{1,3}\Phi^{2,1,3}R^{1,2}$, and the same with $(R^{2,1})^{-1}$ instead of R, in $\mathsf{End}(\otimes^3)$.

[2] Strictly speaking, one should fix a choice of parenthesization on $\otimes^n$.

[3] Following footnotes 1 and 2, stricly speaking, one shall have $\Phi \in \mathsf{Aut}((- \otimes -) \otimes -)$.

This naturally leads to the following universal version of Problem 1:

Problem 2 Find an invertible element $\Phi \in 1 + \hat{\mathfrak{U}}(\mathfrak{t}_3)^{\geq 1}$ such that

(1) Unit condition: $\Phi^{1,\emptyset,2} = 1$, in $\hat{\mathfrak{U}}(\mathfrak{t}_2)$.
(2) Inverse condition: $\Phi^{-1} = \Phi^{3,2,1}$, in $\hat{\mathfrak{U}}(\mathfrak{t}_3)$.
(3) Pentagon equation: $\Phi^{2,3,4}\Phi^{1,23,4}\Phi^{1,2,3} = \Phi^{1,2,34}\Phi^{12,3,4}$, in $\hat{\mathfrak{U}}(\mathfrak{t}_4)$.
(4) Hexagon equations: $\Phi^{2,3,1}e^{\pm\frac{t_{12}+t_{13}}{2}}\Phi^{1,2,3} = e^{\pm\frac{t_{13}}{2}}\Phi^{2,1,3}e^{\pm\frac{t_{12}}{2}}$ in $\hat{\mathfrak{U}}(\mathfrak{t}_3)$.

It is clear from the above discussion that every solution to Problem 2 is sent, using the morphism $\varphi^{\bullet}_{\hbar t}$ of presheaves of algebras on Fin$_*$, to a solution of Problem 1 such that $R = e^{\frac{\hbar t}{2}}$.

1.3 Drinfeld Associators

Observe that the center of $\mathfrak{t}_3$ is one dimensional and generated by $c := t_{12} + t_{13} + t_{23}$. Hence we have a Lie algebra isomorphism

$$\mathfrak{t}_3 \cong \mathfrak{f}_2 \oplus \Bbbk c\,,$$

where $\mathfrak{f}_2$ is the free Lie algebra on two generators $x = t_{12}$ and $y = t_{23}$. Through this identification, the Lie algebra morphism $(-)^{1,\emptyset,2} : \mathfrak{t}_3 \to \mathfrak{t}_2$ sends x and y to 0 and c to t_{12}.

Given a solution Φ of Problem 2, the unit condition (1) is equivalent to requiring that

$$\Phi = \Phi(x, y) \in \hat{\mathfrak{U}}(\mathfrak{f}_2) \subset \hat{\mathfrak{U}}(\mathfrak{t}_3)\,.$$

Then the inverse condition (2) reads as $\Phi(x, y)^{-1} = \Phi(y, x)$, and the pentagon equation (3) as

$$\begin{aligned}\Phi(t_{23}, t_{34})\Phi(t_{12} + t_{13}, t_{24} + t_{34})\Phi(t_{12}, t_{23}) \\ = \Phi(t_{12}, t_{23} + t_{24})\Phi(t_{13} + t_{23}, t_{34})\,.\end{aligned} \qquad (\pentagon)$$

Finally, the hexagon equations (4) become equivalent to the single equation

$$e^{x/2}\Phi(y, x)e^{y/2}\Phi(z, y)e^{z/2}\Phi(x, z) = 1 \qquad (\hexagon)$$

in the complete associative algebra $\Bbbk\langle\langle x, y, z\rangle\rangle/(x + y + z)$.

Definition 1.7 (Drinfeld Associator) A Drinfeld 1-associator (with coefficients in $\Bbbk$) is a group-like element $\Phi(x, y) \in \exp(\widehat{\mathfrak{f}_2})$ such that $\Phi(x, y)^{-1} = \Phi(y, x)$ and sastisfying ($\pentagon$) and ($\hexagon$). The set of Drinfeld 1-associators (with coefficients in $\Bbbk$) is denoted $\mathrm{Assoc}_1(\Bbbk)$.

From the above discussion, one sees that Drinfeld 1-associators are in bijection with group-like solutions to Problem 2.

Remark 1.8 One can more generally define Drinfeld λ-associators for any $\lambda \in \Bbbk$: all conditions remain the same except for Eq. ($\bigcirc$), that becomes

$$e^{\lambda x/2}\Phi(y, x)e^{\lambda y/2}\Phi(z, y)e^{\lambda z/2}\Phi(x, z) = 1 .$$

For every $\lambda \neq 0$, rescaling x and y (to λx and λy) gives a bijection between $\mathrm{Assoc}_1(\Bbbk)$ and the set $\mathrm{Assoc}_\lambda(\Bbbk)$ of Drinfeld λ-associators.

Theorem 1.9 ([Dri91]) *Let $\Bbbk$ be a field of characteristic zero. The set of Drinfeld 1-associators with coefficients in $\Bbbk$ is non empty.*

In [Dri91], Drinfeld first constructs a 2πi-associator Φ_{KZ} with coefficients in $\mathbb{C}$, given as the renormalized holonomy from 0 to 1 of a differential equation known as the Knizhnik–Zamolodchikov equation. Then, one can rescale it to get a 1-associator with coefficients in $\mathbb{C}$. One can finally use descent methods to prove existence over $\mathbb{Q}$ (this uses the fact that associators actually form a torsor).

1.4 Motivations and Perspectives

1.4.1 The Operadic Approach

The insertion-coproduct morphisms defined above can be used to endow the family $(\mathfrak{t}_n)_{n\geq 0}$ with an operad structure in the symmetric monoidal category of Lie algebras with the direct sum $(\mathsf{Lie}, \oplus, 0)$. This will allow us to restate the above definition in an operadic fashion (see Sect. 2). Indeed, using this operad structure, one can write the set of associators as a set of isomorphisms between two different operads. This set is thus a torsor over the respective automorphism groups, which are none other than the *Grothendieck–Teichmüller* group and the *graded Grothendieck–Teichmüller* group. It is Tamarkin who, inspired by Bar-Natan's work [Bar98], first came up with this operadic picture in his proof [Tam03] of the rationnal formality of the little disks operad se also [Tam02]).

The work of Tamarkin [Tam03, Tam02] (see also [Kon99a]) actually suggests a homotopical version of this operadic approach to the torsor of associators and the *Grothendieck–Teichmüller* groups, that has been achieved in [Fre17b].

1.4.2 Motivic Aspects

Drinfeld's proof [Dri91] of the existence of an associator with complex coefficients is of motivic nature, in the following sense. An associator, called the Knizhnik–Zamolodchikov (KZ) associator, is constructed as the regularized holonomy of an algebraic flat connection defined on $\mathbb{P}^1 - \{0, 1, +\infty\}$ (the KZ connection). The coefficients of the KZ associator Φ_{KZ} are therefore periods (in the sense of [KZ01]); in fact, Φ_{KZ} is a generating series for very specific periods known as multiple zeta values. We refer to [LM96] for a proof, and [Zag94] for generalities on multiple zeta values (MZV). The proof that Φ_{KZ} satisfies the defining equations of a Drinfeld ($2\pi i$-)associator only involves "natural" relations between these periods (because they are all proven by saying that the holonomy of a flat connection along a contractible loop is trivial); natural (actually, *motivic*) relations are linearity, change of variables and Stokes formula (see e.g. [Kon99b]).

Brown [Bro12] proved that MZV linearly span (over $\mathbb{Q}[\frac{1}{2\pi i}]$) all periods of mixed Tate motives that are unramified over $\mathbb{Z}$. This implies that there is an injective morphism $G_{\mathcal{M}_T(\mathbb{Z})_{\mathbb{Q}}} \hookrightarrow \mathrm{GRT}(\mathbb{Q})$, where

- $G_{\mathcal{M}_T(\mathbb{Z})_{\mathbb{Q}}}$ is the Galois group of the tannakian category $\mathcal{M}_T(\mathbb{Z})_{\mathbb{Q}}$ of mixed Tate motives that are unramified over $\mathbb{Z}$ (see [Gon01b, DG05]);
- $\mathrm{GRT}(\mathbb{Q})$ is the graded Grothendieck–Teichmüller group already mentionned above (see also Sect. 2.5 below).

It turns out that both $G_{\mathcal{M}_T(\mathbb{Z})_{\mathbb{Q}}}$ and $\mathrm{GRT}(\mathbb{Q})$ are semi-direct products of the multiplicative group $\mathbb{G}_m$ with a pro-unipotent $\mathbb{Q}$-group. More importantly, Goncharov [Gon01b] proved that in the case of $G_{\mathcal{M}_T(\mathbb{Z})_{\mathbb{Q}}}$ the graded Lie algebra of the associated pro-unipotent $\mathbb{Q}$-group is free with one generator in each odd degree ≥ 3.

It is conjectured that the morphism $G_{\mathcal{M}_T(\mathbb{Z})_{\mathbb{Q}}} \to \mathrm{GRT}(\mathbb{Q})$ is an isomorphism. This would imply that the relations among MZV given by the defining equations of Drinfeld associators imply all the relations of motivic nature between MZV.

There is another set of relations between MZV, called (regularized) double shuffle relations. They are essentially combinatorial relations obtained from formally manipulating the two standard ways of defining MZV: the one *via* infinite sums and the one *via* iterated integrals. Indeed, the product of two Euler sums is a linear combination of Euler sums (indexed by shuffles) and the product of two iterated integrals is a linear combination of iterated integrals (indexed by quasi-shuffles). These relations were systematically studied by Racinet [Rac02], who proved that they are implied by the motivic relations. Furusho [Fur11] proved they are also implied by the associator relations.

1.4.3 Known Examples of Drinfeld Associators

Apart from the KZ associator Φ_{KZ} (and its "complex conjugate" $\Phi_{\overline{KZ}}$), another associator Φ_{AT} has been constructed by Alekseev and Torossian [AT10, ŠW11] using integrals over compactified configuration spaces of points in the plane. Using similar techniques, a whole family $\{\Phi^t\}_{t\in[0,1]}$ of Drinfeld associators such that $\Phi^0 = \Phi_{KZ}$, $\Phi^{1/2} = \Phi_{AT}$ and $\Phi^1 = \Phi_{\overline{KZ}}$ was constructed by Rossi and Willwacher [RW14].

1.4.4 Generalizations

The whole story of Drinfleld associators is very much related to the geometry of configuration spaces of points on a genus zero surface. Variants of the KZ connection for a higher genus surface Σ have been constructed, as well as twisted versions by a finite group Γ acting on the surface. In several cases, one can guess the correct definition of an associator for such a data by inspecting the natural relations satisfied by holonomies of the corresponding connection. A more systematic approach for defining and studying associators uses operads (operads and their siblings provide a consistent way of organizing abstract relations satisfied by these holonomies). The table below summarizes the state of the art, leaving aside the motivic aspects (that are not our main focus in this survey).

Genus	Group	KZ connect.	Associator type	Operadic def.
0	1	[Dri91]	Drinfeld [Dri91]	[Tam03, Fre17a]
0	$\mathbb{Z}/n\mathbb{Z}$	[Enr07]	cyclotomic [Enr07]	[CG20a]
0	$\Gamma \subset \mathrm{PSU}(2)$	[Maa19]	*unknown*	*unknown*
1	1	[CEE09]	elliptic [Enr14a]	[CG23]
1	$\mathbb{Z}/n\mathbb{Z} \times \mathbb{Z}/m\mathbb{Z}$	[CG20b]	ellipsitomic [CG23]	[CG23]
≥ 1	1	[Enr14b]	higher genus [Gon20]	[Gon20, CIW20]

Note that in genus $g > 1$, since the surface doesn't have a framing, one needs to add the framing data to configuration spaces of points (each point now has a unit tangent vector attached to it). To this day, it is still an open problem to prove the existence of an associator (in the sense of [Gon20] in this case) using the connection from [Enr14b] (or, better, a framed version). It is nevertheless expected that the data of a genuine Drinfeld associator should be enough to construct higher genus associators ; this is the case in genus 1 (see [CEE09, Enr14a]). For higher genus associators, this willl appear in [Bro24]. A similar result is proven in [CIW20] for a more homotopical definition of higher genus associators that should be compared with the one from [Gon20].

The goal of this survey is to review the operadic approach, in genus 0 and in genus 1

1.5 Plan of the Paper

Section 2 is devoted to the first line of the above table. We start with some recollections on braid groups, configurations spaces, operads and groupoids, in Sects. 2.1 and 2.2. We then introduce the two main players in our story:

(1) the (pro-unipotent completion of the) operad of parenthezised braids (Sect. 2.3).
(2) the (degree completion of the) operad of parenthezised chord diagrams (Sect. 2.4).

We move on in Sect. 2.5 with the operadic definition of an associator, which is simply an isomorphism from the first player to the second one. The set of associators naturally becomes a bitorsor, with the two acting groups being the automorphism groups of both players. The presentation by generators and relations of the operad of parenthezised braids given in Sect. 2.6 allows to give an explicit description of the bitorsor of associators, and in particular to show that it is consistent with the Definition 1.7 motivated by Problem 2. Finally, Sect. 2.7 provides a more topological description of the operad of parenthezised braids.

The aim of Sect. 3 is to tell a parallel story for the second line of the above table: Enriquez's cyclotomic associators [Enr07]. We start in Sect. 3.1 with some motivations; the reader can view it as a condensed version of our introductory Sect. 1 for the cyclotomic case. We then review moperads in Sect. 3.2, which were introduced by Willwacher in [Wil16], and which play an analogue role to operads in the cyclotomic picture. Furthermore, we prove that this structure naturally appears when one considers the derivative (in the sense of species) of an operad. This construction recovers many of the examples that appear in this text. In Sect. 3.3 we construct the moperad of parenthesized cyclotomic braids, which will be the first player of this story. And in Sect. 3.4, we construct the moperad of infinitesimal cyclotomic braids, which is our second player. This allows us to define cyclotomic associators purely in terms of isomorphisms of moperads in Sect. 3.5. In Sect. 3.6, we give an explicit description of these cyclotomic associators and finally in Sect. 3.7 we give a topological interpretation of some of the constructions performed.

Finally, in Sect. 4, we tell the story of the fourth line of the table and give a quick overview of the fifth line. The structure remains the same. We start with some motivation in Sect. 4.1. We introduce the first player, the right module of parenthesized elliptic braids in Sect. 4.2; then we introduce the second player, the right module of infinitesimal elliptic braids in Sect. 4.3. We then define elliptic associators as right module isomorphisms of the two in Sect. 4.4. Using a presentation of the right module of parenthesized elliptic braids, we give a more concrete description of elliptic associators in Sect. 4.5. We then present the topological point of view on these objects in Sect. 4.6. Finally, we give a quick overview on the ellipsitomic case in Sect. 4.7: this case can be essentially thought as a combination of the elliptic case with the cyclotomic case.

1.6 Conventions

Let us state here some of the conventions we will adopt throughout this survey.

(1) *Convention for the generators of the pure braid groups:* We choose the following generators for the pure braid groups PB_n: x_{ij} is the pure braid that goes *from* the i-th strand *in front* of the others, does a loop around the j-th strand, and comes back *in front* of the other strands:

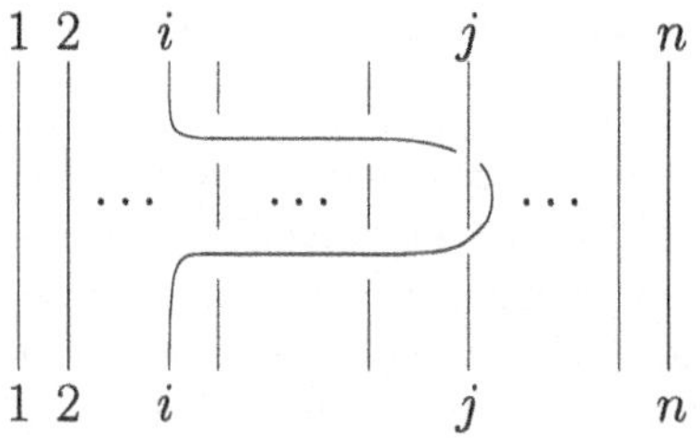

(2) *Topological convention for categories:* We view braid groups as homotopy groups, where the multiplication is given by the concatenation of paths.

This entails that, when working with categories, we *concatenate* arrows instead of composing them. If a, b, c are objects in a category C, we denote

$$a \xrightarrow{f} b \xrightarrow{g} c$$

by $f\,g$ instead of $g\,f$. In other words, $f\,g := g \circ f$.

(3) *Symmetric group actions:* If M is a symmetric sequence, every $M(n)$ is endowed with a *right* $\mathbb{S}_n$-action, where $\mathbb{S}_n$ stands for the n-th symmetric group. Nevertheless, we will often consider a *left* $\mathbb{S}_n$-action, as it simplifies diagrams. In order to pass from one to the other, one merely needs to pre-compose the action by the involution which sends an element to its inverse.

2 Operadic Approach to Drinfeld Associators

2.1 Braid Groups and Configuration Spaces

The space of configurations of n ordered points in the complex plane $\mathrm{Conf}_n(\mathbb{C})$ is given by

$$\mathrm{Conf}_n(\mathbb{C}) := \left\{(x_1, \cdots, x_n) \in \mathbb{C}^n \mid x_i \neq x_j \text{ if } i \neq j\right\}.$$

for $n \geq 0$. It is a path-connected topological space, equipped with an action of the symmetric group $\mathfrak{S}_n$ (which permutes the indices of the points) for all n in $\mathbb{N}$.

Definition 2.1 (Pure Braid Groups) The *pure braid group* PB_n on n strands is given by the fundamental group

$$\mathrm{PB}_n := \pi_1(\mathrm{Conf}_n(\mathbb{C})).$$

Elements of the pure braid group PB_n can be represented as braids. One can choose a point $(x_1, \cdots, x_n)$ in $\mathrm{Conf}_n(\mathbb{C})$ (since it is path-connected, two choices yield isomorphic groups), and represent elements in PB_n as braids with n strands. Notice that the i-th strand of a braid in the pure braid group starts at i-th spot and must also finish at i-th spot. For

instance, the following braid

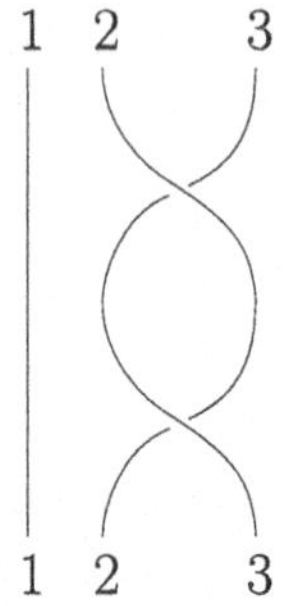

is an element of PB_3.

Theorem 2.2 *The pure braid group* PB_n *on n strands admits the following presentation.*

(1) *It is generated by* $\{x_{ij}\}$ *for* $1 \leq i < j \leq n$. *These elementary pure braids can be represented as*

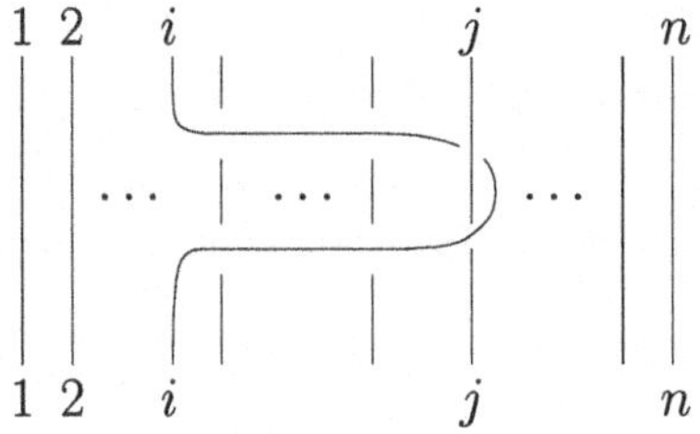

(2) *The generators are subject to the following relations*

(a) $[x_{ij}, x_{kl}] = 1$ *for* $i < j < k < l$,
(b) $[x_{il}, x_{jk}] = 1$ *for* $i < j < k < l$,
(c) $x_{ik}x_{jk}x_{ij} = x_{jk}x_{ij}x_{ik} = x_{ij}x_{ik}x_{jk}$ *for* $i < j < k$,
(d) $[x_{kl}x_{ik}x_{kl}^{-1}, x_{jl}] = 1$ *for* $i < j < k < l$.

There is a fiber sequence of topological spaces

$$\mathbb{C} - \{x_1, \cdots, x_n\} \hookrightarrow \mathrm{Conf}_{n+1}(\mathbb{C}) \twoheadrightarrow \mathrm{Conf}_n(\mathbb{C}) ,$$

given by forgetting the last point of the configuration. It is split by the morphism $(x_1, \ldots, x_n) \mapsto (x_1, \ldots, x_n, 1 + \sum_{i=1}^n |x_i|)$. This induces a split short exact sequence of homotopy groups

$$1 \longrightarrow \mathbb{F}_n \longrightarrow \mathrm{PB}_{n+1} \longrightarrow \mathrm{PB}_n \longrightarrow 1 ,$$

since all the higher homotopy groups are trivial. Here and below, $\mathbb{F}_n$ denotes the free group on n generators. In particular, there is the following decomposition

$$\mathrm{PB}_n \cong \mathbb{F}_n \rtimes (\mathbb{F}_{n-1} \rtimes \cdots (\mathbb{F}_2 \rtimes \mathbb{F}_1))) ,$$

which follows inductively from the previous split short exact sequence.

Lemma 2.3 *There is an isomorphism of groups*

$$\mathrm{PB}_3 \cong \mathbb{F}_2 \times \mathbb{F}_1 ,$$

where the generators of $\mathbb{F}_2$ *are sent to* x_{12} *and* x_{23} *and where the generator of* $\mathbb{F}_1$ *is sent to* $x_{12}x_{13}x_{23}$.

Proof In this particular case, one can choose another split fibration

$$\mathbb{C} - \{x_1, x_3\} \hookrightarrow \mathrm{Conf}_3(\mathbb{C}) \twoheadrightarrow \mathrm{Conf}_2(\mathbb{C}) .$$

Here we forget the second point, and the splitting is given by sending a configuration in $\mathrm{Conf}_2(\mathbb{C})$ to the configuration given by keeping the two original points and adding a point in the middle of the two. Given this splitting, one can directly check that the corresponding semi-direct product

$$\mathrm{PB}_3 \cong \mathbb{F}_2 \rtimes \mathbb{F}_1$$

is in fact a direct product (this is because the full twist $x_{12}x_{13}x_{23}$ is central). □

2.2 Conventions on Operads and Groupoids

Let $(\mathcal{E}, \otimes, \mathbb{1})$ be symmetric monoidal category where the tensor product commutes with colimits. The category of symmetric sequences $\mathbb{S}\text{-}\mathsf{mod}_{\mathcal{E}}$ is given by collections of objects $(e(n))_{n\geq 0}$ for $n \geq 0$, where $e(n)$ is endowed with a right action of $\mathfrak{S}_n$.

The category $\mathbb{S}\text{-}\mathsf{mod}_{\mathcal{E}}$ can be endowed with a monoidal product $\circ$ called the plethysm product. The category $(\mathbb{S}\text{-}\mathsf{mod}_{\mathcal{E}}, \circ, \mathbb{1}_0)$ forms a monoidal category, where $\mathbb{1}_\circ$ is the symmetric sequence given by $\mathbb{1}$ in arity one and the initial object $\emptyset$ elsewhere. *Operads* are defined to be unital monoids in $(\mathbb{S}\text{-}\mathsf{mod}_{\mathcal{E}}, \circ, \mathbb{1}_\circ)$.

One can also consider the category of sequences in $\mathcal{E}$, indexed by $\mathbb{N}$. This category, denoted by $\mathbb{N}\text{-}\mathsf{mod}_{\mathcal{E}}$, also has a plethysm product which endows it with a monoidal structure. Monoids in this category are called *non-symmetric operads*.

Example 2.4 The category of small sets Sets together with the cartesian product of sets satisfies these hypothesis. Similarly, the categories of small groups Grp, small groupoids Grpd, or topological spaces Top all satisfy them with the cartesian product again. △

Remark 2.5 Notice that the hypothesis that the tensor product commutes with colimits is too strong in order to include the symmetric monoidal category $(\mathsf{Lie}, \oplus, 0)$. Nevertheless, since it is cocomplete, one can still define operads as monoids in the normal oplax monoidal category $(\mathbb{S}\text{-}\mathsf{mod}_{\mathsf{Lie}}, \circ, \mathbb{1}_0)$. We refer to [Chi12] for more details.

From now on, we restrict to operads $\mathcal{P}$ which are *pointed reduced*, meaning $\mathcal{P}(1) = \mathcal{P}(0) = \mathbb{1}$. Let $\mathcal{U}nit$ be the symmetric sequence given by $\mathcal{U}nit(0) = \mathcal{U}nit(1) = \mathbb{1}$ and $\emptyset$ elsewhere. It has a unique operad structure which is given by the obvious compositions maps. An operad $\mathcal{P}$ is pointed reduced if and only if it admits a morphism of operads $f : \mathcal{U}nit \longrightarrow \mathcal{P}$ which is an isomorphism in arities zero and one.

Example 2.6

(1) The collection of the pure braid groups $(\mathrm{PB}_n)_{n\geq 0}$ forms a non symmetric operad in the category of $(\mathsf{Grp}, \times, \{*\})$, where the structure is given by the insertion of braids. Notice that the insertion of the empty braid in $\mathrm{PB}_0 \cong \{*\}$ suppresses the strand into which it is inserted. One can check that the natural action of $\mathfrak{S}_n$ is *not* compatible with the insertion of braids.
(2) The *little disks operad* $\mathbb{E}_2$ forms an operad in the category of topological spaces $(\mathsf{Top}, \times, \{*\})$ (see [May72]). It does not form an operad in the category of *pointed* topological spaces $(\mathsf{Top}_*, \times, \{*\})$. Indeed, if it did, since we know that for all n there is a weak equivalence $\mathbb{E}_2(n) \xrightarrow{\sim} \mathrm{Conf}_n(\mathbb{C})$, then one could apply the strong monoidal functor $\pi_1(-)$ and obtain a symmetric operad structure on the collection $(\mathrm{PB}_n)_{n\geq 0}$. △

One can "try to symmetrize" the non-symmetric operad of pure braid groups $(\mathrm{PB}_n)_{n\geq 0}$ by passing to the larger category of groupoids and defining the following operad in it.

Example 2.7 Let $\mathcal{CoB}$ be the operad in groupoids given by the following description:

(1) The objects of $\mathcal{CoB}(n)$ are elements σ of the symmetric group $\mathbb{S}_n$. For example, (321) is in $\mathrm{Ob}(\mathcal{CoB}(3))$.
(2) The set of morphisms between two permutations σ and τ in $\mathbb{S}_n$ is given by the set of braids going from σ to τ. Pictorially, we have:

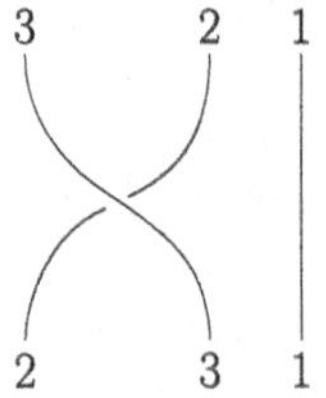

$$\in \mathrm{Hom}_{\mathcal{CoB}(3)}(321, 231)$$

(3) The operadic composition is given by the insertion of braids together with the relabeling of strands according to the insertion spot. Pictorially, we have:

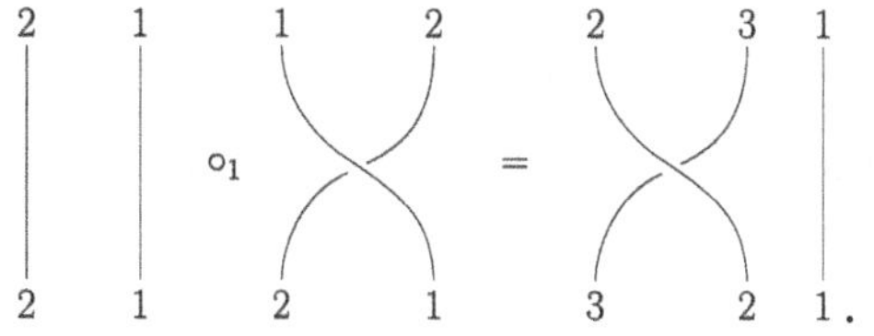

△

Remark 2.8 Since the operad $\mathbb{E}_2$ is not pointed, one could take the *fundamental groupoid* $\Pi_1(\mathbb{E}_2)$ instead of the fundamental group as before. Since the functor $\Pi_1(-)$ is strong monoidal, this defines an operad in the category of groupoids. Nevertheless, the operad of objects of $\Pi_1(\mathbb{E}_2)$ is huge.

One can ask oneself if the operad $Co\mathcal{B}$ defined above is *a model* for $\Pi_1(\mathbb{E}_2)$, that is, if these two operads are equivalent as operads in the category of groupoids. It is indeed a model for $\Pi_1(\mathbb{E}_2)$. Nevertheless, it is not a *cofibrant* operad in the category of groupoids, as the operad of objects of $Co\mathcal{B}$ is not free.

2.3 *The Operad of Parenthesized Braids*

Definition 2.9 (Fake Pull-Back) Let $\mathcal{O}, \mathcal{P}$ be two operads in $(\mathsf{Grpd}, \times, \{*\})$ and suppose there is a morphism $f : \mathrm{Ob}(\mathcal{O}) \longrightarrow \mathrm{Ob}(\mathcal{P})$ of operads in the category $(\mathsf{Sets}, \times, \{*\})$. The *fake pull-back* of $\mathcal{P}$ along f, denoted $f^*\mathcal{P}$, is the operad in the category of groupoids defined as follows:

(1) The objects of $f^*\mathcal{P}(n)$ are given by $\mathrm{Ob}(f^*\mathcal{P}(n)) := \mathrm{Ob}(\mathcal{O}(n))$.
(2) The morphisms between a, b in $\mathrm{Ob}(f^*\mathcal{P}(n))$ are given by

$$\mathrm{Hom}_{f^*\mathcal{P}(n)}(a, b) := \mathrm{Hom}_{\mathcal{P}(n)}(f(a), f(b)) .$$

Remark 2.10 It is straightforward to check that $f^*\mathcal{P}$ forms an operad in the category of groupoids endowed with the composition of the operad $\mathcal{O}$ on objects and the composition of the operad $\mathcal{P}$ on morphisms.

Let $\mathcal{P}a$ be the free operad in sets generated by an arity two operation. The set $\mathcal{P}a(n)$ is the set of maximally parenthesized permutations of $\mathfrak{S}_n$. One can view this operad in sets inside the category of operads in groupoids by declaring that all sets of morphisms are trivial (the empty set between two different objects, the identity as the unique endomorphism). There is an obvious morphism of operads in sets $\varphi : \mathrm{Ob}(\mathcal{P}a) \longrightarrow \mathrm{Ob}(Co\mathcal{B})$ given by forgetting the parenthesis on permutations.

Definition 2.11 (Operad of Parenthesized Braids) The operad in the category of groupoids of parenthesized braids $\mathcal{P}a\mathcal{B}$ is defined to be the fake pull-back of $\mathcal{C}o\mathcal{B}$ along the morphism φ; $\mathcal{P}a\mathcal{B} := \varphi^*\mathcal{C}o\mathcal{B}$.

Example 2.12 Objects in $\mathcal{P}a\mathcal{B}(n)$ are fully parenthesized permutations σ in $\mathbb{S}_n$. Morphisms are again given by braids going from one permutation to another. Composition is given by the insertion of braids together with the composition of parenthesized permutations. Pictorially, we have for instance

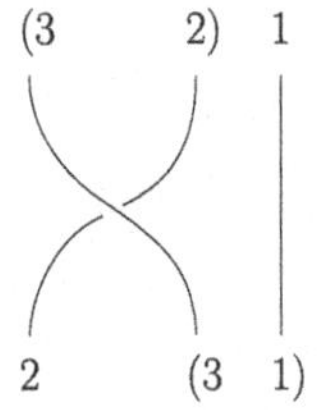

$$\in \mathrm{Hom}_{\mathcal{P}a\mathcal{B}(3)}((32)1, 2(31)) \quad .$$

The operadic composition is given by the insertion of braids together with the insertion of parenthesis

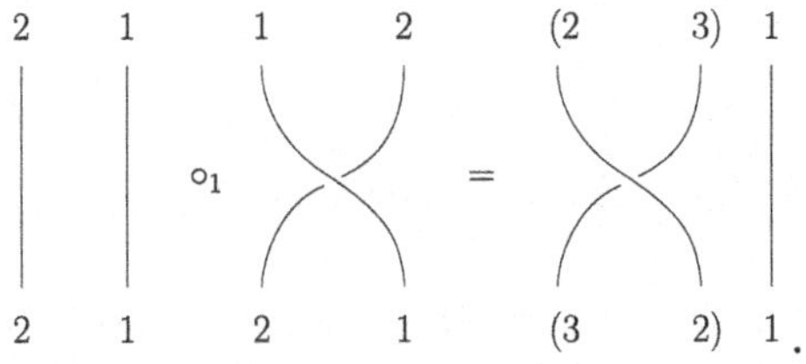

△

Remark 2.13 The operad of objects of $\mathcal{P}a\mathcal{B}$ is free. It provides us with a cofibrant model for $\Pi_1(\mathbb{E}_2)$ in the category of operads in groupoids.

Since the Malcev completion $\widehat{(-)}(\Bbbk)$ is a strong monoidal functor, we can complete the operad $\mathcal{P}a\mathcal{B}$ and still get an operad. See the Appendix A.3 for more details on the Malcev completions and pro-unipotent $\Bbbk$-groupoids.

The resulting operad $\widehat{\mathcal{P}a\mathcal{B}}(\Bbbk)$ in the category of pro-unipotent $\Bbbk$-groupoids will be the *first player* in our story.

2.4 The Operad of Chord Diagrams

Recall the Drinfeld–Kohno family of Lie algebras $(\mathfrak{t}_n)_{n\geq 0}$ from Definition 1.3. The insertion-coproduct maps will endow this family of Lie algebras with an operad structure. This operad can be understood as the holonomy Lie operad (see [MV19, §5.3]) of the little disks operad $\mathbb{E}_2$.

Proposition 2.14 *The family of Lie algebras* $(\mathfrak{t}_n)_{n\geq 0}$ *can be endowed with the following operad structure in the symmetric monoidal category* $(\mathsf{Lie}, \oplus, 0)$*:*

(1) *The action of* σ *in* $\mathbb{S}_n$ *on* $\mathfrak{t}_n$ *is given by* $\sigma \bullet t^n_{ij} := t^n_{\sigma(i)\sigma(j)}$.
(2) *The partial composition maps*

$$\{\circ_p : \mathfrak{t}_n \oplus \mathfrak{t}_m \longrightarrow \mathfrak{t}_{n+m-1}\}_{1\leq p\leq n}$$

only needs to be specified on each component of the direct sum. For $i < j$*, they are given as*

$$\mathfrak{t}_n \ni t_{ij} \longmapsto \begin{cases} t_{i+m-1\, j+m-1} & \text{if } p < i\,. \\ \displaystyle\sum_{k=i}^{i+m-1} t_{k\, j+m-1} & \text{if } p = i\,. \\ t_{i j+m-1} & \text{if } i < p < j\,. \\ \displaystyle\sum_{k=j}^{j+m-1} t_{i k} & \text{if } p = j\,. \\ t_{ij} & \text{if } j < p\,. \end{cases}$$

$$\mathfrak{t}_m \ni t_{ij} \longmapsto t_{i+p-1\, j+p-1}$$

Proof Let us construct the restriction of $\circ_p$ to each of the sources components using the insertion-coproduct morphisms of Definition 1.6. There is a "coproduct type" Lie algebra morphism

$$\circ_p(1) : \mathfrak{t}_n \longrightarrow \mathfrak{t}_{n+m-1}$$

associated to the well-defined map $\{1, \ldots, n + m - 1\} \to \{1, \ldots, n\}$ which sends $\{p, \ldots, p + m - 1\}$ to p, and there is a "insertion type" Lie algebra morphism

$$\circ_p(2) : \mathfrak{t}_m \longrightarrow \mathfrak{t}_{n+m-1}$$

associated to the partially defined map $\{1, \ldots, n + m - 1\} \to \{1, \ldots m\}$ which sends $k \in \{p, \ldots, p + m - 1\}$ to $k - p + 1$. This induces by universal property a Lie algebra morphism

$$\circ_p(1) \amalg \circ_p(2) : \mathfrak{t}_n \amalg \mathfrak{t}_m \longrightarrow \mathfrak{t}_{n+m-1}\,.$$

This induces an operad structure on $(\mathfrak{t}_n)_{n\geq 0}$ for the coproduct of Lie algebras, see Appendix B.1. One can notice that the images of these two maps commute, hence this map descends to the direct sum and gives the partial composition map $\circ_p$ defined in the proposition, which satisfy the axioms of an operad. □

Remark 2.15 Notice that in the above proposition, we have written the *left* $\mathbb{S}_n$-action on $\mathfrak{t}_n$. The right $\mathbb{S}_n$-action is given by

$$t_{ij}^n \bullet \sigma := t^n_{\sigma^{-1}(i)\sigma^{-1}(j)} ,$$

and in general one can pass from a left to a right action by applying the involution $(-)^{-1}$. We have chosen to do so since it simplifies the diagrams in Theorem 2.30. The same will apply for Proposition 3.23 and Theorem 3.33; and for Proposition 4.6 and Theorem 4.15.

Remark 2.16 The above proof is a particular instance of a general phenomenon that we describe in Appendix A.4.

Since the universal enveloping algebra functor $\mathfrak{U}(-)$ is a strong monoidal functor, it sends operads in (Lie, $\oplus$, 0) to operads in (Hopf-alg, $\otimes$, $\Bbbk$).

Definition 2.17 (Chord Diagrams Operad) The operad of *chord diagrams $\mathcal{CD}$* is the operad given by applying the universal enveloping algebra functor to the Drinfeld–Kohno operad.

Unwrapping this definition, the space of arity n operations of $\mathcal{CD}$ is given by the universal enveloping algebra of $\mathfrak{t}_n$, the n-th Drinfeld–Kohno algebra. Since the latter admits an explicit presentation, the former also admits the following presentation

$$\mathfrak{U}(\mathfrak{t}_n) = \frac{\Bbbk\langle t_{ij} \mid 1 \leq i, j \leq n, i \neq j\rangle}{\left(t_{ij} = t_{ji} \ , \ [t_{ij}, t_{ik} + t_{kj}] = 0 \ , \ [t_{ij}, t_{kl}] = 0\right)} .$$

The algebras $\left(\mathfrak{U}(\mathfrak{t}_n)\right)_{n\geq 0}$ are called the *chord diagram algebras* since pictorially one can represent the generators t_{ij} as chords (or infinitesimal braids) in the following way

$t_{ij} =$ 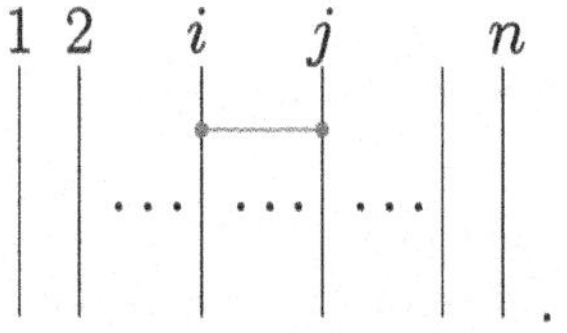

This way, one can represent the relations between the generators as the following relations between chord diagrams.

(1) The first relation can be depicted as

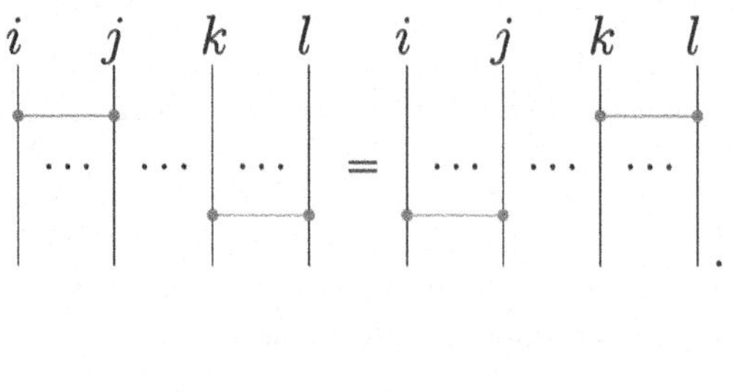

(2) The second relation can be depicted as

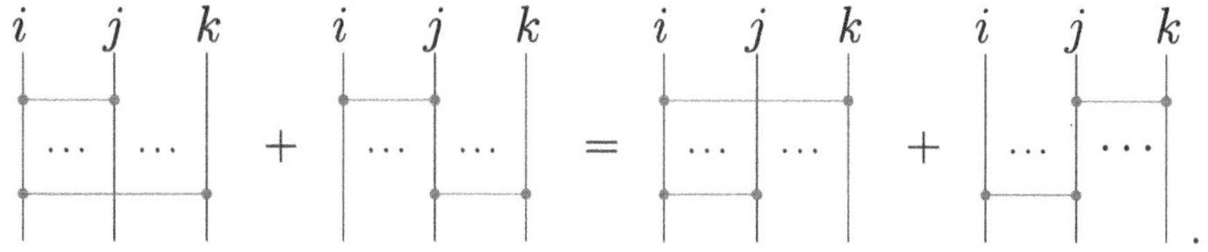

Remark 2.18 (Weight-Filtration) The relations that define the Lie algebras $\mathfrak{t}_n$ are homogeneous for all $n \geq 2$. Hence the algebras $\mathfrak{U}(\mathfrak{t}_n)$ are endowed with a weight filtration where the generators t_{ij} are in weight one.

Let $\mathsf{Cat}(\mathsf{Coalg}_{\Bbbk})$ denote the category of small categories enriched over cocommutative $\Bbbk$-coalgebras. It comes equipped with a symmetric monoidal structure given as follows: let $\mathcal{E}$ and $\mathcal{F}$ be two such categories, their product $\mathcal{E} \otimes \mathcal{F}$ is the category given by the following presentation

(1) The objects of $\mathcal{E} \otimes \mathcal{F}$ are given by pairs in $\mathrm{Ob}(\mathcal{E}) \times \mathrm{Ob}(\mathcal{F})$.
(2) The set of morphisms between two pairs (e, f) and (e', f') is

$$\mathrm{Hom}_{\mathcal{E}\otimes\mathcal{F}}((e, f), (e', f')) := \mathrm{Hom}_{\mathcal{E}}(e, e') \otimes \mathrm{Hom}_{\mathcal{F}}(f, f') ,$$

where the last tensor denotes simply the tensor product of two cocommutative $\Bbbk$-coalgebras.

One checks that it endows $\mathsf{Cat}(\mathsf{Coalg}_{\Bbbk})$ with a symmetric monoidal structure. The unit is given by the category with one object whose endomorphisms are given by $\Bbbk$, which we denote by $\{*\}$. Thus one can define operads in the symmetric monoidal category $(\mathsf{Cat}(\mathsf{Coalg}_{\Bbbk}), \otimes, \{*\})$.

Example 2.19 The chord diagram operad $\mathcal{CD}$ is an operad in $\mathsf{Cat}(\mathsf{Coalg}_{\Bbbk})$ where the category $\mathcal{CD}(n)$ only has one object whose algebra of endomorphisms is given by $\mathfrak{U}(\mathfrak{t}_n)$. △

Remark 2.20 Let $\mathcal{E}$ be in $\mathsf{Cat}(\mathsf{Coalg}_{\Bbbk})$. Notice that for any object e in $\mathcal{E}$, its endomorphisms $\mathrm{Hom}_{\mathcal{E}}(e, e)$ form a cocommutative bialgebra, where the multiplication is given by the composition of endomorphisms.

Definition 2.21 ($\Bbbk$-Linear Extension) The $\Bbbk$-*extension* is a strong monoidal functor

$$\begin{array}{ccc}(\mathsf{Set}, \times, \{*\}) & \longrightarrow & (\mathsf{Cat}(\mathsf{Coalg}_{\Bbbk}), \otimes, \{*\}) \\ S & \longmapsto & S_{\Bbbk} .\end{array}$$

It associates to a set S the category $S_{\Bbbk}$ which is defined as follows:

(1) The set of objects of $S_{\Bbbk}$ is given by S.

(2) Let s, s' be in S, the space of morphisms is given by

$$\mathrm{Hom}_{S_{\Bbbk}}(s, s') := \Bbbk \ .$$

The set-theoretical operad $\mathcal{P}a$ can therefore be promoted into an operad in $\mathsf{Cat}(\mathsf{Coalg}_{\Bbbk})$ by considering its $\Bbbk$-linear extension $\mathcal{P}a_{\Bbbk}$.

Definition 2.22 (Parenthesized Chord Diagrams) The *parenthesized chord diagrams* operad $\mathcal{P}a\mathcal{CD}$ is given by the Hadamard product of $\mathcal{P}a_{\Bbbk}$ and $\mathcal{CD}$ in the category of operads defined above $\mathsf{Cat}(\mathsf{Coalg}_{\Bbbk})$. That is, we have:

$$\mathcal{P}a\mathcal{CD}(n) := \mathcal{P}a_{\Bbbk}(n) \otimes \mathcal{CD}(n) \ ,$$

where $\otimes$ denotes the tensor product in $\mathsf{Cat}(\mathsf{Coalg}_{\Bbbk})$.

Example 2.23 The object of $\mathcal{P}a\mathcal{CD}(n)$ are therefore given by parenthesized permutations in $\mathfrak{S}_n$. The morphisms in $\mathrm{Hom}_{\mathcal{P}a\mathcal{CD}(n)}(\sigma, \tau)$ between two parenthesized permutations can be depicted as chord diagrams on the unique strand that goes from σ to τ. For instance, the element $1 \otimes (t_{24}.t_{34})$ in $\Bbbk \otimes \mathfrak{U}(\mathfrak{t}_4) \cong \mathrm{Hom}_{\mathcal{P}a\mathcal{CD}(4)}(((31)2)4, (4(13))2)$ can be depicted as:

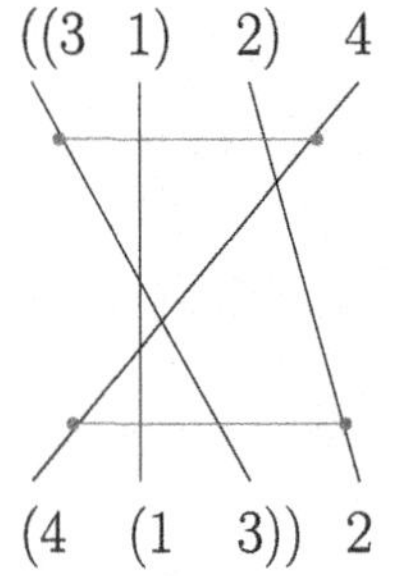

$\in \mathrm{Hom}_{\mathcal{P}a\mathcal{CD}}(((31)2)4, (4(13))2)$.

Composition operations in the operad

$$\circ_i : \mathcal{P}a\mathcal{CD}(n) \otimes \mathcal{P}a\mathcal{CD}(m) \longrightarrow \mathcal{P}a\mathcal{CD}(n+m-1)$$

are defined as follows. On objects, the composition of two parenthesized permutations is given by the insertion the i-th spot. On morphisms, that is, on chord diagrams, its given by the sum of possible distributions of the existing chords along the inserted strands. For instance, the composition of $1 \otimes t_{12} \circ_2 1 \otimes \mathrm{id}$ gives

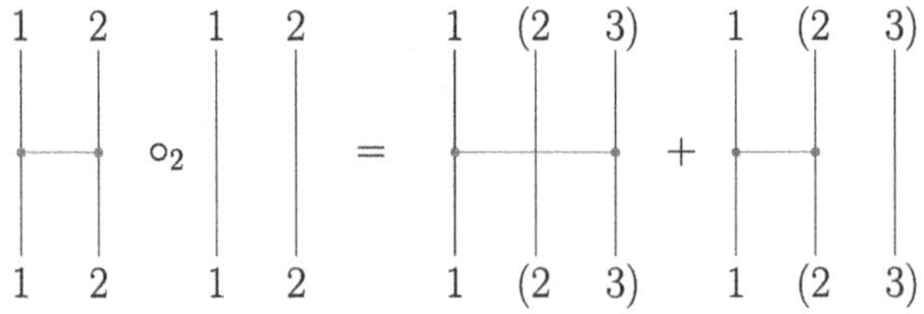

which is an endomorphism of $1(23)$ in $\mathcal{P}a\mathcal{CD}(3)$. Here are some examples of distinguished morphisms in the operad $\mathcal{P}a\mathcal{CD}$:

(1) The element $H := 1 \otimes t_{12}$ in $\mathrm{Hom}_{\mathcal{P}a\mathcal{CD}(2)}(12, 12)$ depicted as

$H :=$
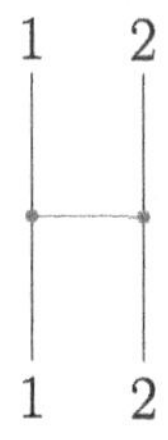

(2) The element $X := 1 \otimes 1$ in $\mathrm{Hom}_{\mathcal{P}a\mathcal{CD}(2)}(12, 21)$ depicted as

$X :=$
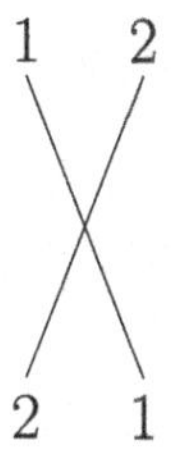

(3) The element $\alpha := 1 \otimes 1$ in $\mathrm{Hom}_{\mathcal{P}a\mathcal{CD}(3)}((12)3, 1(23))$ depicted as

$\alpha :=$
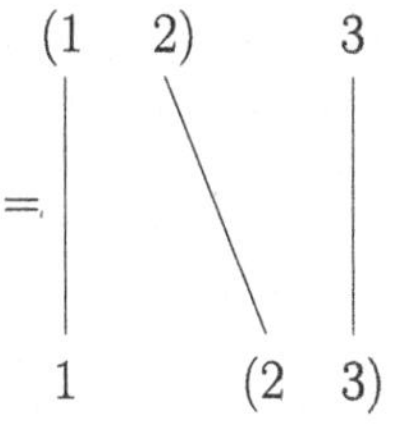

△

Remark 2.24 The operad $\mathcal{P}a\mathcal{CD}$ is in fact generated, as an operad in the category $\mathsf{Cat}(\mathsf{Coalg}_{\Bbbk})$, by the object 12 in $\mathcal{P}a\mathcal{CD}(2)$ and by the morphisms α, X and H. They satisfy some relations but, to the best of our knowledge, there is no explicit presentation for $\mathcal{P}a\mathcal{CD}$. See [Fre17a, Section 10.2] for more details.

We define $\widehat{\mathcal{P}a\mathcal{CD}}$ as the operad in categories enriched in complete cocommutative coalgebras obtained by completing $\mathcal{P}a\mathcal{CD}$ with respect to the filtration defined in Remark 2.18. Indeed, the weigh filtration on each algebra $\mathfrak{U}(\mathfrak{t}_n)$ induces a weight filtration on each hom-coalgebra $\mathrm{Hom}_{\mathcal{P}a\mathcal{CD}(n)}(\sigma, \tau)$. This operad forms an operad in complete Hopf groupoids. See Appendix A.3 for more details on these notions.

The operad $\mathrm{Grp}(\widehat{\mathcal{P}a\mathcal{CD}})$, obtained by applying the group-like element functor to each complete Hopf groupoid, forms an operad in the category of pro-unipotent $\Bbbk$-groupoids. This will be our *second player*.

2.5 Operadic Definition of Associators

Now we may give a fully operadic definition of an associator.

Definition 2.25 (Operadic Associators) The set of *operadic associators* is given by

$$\mathrm{OpAssoc}(\Bbbk) := \mathrm{Iso}^{+}_{\mathsf{Op}(\mathsf{p.u\text{-}Grpd}_{\Bbbk})}\left(\widehat{\mathcal{P}a\mathcal{B}}(\Bbbk), \mathrm{Grp}(\widehat{\mathcal{P}a\mathcal{CD}})\right),$$

where we consider only isomorphisms of operads in the category of groupoids which are given by the identity morphism on the objects of these operads.

This definition entails that the set of operadic associators is a bitorsor over the automorphism groups of each operad that are the identity morphism on objects. These automorphism groups are of primordial importance.

Definition 2.26 (Grothendieck–Teichmüller Group) The *Grothendieck–Teichmüller group* over $\Bbbk$ is given by

$$\mathrm{GT}(\Bbbk) := \mathrm{Aut}^{+}_{\mathsf{Op}(\mathsf{p.u\text{-}Grpd}_{\Bbbk})}\left(\widehat{\mathcal{P}a\mathcal{B}}(\Bbbk)\right),$$

where we consider only automorphisms of operads in the category of groupoids which are given by the identity morphism on the objects.

Remark 2.27 (Fifty Shades of Grothendieck–Teichmüller) One can define many other versions of the Grothendieck–Teichmüller group. There is a *pro-finite* version, denoted $\widehat{\mathrm{GT}}$, where one considers the automorphisms of operads that fix the objects of the pro-finite completion of $\mathcal{P}a\mathcal{B}$. It was shown in [Iha94] that the absolute Galois group $\mathrm{Gal}(\overline{\mathbb{Q}}/\mathbb{Q})$ embeds into $\widehat{\mathrm{GT}}$. These two groups are conjecturally isomorphic. For more on this, see [BR24]. In the same spirit, one can defined a *pro-ℓ* version of the Grothendieck–Teichmüller group. Notice that if one does not complete in any way $\mathcal{P}a\mathcal{B}$, its automorphism group is simply $\mathbb{Z}/2\mathbb{Z}$. This corresponds heuristically to the complex conjugation in $\mathrm{Gal}(\overline{\mathbb{Q}}/\mathbb{Q})$.

Definition 2.28 (Graded Grothendieck–Teichmüller Group) The *graded Grothendieck–Teichmüller group* over $\Bbbk$ is given by

$$\mathrm{GRT}(\Bbbk) := \mathrm{Aut}^{+}_{\mathsf{Op}(\mathsf{p.u\text{-}Grpd}_{\Bbbk})}\left(\mathrm{Grp}(\widehat{\mathcal{P}a\mathcal{CD}})\right) \cong \mathrm{Aut}^{+}_{\mathsf{OpCat}(\mathsf{Coalg}_{\Bbbk})}\left(\widehat{\mathcal{P}a\mathcal{CD}}\right),$$

where we consider only automorphisms of operads in the category of groupoids (or equivalently, of operad in the category of categories enriched in counital cocommutative coalgebras) which are given by the identity morphism on the objects.

Remark 2.29 In the three above definitions, one could replace isomorphisms that are the identity on objects by isomorphisms in the homotopy category of a suitable model category of operads in pro-unipotent groupoids. It appears that this gives the same result, as is very well-explained (for the pro-finite case) in [Hor17]. It roughly relies on the fact that any morphism between these operads is homotopic to one being the identity on objects.

2.6 More Concrete Descriptions

Theorem 2.30 *The operad $\mathcal{P}a\mathcal{B}$, as an operad in groupoids, admits the following presentation. It is generated by the object* 12 *in $\mathcal{P}a\mathcal{B}(2)$ and by the following morphisms:*

(1) *The braiding R which pictorially is given by the following braid*

$R :=$ 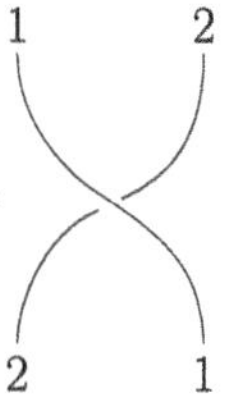

$\in \mathrm{Hom}_{\mathcal{P}a\mathcal{B}(3)}(12, 21)$

(2) *The associator Φ which pictorially is given by the following braid*

$\Phi :=$ 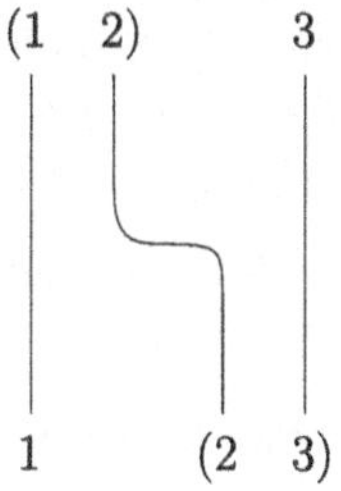

$\in \mathrm{Hom}_{\mathcal{P}a\mathcal{B}(3)}((12)3, 1(23))$

They satisfy the following relations

(1) *Recall that $\mathcal{P}a\mathcal{B}(0)$ is the trivial groupoid $\{*\}$. The unital relation states that*

$$\Phi \circ_2 id_{\{*\}} = id_{12} .$$

(2) *The pentagon relation amounts to the commutativity of the following diagram*

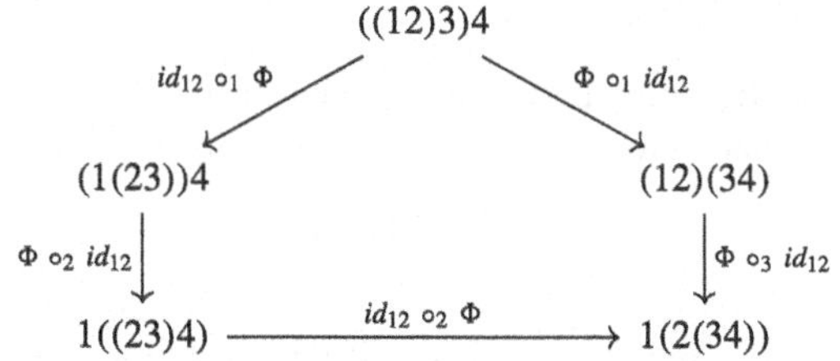

(3) *The first hexagon relation amounts to the commutativity of the following diagram*

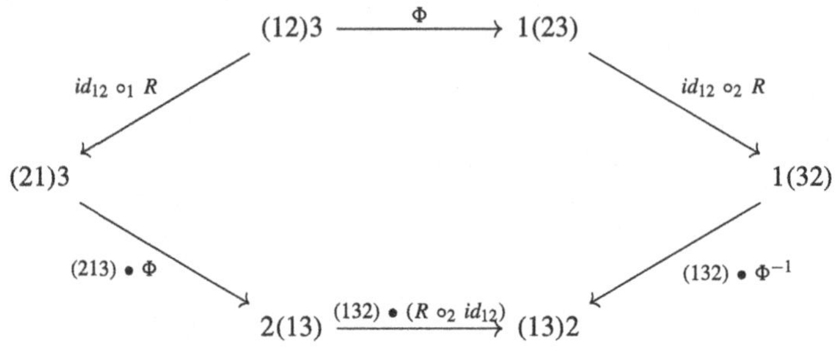

(4) *The second hexagon relation amounts to the commutativity of the above diagram when replacing* R *by* $\tilde{R} := ((21) \bullet R)^{-1}$.

The proof can be found in [Fre17a, Theorem 6.2.4] (although originally proven in [Bar98] using a different language).

Remark 2.31 Notice that the unitality relation, together with the pentagon relation, implies the two other unitality relations $\Phi \circ_1 \mathrm{id}_{\{*\}} = \Phi \circ_3 \mathrm{id}_{\{*\}} = \mathrm{id}_{12}$ as well. Indeed, if one applies twice $- \circ_1 \mathrm{id}_{\{*\}}$ to the pentagon relation, one finds $(\Phi \circ_1 \mathrm{id}_{\{*\}})^2 = \Phi \circ_1 \mathrm{id}_{\{*\}}$; this implies that $\Phi \circ_1 \mathrm{id}_{\{*\}} = \mathrm{id}_{12}$. One proves along the same lines that $\Phi \circ_3 \mathrm{id}_{\{*\}} = \mathrm{id}_{12}$.

Corollary 2.32 *There is a one to one correspondence between classical associators as in Definition 1.7 and operadic associators as in Definition 2.25.*

Sketch of Proof Let $F : \widehat{\mathcal{P}a\mathcal{B}}(\Bbbk) \xrightarrow{\sim} \mathrm{Grp}(\widehat{\mathcal{P}a\mathcal{CD}})$ be an isomorphism which is the identity on objects. By the above theorem, it is completely determined by the image of the generators R and Φ.

The image of R by F has to be of the form

$$F(R) = e^{\lambda . t_{12}/2} X\ , \quad \text{in} \quad \mathrm{Hom}_{\mathcal{P}a\mathcal{CD}(2)}(12, 21)\ ,$$

where $e^{\lambda . t_{12}/2}$ is in $\exp(\widehat{\mathfrak{t}_2}) \cong \Bbbk$ and where X is the element of $\mathcal{P}a\mathcal{CD}$ given in Example 2.23. Furthermore, F is an isomorphism if and only if λ is in $\Bbbk^\times$.

The image of Φ by F has to be of the form

$$F(\Phi) = f(\Phi)\alpha\ , \quad \text{in} \quad \mathrm{Hom}_{\mathcal{P}a\mathcal{CD}(3)}((12)3, 1(23))\ ,$$

where $f(\Phi)$ is in $\exp(\widehat{\mathfrak{t}_3})$ and where α is the element of $\widehat{\mathcal{P}a\mathcal{CD}}$ given in Example 2.23. Recall from Sect. 1.3 that there is a canonical isomorphism of Lie algebras

$$\mathfrak{t}_3 \cong \Bbbk c \oplus \mathfrak{f}_2\ ,$$

where $c = t_{12} + t_{23} + t_{13}$ and where $\mathfrak{f}_2$ is the free Lie algebra generated by t_{12} and t_{23}.

This gives a canonical isomorphism $\exp(\widehat{\mathfrak{t}_3}) \cong \Bbbk \times \widehat{\mathbb{F}_2}(\Bbbk)$. Therefore we can decompose $f(\Phi)$ as

$$f(\Phi) = (\Xi(c), \Phi(t_{12}, t_{23})) \quad \text{in} \quad \exp(\widehat{\mathfrak{t}_3}) \cong \Bbbk \times \widehat{\mathbb{F}_2}(\Bbbk) .$$

Now one can check that the morphism

$$- \circ_2 \mathrm{id}_{\{*\}} : \mathrm{Hom}_{\mathcal{P}a\mathcal{CD}(3)}((12)3, 1(23)) \longrightarrow \mathrm{Hom}_{\mathcal{P}a\mathcal{CD}(2)}(12, 12)$$

is the one determined by the morphism $\mathfrak{t}_3 \twoheadrightarrow \mathfrak{t}_2$ which sends t_{13} to t_{12} and t_{12}, t_{23} to zero. By pre-composing it with the isomorphism $\Bbbk c \oplus \mathfrak{f}_2 \cong \mathfrak{t}_3$, the resulting morphism sends t_{12}, t_{23} to zero and c to t_{12}. The unital relation

$$\Phi \circ_2 \mathrm{id}_{\{*\}} = \mathrm{id}_{12}$$

therefore gives that $\Xi(c) = 1$, and thus that $f(\Phi)$ is given by $\Phi(t_{12}, t_{23})$ in $\exp(\widehat{\mathfrak{f}_2})$.

So far, we have identified the data present in Definition 1.7 and in Definition 2.25. Let us briefly illustrate how the relations in both definitions are sent to each other for the particular case of the pentagon equation. Notice that there is a dictionary between the notation used in the introduction and the operadic notation. The morphism $\Phi \circ_2 \mathrm{id}_{12}$ rewrites as $\Phi^{1,23,4}$ since it is the associator where 2 and 3 are considered as "a single object". The morphism $\mathrm{id}_{12} \circ_2 \Phi$ rewrites as $\Phi^{2,3,4}$ since the object 1 "plays no role". The pentagon relation imposed on Φ then rewrites in this particular notation as

$$\Phi^{1,2,3}\Phi^{1,23,4}\Phi^{2,3,4} = \Phi^{12,3,4}\Phi^{1,2,34} ,$$

which is the same equation as in the introduction, but where the products are in the opposite order. More generally, one can check (see [Fre17a, Theorem 10.2.9] for a detailled proof) that this dictionary identifies the relations imposed on Drinfeld associators with the relations imposed on the operadic associators (with reversed order). The one to one correspondence between classical associators as in Definition 1.7 and operadic associators as in Definition 2.25 is therefore given by $\Phi \mapsto \Phi^{-1}$. □

Remark 2.33 (Algebraic and Topological Conventions) As observed in the above proof, the pentagon equation for a Drinfeld associator is written as

$$\Phi^{2,3,4}\Phi^{1,23,4}\Phi^{1,2,3} = \Phi^{1,2,34}\Phi^{12,3,4}$$

in the Problem 2, whereas the pentagon equation of Theorem 2.30, translated into the same notation, are in the opposite order

$$\Phi^{1,2,3}\Phi^{1,23,4}\Phi^{2,3,4} = \Phi^{12,3,4}\Phi^{1,2,34} .$$

This is due to our unusual choice of convention, that leads us to view the product ab of two elements of an algebra as a concatenation of arrows, and thus as the *opposite* of their composition (when one sees the algebra as a category with only one object).

Similarly, this presentation of $\mathcal{P}a\mathcal{B}$ allows us to give an explicit description of the Grothendieck–Teichmüller group.

Theorem 2.34 *Elements of the Grothendieck–Teichmüller group* GT($\Bbbk$) *are in bijection with pairs* (μ, f) *in* $\Bbbk^\times \times \widehat{\mathbb{F}}_2(\Bbbk)$, *where* $\widehat{\mathbb{F}}_2(\Bbbk)$ *denotes the Malcev completion of the free group on two generators, which satisfy the following conditions*

(1) $f(x,y)^{-1} = f(y,x)$ *in* $\widehat{\mathbb{F}}_2(\Bbbk)$.
(2) $x^\nu f(x,y) y^\nu f(y,z) z^\nu f(z,x) = 1$ *if* $xyz = 1$ *in* $\widehat{\mathbb{F}}_3(\Bbbk)$, *where* $\nu = \frac{\mu-1}{2}$.
(3) *Let* x_{ij} *denote the standard generator of* $\widehat{\mathrm{PB}}_4(\Bbbk)$, *the Malcev completion of the pure braid group on four strands, defined in Theorem 2.2. Then the element* f *satisfies*

$$f(x_{12}, x_{23}) f(x_{12}x_{13}, x_{24}x_{34}) f(x_{23}, x_{34}) = f(x_{13}x_{23}, x_{34}) f(x_{12}, x_{23}x_{24})$$

in $\widehat{\mathrm{PB}}_4(\Bbbk)$.

Under this bijection, the group structure of the Grothendieck–Teichmüller group is given by:

$$(\mu_1, f_1) \star (\mu_2, f_2) = \left(\mu_1\mu_2, f_1\left(x^{\mu_2}, f_2(x,y) y^{\mu_2} f_2(y,x)\right) f_2(x,y)\right).$$

Under this bijection, its action on the set of associators is given by:

$$(\mu, f) \bullet (\lambda, \Phi) = \left(\mu\lambda, f\left(e^{\mu t_{12}}, \Phi(t_{12}, t_{23}) e^{\mu t_{23}} \Phi(t_{23}, t_{12})\right) \Phi(t_{12}, t_{23})\right) .$$

Proof An explicit proof of this can be found in [Fre17a, Theorem 11.1.7]. □

Remark 2.35 Beware that [Fre17a] uses a different presentation of the pure braid group. In *op.cit.*, x_{ij} denotes the pure braid that goes *from* the j-th strand, passing *behind* other strands, does a loop around the i-th strand, and comes back passing *behind* again. Hence the formula given in condition (3) of Theorem 2.34 does not coincide with the one appearing in [Fre17a, Theorem 11.1.7].

Remark 2.36 Using Remark 2.24, one can also give an explicit description of the *graded Grothendieck–Teichmüller group* GRT($\Bbbk$), which coincides with the original definition of Drinfeld.

2.7 *Topological Description of* $\mathcal{P}a\mathcal{B}$

The operad in the category of groupoids $\mathcal{P}a\mathcal{B}$ can be obtained via a topological construction. For this, we consider the configuration spaces

$$\mathrm{Conf}(\mathbb{C}, \mathrm{I}) := \left\{(x_1, \cdots, x_n) \in \mathbb{C}^{\mathrm{I}} \mid x_i \neq x_j \text{ if } i \neq j\right\} ,$$

for any finite set I. We then consider the space

$$\mathrm{C}(\mathbb{C},\mathrm{I}) := \mathrm{Conf}(\mathbb{C},\mathrm{I})/\mathbb{C} \ltimes \mathbb{R}_{>0} ,$$

of configurations of (pairwise distinct) points modulo translations and dilations. Their Fulton–MacPherson compactifications $\overline{\mathrm{C}}(\mathbb{C},\mathrm{I})$ can be endowed with a canonical operad structure. Indeed, one can compute that the boundary of

$$\partial\, \overline{\mathrm{C}}(\mathbb{C},\mathrm{I}) \cong \bigcup_{k\geq 0} \bigcup_{\mathrm{J}_1\sqcup\cdots\sqcup\mathrm{J}_k=\mathrm{I}} \overline{\mathrm{C}}(\mathbb{C},[k]) \times \left(\prod_{j=1}^{k} \overline{\mathrm{C}}(\mathbb{C},\mathrm{J}_j)\right)$$

where $[k] = \{1, \cdots, k\}$. Therefore the operad structure is simply given by the inclusions

$$\gamma_{\mathrm{J}_1,\cdots,\mathrm{J}_k} : \overline{\mathrm{C}}(\mathbb{C},[k]) \times \left(\prod_{j=1}^{k} \overline{\mathrm{C}}(\mathbb{C},\mathrm{J}_j)\right) \hookrightarrow \partial\, \overline{\mathrm{C}}(\mathbb{C},\mathrm{J}_1\sqcup\cdots\sqcup\mathrm{J}_k) \hookrightarrow \overline{\mathrm{C}}(\mathbb{C},\mathrm{J}_1\sqcup\cdots\sqcup\mathrm{J}_k) .$$

We denote this operad in topological spaces by $\overline{\mathrm{C}}(\mathbb{C})$.

Remark 2.37 There exists no choice of a family of points in $\overline{\mathrm{C}}(\mathbb{C})$ compatible with the operad structure. In other words, this operad in topological spaces can not be promoted into an operad in pointed topological spaces.

Remark 2.38 There is a direct weak-equivalence of operads $\overline{\mathrm{C}}(\mathbb{C}) \xrightarrow{\sim} \mathbb{E}_2$, where $\mathbb{E}_2$ is the little disks operad. See [Hoe12].

Recall $\mathcal{P}a$, the operad in the category of sets generated by a single arity two operation. Elements in $\mathcal{P}a(n)$ are in bijection with maximally parenthesized permutations of $\mathfrak{S}_n$. Now lets view this operad as a topological operad where we impose the discrete topology on the sets $\mathcal{P}a(n)$.

Lemma 2.39 *There is an inclusion of topological operads*

$$\iota : \mathcal{P}a \hookrightarrow \overline{\mathrm{C}}(\mathbb{C}) .$$

Pictorially, the morphism ι establishes the following correspondence

$$((13)(24)\cdots n) \longrightarrow$$

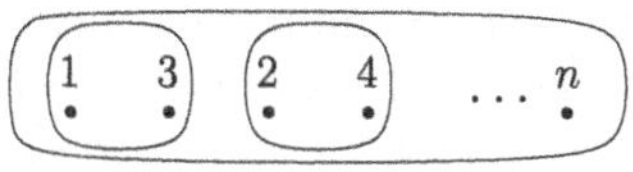

between maximally parenthesized permutations and configurations of labeled n points inside the real line $\mathbb{R} \subset \mathbb{C}$, where the parenthesis on the right denote points which are infinitesimally close in the Fulton–MacPherson compactification.

Theorem 2.40 *There is an isomorphism of operads in the category of groupoids*

$$\mathcal{P}a\mathcal{B} \cong \Pi_1(\overline{C}(\mathbb{C}), \mathcal{P}a) .$$

Sketch of Proof Both of these operads in groupoids have the same underlying operad of objects, so we can consider the identity morphism on objects. Therefore what is left is to identify the automorphism groups on each side. Take the leftmost parenthesization of the trivial permutation $(((12)3)\cdots n)$ as an object, one can construct an explicit isomorphism

$$\mathrm{PB}_n \cong \mathrm{Aut}_{\mathcal{P}a\mathcal{B}(n)}((((12)3)\cdots n)) \cong \pi_1(\overline{\mathrm{C}}(\mathbb{C}), \iota(((12)3)\cdots n))$$

by sending the generator x_{ij} to the loop where the i-th point travels around only the j-th point and then goes back to its original position. □

3 Cyclotomic Associators

3.1 Motivation

3.1.1 A Variation on the Quantization Problem

Let $\mathfrak{g}$ be a Lie algebra and let $\sigma : \mathfrak{g} \to \mathfrak{g}$ be an automorphism of order N. It defines a Lie subalgebra of fixed points $\mathfrak{h} := \mathfrak{g}^\sigma$. One can always get a decomposition $\mathfrak{g} \cong \mathfrak{h} \oplus \mathfrak{m}$ as an $\mathfrak{h}$-module, where $\mathfrak{m}$ is the direct sum of the other eigenspaces of σ. Let t be a σ-invariant Casimir element of $\mathfrak{g}$; it decomposes as $t = t_{\mathfrak{h}} + t_{\mathfrak{m}}$, where $t_{\mathfrak{h}} \in S^2(\mathfrak{h})$ and $t_{\mathfrak{m}} \in S^2(\mathfrak{m})$.

The category $\mathsf{Rep}\,(\mathfrak{U}(\mathfrak{g}) \rtimes \Gamma)$, where $\Gamma := \mathbb{Z}/N\mathbb{Z}$, is a symmetric monoidal category, thus a braided monoidal category. One can show that the category $\mathsf{Rep}(\mathfrak{h})$ is in fact a *braided module category* over the braided monoidal category $\mathsf{Rep}\,(\mathfrak{U}(\mathfrak{g}) \rtimes \Gamma)$. See [Enr07] and [Bro13] for more on this notion. Equivalently, one can say that $\mathsf{Rep}(\mathfrak{h})$ is a σ-braided module category over $\mathsf{Rep}(\mathfrak{g})$ in the sense of [DCNTY19].

We have already seen that t can be used to define a first order deformation of $\mathsf{Rep}(\mathfrak{g})$ as a braided module category. Similarly, following [Enr07, Bro13, DCNTY19], $t_{\mathfrak{h}}$ can be used to define a first order deformation of $\mathsf{Rep}(\mathfrak{h})$ as a σ-braided module category over (the above mentionned first order deformation of) $\mathsf{Rep}(\mathfrak{g})$. Thus, Problem 1 of the introduction can be extended to this setting, and stated very shortly as follows: can one extend this first order deformation into a formal deformation over $\Bbbk[[\hbar]]$?

Cyclotomic associators provide a universal answer to this type of deformation quantization problems analogously to classical associators before.

3.1.2 Universal Reformulation

Let us be a bit more specific about the meaning of "a universal answer" in the previous paragraph. Recall the morphisms

$$\varphi_t^n : \mathfrak{U}(\mathfrak{t}_n) \to \left(\mathfrak{U}(\mathfrak{g})^{\otimes n}\right)^{\mathfrak{g}} \cong \mathsf{End}(\otimes^n)\,,$$

that are compatible with insertion-coproduct morphisms. There is a cyclotomic version $\mathfrak{t}_n^\Gamma$, $\Gamma := \mathbb{Z}/N\mathbb{Z}$, of the Drinfeld–Kohno graded Lie algebra (see Definition 3.22 below). It comes along with algebra morphisms

$$\varphi_{t,\sigma}^n : \mathfrak{U}(\mathfrak{t}_n^\Gamma) \to \left(\mathfrak{U}(\mathfrak{h}) \otimes \left(\mathfrak{U}(\mathfrak{g}) \rtimes \Gamma\right)^{\otimes n}\right)^{\mathfrak{h}} \cong \mathsf{End}(\odot^n)\,,$$

where the functor $\odot^n$: $\mathsf{Rep}(\mathfrak{h}) \times \mathsf{Rep}\,(\mathfrak{U}(\mathfrak{g}) \rtimes \mathbb{Z}/N\mathbb{Z})^n \to \mathsf{Rep}(\mathfrak{h})$ is the n-fold module structure of $\mathsf{Rep}(\mathfrak{h})$.

It turns out that the natural transformations and relations defining a braided module category sit in $\mathsf{End}(\odot^n)$ and can be expressed using insertion-coproduct type morphisms. Such insertion-coproduct morphisms also exist on the cyclotomic Drinfeld–Kohno Lie agebras $\mathfrak{t}_n^\Gamma$, and the morphisms $\varphi_{t,\sigma}^n$ are compatible with them. Therefore, a universal solution to the quantization problem is a solution that lives at the level of $\mathfrak{U}(\mathfrak{t}_n^\Gamma)$ (in fact, it lives in its degree completion, see Remark 1.5).

3.1.3 Number Theoretic Aspects

As mentioned in the previous Section, the absolute Galois group $\mathrm{Gal}(\overline{\mathbb{Q}}/\mathbb{Q})$ embeds into the pro-finite version $\widehat{\mathrm{GT}}$ of the Grothendieck–Teichmüller group. Cyclotomic associators provide new versions of the pro-finite Grothendieck–Teichmüller group, and the absolute Galois groups $\mathrm{Gal}(\overline{\mathbb{Q}}/\mathbb{Q}(\mu_N))$ embed into these. Here $\mathbb{Q}(\mu_N)$ stands for the N-th cyclotomic extension of $\mathbb{Q}$, and μ_N is the group of N-th roots of 1.

3.1.4 Motivic Aspects

Enriquez's proof [Enr07] of the existence of a cyclotomic associator is also motivic, as an explicit cyclotomic associator Ψ_{KZ} is constructed as the regularized holonomy of an algebraic flat connection defined on $\mathbb{P}^1 - (\{0,+\infty\} \cup \mu_N)$ that generalizes the KZ connection. The cyclotomic associator Ψ_{KZ} is in fact a generating series for special values of multiple polylogarithms at N-th roots of 1 (see *loc.cit*). These periods, sometimes called N-coloured MZV, have very interesting combinatorial [Rac02] and motivic [Gon98, Gon01a, Gon01b, Gon01c, DG05] features.

3.2 Moperads

Let $(\mathcal{E}, \otimes, \mathbb{1})$ be a cocomplete symmetric monoidal category such that the monoidal product $\otimes$ commutes with small colimits in each variable. The category of symmetric sequences $\mathbb{S}\text{-mod}_{\mathcal{E}}$ has a monoidal structure other than the pleythism, which is symmetric and is given by

$$(M \circledast N)(n) := \coprod_{p+q=n} (M(p) \otimes N(q))^{\mathbb{S}_n}_{\mathbb{S}_p \times \mathbb{S}_q} ,$$

for M, N two symmetric sequence in $\mathcal{E}$. The unit for this symmetric monoidal structure is given by the symmetric sequence $\mathbb{1}_{\circledast}$ which is $\mathbb{1}$ in arity zero and zero elsewhere. This product is compatible with the $\circ$-product in the following way: there is a natural transformation

$$\varepsilon_{M,N,L} : M \circledast (N \circ L) \longrightarrow (M \circledast N) \circ L ,$$

for any triple of symmetric sequences (M, N, L) which is given by assembling the evident maps

$$\begin{array}{c} M(p) \otimes (N(q) \otimes L(i_1) \otimes \cdots \otimes L(i_q)) \\ \downarrow \\ (M(p) \otimes N(q)) \otimes L(i_1) \otimes \cdots \otimes L(i_q) \otimes L(1) \otimes \cdots \otimes L(1) \end{array}$$

Definition 3.1 (Moperad) Let $\mathcal{P}$ be an operad. A *moperad* $(M, \gamma_M, \eta, \gamma_{\mathcal{P}})$ over $\mathcal{P}$ is the data of a symmetric sequence M such that

(1) (M, γ_M, η) forms a unital monoid for the $\circledast$ product.
(2) $(M, \gamma_{\mathcal{P}}, \eta)$ is a right module over the operad $\mathcal{P}$.

These structures make the following diagram commutes

$$\begin{array}{ccccc} M \circledast (M \circ \mathcal{P}) & & \xrightarrow{\gamma_{\mathcal{P}}} & & M \circledast M \\ \downarrow{\scriptstyle \varepsilon_{M,M,\mathcal{P}}} & & & & \downarrow{\scriptstyle \gamma_M} \\ (M \circledast M) \circ \mathcal{P} & \xrightarrow{\gamma_M} & M \circ \mathcal{P} & \xrightarrow{\gamma_{\mathcal{P}}} & M . \end{array}$$

Remark 3.2 This structure was introduced for the first time in [Wil16]. See *loc.cit* or [CG20a].

This structure can be rephrased in terms of partial composition maps as the data of

$$\begin{cases} \circ_0 : M(k) \otimes M(m) \longrightarrow M(k+m) , \\ \circ_i : M(k) \otimes \mathcal{P}(m) \longrightarrow M(k+m-1) \text{ for } 1 \leq i \leq k, \end{cases}$$

which satisfy analogous compatibility conditions. One can visualize these compositions as follows: operations in $M(k)$ are represented as rooted trees with $k+1$ leaves, where the up most left leaf is fixed and not acted upon by the symmetric group $\mathfrak{S}_k$. This leaf is called the 0-th leaf. The compositions $\circ_0$ amount to inserting operations in M into the 0-th leaf, and the compositions $\circ_i$ amount to inserting operations in $\mathcal{P}$ into the other, non-fixed, leaves. Pictorially it gives

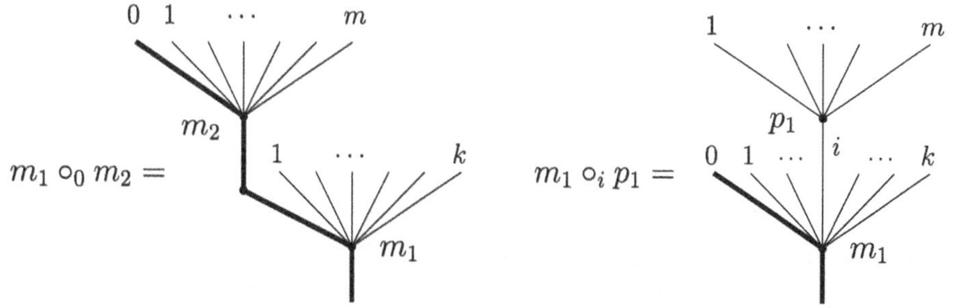

for m_1 an operation in $M(k)$, m_2 an operation in $M(m)$ and p_1 an operation in $\mathcal{P}(m)$.

3.2.1 Moperads as the Derivatives of Operads

Symmetric sequences in $\mathcal{E}$ can equivalently be described as *species of structures*, that is, as functors

$$M : \mathrm{Fin}_{\mathrm{bij}}^{\mathrm{op}} \longrightarrow \mathcal{E}$$

from the category of finite sets with bijections to the category $\mathcal{E}$. The theory of species of structures was introduced by Joyal in [Joy81]. We refer to [BLL98] for more details.

Remark 3.3 In order to extend a symmetric sequence to a species of structures, there is a canonical choice given by

$$M(I) := \left(\coprod_{f:I \longrightarrow [n]} M(n) \right)_{\mathfrak{S}_n} ,$$

where I is a finite set with n elements, $[n] = \{1, \cdots, n\}$ and the sum is over all bijections f. The notion of an operad and a moperad can be extended to species of structures in a straightforward way.

Definition 3.4 (Derivative of a Species of Structures) Let

$$M : \mathrm{Fin}_{\mathrm{bij}}^{\mathrm{op}} \longrightarrow \mathcal{E}$$

be a species of structure in $\mathcal{E}$. Its *derivative* M' is the species of structures whose image is given by

$$M'(I) := M(I \sqcup \{*\})$$

for any finite set I.

The derivatives are a classical operation on species of structures: if the species happens to be an operad, then its derivative is in fact a moperad over it.

Proposition 3.5 *Let $\mathcal{P}$ be a species of structure in $\mathcal{E}$ endowed with an operad structure. Then $\mathcal{P}'$ is canonically endowed with a moperad structure over $\mathcal{P}$ given as follows.*

(1) *The monoid structure*

$$\circ_0 : \mathcal{P}'(I) \otimes \mathcal{P}'(J) \longrightarrow \mathcal{P}'(I \sqcup J)$$

is given by the operadic composition

$$\circ_{\{*\}} : \mathcal{P}(I \sqcup \{*\}) \otimes \mathcal{P}(J \sqcup \{*\}) \longrightarrow \mathcal{P}(I \sqcup J \sqcup \{*\})$$

at the added point $\{*\}$.

(2) *The right module structure*

$$\circ_i : \mathcal{P}'(I) \otimes \mathcal{P}(J) \longrightarrow \mathcal{P}'((I - \{i\}) \sqcup J)$$

is given by the operadic composition

$$\circ_i : \mathcal{P}(I \sqcup \{*\}) \otimes \mathcal{P}(J) \longrightarrow \mathcal{P}((I - \{i\}) \sqcup J \sqcup \{*\})$$

at the point i in I.

Proof It is straightforward to check that the associativity conditions on the operad structure of $\mathcal{P}$ induce those of the moperad structure on $\mathcal{P}'$. □

Example 3.6 Let $\mathcal{MU}nit$ be the symmetric sequence given for all n by $\mathcal{MU}nit(n) := \mathbb{1}$, then $\mathcal{MU}nit$ forms a moperad over $\mathcal{U}nit$ with the obvious compositions maps. One can easily see that $\mathcal{MU}nit$ is the derivative of the operad $uCom$, defined by $uCom(n) := \mathbb{1}$, encoding unital commutative algebras in $\mathcal{E}$. △

Recall our convention that all operads $\mathcal{P}$ come equipped with a morphism of operads $f : \mathcal{U}nit \longrightarrow \mathcal{P}$ which is an isomorphism in arities zero and one, that is, our operads are *pointed reduced*. We will use the same convention for moperads: any moperad M over an operad $\mathcal{P}$ will come with a morphism of moperads $\varphi_f : \mathcal{MU}nit \longrightarrow M$ over the above mentioned morphism f. This implies that there is a distinguished morphism of symmetric sequences $\mathcal{P} \longrightarrow M$ for all the moperads considered from now on.

Example 3.7 The moperad $\mathcal{P}a_0$ over the operad of parenthesized permutations $\mathcal{P}a$ in the category of sets is defined as follows: the set $\mathcal{P}a_0(n)$ is given by the set of maximal parenthesization of $0\sigma_1 \cdots \sigma_n$, where $\sigma_1 \cdots \sigma_n$ is a permutation of $\mathfrak{S}_n$. The right module structure of $\mathcal{P}a_0$ over $\mathcal{P}a$ is given by the insertion of parenthesized permutations on the right of the zero element: $(0\sigma) \circ_i \tau := 0\sigma \circ_i \tau$, for any parenthesized permutation τ. For example,

$$(01)2 \circ_1 (13)2 = (0(13)2)4 .$$

The moperad structure is given by

$$0\sigma \circ_0 0\tau := 0\tau\sigma ,$$

where one inserts 0τ together with its maximal parenthesization into the zero spot of the fist maximally parenthesized word 0σ, giving another maximally parenthesized word. For example,

$$(01)2 \circ_0 0(3(21)) = (0(3(21))4)5 .$$

The inclusion of $\mathcal{MU}nit$ into $\mathcal{P}a_0$ sends $\mathcal{MU}nit(n) = \{*\}$ into the leftmost parenthesization of $0\mathrm{id}_n = 01 \cdots n$, where id_n is the trivial permutation in $\mathfrak{S}_n$. One can directly see that it forms a moperad by applying Proposition 3.5 to $\mathcal{P}a$ (indeed, $\mathcal{P}a_0 = \mathcal{P}a'$). △

Example 3.8 The moperad $\mathcal{C}o\mathcal{B}^{(1)}$ over the operad $\mathcal{C}o\mathcal{B}$ from Example 2.7, is defined as follows:

(1) The objects of $\mathcal{C}o\mathcal{B}^{(1)}(n)$ are given by words 0σ, where σ is a permutation of $\mathfrak{S}_n$. For instance, 0321 is in $\mathrm{Ob}(\mathcal{C}o\mathcal{B}^{(1)}(3))$.
(2) The morphisms of $\mathcal{C}o\mathcal{B}^{(1)}(n)$ are given, for two words 0σ and 0τ, by the set of braids going from 0σ to 0τ, where the zeroth strand is frozen. Pictorially, we have:

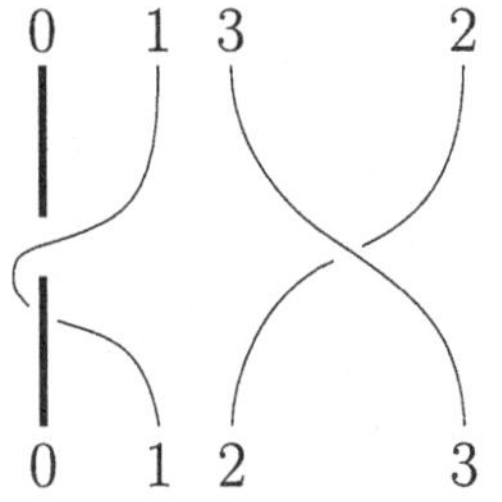

$$\in \mathrm{Hom}_{\mathcal{C}o\mathcal{B}^{(1)}(3)}(0132, 0123)$$

(3) The right module structure of $\mathcal{C}o\mathcal{B}^{(1)}$ over $\mathcal{C}o\mathcal{B}$ is given by the composition of braids on the last n spots. The monoid structure of $\mathcal{C}o\mathcal{B}^{(1)}$ is given by the insertion of the second braid into the zeroth spot of the first braid. The morphism from $\mathcal{MU}nit$ to $\mathcal{C}o\mathcal{B}^{(1)}$ sends $\{*\}$ to $0\mathrm{id}_n$.

One can directly see that it forms a moperad by applying Proposition 3.5 to $\mathcal{C}o\mathcal{B}$ (indeed, $\mathcal{C}o\mathcal{B}^{(1)} = \mathcal{C}o\mathcal{B}'$). △

Consider $\mathcal{P}a_0$ as a moperad over $\mathcal{P}a$ in groupoids where there are no non-trivial morphisms between objects. There is a morphism of moperads in sets $\psi : \mathrm{Ob}(\mathcal{P}a_0) \longrightarrow$

$\mathrm{Ob}(\mathcal{CoB}^{(1)})$ which lies above the morphism of operads in sets $\varphi : \mathrm{Ob}(\mathcal{P}a) \longrightarrow \mathrm{Ob}(\mathcal{CoB})$. Both are given by forgetting the parenthesization.

Definition 3.9 (Moperad of Parenthesized Braids) The *moperad of parenthesized braids* $\mathcal{P}a\mathcal{B}^{(1)}$ over the operad in groupoids $\mathcal{P}a\mathcal{B}$ is defined to be the fake pull-back of $\mathcal{C}o\mathcal{B}^{(1)}$ along the morphism ψ.

Remark 3.10 It is straightforward to check that the fake pull-back of Definition 2.9 generalizes to moperads.

Remark 3.11 The moperad $\mathcal{P}a\mathcal{B}^{(1)}$ can also directly be obtained by applying Proposition 3.5 to $\mathcal{P}a\mathcal{B}$: $\mathcal{P}a\mathcal{B}^{(1)} = \mathcal{P}a\mathcal{B}'$.

3.3 Moperads of Parenthesized Cyclotomic Braids

3.3.1 Group Actions on Moperads

Consider the terminal operad $uCom$ in the category of sets, given by $uCom(n) = \{*\}$ for all $n \geq 0$ together with the obvious compositions. For any group G, one can defined a moperad $\underline{G}$ over $uCom$ as follows

(1) The set $\underline{G}(n) := G^{\times n}$, together with the evident $\mathfrak{S}_n$ action.
(2) The right module structure over $uCom$ is given by

$$\begin{array}{ccc} G^{\times n} \cong G^{\times n} \times uCom(m) & \xrightarrow{\quad\circ_i^G\quad} & G^{\times n+m-1} \\ (g_1, \cdots, g_n) & \longmapsto & (g_1, \cdots, g_{i-1}, g_i, \cdots, g_i, g_{i+1}, \cdots, g_n)\,, \end{array}$$

where we have m copies of the element g_i, for $1 \leq i \leq n$.
(3) The monoid structure maps $\circ_0^G : G^{\times n} \times G^{\times m} \longrightarrow G^{\times(n+m)}$ are given by the evident isomorphism.

The data of $\underline{G}$ over $uCom$ actually forms a moperad in the category of groups, as all the above maps are compatible with the group structures.

Definition 3.12 (G-Action on Moperads) Let M be a moperad over an operad $\mathcal{P}$ in the category of sets. The data of a G-action on the moperad M is the data of a morphism of moperads over $\mathcal{P}$

$$\gamma_G : \underline{G} \times M \longrightarrow M\,,$$

where $\underline{G} \times M$ is a moperad over $uCom \times \mathcal{P} \cong \mathcal{P}$ under the natural identification. The structural morphism γ_G satisfies the condition imposed by the commutativity of the following diagram

$$\begin{array}{ccc} \underline{G} \times \underline{G} \times M & \xrightarrow{\mathrm{id} \times \gamma_G} & \underline{G} \times M \\ \Big\downarrow{\scriptstyle \mathrm{Mult}_{\underline{G}} \times \mathrm{id}_n} & & \Big\downarrow{\scriptstyle \gamma_G} \\ \underline{G} \times M & \xrightarrow{\gamma_G} & M\,, \end{array}$$

where the morphism $\mathrm{Mult}_{\underline{G}}$ is given in arity n by the group multiplication of $G^{\times n}$.

Remark 3.13 The data of a G-action on a moperad M over $\mathcal{P}$ amounts to the data of an $\mathfrak{S}_n$-equivariant left $G^{\times n}$-action on $M(n)$ such that the right module structure maps $\{\circ_i\}$ are $\{\circ_i^G\}$-equivariant and such that the monoid maps $\{\circ_0\}$ is $G^{\times n+m}$-equivariant.

3.3.2 Semi-Direct Products

Let G be a group. A G-groupoid is the data of a groupoid $\mathcal{D}$ together with a morphism of groups from G to the automorphism group $\mathrm{Aut}_{\mathsf{Grpd}}(\mathcal{D})$ of the groupoid $\mathcal{D}$; hence each element g of G induces an automorphism of groupoids of $\mathcal{D}$ which will be denoted by $g \star -$. Together with G-equivariant morphisms, G-groupoids form a category, denoted by G-Grpd. We denote $\mathrm{B}G$ the classifying groupoid of G, which has only one object $\{*\}$, whose automorphism group is precisely G.

Definition 3.14 (Semi-Direct Product) The *semi-direct product* is a functor

$$\begin{aligned} G\text{-}\mathsf{Grpd} &\longrightarrow \mathsf{Grpd}/\mathrm{B}G \\ \mathcal{D} &\longmapsto \mathcal{D} \rtimes G\,. \end{aligned}$$

The semi-direct product $\mathcal{D} \rtimes G$, which is a groupoid above the groupoid $\mathrm{B}G$, is the groupoid defined as follows:

(1) The object of $\mathcal{D} \rtimes G$ are the objects of $\mathcal{D}$.
(2) Let d be in $\mathrm{Ob}(\mathcal{D} \rtimes G)$ and g be in G, then

$$\mathrm{Hom}_{\mathcal{D}\rtimes G}(g \star d, d) := \mathrm{Hom}_{\mathcal{D}}(g \star d, d) \amalg \{\varepsilon_g\}\,,$$

that is, there is a free added arrow $\varepsilon_g : g \star d \longrightarrow d$ for all g and all d. The set of free added arrows satisfies the following conditions:

(a) For any g, h in G and object d:

$$\left(g \star (h \star d) \xrightarrow{\varepsilon_g} h \star d \xrightarrow{\varepsilon_h} d \right) = \left((g.h) \star d \xrightarrow{\varepsilon_{g.h}} d\,, \right)$$

where $g \star d$ denotes the image of the object d via the functor $g \star -$.

(b) For any morphism $f : d \longrightarrow e$ in $\mathcal{D}$ and any g in G:

$$\left(g \star d \xrightarrow{\varepsilon_g} d \xrightarrow{f} e \xrightarrow{(\varepsilon_g)^{-1}} g \star e \right) = \left(g \star d \xrightarrow{g \star f} g \star e \,, \right)$$

where $g \star f$ denotes the image of the arrow f via the functor $g \star -$.

(3) The structural morphism $\mathcal{D} \rtimes G \longrightarrow \mathrm{B}G$ sends every object to $\{*\}$ and every arrow to id_* except the arrows $\{\varepsilon_g\}$, which are sent to g, for every g in G.

There is another functor, which associates to any groupoid above $\mathrm{B}G$ a groupoid endowed with a G-action.

Definition 3.15 (Pseudo-Free G-Groupoid) The *pseudo-free G-groupoid* is a functor

$$\begin{aligned} \mathsf{Grpd}/\mathrm{B}G &\longrightarrow G\text{-}\mathsf{Grpd} \\ \mathcal{Q} &\longmapsto \mathcal{G}(\mathcal{Q}) \,. \end{aligned}$$

Let $p : \mathcal{Q} \longrightarrow \mathrm{B}G$ be a groupoid above $\mathrm{B}G$, the pseudo-free G-groupoid generated by $p : \mathcal{Q} \longrightarrow \mathrm{B}G$, denoted by $\mathcal{G}(\mathcal{Q})$, is defined as follows:

(1) The set of objects of $\mathcal{G}(\mathcal{Q})$ is given by $\mathrm{Ob}(\mathcal{Q}) \times G$.
(2) The set of morphisms

$$\mathrm{Hom}_{\mathcal{G}(\mathcal{Q})}((x, g), (y, h)) := \{f \in \mathrm{Hom}_{\mathcal{Q}}(x, y) \mid g \bullet p(f) = h\} \,,$$

for x, y in $\mathrm{Ob}(\mathcal{Q})$ and g, h in G.
(3) The G-action $g \star -$ sends (x, h) to (x, gh) and sends any arrow $f : (x_1, h_1) \longrightarrow (x_2, h_2)$ to $f : (x_1, gh_1) \longrightarrow (x_2, gh_2)$.

Remark 3.16 The pseudo-free G-construction and the semi-direct product construction do *not* define adjoint functors.

Remark 3.17 Let (x, g) be an object of $\mathcal{G}(\mathcal{Q})$, the automorphism group $\mathrm{Aut}_{\mathcal{G}(\mathcal{Q})}((x, g))$ is given by the automorphisms $f : x \longrightarrow x$ in $\mathcal{Q}$ such that $p(f) = \mathrm{id}_{p(x)}$.

Example 3.18 Let B_n be the braid group on n strands. The morphism $\mathrm{B}_n \twoheadrightarrow \mathfrak{S}_n$ gives a morphism of groupoids $p : \mathrm{BB}_n \twoheadrightarrow \mathrm{B}\mathfrak{S}_n$. The pseudo-free $\mathfrak{S}_n$-groupoid generated by p is given by the groupoid $CoB(n)$, with its corresponding $\mathfrak{S}_n$-action. △

We denote again $uCom$ the terminal operad in the category of groupoids, given by the trivial groupoid in each arity. Since the classifying groupoid functor $\mathrm{B}(-)$ is strong monoidal, the moperad $\underline{G}$ in sets is sent to a moperad $\mathrm{B}\underline{G}$ over $uCom$ in the category of groupoids.

Let $\mathcal{P}$ be an operad in groupoids. Since the above functor is lax symmetric monoidal, any moperad M over $\mathcal{P}$ together with a morphism of moperads $\phi : M \longrightarrow \mathrm{B}\underline{G}$ above the trivial morphism $\mathcal{P} \twoheadrightarrow uCom$ will give, under the functor $\mathcal{G}(-)$, a moperad $\mathcal{G}(M)$ over $\mathcal{P}$

which comes equipped with a $\mathsf{B}\underline{G}$-action (in the sense of Definition 3.12 extended to the groupoid case).

Example 3.19 There is a morphism of moperads $\iota : Co\mathcal{B}^{(1)} \longrightarrow \mathsf{B}\underline{\mathbb{Z}/N\mathbb{Z}}$ above the trivial morphism $Co\mathcal{B} \twoheadrightarrow uCom$. On objects, it sends any word 0σ in $Co\mathcal{B}^{(1)}(n)$ to the single object $\{*\}$. On morphisms, it sends a braid between two words 0σ and 0τ in $Co\mathcal{B}^{(1)}(n)$ to the element $(\gamma_1, \cdots, \gamma_n)$ in $(\mathbb{Z}/N\mathbb{Z})^{\times n}$. The value of γ_i is given by the number of loops that the i-th strand of the braid makes around the zeroth strand, counted module N. For instance,

$$Co\mathcal{B}(3) \longrightarrow \mathsf{B}(\mathbb{Z}/N\mathbb{Z})^3$$

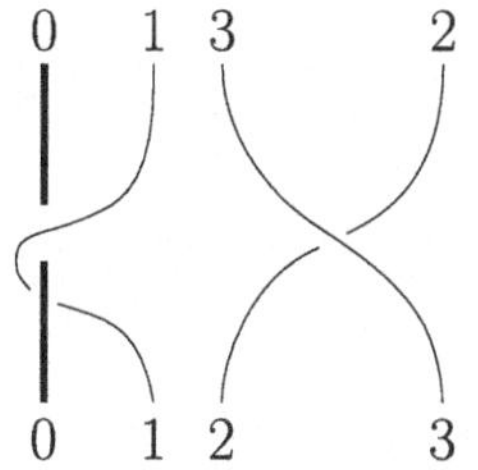

$$\mapsto \quad (\bar{1}, \bar{0}, \bar{0})$$

This allows us to define $Co\mathcal{B}^{(N)}$ as the moperad over $Co\mathcal{B}$ given, in arity n, by the pseudo-free $(\mathbb{Z}/N\mathbb{Z})^n$-groupoid on the morphism $\iota(n) : Co\mathcal{B}^{(1)}(n) \longrightarrow \mathsf{B}((\mathbb{Z}/N\mathbb{Z})^n)$. Objects of $Co\mathcal{B}^{(N)}(n)$ are given by words $0\sigma_1 \cdots \sigma_n$ labeled by $(\gamma_1, \cdots, \gamma_n)$ in $(\mathbb{Z}/N\mathbb{Z})^n$. We will denote them $0(\sigma_1)_{\gamma_1} \cdots (\sigma_n)_{\gamma_n}$. Morphisms between two labeled words $0(\sigma_1)_{\gamma_1} \cdots (\sigma_n)_{\gamma_n}$ and $0(\tau_1)_{\lambda_1} \cdots (\tau_n)_{\lambda_n}$ are given by braids which satisfy the following condition: the number of loops that the i-th strand does around the zeroth strand is $\lambda_i - \gamma_i$ modulo N for all $1 \leq i \leq n$. △

Similarly to the previous example, there is also a morphism of moperads $\jmath$: $\mathcal{P}a\mathcal{B}^{(1)} \longrightarrow \mathsf{B}\underline{\mathbb{Z}/N\mathbb{Z}}$ above the trivial morphism $\mathcal{P}a\mathcal{B} \twoheadrightarrow uCom$. It again sends parenthesized permutations to $\{*\}$ and braids between them to the numbers of loops around the zeroth strand modulo N.

Definition 3.20 (Moperad of Parenthesized Cyclotomic Braids) The *moperad of parenthesized braids* $\mathcal{P}a\mathcal{B}^{(N)}$ is the moperad over $\mathcal{P}a\mathcal{B}$ defined in each arity by

$$\mathcal{P}a\mathcal{B}^{(N)}(n) := \mathcal{G}\left(\jmath(n) : \mathcal{P}a\mathcal{B}^{(1)}(n) \longrightarrow \mathsf{B}((\mathbb{Z}/N\mathbb{Z})^n)\right) .$$

Its moperad structure comes from the fact that $\jmath : \mathcal{P}a\mathcal{B}^{(1)} \longrightarrow \mathsf{B}\underline{\mathbb{Z}/N\mathbb{Z}}$ is a morphism of moperads.

Remark 3.21 The description of $\mathcal{C}o\mathcal{B}^{(N)}$ given in Example 3.19 applies *mutatis mutandis* to $\mathcal{P}a\mathcal{B}^{(N)}$ as well. In this case, the only real difference is that objects of $\mathcal{C}o\mathcal{B}^{(N)}(n)$ are words $0\sigma_1 \cdots \sigma_n$ while objects in $\mathcal{P}a\mathcal{B}^{(N)}(n)$ are words $0\sigma_1 \cdots \sigma_n$ together with a *maximal parenthesization*.

The moperad $\widehat{\mathcal{P}a\mathcal{B}}^{(N)}(\mathbb{k})$ over $\widehat{\mathcal{P}a\mathcal{B}}(\mathbb{k})$ is obtained by applying the Malcev completion functor to the aforementioned moperad. It gives a moperad in the category of pro-unipotent $\mathbb{k}$-groupoids. This will be *our first cyclotomic player*.

3.4 Infinitesimal Cyclotomic Braids

In the last section, we defined the moperad $\mathcal{P}a\mathcal{B}^{(N)}$ over $\mathcal{P}a\mathcal{B}$, which will be the first player necessary to define cyclotomic associators. Now we need the "infinitesimal version" of this moperad. From now on, we fix $\Gamma := \mathbb{Z}/N\mathbb{Z}$.

Definition 3.22 (Cyclotomic Drinfeld–Kohno Algebras) The n-th *cyclotomic Drinfeld–Kohno* Lie algebra $\mathfrak{t}_n^\Gamma$ is given by the following presentation:

(1) It is generated by elements t_{0i} for $1 \leq i \leq n$ and by elements t_{ij}^α for $1 \leq i, j \leq n$, where $i \neq j$, and for α in Γ.
(2) These generators are subject to the following relations:

  (a) $t_{ij}^\alpha = t_{ji}^{-\alpha}$.
  (b) $[t_{0i}, t_{jk}^\alpha] = 0$ and $[t_{ij}^\alpha, t_{kl}^\beta] = 0$.
  (c) $[t_{ij}^\alpha, t_{ik}^{\alpha+\beta} + t_{jk}^\beta] = 0$.
  (d) $[t_{0i}, t_{0j} + \sum_{\alpha\in\Gamma} t_{ij}^\alpha] = 0$.
  (e) $[t_{0i} + t_{0j} + \sum_{\beta\in\Gamma} t_{ij}^\beta, t_{ij}^\alpha] = 0$.

  for all distinct i, j, k, l and for all α, β in Γ.

Proposition 3.23 ([CG20a]) *The collection* $(\mathfrak{t}_n^\Gamma)_{n\geq 0}$ *can be endowed with the following moperad structure over the operad* $(\mathfrak{t}_n)_{n\geq 0}$ *in the symmetric monoidal category* $(\mathsf{Lie}, \oplus, 0)$*:*

(1) *The action of the symmetric group* $\mathfrak{S}_n$ *on* $\mathfrak{t}_n^\Gamma$ *is given by :*

$$\sigma \bullet t_{ij}^\alpha := t_{\sigma(i)\sigma(j)}^\alpha \text{ and } \sigma \bullet t_{0i} := t_{0\sigma(i)} .$$

(2) *The partial composition maps* $\{\circ_p : \mathbf{t}_n^\Gamma \oplus \mathbf{t}_m \longrightarrow \mathbf{t}_{n+m-1}^\Gamma\}$ *of the right module structure are given by the following components (we assume that* $i < j$*):*

$$\mathbf{t}_m \ni t_{ij} \longmapsto t^0_{i+p-1\, j+p-1}$$

$$\mathbf{t}_n^\Gamma \ni t_{ij}^\alpha \longmapsto \begin{cases} t^\alpha_{i+m-1\, j+m-1} & \text{if } p < i\,. \\ \displaystyle\sum_{k=i}^{i+m-1} t^\alpha_{k\, j+m-1} & \text{if } p = i\,. \\ t^\alpha_{i j+m-1} & \text{if } i < p < j\,. \\ \displaystyle\sum_{k=j}^{j+m-1} t^\alpha_{ik} & \text{if } p = j\,. \\ t^\alpha_{ij} & \text{if } j < p\,. \end{cases}$$

$$\mathbf{t}_n^\Gamma \ni t_{0j} \longmapsto \begin{cases} t_{0\, j+m-1} & \text{if } p < j\,. \\ \displaystyle\sum_{k=j}^{j+m-1} t_{0k} + \sum_{\gamma\in\Gamma}\sum_{j\leq l<r<j+m} t^\gamma_{lr} & \text{if } p = j\,. \\ t_{0j} & \text{if } j < p\,. \end{cases}$$

(3) *The monoidal structure maps* $\{\circ_0 : \mathbf{t}_n^\Gamma \oplus \mathbf{t}_m^\Gamma \longrightarrow \mathbf{t}_{n+m}^\Gamma\}$ *are given by the following components:*

$$\mathbf{t}_m^\Gamma \ni t_{0k} \longmapsto t_{0k}\,.$$

$$\mathbf{t}_m^\Gamma \ni t_{kl}^\alpha \longmapsto t_{kl}^\alpha\,.$$

$$\mathbf{t}_n^\Gamma \ni t_{0i} \longmapsto t_{0i} + \sum_{\gamma\in\Gamma}\sum_{r\geq 1}^{m} t^\gamma_{ri}\,.$$

$$\mathbf{t}_n^\Gamma \ni t_{ij}^\alpha \longmapsto t_{ij}^\alpha\,.$$

It comes equipped with the following action of Γ*: let* $(\gamma_1, \cdots, \gamma_i, \cdots, \gamma_n)$ *be in* Γ^n*, then its action on the generators of* $\mathfrak{t}_n^\Gamma$ *is given by*

$$\begin{cases} \gamma_i \bullet t_{0j} = t_{0j} \text{ for } 1 \leq k \leq n\,, \\ \gamma_i \bullet t_{jk}^\alpha = t_{jk}^\alpha \text{ if } i \neq j, k\,, \\ \gamma_i \bullet t_{jk}^\alpha = t_{jk}^{\alpha+\gamma_i} \text{ if } i = j\,, \\ \gamma_i \bullet t_{jk}^\alpha = t_{jk}^{\alpha-\gamma_i} \text{ if } i = k\,. \end{cases}$$

Proof We have already seen in the Introduction that the insertion-coproduct morphisms gives $t_\bullet$ the structure of a presheaf of graded Lie algebras on Fin_*. Following [Enr07] one can extend it to a presheaf of graded Lie algebras on $\mathrm{bFin}_{*,*}$ (defined in Appendix B.3) in the following way (without loss of generality, we allow ourselves to work with a skeleton of $\mathrm{bFin}_{*,*}$):

- The Lie algebra associated with $\{* = 0, 1, \ldots, n\}$ is $\mathfrak{t}_n$.
- The Lie algebra associated with $\{*, 0, 1, \ldots, n\}$ is $\mathfrak{t}_n^\Gamma$.
- For every doubly pointed map $f : \{*, 0, 1, \ldots, m\} \to \{*, 0, 1, \ldots, n\}$, the corresponding map $(-)^f :\to \mathfrak{t}_n^\Gamma \to \mathfrak{t}_m^\Gamma$ is defined by

$$(t_{ij}^\alpha)^f := \sum_{\substack{k \in f^{-1}(i) \\ l \in f^{-1}(j)}} t_{kl}^\alpha$$

  and

$$(t_{0i})^f := \sum_{j \in f^{-1}(i)} t_{0j} + \sum_{\substack{j,k \in f^{-1}(i) \\ j<k}} \sum_{\gamma \in \Gamma} t_{jk}^\gamma + \sum_{\substack{j \in f^{-1}(0)\setminus\{0\} \\ k \in f^{-1}(i)}} \sum_{\gamma \in \Gamma} t_{jk}^\gamma\,.$$

- For every doubly pointed map $f : \{*, 0, 1, \ldots, m\} \to \{* = 0, 1, \ldots, n\}$, the corresponding map $(-)^f : \mathfrak{t}_n \to \mathfrak{t}_m^\Gamma$ is defined by

$$(t_{ij})^f := \sum_{\substack{k \in f^{-1}(i) \\ l \in f^{-1}(j)}} t_{kl}^0\,.$$

- For every pointed map $f : \{* = 0, 1, \ldots, m\} \to \{* = 0, 1, \ldots, n\}$, the corresponding map $(-)^f : \mathfrak{t}_n \to \mathfrak{t}_m$ is the usual insertion-coproduct morphism from Definition 1.6. △

This defines a presheaf on $\mathrm{bFin}_{*,*}$, and therefore by Appendix B.3, a moperad structure in the category of Lie algebras with the coproduct. Like in Proposition 2.14, the images of theses maps commute and therefore factor through the product of Lie algebras. □

Definition 3.24 (**N-Chord Diagram Moperad**) The *N-chord diagram moperad* $\mathcal{CD}_0^\Gamma$ is the moperad over $\mathcal{CD}$ given by applying the universal enveloping algebra to the cyclotomic Drinfeld–Kohno moperad.

In each arity, the moperad $\mathcal{CD}_0^\Gamma$ is given by

$$\mathcal{CD}_0^\Gamma(n) = \mathfrak{U}(\mathfrak{t}_n^\Gamma)\,,$$

where $\mathfrak{U}(\mathfrak{t}_n^\Gamma)$ is the algebra of N-chord diagrams on $n+1$ strands. Following [Bro13], one can represent the elements of this algebra as chord diagrams on $n+1$ strands, numbered from 0 to n, with labels on the last n strands: each strand outside the zeroth one is labeled with elements of $\mathbb{Z}/N\mathbb{Z}$ such that their sum is 0 in $\mathbb{Z}/N\mathbb{Z}$. For instance, the element $t_{0i}.t_{jk}^{\alpha}.t_{jl}^{\beta}$ in $\mathfrak{U}(\mathfrak{t}_n^\Gamma)$ can be represented as

$$t_{0i}.t_{jk}^{\alpha}.t_{jl}^{\beta} =$$

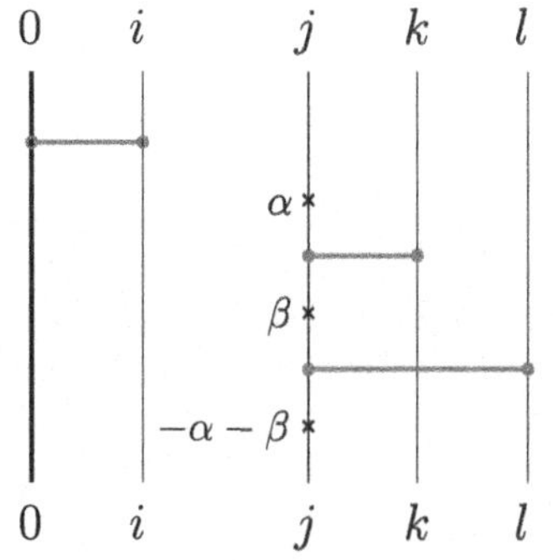

One can also represent all the relations that define the algebra $\mathfrak{U}(\mathfrak{t}_n^\Gamma)$ in terms of equalities of such N-chord diagrams (see [CG20a]).

We have that $\mathcal{CD}_0^\Gamma$ is a moperad over $\mathcal{CD}$ in the category $\mathsf{Cat(Coalg_{\Bbbk})}$ of small categories enriched over counital cocommutative $\Bbbk$-coalgebras, where both only have one object. In order to obtain a moperad over the operad $\mathcal{PaCD}$, we need to add objects to $\mathcal{CD}_0^\Gamma$ first. Consider the moperad $\underline{\Gamma}$ over $uCom$ in the category of sets. By applying the $\Bbbk$-linear extension functor introduced in Definition 2.21, we obtain a moperad $\underline{\Gamma}_{\Bbbk}$ over the terminal operad $uCom$, this time in the category $\mathsf{Cat(Coalg_{\Bbbk})}$. The category $\underline{\Gamma}_{\Bbbk}(n)$ has Γ^n a set of objects and $\Bbbk$ as its cocommutative coalgebra of morphisms for any two objects.

Definition 3.25 (Cyclotomic Chord Moperad) The *cyclotomic chord moperad* $\mathcal{CD}^\Gamma$ is defined as the Hadamard product

$$\mathcal{CD}^\Gamma := \mathcal{CD}_0^\Gamma \otimes \underline{\Gamma}_{\Bbbk}$$

of moperads in the category $\mathsf{Cat(Coalg_{\Bbbk})}$. It is naturally a moperad over $\mathcal{CD} \otimes uCom \cong \mathcal{CD}$. It comes with a natural Γ-action given by the product of the Γ-actions on each component of the product. (We view Γ acting on $\underline{\Gamma}_{\Bbbk}$ via translation on objects).

Remark 3.26 Let $(\alpha_1, \cdots, \alpha_n)$ and $(\beta_1, \cdots, \beta_n)$ be two objects in $\mathcal{CD}^\Gamma(n)$. The elements of the cocommutative coalgebra $\mathfrak{U}(\mathfrak{t}_n^\Gamma)$ of morphisms between them can be represented as chord diagrams on $n+1$ strands, where the last n strands are labeled by elements of $\mathbb{Z}/N\mathbb{Z}$ in the following way: the sum of the labels on the i-th strand must be equal to $\beta_i - \alpha_i$, for all $1 \leq i \leq n$. For example, the following chord diagram

$L^\alpha :=$

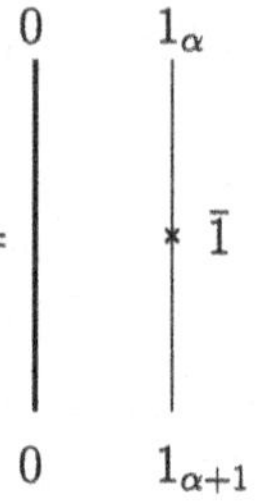

defines a morphism between α and $\alpha + \bar{1}$, for every α in $\mathcal{CD}^\Gamma(1) = \mathbb{Z}/N\mathbb{Z}$.

Now recall the set-theoretical moperad $\mathcal{P}a_0$ over $\mathcal{P}a$ given in Example 3.7. By applying the $\Bbbk$-linear extension functor to both, we obtain a moperad $(\mathcal{P}a_0)_\Bbbk$ over $\mathcal{P}a_\Bbbk$, this time in the category $\mathsf{Cat}(\mathsf{Coalg}_\Bbbk)$.

Definition 3.27 (Cyclotomic Parenthesized Chord Moperad) The *cyclotomic parenthesized chord moperad* $\mathcal{P}a\mathcal{CD}^\Gamma$ is defined by the Hadamard product

$$\mathcal{P}a\mathcal{CD}^\Gamma := (\mathcal{P}a_0)_\Bbbk \otimes \mathcal{CD}^\Gamma$$

of moperads in the category $\mathsf{Cat}(\mathsf{Coalg}_\Bbbk)$. It is naturally a moperad over $\mathcal{P}a_\Bbbk \otimes \mathcal{CD} = \mathcal{P}a\mathcal{CD}$. The Γ-action on $\mathcal{CD}^\Gamma$ endows this product with a Γ-action as well.

Example 3.28 It follows from this definition that objects in $\mathcal{P}a\mathcal{CD}^\Gamma(n)$ are given by maximal parenthesizations on words $0(\sigma_1)_{\alpha_1} \cdots (\sigma_n)_{\alpha_n}$, where σ is a permutation in $\mathfrak{S}_n$ and $(\alpha_1, \cdots, \alpha_n)$ is an element of Γ^n. Morphisms between two maximally parenthesized words $0(\sigma_1)_{\alpha_1} \cdots (\sigma_n)_{\alpha_n}$ and $0(\tau_1)_{\gamma_1} \cdots (\tau_n)_{\gamma_n}$ are given by chord diagrams on the unique strands that go from one permutation to the other, where each strand is labeled by elements of Γ, such that the sums of the labels of the i-th strand is equal to $\gamma_i - \alpha_i$, for all $1 \leq i \leq n$. Here are two distinguished morphisms of $\mathcal{P}a\mathcal{CD}^\Gamma$:

$K^\alpha :=$

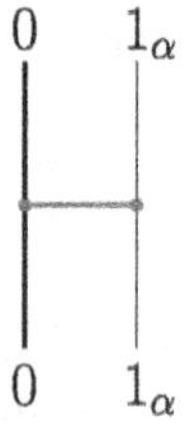

$\beta :=$

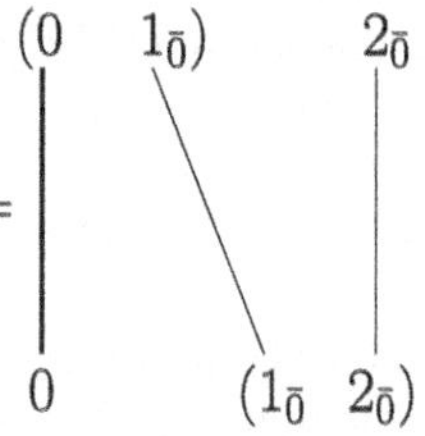

Here K^α is the endomorphism of α in $\mathcal{P}a\mathcal{CD}^\Gamma(1)$ which corresponds to t_{01} and β is the isomorphism between $(0 1_{\bar{0}}) 2_{\bar{0}}$ and $0(1_{\bar{0}} 2_{\bar{0}})$ in $\mathcal{P}a\mathcal{CD}^\Gamma(2)$ which corresponds to the trivial chord diagram. △

Remark 3.29 The elements β, L^α and K^α for all α in Γ generate $\mathcal{P}a\mathcal{CD}^\Gamma$ as a moperad over $\mathcal{P}a\mathcal{CD}$ with a Γ-action.

Notice that the relations defining the cyclotomic Drinfeld–Kohno Lie algebras $\mathfrak{t}_n^\Gamma$ are homogeneous, therefore they inherit a weight filtration with generators in weight one. See Remark 2.18 for the standard case. For all $n \geq 0$, $\mathcal{P}a\mathcal{CD}^\Gamma(n)$ forms a Hopf groupoid, therefore if we want to obtain a pro-unipotent groupoid we ought to complete it.

We define $\widehat{\mathcal{P}a\mathcal{CD}}^\Gamma$ as the moperad over $\widehat{\mathcal{P}a\mathcal{CD}}$ in the category of categories enriched over complete cocommutative coalgebras obtained by completing each hom-set of $\mathcal{P}a\mathcal{CD}^\Gamma(n)$ with respect to the aforementioned weight filtration. It is a moperad in complete Hopf groupoids.

We obtain a moperad $\mathrm{Grp}(\widehat{\mathcal{P}a\mathcal{CD}}^\Gamma)$ over $\mathrm{Grp}(\widehat{\mathcal{P}a\mathcal{CD}})$ by taking the group-like elements functor to both. This gives a moperad in the category of pro-unipotent $\Bbbk$-groupoids. This will be *our second cyclotomic player*.

3.5 Cyclotomic Associators and Grothendieck–Teichmüller Groups

We are again able to present a purely (m)operadic version of cyclotomic associators.

Definition 3.30 (Cyclotomic Associators) The set of *cyclotomic associators* is given by

$$\mathrm{Assoc}^\Gamma(\Bbbk) :=$$

$$\mathrm{Iso}^+_{\mathsf{Mop(p.u\text{-}Grpd_\Bbbk)}}\left(\left(\widehat{\mathcal{P}a\mathcal{B}}^\Gamma(\Bbbk), \widehat{\mathcal{P}a\mathcal{B}}(\Bbbk)\right), \left(\mathrm{Grp}(\widehat{\mathcal{P}a\mathcal{CD}}^\Gamma), \mathrm{Grp}(\widehat{\mathcal{P}a\mathcal{CD}})\right)\right)^\Gamma.$$

An element of $\mathrm{Assoc}^\Gamma(\Bbbk)$ amounts to the data of:

(1) an isomorphism F between $\widehat{\mathcal{P}a\mathcal{B}}(\Bbbk)$ and $\mathrm{Grp}(\widehat{\mathcal{P}a\mathcal{CD}})$ which is the identity on objects (an associator) ,

(2) an isomorphism G between $\widehat{\mathcal{P}a\mathcal{B}}^\Gamma(\Bbbk)$ and $\mathrm{Grp}(\widehat{\mathcal{P}a\mathcal{CD}}^\Gamma)$ of moperads which is Γ-equivariant and which is the identity on objects, lying above the isomorphism F.

This definition entails that the set of cyclotomic associators is a bitorsor over the automorphism groups of each moperad that are the identity morphism on objects. This allows us to define the cyclotomic versions of the Grothendieck–Teichmüller group and of the graded Grothendieck–Teichmüller group.

Definition 3.31 (Cyclotomic Grothendieck–Teichmüller Group) The *cyclotomic Grothendieck–Teichmüller group* over $\Bbbk$ is given by

$$\mathrm{GT}^\Gamma(\Bbbk) := \mathrm{Aut}^+_{\mathsf{Mop(p.u\text{-}Grpd_\Bbbk)}}\left(\widehat{\mathcal{P}a\mathcal{B}}^\Gamma(\Bbbk), \widehat{\mathcal{P}a\mathcal{B}}(\Bbbk)\right)^\Gamma,$$

that is, the group of Γ-equivariant automorphisms of the moperads of $\widehat{\mathcal{P}a\mathcal{B}}^\Gamma(\Bbbk)$ which are the identity on objects.

Definition 3.32 (Cyclotomic Graded Grothendieck–Teichmüller Group) The *cyclotomic graded Grothendieck–Teichmüller group* over $\Bbbk$ is given by

$$\mathrm{GRT}^{\Gamma}(\Bbbk) := \mathrm{Aut}^{+}_{\mathsf{MopCat(Coalg_{\Bbbk})}}\left(\widehat{\mathcal{PaCD}}^{\Gamma}, \widehat{\mathcal{PaCD}}\right)^{\Gamma},$$

that is, the group of Γ-equivariant automorphisms of the moperads of $\widehat{\mathcal{PaCD}}^{\Gamma}$ which are the identity on objects.

3.6 More Concrete Descriptions

Theorem 3.33 ([CG20a, Theorem 4.6]) *The moperad $\mathcal{PaB}^{\Gamma}$, as a moperad over $\mathcal{PaB}$ in groupoids endowed with a Γ-action, and having $\mathcal{Pa}_0$ as its moperad of objects, admits the following presentation. The generating morphisms are*

(1) *The loop E around the frozen zeroth strand, pictorially given by*

$E :=$ 

$\in \mathrm{Hom}_{\mathcal{PaB}^{\Gamma}(1)}(01_{\bar{0}}, 01_{\bar{1}})$

(2) *The associator Ψ which pictorially is given by the following braid*

$\Psi :=$ 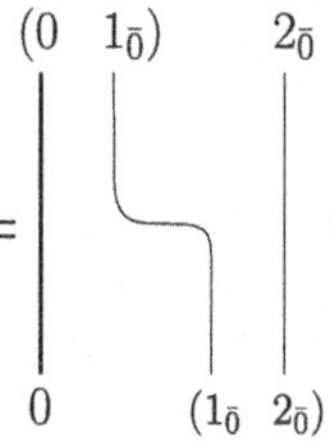

$\in \mathrm{Hom}_{\mathcal{PaB}^{\Gamma}(2)}((01_{\bar{0}})2_{\bar{0}}, 0(1_{\bar{0}}2_{\bar{0}}))$

They satisfy the following relations

(1) *Recall that $\mathcal{PaB}(0)$ is the trivial groupoid $\{*\}$. The unital relation states that*

$$\Psi \circ_1 id_{\{*\}} = id_{01_{\bar{0}}}\,.$$

Notice that it implies that $\Psi \circ_2 id_{\{*\}} = id_{01_{\bar{0}}}$.

(2) *The mixed pentagon relation amounts to the commutativity of the following diagram*

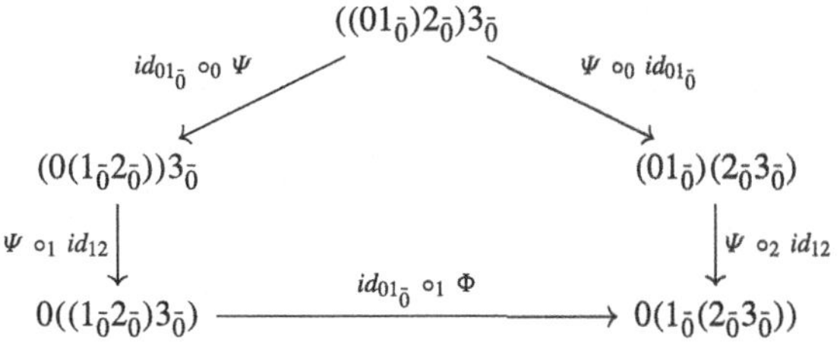

(3) *The twisted ribbon relation amounts to the commutativity of the following diagram*

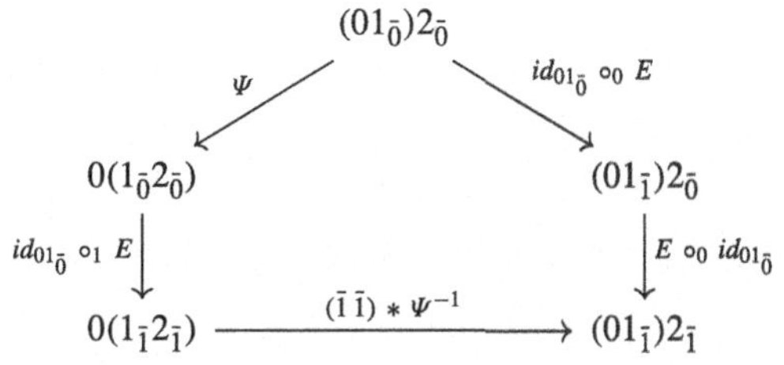

(4) *The twisted octagon relation amounts to the commutativity of the following diagram*

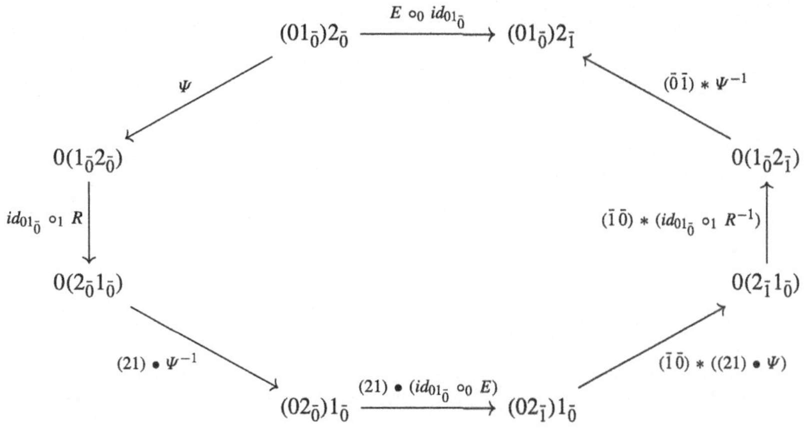

Remark 3.34 For the sake of completeness, let us rewrite the above relations in the "insertion-coproduct" notation:

(1) The unital relation states that $\Psi^{0,\emptyset,1_{\bar{0}}} = \mathrm{id}$.
(2) The mixed pentagon relation can be written as

$$\Psi^{01_{\bar{0}},2_{\bar{0}},3_{\bar{0}}}\,\Psi^{0,1_{\bar{0}},2_{\bar{0}}3_{\bar{0}}} = \Psi^{0,1_{\bar{0}},2_{\bar{0}}}\,\Psi^{0,1_{\bar{0}}2_{\bar{0}},3_{\bar{0}}}\,\Phi^{1_{\bar{0}},2_{\bar{0}},3_{\bar{0}}}\Psi^{0,1_{\bar{0}},2_{\bar{0}}}\,.$$

(3) The twisted ribbon relation can be written as

$$\Psi^{0,1_{\bar{0}},2_{\bar{0}}}\ E^{0,1_{\bar{0}}2_{\bar{0}}}\ (\Psi^{0,1_{\bar{1}},2_{\bar{1}}})^{-1} = E^{0,1_{\bar{0}}}\ E^{01_{\bar{1}},2_{\bar{0}}}\ .$$

(4) The twisted octagon relation can be written as

$$E^{01_{\bar{1}},2_{\bar{0}}} = \Psi^{0,1_{\bar{0}},2_{\bar{0}}}\ R^{1_{\bar{0}},2_{\bar{0}}}\ (\Psi^{0,2_{\bar{0}},1_{\bar{0}}})^{-1}\ E^{0,2_{\bar{0}}}\ \Psi^{0,2_{\bar{1}},1_{\bar{0}}}\ R^{2_{\bar{1}},1_{\bar{0}}}\ (\Psi^{0,1_{\bar{0}},2_{\bar{1}}})^{-1}\ .$$

3.6.1 Relative Malcev Completions

Using this presentation of the moperad $\mathcal{P}a\mathcal{B}^{\Gamma}$, our goal is going to be to describe cyclotomic associators. Recall that $\Gamma = \mathbb{Z}/N\mathbb{Z}$. First consider the free group $\mathbb{F}_2$ generated by two elements x and y. There is a group morphism

$$\begin{cases} \mathbb{F}_2 \xrightarrow{\ p\ } \mathbb{Z} \\ x \longmapsto 1 \\ y \longmapsto 0 \end{cases}$$

This gives a morphism $p_N : \mathbb{F}_2 \twoheadrightarrow \mathbb{Z}/N\mathbb{Z}$ by composing p with the obvious projection.

Lemma 3.35 *There is an isomorphism of groups*

$$Ker(p_N) \cong \mathbb{F}_{N+1}\ ,$$

where $\mathbb{F}_{N+1}$ *is a free group generated by* x^N *and* $x^a y x^{-a}$ *for* $a = 0, \cdots, N-1$.

This gives an exact sequence of groups:

$$1 \longrightarrow \mathbb{F}_{N+1} \hookrightarrow \mathbb{F}_2 \xrightarrow{p_N} \mathbb{Z}/N\mathbb{Z} \longrightarrow 1\ .$$

Let $f : G \twoheadrightarrow H$ be a surjective morphism of groups, and suppose that $\mathrm{Ker}(f)$ is finitely generated as a group.

The *relative* $\mathbb{k}$*-Malcev completion* $\widehat{f} : \widehat{G}(\mathbb{k}, f) \twoheadrightarrow H$ of the morphism f can be defined as the initial $\mathbb{k}$*-pro-unipotent group* whose $\mathbb{k}$-points lie above H. See [Enr07, Section 1.1] for more details or [Hai98] for the original construction. One can show that it fits into the following exact sequence:

$$1 \longrightarrow \widehat{\mathrm{Ker}(f)}(\mathbb{k}) \hookrightarrow \widehat{G}(\mathbb{k}, f) \xrightarrow{\widehat{f}} H \longrightarrow 1\ ,$$

where $\widehat{\mathrm{Ker}(f)}(\mathbb{k})$ is the classical $\mathbb{k}$-pro-unipotent completion.

Remark 3.36 There is always a monomorphism of groups $\widehat{G}(\mathbb{k}, f) \hookrightarrow \widehat{G}(\mathbb{k})$.

Lemma 3.37 *There is an isomorphism of groups*

$$\widehat{\mathbb{F}_2}(\mathbb{k}, p_N) \cong \widehat{\mathbb{F}_{N+1}}(\mathbb{k}) \ltimes \mathbb{Z}/N\mathbb{Z}$$

where the semi-direct product on the right is given by the following action

$$\alpha * x^N = 0 \quad \text{and} \quad \alpha * x^a y x^{-a} = x^{a+\alpha} y x^{-a-\alpha} ,$$

for α *in* $\mathbb{Z}/N\mathbb{Z}$ *.*

Proof The exact sequence

$$1 \longrightarrow \widehat{\mathbb{F}_{N+1}}(\Bbbk) \hookrightarrow \widehat{\mathbb{F}}_2(\Bbbk, p_N) \xrightarrow{\widehat{p_N}} \mathbb{Z}/N\mathbb{Z} \longrightarrow 1$$

admits a splitting since the group $\mathbb{Z}/N\mathbb{Z}$ is finite. One can check the formula by hand. □

3.6.2 Cyclotomic Braid Groups

Recall that the *cyclotomic pure braid groups* PB_n^Γ are given by the kernel of the following morphism

$$\begin{cases} \mathrm{PB}_n^1 \xrightarrow{\quad c_N \quad} \Gamma^n \\ x_{0j} \longmapsto (\overline{0}, \cdots, \overline{0}, \overline{1}, \overline{0}, \cdots, \overline{0}) \\ x_{ij} \longmapsto (\overline{0}, \cdots, \overline{0}) \end{cases}$$

where $(\overline{0}, \cdots, \overline{0}, \overline{1}, \overline{0}, \cdots, \overline{0})$ has $\overline{1}$ in the j-th position, and where PB_n^1 denotes the pure braid group with one fixed strand and n free strands. We therefore have a morphism of short exact sequences

$$\begin{array}{ccccccccc} 0 & \longrightarrow & \mathbb{Z} & \hookrightarrow & \mathrm{PB}_2^1 & \twoheadrightarrow & \mathbb{F}_2 & \longrightarrow & 1 \\ & & \downarrow & & \downarrow c_N & & \downarrow p_N & & \\ 0 & \longrightarrow & \Gamma & \hookrightarrow & \Gamma^2 & \twoheadrightarrow & \Gamma & \longrightarrow & 0 . \end{array}$$

Recall the splitting[4] $\mathbb{F}_2 \longrightarrow \mathrm{PB}_2^1$ given by $x \mapsto x_{01}$ and $y \mapsto x_{12}$ of the upper short exact sequence, that lifts the splitting of the lower one given by $\Gamma \ni \alpha \mapsto (\alpha, \overline{0})$. Passing to relative Malcev completions, we get a splitting $\widehat{\mathbb{F}}_2(\Bbbk, p_N) \longrightarrow \widehat{\mathrm{PB}_2^1}(\Bbbk, c_N)$ of the short exact sequence

$$0 \longrightarrow \Bbbk \hookrightarrow \widehat{\mathrm{PB}_2^1}(\Bbbk, c_N) \twoheadrightarrow \widehat{\mathbb{F}}_2(\Bbbk, p_N) \longrightarrow 1 ,$$

[4] Up to a shift of the indices/strand labels (that corresponds to the canonical identification $\mathrm{PB}_2^1 = \mathrm{PB}_3$) this is the splitting from Lemma 2.3.

and, taking kernels, a splitting $\widehat{\mathbb{F}_{N+1}}(\Bbbk) \longrightarrow \widehat{\mathrm{PB}_2^\Gamma}(\Bbbk)$ of the short exact sequence

$$0 \longrightarrow \Bbbk \hookrightarrow \widehat{\mathrm{PB}_2^\Gamma}(\Bbbk) \twoheadrightarrow \widehat{\mathbb{F}_{N+1}}(\Bbbk) \longrightarrow 1\,.$$

It can be checked that this induces an isomorphism of groups

$$\widehat{\mathrm{PB}_2^\Gamma}(\Bbbk) \cong \Bbbk \times \widehat{\mathbb{F}_{N+1}}(\Bbbk)\,.$$

Similarly, one can check that the center of $\mathfrak{t}_2^\Gamma$ is a one dimensional vector space generated by $t_{01} + t_{02} + \sum_{\alpha\in\Gamma} t_{12}^\alpha$, and that furthermore

$$\mathfrak{t}_2^\Gamma \cong \frac{\mathcal{L}ie(t_{01}, t_{02}, t_{12}^\alpha \mid \alpha = 0, \cdots, N-1)}{(t_{01} + t_{02} + \sum_{\alpha\in\Gamma} t_{12}^\alpha \text{ is central})}\,.$$

There is a canonical isomorphism of Lie algebras

$$\mathfrak{t}_2^\Gamma \cong \Bbbk.c \oplus \mathfrak{f}_{N+1}\,,$$

where $c = t_{01} + t_{02} + \sum_{\alpha\in\Gamma} t_{12}^\alpha$ and where $\mathfrak{f}_{N+1}$ is generated by t_{01} and t_{12}^α, for $\alpha = 0, \cdots, N-1$.

Therefore an element in $\exp(\widehat{\mathfrak{t}_2^\Gamma})$ can by written as a pair

$$\left(\Xi(c), \Psi(t_{01}, t_{12}^{\bar{0}}, \cdots, t_{12}^{\overline{N-1}})\right) \quad \text{in} \quad \Bbbk \times \Bbbk\langle\langle t_{01}, t_{02}, t_{12}^\alpha \mid \alpha = 0, \cdots, N-1\rangle\rangle\,,$$

where Ψ is a non-commutative formal power series in variables t_{01} and t_{12}^α for $\alpha = 0, \cdots, N-1$.

Theorem 3.38 ([CG20a, Theorem 5.5]) *There is a bijection between the set of cyclotomic associators* Assoc^Γ *and the set of triples* (λ, Φ, Ψ) *where*

(1) (λ, Φ) *is a Drinfeld associator;*

(2) Ψ *is a non-commutative formal series* $\Psi(t_{01}, t_{12}^{\bar{0}}, \cdots, t_{12}^{\overline{N-1}})$ *such that*

  (a) *the following equation is satisfied in* $\exp(\widehat{\mathfrak{t}_3^\Gamma})$*:*

$$\Psi\left(t_{01} + \sum_{\alpha\in\Gamma} t_{12}^\alpha + \sum_{\alpha\in\Gamma} t_{13}^\alpha, t_{12}^{\bar{0}}, \cdots, t_{12}^{\overline{N-1}}\right) \Psi\left(t_{01}, t_{12}^{\bar{0}} + t_{13}^{\bar{0}}, \cdots, t_{12}^{\overline{N-1}} + t_{13}^{\overline{N-1}}\right) =$$

$$\Psi\left(t_{01}, t_{12}^{\bar{0}}, \cdots, t_{12}^{\overline{N-1}}\right) \Psi\left(t_{01} + t_{02} + \sum_{\alpha\in\Gamma} t_{12}^\alpha, t_{12}^{\bar{0}} + t_{13}^{\bar{0}}, \cdots, t_{12}^{\overline{-N+1}} + t_{13}^{\overline{-N+1}}\right) \Phi(t_{12}, t_{23})$$

(b) *the following equation is satisfied in* $\exp(\widehat{\mathfrak{t}}_2^\Gamma)$*:*

$$e^{\frac{\lambda}{N}t_{01}}\,\Psi\left(t_{01},t_{12}^{\bar{0}},\cdots,t_{12}^{\overline{N-1}}\right)\,e^{\frac{\lambda}{2}t_{12}^{\bar{0}}}\,\Psi\left(t_{01},t_{12}^{\bar{0}},\cdots,t_{12}^{-\alpha},\cdots,t_{12}^{\overline{-N+1}}\right)^{-1}\,e^{\frac{\lambda}{N}t_{02}}$$

$$(\bar{0},\bar{1})*\left(\Psi\left(t_{01},t_{12}^{\bar{0}},\cdots,t_{12}^{-\alpha},\cdots,t_{12}^{\overline{-N+1}}\right)\,e^{\frac{\lambda}{2}t_{12}^{\bar{0}}}\,\Psi\left(t_{01},t_{12}^{\bar{0}},\cdots,t_{12}^{\overline{N-1}}\right)^{-1}\right)=1\,.$$

Sketch of Proof The data of an isomorphism $F:\widehat{\mathcal{P}a\mathcal{B}}(\Bbbk)\xrightarrow{\sim}\mathrm{Grp}(\widehat{\mathcal{P}a\mathcal{CD}})$ which is the identity on objects is equivalent to the data of a pair (λ_1,Φ) which is a Drinfeld associator.

Let $G:\widehat{\mathcal{P}a\mathcal{B}}^{\Gamma}(\Bbbk)\xrightarrow{\sim}\mathrm{Grp}(\widehat{\mathcal{P}a\mathcal{CD}}^{\Gamma})$ be an Γ-equivariant isomorphism of moperads above an associator F, which is also the identity on objects. The morphism G is completely determined by the image of E and Ψ.

The image of E by G has to be of the form

$$G(E)=e^{\lambda_2 t_{01}}L^{\bar{1}}\quad\text{in}\quad\mathrm{Hom}_{\mathcal{P}a\mathcal{CD}^{\Gamma}(1)}(01_{\bar{0}},01_{\bar{1}})\,,$$

where $e^{\lambda_2 t_{01}}$ is in $\exp(\widehat{\mathfrak{t}}_1^\Gamma)\cong\Bbbk$ and where is the $L^{\bar{1}}$ element described in Remark 3.26. One can show that $\lambda_2=\lambda_1/N$, see [CG20a, Lemma 5.7].

The image of Ψ by G has to be of the form

$$G(\Psi)=g(\Psi)\beta\,,\quad\text{in}\quad\mathrm{Hom}_{\mathcal{P}a\mathcal{CD}^{\Gamma}(2)}((01_{\bar{0}})2_{\bar{0}},0(1_{\bar{0}}2_{\bar{0}}))\,,$$

where $g(\Psi)$ is in $\exp(\widehat{\mathfrak{t}}_2^\Gamma)$ and where β is the element of $\widehat{\mathcal{P}a\mathcal{CD}}^{\Gamma}$ described in Example 3.28. The isomorphism

$$\mathfrak{t}_2^\Gamma\cong\Bbbk.c\oplus\mathfrak{f}_{N+1}\,,$$

allows us to canonically decompose

$$g(\Psi)=\left(\Xi(c),\Psi(t_{01},t_{12}^{\bar{0}},\cdots,t_{12}^{\overline{N-1}})\right)\quad\text{in}\quad\exp(\widehat{\mathfrak{t}}_2^\Gamma)\cong\Bbbk\times\widehat{\mathbb{F}_{N+1}}(\Bbbk)\,,$$

where $c=t_{01}+t_{02}+\sum_{\alpha\in\Gamma}t_{12}^\alpha$. The morphism

$$-\circ_1\mathrm{id}_{\{*\}}:\mathrm{Hom}_{\mathcal{P}a\mathcal{CD}^{\Gamma}(2)}((01_{\bar{0}})2_{\bar{0}},0(1_{\bar{0}}2_{\bar{0}}))\longrightarrow\mathrm{Hom}_{\mathcal{P}a\mathcal{CD}^{\Gamma}(1)}(01_{\bar{0}},01_{\bar{0}})$$

is the one determined by the morphism $\mathfrak{t}_2^\Gamma\twoheadrightarrow\mathfrak{t}_1^\Gamma$ which sends t_{01},t_{12}^α to zero and t_{02} to t_{01}. Its pre-composition with the isomorphism $\Bbbk.c\oplus\mathfrak{f}_{N+1}\cong\mathfrak{t}_2^\Gamma$ is given by sending t_{01},t_{12}^α to zero and c to t_{01}. The unital relation

$$\Psi\circ_1\mathrm{id}_{\{*\}}=\mathrm{id}_{01_{\bar{0}}}$$

therefore gives that $\Xi(c) = 1$, and thus that the element $g(\Phi)$ is simply given by the formal power series $\Psi(t_{01}, t_{12}^{\bar{0}}, \cdots, t_{12}^{\overline{N-1}})$ in $\exp(\widehat{\mathfrak{f}}_{N+1})$.

Finally, one shows by direct computations, starting from the above presentation of $\mathcal{P}a\mathcal{B}^{\Gamma}$, that the relations in this presentation impose the relations in the theorem. □

Theorem 3.39 ([Enr07, Section 2.2]) *Let $\mathbb{k}$ be a field of characteristic zero and let $\Gamma = \mathbb{Z}/N\mathbb{Z}$. Then the set of cyclotomic associators* $\mathrm{Assoc}^{\Gamma}(\mathbb{k})$ *is non empty.*

There is an explicit construction of a cyclotomic associator over $\mathbb{k} = \mathbb{C}$ given as the renormalized holonomy from 0 to 1 of a cyclotomic version of the Knizhnik–Zamolodchikov differential equation. One can then use again descent methods to prove existence over $\mathbb{k} = \mathbb{Q}$.

Theorem 3.40 ([CG20a, Section 4.5]) *Elements of the cyclotomic Grothendieck–Teichmüller group* $\mathrm{GT}^{\Gamma}(\mathbb{k})$ *are in bijection with triples*

$$(\mu, f, g) \quad in \quad \mathbb{k}^{\times} \times \widehat{\mathbb{F}}_2(\mathbb{k}) \times \widehat{\mathbb{F}}_{N+1}(\mathbb{k})$$

where (μ, f) is an element of the Grothendieck–Teichmüller group $\mathrm{GT}(\mathbb{k})$ *and where g is seen as a word $g(x^N, y, xyx^{-1}, \ldots, x^{N-1}yx^{1-N})$ which satisfies the following relations:*

(1) *The element g can be seen inside of $\widehat{\mathrm{PB}}_3^{\Gamma}(\mathbb{k})$, the Malcev completion of the cyclotomic pure braid group.*[5] *Let x_{ij} denote the standard generators of this group, then g satisfies*

$$g(x_{01}, x_{12})g(x_{01}x_{02}, x_{13}x_{23})f(x_{12}, x_{23}) = g(x_{02}x_{12}, x_{23})g(x_{01}, x_{12}x_{13})$$

in $\widehat{\mathrm{PB}}_3^{\Gamma}(\mathbb{k})$.

(2) *Let α be equal to $\bar{1}$ in Γ. Then we have*

$$x^{\frac{\mu-1}{N}} g(x, y) y^{\frac{\mu+1}{N}} g(z, y)^{-1} z^{\frac{\mu-1}{N}} \alpha * \left(g(z, y) y^{\frac{\mu-1}{N}} g(x, y)^{-1}\right) = 1 \,,$$

if $xyz = 1$ in $\widehat{\mathbb{F}}_2(\mathbb{k}, p_N)$, where the action of α on g is given by Lemma 3.37.

---

[5] There are four different morphisms of short exact sequences from the short exact sequence where $\widehat{\mathbb{F}}_{N+1}(\mathbb{k})$ appears as the kernel to the short exact sequence where $\widehat{\mathrm{PB}}_3^{\Gamma}(\mathbb{k})$ appears as the kernel. Before applying the relative Malcev completion, these maps are given as follows.

(a) The middle map $\mathbb{F}_2 \longrightarrow \mathrm{PB}_3^1$ is given by $x \mapsto x_{01}$ and $y \mapsto x_{12}$, which lifts the map $\Gamma \longrightarrow \Gamma^3$ given by $n \mapsto (n, 0, 0)$.
(b) The middle map $\mathbb{F}_2 \longrightarrow \mathrm{PB}_3^1$ is given by $x \mapsto x_{01}x_{02}$ and $y \mapsto x_{13}x_{23}$, which lifts the map $\Gamma \longrightarrow \Gamma^3$ given by $n \mapsto (n, n, 0)$.
(c) The middle map $\mathbb{F}_2 \longrightarrow \mathrm{PB}_3^1$ is given by $x \mapsto x_{02}x_{12}$ and $y \mapsto x_{23}$, which lifts the map $\Gamma \longrightarrow \Gamma^3$ given by $n \mapsto (0, n, 0)$.
(d) The middle map $\mathbb{F}_2 \longrightarrow \mathrm{PB}_3^1$ is given by $x \mapsto x_{01}$ and $y \mapsto x_{12}x_{13}$, which lifts the map $\Gamma \longrightarrow \Gamma^3$ given by $n \mapsto (n, 0, 0)$.

In each case, the map $\mathbb{F}_{N+1} \longrightarrow \mathrm{PB}_3^{\Gamma}$ is fully determined by the other two, and provides the corresponding way to see an element $g \in \mathbb{F}_{N+1}$ inside PB_3^{Γ}.

Under this bijection, the group structure of the Grothendieck–Teichmüller group is given by:

$$(\mu_1, f_1, g_1) \star (\mu_2, f_2, g_2) = (\mu, f, g) ,$$

where

$$\mu := \mu_1\mu_2 \quad \text{and} \quad f := f_1\left(x^{\mu_2}, f_2(x, y)y^{\mu_2} f_2(y, x)\right) f_2(x, y) ,$$

and

$$g := g_1\left(x^{\frac{\mu_2-1}{N}}, g_2(x, y)y^{\mu_2} g_2(x, y)^{-1}\right) g_2(x, y).$$

And, under this bijection, its action on the set of cyclotomic associators is given by:

$$(\mu, f, g) \bullet (\lambda, \Phi, \Psi) = \left(\mu\lambda, f\left(e^{\lambda t_{12}}, \Phi(t_{12}, t_{23})e^{\lambda t_{23}}\Phi(t_{23}, t_{12})\right)\Phi(t_{12}, t_{23}), \Upsilon\right) ,$$

where

$$\begin{aligned}\Upsilon := \Psi(t_{12}, t_{23}^0, \cdots, t_{23}^{N-1}) \\ g\Bigg(e^{\lambda t_{12}}, Ad_{\Psi(t_{12}, t_{23}^0, \cdots, t_{23}^{N-1})} e^{\lambda t_{23}^0}, \\ Ad_{e^{(\lambda/N)t_{12}}\,\Psi(t_{12}, t_{23}^1, \cdots, t_{23}^{N})}\, e^{\lambda t_{23}^1}, \cdots, \\ Ad_{e^{(\lambda(N-1)/N)t_{12}}\,\Psi(t_{12}, t_{23}^{N-1}, \cdots, t_{23}^{2N-2})} e^{\lambda t_{23}^{N-1}}\Bigg).\end{aligned}$$

Here by convention $Ad_U V := UVU^{-1}$.

Remark 3.41 There is also an explicit description of the cyclotomic graded Grothendieck–Teichmüller group which can be obtained by computing the relations that elements of Remark 3.29 satisfy.

3.7 Topological Description of $\mathcal{P}a\mathcal{B}^\Gamma$

Analogue statements to those mentioned in Sect. 2.7 hold in the cyclotomic case. First, we introduce the configuration space

$$\mathrm{C}(\mathbb{C}^\times, n) := \left\{(x_1, \cdots, x_n) \in \left(\mathbb{C}^\times\right)^n \mid x_i \neq x_j \text{ if } i \neq j\right\} / \mathbb{R}_{>0} .$$

This space is isomorphic to the space of configurations of $n+1$ points. Here the missing origin in $\mathbb{C}^\times$ is seen as a *fixed* point in the configurations.

We consider their Fulton–MacPherson compactifications $\overline{\mathrm{C}}(\mathbb{C}^\times, n)$, which this time have a *topological moperad structure* over the topological operad $\overline{\mathrm{C}}(\mathbb{C})$. Indeed, one can compute that

$$\partial\, \overline{\mathrm{C}}(\mathbb{C}^\times, n) \cong \bigcup_{k\geq 1}\ \bigcup_{\mathrm{J}_1\sqcup\cdots\sqcup \mathrm{J}_k=[n]} \overline{\mathrm{C}}(\mathbb{C}^\times, [k]) \times \overline{\mathrm{C}}(\mathbb{C}^\times, J_m) \times \left(\prod_{j=1, j\neq m}^{k} \overline{\mathrm{C}}(\mathbb{C}, \mathrm{J}_j) \right) ,$$

where $[n] = \{0, \cdots , n\}$ and where J_m for $1 \leq m \leq k$ is the subset of $[n]$ which contains the element 0. Using this decomposition of the boundary, one can write the moperad structure maps as obvious inclusions. We denote this topological moperad over $\overline{\mathrm{C}}(\mathbb{C})$ by $\overline{\mathrm{C}}(\mathbb{C}^\times)$.

Remark 3.42 The moperad $\overline{\mathrm{C}}(\mathbb{C}^\times)$ is also given by applying Proposition 3.5 to $\overline{\mathrm{C}}(\mathbb{C})$.

Recall $\mathcal{P}a_0$, the moperad over $\mathcal{P}a$ in the category of sets defined in Example 3.7. Now lets view this moperad as a topological moperad over $\mathcal{P}a$ where we impose the discrete topology both on $\mathcal{P}a_0$ and on $\mathcal{P}a$.

Lemma 3.43 *There is an inclusion of topological moperads*

$$\epsilon : \mathcal{P}a_0 \hookrightarrow \overline{C}(\mathbb{C}^\times)$$

over the topological operad $\mathcal{P}a$.

Proof Follows from Lemma 2.39 and Proposition 3.5. □

Proposition 3.44 *There is an isomorphism of moperads in the category of groupoids*

$$\mathcal{P}a\mathcal{B}^{(1)} \cong \Pi_1(\overline{C}(\mathbb{C}^\times), \mathcal{P}a_0) ,$$

over the operad $\mathcal{P}a\mathcal{B}$.

Proof Follows from Theorem 2.40 and Proposition 3.5. □

3.7.1 Twisted Configuration Spaces

Let again $\Gamma = \mathbb{Z}/N\mathbb{Z}$. We consider the following *twisted configuration spaces* given by

$$\mathrm{C}(\mathbb{C}^\times, n, \Gamma) := \left\{(x_1, \cdots , x_n) \in \left(\mathbb{C}^\times\right)^n \mid x_i \neq \zeta.x_j \text{ if } i \neq j \mid \zeta \in \mu_N\right\} /\mathbb{R}_{>0} ,$$

where μ_N denotes the group given by the complex N-roots of the unit.

Remark 3.45 One easily sees that the fundamental group of $\mathrm{C}(\mathbb{C}^\times, n, \Gamma)$ is the cyclotomic pure braid group PB_n^Γ. Actually, the hyperplanes defined by $x_i = \zeta.x_j$ and $x_i = 0$ are the reflection hyperplanes associated with the complex reflection group $\Gamma^n \rtimes \mathbb{S}_n$. Therefore PB_n^Γ is the pure braid group associated with $\Gamma^n \rtimes \mathbb{S}_n$, in the sense of [BMR95].

These spaces admits Fulton–MacPherson compactifications $\overline{\mathrm{C}}(\mathbb{C}^\times, n, \Gamma)$. This family also has a natural moperad structure over the operad $\overline{\mathrm{C}}(\mathbb{C})$, which is again given by a similar decomposition of its boundary. We denote by $\overline{\mathrm{C}}(\mathbb{C}^\times, \Gamma)$ this moperad, which is naturally endowed, as a moperad, with a Γ-action.

The moperad $\mathcal{P}a^\Gamma$ is the set theoretical moperad over $\mathcal{P}a$, where $\mathcal{P}a^\Gamma(n)$ is given by maximal parenthesizations on words $0(\sigma_1)_{\alpha_1}\cdots(\sigma_n)_{\alpha_n}$, where σ is a permutation in $\mathfrak{S}_n$ and $(\alpha_1, \cdots, \alpha_n)$ is an element of Γ^n. It is naturally endowed, as a moperad, with a Γ-action. We view this construction in the category of topological spaces by endowing it with the discrete topology.

Lemma 3.46 *There is a Γ-equivariant inclusion of topological moperads*

$$\epsilon : \mathcal{P}a^\Gamma \hookrightarrow \overline{\mathcal{C}}(\mathbb{C}^\times, \Gamma)$$

over the topological operad $\mathcal{P}a$.

Proposition 3.47 *There is a Γ-equivariant isomorphism of moperads in the category of groupoids*

$$\mathcal{P}a\mathcal{B}^{(N)} \cong \Pi_1(\overline{\mathcal{C}}(\mathbb{C}^\times, \Gamma), \mathcal{P}a^\Gamma) ,$$

over the operad $\mathcal{P}a\mathcal{B}$.

Remark 3.48 Proofs of the above results can be found in [CG20a], in particular inside the proof of Theorem 3.4 and inside the proof of Theorem 4.6.

4 Elliptic and Ellipsitomic Associators

4.1 Motivations and General Context

One can phrase a deformation problem extending Problem 1, making use of the notion of an *elliptic structure* over a braided monoidal category [CEE09, Enr14a]. According to [Enr14a] an elliptic structure over a braided monoidal category $\mathcal{C}$ is a functor $F : \mathcal{C} \to \mathcal{C}_1$ together with two natural automorphisms of $F(- \otimes -)$ satisfying certain relations that we won't specify here. These relations naturally live in $\mathrm{End}\big(F(\otimes^n)\big)$, that is an operadic module over $\mathrm{End}(\otimes^n)$.

For a finite dimensional Lie algebra $\mathfrak{g}$, following [CEE09] we consider the algebra morphism from $\mathfrak{U}(\mathfrak{g})$ to the algebra of $\mathsf{D}(\mathfrak{g})$ of differential operators on $\mathfrak{g}$ sending a generator $x \in \mathfrak{g}$ to the linear vector field on $\mathfrak{g}$ given by $\mathrm{ad}_x = [x, -]$. The induction functor $F := \mathsf{D}(\mathfrak{g}) \otimes_{\mathfrak{U}(\mathfrak{g})} -$ then goes from $\mathcal{C} := \mathsf{Rep}(\mathfrak{g})$ to $\mathcal{C}_1 := \mathsf{Rep}\big(\mathsf{D}(\mathfrak{g})\big)$. One then gets that

$$\mathrm{End}\big(F(\otimes^n)\big) \cong \mathrm{Hom}\big(\otimes^n, \mathrm{Res} \circ F(\otimes^n)\big) \cong \Big(\mathsf{D}(\mathfrak{g}) \otimes_{\mathfrak{U}(\mathfrak{g})} \big(\mathfrak{U}(\mathfrak{g})^{\otimes n}\big)\Big)^{\mathfrak{g}},$$

where Res : $C_1 \to C$ is the restriction functor, that is right adjoint to F, and the algebra map $\mathfrak{U}(\mathfrak{g}) \to \mathfrak{U}(\mathfrak{g})^{\otimes n}$ is the iterated coproduct.[6] In this context we have a (quite trivial) elliptic structure, and any $t \in S^2(\mathfrak{g})^{\mathfrak{g}}$ that is non-degenerate (meaning that the induced map $\mathfrak{g}^* \to \mathfrak{g}$ is an isomorphism) provides a first order deformation of it. A natural deformation problem is the following: can one extend this first order deformation to a formal one?

Just as in Sect. 1.2 and 3.1, there is a collection of graded Lie algebras $(\bar{\mathfrak{t}}_n^{\mathrm{ell}})_{n\geq 0}$, see Sect. 4.3, together with insertion-coproduct type morphisms that makes an operadic module over the Drinfeld–Kohno operad $(\mathfrak{t}_n)_{n\geq 0}$. It comes with an operadic module morphism $\bar{\mathfrak{t}}_\bullet^{\mathrm{ell}} \longrightarrow \mathrm{End}\big(F(\otimes^\bullet)\big)$, making it a perfect location for finding "universal" solutions to our deformation problem, that we call *elliptic associators*.

One can construct such universal solutions as holonomies of a genus one version of the KZ connection (known as the Knizhnik–Zamolodchikov–Bernard connection, see [CEE09]). Such a connection depends on the choice of a complex structure on the torus, and actually extends to the whole moduli space of marked elliptic curves. As a result, we get that these holonomies have many interesting features:

- They satisfy modular invariance properties (see [Enr14a]).
- Their coefficients are interesting numbers, involving iterated integrals of Eiseinstein series, that one can view as elliptic analogs of MZV [Enr16]. As a consequence (of the defining relations of an elliptic associator), one gets relations between iterated integrals of Eiseinstein series and ordinary MZV.
- All of the above shall lift to some appropriate motivic context (presumbably the one of [HM20]).

Note that the notion of an elliptic structure has been generalized to higher genus in [Hum12], under the name *genus g structure* ($g = 1$ corresponding to the elliptic case), in relation with the study of invariants for tangles in a thickened surface.

4.2 The Module of Parenthesized Elliptic Braids

Our goal is to build an elliptic version of $\mathcal{P}a\mathcal{B}$, much in the same way as in Sect. 2.3. The elliptic version $\mathcal{P}a\mathcal{B}_{\mathrm{ell}}$ is going to be a right module over the operad $\mathcal{P}a\mathcal{B}$. In order to construct it, we will need to consider the *pure braid group of the torus*.

Let $\mathbb{T}$ be the topological 2-torus. The configuration space of n points in the torus $\mathbb{T}$ is defined as follows:

$$\mathrm{Conf}_n(\mathbb{T}) := \left\{(x_1, \cdots, x_n) \in \mathbb{T}^n \mid x_i \neq x_j \text{ if } i \neq j\right\} .$$

[6] Note that, according to [CEE09, Remark 5.7], there is an isomorphism

$$\Big(\mathsf{D}(\mathfrak{g}) \otimes_{\mathfrak{U}(\mathfrak{g})} \big(\mathfrak{U}(\mathfrak{g})^{\otimes n}\big)\Big)^{\mathfrak{g}} \cong \Big(\mathsf{D}(\mathfrak{g}) \otimes \big(\mathfrak{U}(\mathfrak{g})^{\otimes (n-1)}\big)\Big)^{\mathfrak{g}}.$$

Its *reduced version*

$$\mathrm{C}(\mathbb{T}, n) := \mathrm{Conf}_n(\mathbb{T})/\mathbb{T} ,$$

is obtained by considering configurations of n points modulo the action of the torus on itself by multiplication, using its abelian group structure. It is a path-connected topological space for $n \geq 0$. It comes equipped with an action of the symmetric group $\mathfrak{S}_n$, given by the permutation of the indices of the points.

Definition 4.1 (Reduced Braid Group of the Torus) The *reduced pure braid group of the torus group* $\overline{\mathrm{PB}}_n(\mathbb{T})$ on n strands is given by the fundamental group

$$\overline{\mathrm{PB}}_n(\mathbb{T}) := \pi_1(\mathrm{C}(\mathbb{T}, n)) .$$

Let us describe this group. It is generated by elements $\{a_i\}$ and $\{b_i\}$ for $1 \leq i \leq n$, which correspond to i-th point following the trajectory given by each of the generators of the fundamental group of the torus $\pi_1(\mathbb{T})$. They can be depicted as follows:

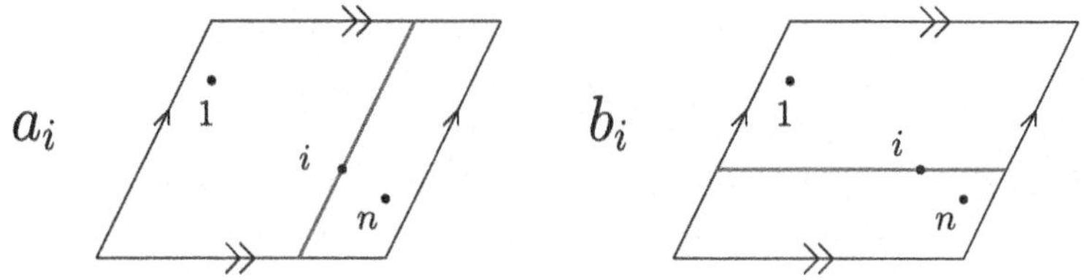

For all n, there is a canonical morphism of groups

$$\begin{aligned} \mathrm{PB}_n &\longrightarrow \overline{\mathrm{PB}}_n(\mathbb{T}) \\ x_{ij} &\longmapsto \left[a_j^{-1}, b_i^{-1}\right] , \end{aligned}$$

induced by the inclusion of the plane into the fundamental domain of the torus. Thus any generator x_{ij} of the pure braid group can be seen as an element in the reduced pure braid group of the torus, and it still corresponds to the i-th point moving around the j-th point.

Without giving a full presentation of $\overline{\mathrm{PB}}_n(\mathbb{T})$, let us mention some of the important relations that the generators $\{a_i\}$ and $\{b_i\}$ satisfy. First notice that these generators commute, that is

$$[a_i, a_j] = [b_i, b_j] = 1$$

for $i < j$, and we have that $a_1 \cdots a_n = b_1 \cdots b_n = 1$. Finally, for all $1 \leq i \leq n$, we have the relations

$$a_{i+1} = \sigma_i^{-1} a_i \sigma_i^{-1} \quad \text{and} \quad b_{i+1} = \sigma_i^{-1} b_i \sigma_i^{-1} ,$$

which are satisfied in the *reduced braid group of the torus* $\overline{\mathrm{B}}_n(\mathbb{T})$, defined as

$$\overline{\mathrm{B}}_n(\mathbb{T}) := \pi_1(\mathrm{C}(\mathbb{T}, n)/\mathfrak{S}_n) ,$$

and where σ_i is the following generator:

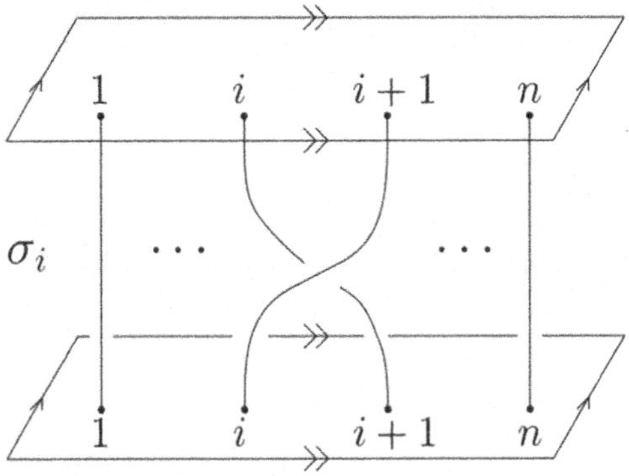

There is an elliptic analogue of the operad in the category of groupoids $\mathcal{C}o\mathcal{B}$, defined in Example 2.7. The symmetric sequence in the category of groupoids $(\mathcal{C}o\mathcal{B}_{\text{ell}}(n))_{n\geq 0}$ is given as follows.

(1) The objects of $\mathcal{C}o\mathcal{B}_{\text{ell}}(n)$ are elements of the symmetric group $\mathfrak{S}_n$.
(2) Morphisms between two permutations σ and τ in $\mathfrak{S}_n$ are given by reduced braids on the torus going from σ to τ.

The $\mathfrak{S}_n$-module $\mathcal{C}o\mathcal{B}_{\text{ell}}(n)$ has a canonical right module structure over the operad $\mathcal{C}o\mathcal{B}$, that is, there are maps

$$\{\circ_i : \mathcal{C}o\mathcal{B}_{\text{ell}}(n) \times \mathcal{C}o\mathcal{B}(k) \longrightarrow \mathcal{C}o\mathcal{B}_{\text{ell}}(n+k-1)\}$$

for $1 \leq i \leq n$ which satisfy the obvious axioms. These functors are given on objects by the insertion of permutations and on morphisms by the inserting of a pure braid going from σ' to τ' into the i-th spot of a reduced pure braid on the torus going from σ to τ.

There is an obvious morphism $\varphi : \mathrm{Ob}(\mathcal{P}a) \longrightarrow \mathrm{Ob}(\mathcal{C}o\mathcal{B}_{\text{ell}})$ of symmetric sequences in sets which simply forgets the parenthesis on permutations.

Definition 4.2 (The Right Module $\mathcal{P}a\mathcal{B}_{\text{ell}}$) The symmetric sequence in the category of groupoids $\mathcal{P}a\mathcal{B}_{\text{ell}}$ is defined to be the fake pull-back of $\mathcal{C}o\mathcal{B}_{\text{ell}}$ along the morphism φ. It inherits a right module structure over the operad $\mathcal{P}a\mathcal{B}$ coming from the right module structure of $\mathcal{C}o\mathcal{B}_{\text{ell}}$.

Remark 4.3 In addition to the elements R and Φ described in Theorem 2.30 which are also present in $\mathcal{P}a\mathcal{B}_{\mathrm{ell}}$, there are two other distinguished elements. The object (12) in $\mathcal{P}a\mathcal{B}_{\mathrm{ell}}(2)$ has this time two non-trivial automorphisms, denoted by A and B: A, resp. B, is induced by the element $a_1 = a_2^{-1}$, resp. $b_1 = b_2^{-1}$, in $\overline{\mathrm{PB}}_2(\mathbb{T})$.

The Malcev completion $\widehat{\mathcal{P}a\mathcal{B}}_{\mathrm{ell}}(\Bbbk)$ is a right module over $\widehat{\mathcal{P}a\mathcal{B}}(\Bbbk)$ in the underlying category of pro-unipotent $\Bbbk$-groupoids, since the Malcev completion functor is strong monoidal. This gives our *first elliptic player*.

4.3 The Module of Elliptic Chord Diagrams

Here we define the "infinitesimal version" of the right module $\mathcal{P}a\mathcal{B}_{\mathrm{ell}}$, given in terms of elliptic chord diagrams.

Definition 4.4 (Elliptic Drinfeld–Kohno Algebras) The n-th *elliptic Drinfeld–Kohno algebra* $\mathfrak{t}_n^{\mathrm{ell}}$ is the bigraded Lie algebra given by the following presentation:

(1) It is generated by elements $\{t_{ij}\}$ for $1 \leq i \neq j \leq n$ in bidegree $(1, 1)$, by elements α_i for $1 \leq i \leq n$ in bidegree $(1, 0)$ and by elements β_i for $1 \leq i \leq n$ in bidegree $(0, 1)$.
(2) The generators are subject to the following relations:

   (a) $t_{ij} = t_{ji}$.
   (b) $[t_{ij}, t_{kl}] = 0$.
   (c) $[t_{ij}, t_{ik} + t_{jk}] = 0$.
   (d) $[\alpha_i, \beta_j] = t_{ij}$.
   (e) $[\alpha_i, \alpha_j] = [\beta_i, \beta_j] = 0$.
   (f) $[\alpha_i, \beta_i] = -\sum_{j \mid j \neq i} t_{ij}$.
   (g) $[\alpha_i, t_{jk}] = [\beta_i, t_{jk}] = 0$.
   (h) $[\alpha_i + \alpha_j, t_{ij}] = [\beta_i + \beta_j, t_{ij}] = 0$.

Remark 4.5 The elements $\sum_i \alpha_i$ and $\sum_i \beta_i$ are central elements in $\mathfrak{t}_n^{\mathrm{ell}}$.

Proposition 4.6 *The family $(\mathfrak{t}_n^{\mathrm{ell}})_{n \geq 0}$ can be endowed with a right module structure over the operad $(\mathfrak{t}_n)_{n \geq 0}$ in the category of graded Lie algebras.*

(1) *The action of the symmetric group $\mathfrak{S}_n$ on $\mathfrak{t}_n^{\mathrm{ell}}$ is given by :*

$$\sigma \bullet t_{ij} := t_{\sigma(i)\sigma(j)} \ , \quad \sigma \bullet \alpha_i := \alpha_{\sigma(i)} \ \text{and} \ \sigma \bullet \beta_i := \beta_{\sigma(i)} \ .$$

(2) *The partial composition maps*

$$\{\circ_p : \mathfrak{t}_n^{\mathrm{ell}} \oplus \mathfrak{t}_m \longrightarrow \mathfrak{t}_{n+m-1}^{\mathrm{ell}}\}$$

of the right module structure are given by the following components (where we assume $i < j$):

$$\mathfrak{t}_m \ni t_{ij} \longmapsto t_{i+p-1\, j+p-1}\,.$$

$$\mathfrak{t}_n^{\mathrm{ell}} \ni t_{ij} \longmapsto \begin{cases} t_{i+m-1\, j+m-1} & \text{if } p < i\,. \\ \sum_{k=i}^{i+m-1} t_{k\, j+m-1} & \text{if } p = i\,. \\ t_{i\, j+m-1} & \text{if } i < p < j\,. \\ \sum_{k=j}^{j+m-1} t_{i\,k} & \text{if } p = j\,. \\ t_{i\,j} & \text{if } j < p\,. \end{cases}$$

$$\mathfrak{t}_n^{\mathrm{ell}} \ni \alpha_i \longmapsto \begin{cases} \alpha_{i+m-1} & \text{if } p < i\,. \\ \sum_{k=i}^{i+m-1} \alpha_k & \text{if } p = i\,. \\ \alpha_i & \text{if } i < p\,. \end{cases}$$

$$\mathfrak{t}_n^{\mathrm{ell}} \ni \beta_i \longmapsto \begin{cases} \beta_{i+m-1} & \text{if } p < i\,. \\ \sum_{k=i}^{i+m-1} \beta_k & \text{if } p = i\,. \\ \beta_i & \text{if } i < p\,. \end{cases}$$

Proof Using the formulas above, one can define a presheaf on the category $\mathrm{Fin}_{*,\not *}$. As explained in the Appendix B.2, this induces a right module structure for the cocartesian monoidal structure, which descends to the cartesian monoidal structure on Lie algebras since the images of the insertion morphisms commute with the images of the coproduct morphisms. See proofs of Propositions 2.14 and 3.23 for analogue arguments. □

Definition 4.7 (Reduced Elliptic Drinfeld–Kohno Algebras) The n-th *reduced elliptic Drinfeld–Kohno algebra* $\bar{\mathfrak{t}}_n^{\mathrm{ell}}$ is given by

$$\bar{\mathfrak{t}}_n^{\mathrm{ell}} := \frac{\mathfrak{t}_n^{\mathrm{ell}}}{\left(\sum_i \alpha_i\,,\ \sum_i \beta_i\right)}\,.$$

The right module structure defined in Proposition 4.6 descends to a right module structure on the symmetric sequence $(\bar{\mathfrak{t}}_n^{\mathrm{ell}})_{n\geq 0}$ in the category of graded Lie algebras given by the *reduced* elliptic Drinfeld–Kohno algebras.

Definition 4.8 (The Right Module of Reduced Elliptic Chord Diagrams) The *right module of reduced elliptic chord diagrams* $\mathcal{CD}_{\mathrm{ell}}$ is the right module over the operad of chord diagrams $\mathcal{CD}$ obtained by applying the universal enveloping algebra functor to the right module of *reduced* elliptic Drinfeld–Kohno algebras.

Morphisms in $\mathcal{CD}_{\mathrm{ell}}(n)$ can be represented as chord diagrams on n strands, with two extra types of chords which correspond to the added generators α_i and β_i. These extra chords can be depicted as follows:

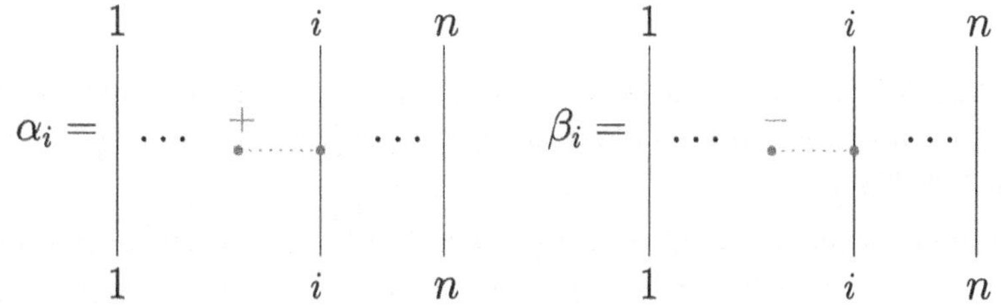

As an example, the relation (f) in the Definition 4.4 can be represented in terms of chord diagrams as follows:

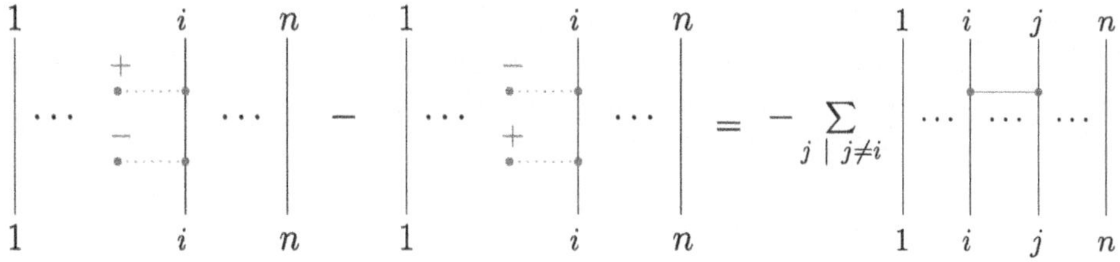

Recall the set-theoretical operad $\mathcal{P}a$ of parenthesized permutations. We consider its $\mathbb{k}$-linear extension $(\mathcal{P}a)_{\mathbb{k}}$ to the underlying category of small categories enriched in cocommutative $\mathbb{k}$-coalgebras, constructed in Definition 2.21. It is canonically a right module over itself.

Definition 4.9 (Right Module of Parenthesized Reduced Elliptic Chord Diagrams) The *right module of parenthesized elliptic chord diagrams* $\mathcal{P}a\mathcal{CD}_{\mathrm{ell}}$ is defined as the Hadamard tensor product

$$\mathcal{P}a\mathcal{CD}_{\mathrm{ell}} := (\mathcal{P}a)_{\mathbb{k}} \otimes \mathcal{CD}_{\mathrm{ell}} ,$$

in the category of symmetric sequences in the category $\mathsf{Cat}(\mathsf{Coalg}_{\mathbb{k}})$. It comes equipped with a right module structure over the operad $\mathcal{P}a\mathcal{CD}$ since this operad is given by the Hadamard product of $(\mathcal{P}a)_{\mathbb{k}}$ and $\mathcal{CD}$.

Example 4.10 There are two distinguished endomorphisms of the object (12) in $\mathcal{P}a\mathcal{CD}_{\mathrm{ell}}(2)$, called α and β, which are given by $1 \otimes \alpha_1$ and $1 \otimes \beta_1$, and which can be represented as follows:

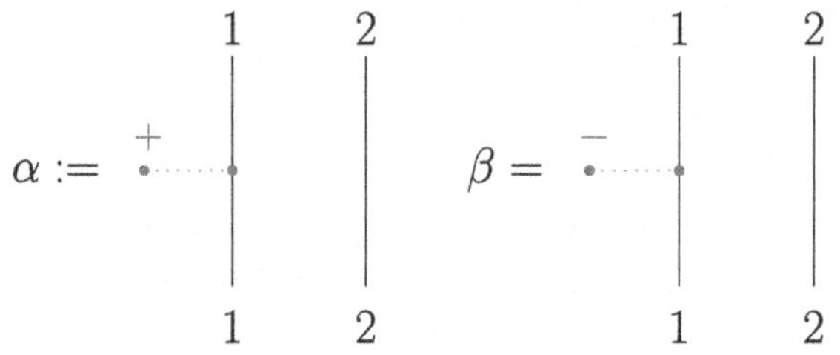

△

Remark 4.11 The morphisms α and β generate $\mathcal{P}a\mathcal{CD}_{\mathrm{ell}}$ as a right module over $\mathcal{P}a\mathcal{CD}$ the underlying category of $\mathsf{Cat}(\mathsf{Coalg}_{\Bbbk})$, and in fact one can give an explicit presentation of this right module.

The relations that define the elliptic Drinfeld–Kohno Lie algebras $\mathfrak{t}_n^{\mathrm{ell}}$ are homogeneous with respect to the bigrading imposed on generators. Hence it has a canonical filtration given by the total degree of the bifiltration.

We define $\widehat{\mathcal{P}a\mathcal{CD}}_{\mathrm{ell}}$ as the right module over $\widehat{\mathcal{P}a\mathcal{CD}}$ in the category of categories enriched over complete cocommutative coalgebras obtained by completing each hom-set of $\mathcal{P}a\mathcal{CD}_{\mathrm{ell}}(n)$ with respect to the aforementioned filtration. It defines a right module in the underlying category of complete Hopf groupoids.

We get a right module $\mathrm{Grp}(\widehat{\mathcal{P}a\mathcal{CD}}_{\mathrm{ell}})$ over $\mathrm{Grp}(\widehat{\mathcal{P}a\mathcal{CD}})$ in the underlying category of pro-unipotent $\Bbbk$-groupoids by applying the group-like element functor on both sides. This will be our *second elliptic player*.

4.4 Elliptic Associators

These two elliptic players enable us to define elliptic associators purely in operadic terms.

Definition 4.12 (Elliptic Associators) The set of *elliptic associators* is given by

$$\mathrm{Assoc}_{\mathrm{ell}}(\Bbbk) := \mathrm{Iso}^{+}_{\mathsf{RMod(p.u\text{-}Grpd_{\Bbbk})}}\left(\left(\widehat{\mathcal{P}a\mathcal{B}}_{\mathrm{ell}}(\Bbbk), \widehat{\mathcal{P}a\mathcal{B}}(\Bbbk)\right), \left(\mathrm{Grp}(\widehat{\mathcal{P}a\mathcal{CD}}_{\mathrm{ell}}), \mathrm{Grp}(\widehat{\mathcal{P}a\mathcal{CD}})\right)\right).$$

An elements in $\mathrm{Assoc}_{\mathrm{ell}}(\Bbbk)$ amounts to the data of:

(1) an isomorphism F between $\widehat{\mathcal{P}a\mathcal{B}}(\Bbbk)$ and $\mathrm{Grp}(\widehat{\mathcal{P}a\mathcal{CD}})$ which is the identity on objects (an associator) ,

(2) an isomorphism S between $\widehat{\mathcal{P}a\mathcal{B}}_{\mathrm{ell}}(\Bbbk)$ and $\mathrm{Grp}(\widehat{\mathcal{P}a\mathcal{CD}}_{\mathrm{ell}})$ of right modules which is the identity on objects, and which lies above the isomorphism F.

This definition entails that the set of elliptic associators is a bitorsor over the automorphism groups of each right module that are the identity morphism on objects. This allows us to define, as in the previous cases, the elliptic Grothendieck–Teichmüller group and the elliptic graded Grothendieck–Teichmüller group.

Definition 4.13 (Elliptic Grothendieck–Teichmüller Group) The *elliptic Grothendieck–Teichmüller group* over $\Bbbk$ is given by

$$\mathrm{GT}_{\mathrm{ell}}(\Bbbk) := \mathrm{Aut}^{+}_{\mathsf{RMod(p.u\text{-}Grpd_{\Bbbk})}}\left(\left(\widehat{\mathcal{P}a\mathcal{B}}_{\mathrm{ell}}(\Bbbk), \widehat{\mathcal{P}a\mathcal{B}}(\Bbbk)\right)\right),$$

that is, the group of automorphisms of right modules of $\widehat{\mathcal{P}a\mathcal{B}}_{\mathrm{ell}}(\Bbbk)$ which are the identity on objects.

Definition 4.14 (Elliptic Graded Grothendieck–Teichmüller Group) The *elliptic graded Grothendieck–Teichmüller group* over $\Bbbk$ is given by

$$\mathrm{GRT}_{\mathrm{ell}}(\Bbbk) := \mathrm{Aut}^{+}_{\mathsf{RModCat(Coalg_{\Bbbk})}}\left(\left(\widehat{\mathcal{P}a\mathcal{CD}}_{\mathrm{ell}}, \widehat{\mathcal{P}a\mathcal{CD}}\right)\right),$$

that is, the group of automorphisms of right modules of $\widehat{\mathcal{P}a\mathcal{CD}}_{\mathrm{ell}}$ which are the identity on objects.

4.5 More Concrete Descriptions

Theorem 4.15 ([CG23, Theorem 3.3]) *The right module $\mathcal{P}a\mathcal{B}_{\mathrm{ell}}$, as a right module over $\mathcal{P}a\mathcal{B}$ in groupoids having $\mathcal{P}a$ as a $\mathcal{P}a$-right module of objects, admits the following presentation. Thegenerating morphisms are the endomorphisms A and B of* (12) *in $\mathcal{P}a\mathcal{B}_{\mathrm{ell}}(2)$. They satisfy the following relations*

(1) *The nonagon relation for A amounts to the commutativity of the following diagram*

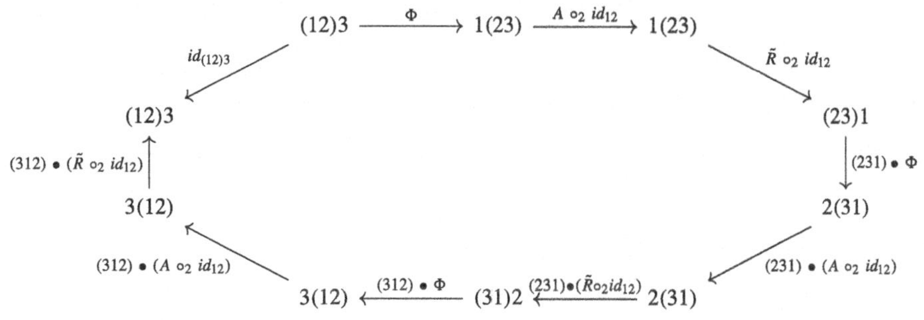

where $\tilde{R}$ is given by $((21) \bullet R)^{-1}$.

(2) *The nonagon relation for B amounts to the above diagram, where one replaces A with B.*
(3) *The mixed relation amounts to the commutativity of the following diagram*

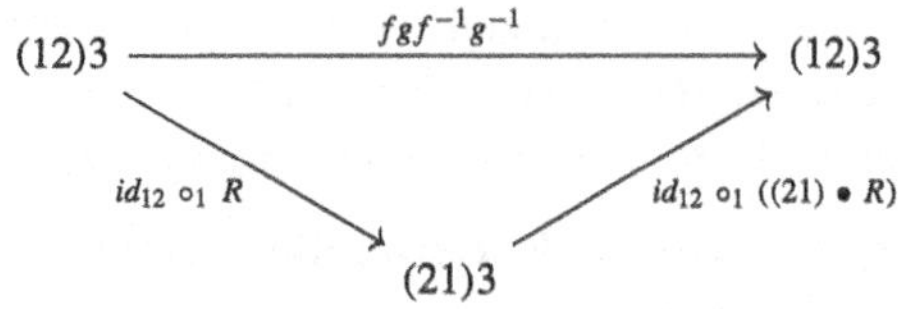

where the arrow f is given by the following composition

$$(12)3 \xrightarrow{\Phi} 1(23) \xrightarrow{A \circ_2 id_{12}} 1(23) \xrightarrow{\Phi^{-1}} (12)3\,,$$

and where the arrow g is given by the following composition

$$(12)3 \xrightarrow{id_{12} \circ_1 \tilde{R}} (21)3 \xrightarrow{(213) \bullet \Phi} 2(13) \xrightarrow{(213)\bullet(B \circ_2 id_{12})} 2(13) \xrightarrow{((213) \bullet \Phi)^{-1}} (21)3 \xrightarrow{id_{12} \circ_1 ((21) \bullet} (21)3.$$

We now deduce from this presentation a more explicit description of elliptic associators.

Observe that the reduced elliptic Drinfeld–Kohno algebra $\bar{t}_2^{\text{ell}}$ is a free Lie algebra on two generators, $\alpha := \alpha_1$ and $\beta := \beta_2$. Therefore an element in $\widehat{\mathfrak{U}}(\bar{t}_2^{\text{ell}})$ can be written a non-commutative formal power series in variables α_1 and β_2.

Theorem 4.16 ([CG23, Theorem 3.9]) *There is a bijection between the set of elliptic associators* $\mathrm{Assoc}_{\mathrm{ell}}(\Bbbk)$ *and the set of quadruples* $(\lambda, \Phi, A_+, A_-)$ *where*

(1) (λ, Φ) *is a Drinfeld associator;*
(2) A_+ *and* A_- *are group-like formal power series in two non-commutative variables, which satisfy the following equations in* $\widehat{\mathfrak{U}}(\bar{t}_3^{\text{ell}})$*:*

(a)
$$\begin{aligned} &\Phi(t_{12}, t_{23})\, A_\pm(\alpha_1, \beta_2 + \beta_3)\, e^{-\lambda(t_{12}+t_{13})/2} \\ &\Phi(t_{23}, t_{13})\, A_\pm(\alpha_2, \beta_3 + \beta_1)\, e^{-\lambda(t_{23}+t_{12})/2} \\ &\Phi(t_{13}, t_{12})\, A_\pm(\alpha_3, \beta_1 + \beta_2)\, e^{-\lambda(t_{13}+t_{23})/2} = 1 \end{aligned}$$

and

(b)
$$\begin{aligned} e^{\lambda . t_{12}} = \Big[&\Phi(t_{12}, t_{23})\, A_+(\alpha_1, \beta_2 + \beta_3)\, \Phi(t_{12}, t_{23})^{-1}\,, \\ &e^{-\lambda . t_{12}/2}\, \Phi(t_{12}, t_{13})\, A_-(\alpha_2, \beta_1 + \beta_3)\, \Phi(t_{12}, t_{13})^{-1}\, e^{-\lambda . t_{12}/2} \Big] \end{aligned}$$

where the bracket denotes the commutator.

Remark 4.17 The above description of an elliptic associator corresponds to the original definition given by Enriquez in [Enr14a], up to taking inverses (because of our unusual convention for products).

Sketch of Proof An isomorphism S between $\widehat{\mathcal{P}a\mathcal{B}}_{\mathrm{ell}}(\Bbbk)$ and $\mathrm{Grp}(\widehat{\mathcal{P}a\mathcal{C}\mathcal{D}}_{\mathrm{ell}})$ is completely determined by the images of the generators A and B. These images give two group-like elements in $\widehat{\mathfrak{U}}(\bar{t}_2^{\,\mathrm{ell}})$, that is to say two group-like non-commutative formal power series in two variables. The relations above correspond precisely to the relations induced by the presentation of Theorem 4.15 (more precisely, (a) corresponds to the two nonagon relations, and (b) corresponds to the mixed relation). □

Theorem 4.18 ([Enr14a]) *Let $\Bbbk$ be a field of characteristic zero. The set of elliptic associators is non empty.*

For every choice of a complex structure on the torus (i.e. for every elliptic curve), there is an explicit construction of an elliptic associator over $\Bbbk = \mathbb{C}$ given as renormalized holonomies (along A- and B-cycles) of an elliptic version of the Knizhnik–Zamolodchikov differential equation, studied in [CEE09]. One can then use descent methods, once again, to prove existence over $\Bbbk = \mathbb{Q}$.

Remark 4.19 The above still makes sense for the nodal curve, where one gets back the usual Knizhnik–Zamolodchikov differential equation (up to a change of coordinate, see [CEE09]). The holonomies can therefore be expressed using a Drinfeld associator. Explicit formulas are given in [CEE09, Enr14a] and provide an elliptic associator for *any* given Drinfeld associator. In other words, the forgetful map $\mathrm{Assoc}_{\mathrm{ell}}(\Bbbk) \to \mathrm{Assoc}(\Bbbk)$ has a section.

The presentation of $\mathcal{P}a\mathcal{B}_{\mathrm{ell}}$ as a right module also allows us to give an explicit description of the elliptic Grothendieck–Teichmüller group.

Theorem 4.20 ([CG23, Section 3.7]) *Elements of the elliptic Grothendieck–Teichmüller group $\mathrm{GT}_{\mathrm{ell}}(\Bbbk)$ are in bijection with quadruples*

$$(\mu, f, g_+, g_-) \quad in \quad \Bbbk^\times \times \widehat{\mathbb{F}}_2(\Bbbk) \times \widehat{\mathbb{F}}_2(\Bbbk) \times \widehat{\mathbb{F}}_2(\Bbbk)\ ,$$

where (μ, f) is an element of the Grothendieck–Teichmüller group $\mathrm{GT}(\Bbbk)$, and where $g_\pm$ are seen as words $g_\pm(A, B)$ which satisfy the following equations

$$(1)\ \left(f(\sigma_1^2, \sigma_2^2) g_\pm(a_1, b_1)(\sigma_1\sigma_2^2\sigma_1)^{\frac{-\mu-1}{2}} \sigma_1^{-1}\sigma_2^{-1}\right)^3 = 1\ ,$$

$$(2)\ u^2 = \left[g_+(a_1, b_1), u^{-1} g_-(a_1, b_1) u^{-1}\right]\ ,$$

in the reduced braid group of the torus $\overline{B}_3(\mathbb{T})$, where the bracket denotes the commutator and where

$$u = f(\sigma_1^2, \sigma_2^2)^{-1} \sigma_1^\mu f(\sigma_1^2, \sigma_2^2)\ .$$

Under this bijection, the group structure of the elliptic Grothendieck–Teichmüller group is given by:

$$(\mu_1, f_1, (g_1)_\pm) \star (\mu_2, f_2, (g_2)_\pm) = (\mu, f, g_\pm)\,,$$

where

$$\mu = \mu_1\mu_2\,, \quad \textit{where} \quad f(x, y) = f_1\left(x^{\mu_2}, f_2(x, y)y^{\mu_2} f_2(y, x)\right) f_2(x, y)\,,$$

and where

$$g_\pm(A, B) = (g_1)_\pm((g_2)_+(A, B), (g_2)_-(A, B))\,.$$

And, under this bijection, its action on the set of elliptic associators is given by:

$$(\mu, f, g_\pm) \bullet (\lambda, \Phi, A_\pm) =$$

$$\left(\mu\lambda, f\left(e^{\lambda t_{12}}, \Phi(t_{12}, t_{23})e^{\lambda t_{23}}\Phi(t_{23}, t_{12})\right)\Phi(t_{12}, t_{23}), g_\pm(A_+, A_-)\right)\,.$$

Main Steps of Proof The proof of the above theorem goes along the following lines:

- Once the data of $(\mu, f) \in \mathrm{GT}(\Bbbk)$ is fixed, lifting the corresponding automorphism of $\widehat{\mathcal{P}a\mathcal{B}}(\Bbbk)$ to an automorphism of the right $\widehat{\mathcal{P}a\mathcal{B}}(\Bbbk)$-module $\widehat{\mathcal{P}a\mathcal{B}}_{\mathrm{ell}}(\Bbbk)$ requires to provide images $g_+(A, B)$ and $g_-(A, B)$ of A and B, respectively;
- According to Theorem 4.15 these images, together with the images of R and Φ determined by (μ, f), should satisfy the two nonagon relations and the mixed reation;
- One can easily check that equation (1) in the above theorem corresponds to the two nonagon relations. Conjugating with $f(\sigma_1^2, \sigma_2^2)$, one can then rewrite equation (2) in an equivalent form that directly corresponds to the mixed relation:

$$\sigma_1^{2\mu} = \Big[f(\sigma_1^2, \sigma_2^2)g_+(a_1, b_1)f(\sigma_1^2, \sigma_2^2)^{-1},$$
$$\sigma_1^{-\mu}f(\sigma_1^2, \sigma_2^2)g_-(a_1, b_1)f(\sigma_1^2, \sigma_2^2)^{-1}\sigma_1^{-\mu}\Big];$$

- The remaining statements (about the group structure and the action on elliptic associators) are explicit calculations.

□

Remark 4.21 Using Remark 4.11, one can also give an explicit description of the *elliptic graded Grothendieck–Teichmüller group* $\mathrm{GRT}_{\mathrm{ell}}$.

4.6 Topological Description of $\mathcal{P}a\mathcal{B}_{\mathrm{ell}}$

We consider the Fulton–MacPherson compactifications $\overline{\mathrm{C}}(\mathbb{T}, \mathrm{I})$ of the reduced configuration spaces of the torus $\mathrm{C}(\mathbb{T}, \mathrm{I})$, where I is a set with n elements. One can compute that the boundary

$$\partial\, \overline{\mathrm{C}}(\mathbb{C}, \mathrm{I}) \cong \bigcup_{k\geq 0} \bigcup_{\mathrm{J}_1 \sqcup \cdots \sqcup \mathrm{J}_k = \mathrm{I}} \overline{\mathrm{C}}(\mathbb{T}, [k]) \times \left(\prod_{j=1}^{k} \overline{\mathrm{C}}(\mathbb{C}, \mathrm{J}_j) \right)$$

where $[k] = \{1, \cdots, k\}$. Therefore the inclusions

$$\gamma_{\mathrm{J}_1, \cdots, \mathrm{J}_k} : \overline{\mathrm{C}}(\mathbb{T}, [k]) \times \left(\prod_{j=1}^{k} \overline{\mathrm{C}}(\mathbb{C}, \mathrm{J}_j) \right) \hookrightarrow \partial\, \overline{\mathrm{C}}(\mathbb{T}, \mathrm{J}_1 \sqcup \cdots \sqcup \mathrm{J}_k) \hookrightarrow \overline{\mathrm{C}}(\mathbb{T}, \mathrm{J}_1 \sqcup \cdots \sqcup \mathrm{J}_k)$$

endow the family $\left(\overline{\mathrm{C}}(\mathbb{T}, \mathrm{I})\right)_{n\geq 0}$ with a right module structure over the operad $\overline{\mathrm{C}}(\mathbb{C})$. We denote this right module by $\overline{\mathrm{C}}(\mathbb{T})$.

Consider again $\mathcal{P}a$, viewed as a right module over itself in the category of topological spaces when endowed with the discrete topology. There is an inclusion of topological right modules over $\mathcal{P}a$

$$\iota : \mathcal{P}a \hookrightarrow \overline{\mathrm{C}}(\mathbb{T}) ,$$

that arrizes from the embedding $(0, 1) \hookrightarrow S^1 \times S^1 = \mathbb{T}$ given by $t \mapsto (\bar{t}, \bar{0})$.

Theorem 4.22 *There is an isomorphism of right modules over $\mathcal{P}a\mathcal{B}$ in the category of groupoids*

$$\mathcal{P}a\mathcal{B}_{\mathrm{ell}} \cong \Pi_1(\overline{\mathrm{C}}(\mathbb{T}), \mathcal{P}a) .$$

Remark 4.23 In [CG23], $\mathcal{P}a\mathcal{B}_{\mathrm{ell}}$ is in fact *defined* using its topological description. In [Enr14a], Enriquez proves that the algebraic definition satisfies a presentation by generators and relations, that is essentially the content of Theorem 4.15 (even though Enriquez doesn't use operads). The paper [CG23] implicitely contains a translation of Enriquez's definition in terms of operads, and the proof that it coincides with the topological definition.

4.7 An Overview of the Ellipsitomic Case

One can think of the ellipsitomic case as a combination of the elliptic and the cyclotomic cases. The constructions performed in this case in order to define ellipsitomic associators follow the same steps as in the previous cases. For precise statements, we refer to [CG23, Sections 4–6].

4.7.1 First Player

Let $\Gamma := \mathbb{Z}/N\mathbb{Z} \times \mathbb{Z}/M\mathbb{Z}$, where $N, M \geq 1$. Recall that the fundamental group of the topological torus $\mathbb{T}$ is $\mathbb{Z} \times \mathbb{Z}$. Therefore Γ can be obtained as a quotient of the fundamental group of the torus, meaning there is an unique covering space

$$p^\Gamma : \tilde{\mathbb{T}} \twoheadrightarrow \mathbb{T} ,$$

associated to the canonical quotient map $\mathbb{Z} \times \mathbb{Z} \twoheadrightarrow \Gamma$. We consider the reduced associated Γ-twisted configuration space

$$\mathrm{C}(\mathbb{T}, n, \Gamma) := \left\{(x_1, \cdots, x_n) \in \left(\tilde{\mathbb{T}}\right)^n \mid p^\Gamma(x_i) \neq p^\Gamma(x_j) \text{ if } i \neq j\right\} / \tilde{\mathbb{T}} .$$

It has a natural Γ-action and a right module structure over the operad of configuration spaces. One can define the right module of *parenthesized ellipsitomic braids* as the fundamental groupoid

$$\mathcal{P}a\mathcal{B}_{\mathrm{ell}}^\Gamma := \Pi_1(\mathrm{C}(\mathbb{T}, n, \Gamma), \mathcal{P}a^\Gamma) ,$$

where $\mathcal{P}a^\Gamma$ is a very similar construction to that of Sect. 3.7. It is a right module over $\mathcal{P}a\mathcal{B}$ endowed with a compatible Γ-action. It is generated as right module with a diagonally trivial Γ-action by two morphisms $A : 1_{(\bar{0},\bar{0})}2_{(\bar{0},\bar{0})} \longrightarrow 1_{(\bar{1},\bar{0})}2_{(\bar{0},\bar{0})}$ and $B : 1_{(\bar{0},\bar{0})}2_{(\bar{0},\bar{0})} \longrightarrow 1_{(\bar{0},\bar{1})}2_{(\bar{0},\bar{0})}$ are in $\mathcal{P}a\mathcal{B}_{\mathrm{ell}}^\Gamma(2)$.

In fact, one can give an explicit presentation of $\mathcal{P}a\mathcal{B}_{\mathrm{ell}}^\Gamma$ analogue Theorem 4.15, see [CG23, Theorem 4.5]. After considering the appropriate Malcev completions, this gives the first ellipsitomic player.

4.7.2 Second Player

One defines the *infinitesimal ellipsitomic braids* Lie algebra $\mathfrak{t}_n^{\mathrm{ell},\Gamma}$ by adding generators $\{t_{ij}^\gamma\}$ for $\gamma \in \Gamma$ in weight $(1, 1)$ to the generators in Definition 4.4 and then slightly modifying the elliptic relations they satisfy.

This symmetric sequence in the category of graded Lie algebras with a Γ-action has a right module structure over the operad $\{\mathfrak{t}_n\}$, given by analogous formulas to those of Proposition 4.6, and which is compatible with the Γ-action.

Once again, one considers a *reduced* version $\bar{\mathfrak{t}}_n^{\mathrm{ell},\Gamma}$ of these algebras by modding out the central elements $\sum_i \alpha_i$ and $\sum_i \beta_i$. One defines the symmetric sequence of *ellipsitomic chord diagrams* $\mathcal{CD}_{\mathrm{ell}}^\Gamma$ by taking the universal enveloping algebra functor.

One then defines the right module $\mathcal{P}a\mathcal{CD}^{\Gamma}_{\text{ell}}$ of *parenthesized ellipsitomic chord diagrams* by performing analogue constructions to those already explained in Sect. 3.4. It has once again a compatible Γ-action. By considering the completion with respect to the total weight filtration, one gets the second ellipsitomic player.

Remark 4.24 Contrary to the cyclotomic case, there is no *frozen strand* neither on ellipsitomic braids nor on ellipsitomic chord diagrams.

4.7.3 Defining Ellipsitomic Associators

Once this machinery is setup, the rest is completely analogue to the previous cases.

Definition 4.25 (Ellipsitomic Associators) The set of *ellipsitomic associators* is given by

$$\mathrm{Assoc}^{\Gamma}_{\text{ell}}(\Bbbk) :=$$

$$\mathrm{Iso}^{+}_{\mathsf{RMod(p.u\text{-}Grpd_{\Bbbk})}}\left(\left(\widehat{\mathcal{P}a\mathcal{B}}^{\Gamma}_{\text{ell}}(\Bbbk), \widehat{\mathcal{P}a\mathcal{B}}(\Bbbk)\right), \left(\mathrm{Grp}(\widehat{\mathcal{P}a\mathcal{CD}}^{\Gamma}_{\text{ell}}), \mathrm{Grp}(\widehat{\mathcal{P}a\mathcal{CD}})\right)\right)^{\Gamma}.$$

This amounts to the data of:

(1) an isomorphism F between $\widehat{\mathcal{P}a\mathcal{B}}(\Bbbk)$ and $\mathrm{Grp}(\widehat{\mathcal{P}a\mathcal{CD}})$ which is the identity on objects (an associator) ,
(2) an isomorphism W between $\widehat{\mathcal{P}a\mathcal{B}}^{\Gamma}_{\text{ell}}(\Bbbk)$ and $\mathrm{Grp}(\widehat{\mathcal{P}a\mathcal{CD}}^{\Gamma}_{\text{ell}})$ of right modules which is Γ-equivariant and is the identity on objects, and which lies above the isomorphism F.

It is naturally a bitorsor over the *ellipsitomic Grothendieck–Teichmüller group* $\mathrm{GT}^{\Gamma}_{\text{ell}}(\Bbbk)$ and the *ellipsitomic Grothendieck–Teichmüller group* $\mathrm{GRT}^{\Gamma}_{\text{ell}}(\Bbbk)$, whose definitions should be straightforward by now.

Using the presentation of $\widehat{\mathcal{P}a\mathcal{B}}^{\Gamma}_{\text{ell}}$ mentioned before, one can give an explicit description of the set of ellipsitomic associators analogue to Theorem 4.16 and an explicit description of the set of elements in the ellipsitomic Grothendieck–Teichmüller group analogue to Theorem 4.20. There is also an explicit description of elements in the *graded* version of the ellipsitomic Grothendieck–Teichmüller group.

Theorem 4.26 ([CG23, Section 6]) *Let* $\Bbbk$ *be a field of characteristic zero. The set of ellipsitomic associators* $\mathrm{Assoc}^{\Gamma}_{\text{ell}}(\Bbbk)$ *is non empty.*

As in all the previous cases ("usual" associators, cyclotomic associators, elliptic associators), the proof of the existence goes into two steps: (1) one proves that ellipsitomic associators exists over $\mathbb{C}$ using analytic methods, (2) one deduces that they exist over $\mathbb{Q}$ using descent methods. The second step isn't very difficult, and relies on the fact that ellipsitomic associators form a torsor over the ellipsitomic (graded) Grothendieck–Teichmüller group. The first step is the most difficult one: in [CG20b], a flat connection is constructed on the Γ-twisted configuration space of the torus, whose renormalized holonomies along appropriate paths gives rise to an ellipsitomic associator.

4.7.4 Concluding Remarks

The ellipsitomic constructions satisfy some compatibilities with surjective morphisms $\Gamma_1 \twoheadrightarrow \Gamma_2$, that could probably be phrased in functorial terms. In would be interesting to investigate this, even in the simplest case of isomorphisms $\Gamma_1 \tilde{\to} \Gamma_2$, this would allow to define and study ellipsitomic associators associated with any finite quotient map $\mathbb{Z}^2 \to \Gamma$.

A question remains open: what is the Lie theoretic deformation problem to which ellipsitomic associators provide a universal solution? Section 5 of [CG20b] already provides some insight: realizations of (some aspects of) the second ellipsitomic player into Lie theoretic terms are given.

The connection involved in the construction of ellipsitomic associators given in [CG20b] depends on the choice of an elliptic curves equipped with a Γ-level structure. Moreover, this connection extends to the whole moduli space of elliptic curves with Γ-level structure and marked points. As a result, the corresponding ellipsitomic associators (obtained as suitable renormalized holonomies) should have many interesting features similar to the ones of elliptic associators:

- Their coefficients are interesting numbers, that should be expressed as iterated integrals of Eiseinstein series associated with a congruence subrgoup $\mathrm{SL}_2^\Gamma \subset \mathrm{SL}_2(\mathbb{Z})$.
- Modular invariance (with respect to the congruence subgroup SL_2^Γ) of these Eiseistein series have been investigated in the last section of [CG20b]. It would be interesting to understand the modular invariance properties of ellipsitomic associators themselves.
- All of this should probably lift to a more abstract motivic context, yet to be determined.

Appendix A: Pro-unipotent Completions

An *unipotent algebraic group G* over $\Bbbk$ is an algebraic group which embeds into $\mathbb{UT}_n$, the algebraic group given by the upper-triangular matrices, for some n in $\mathbb{N}$. A classical result states that the category of affine algebraic groups over $\Bbbk$ is equivalent to the category of commutative Hopf algebras over $\Bbbk$ which are finitely generated as algebras. This gives that the category of pro-affine algebraic groups over $\Bbbk$ is to the category of commutative Hopf algebras over $\Bbbk$, without any finite generation assumptions.

Likewise, one can show that the category of unipotent algebraic groups over $\Bbbk$ is equivalent to the category of commutative Hopf algebras over $\Bbbk$ which are finitely generated as algebras and which are *conilpotent* as coalgebras.

Proposition A.1 *There is a contravariant equivalence of categories between the category of commutative conilpotent Hopf algebras over $\Bbbk$ and the pro-category of unipotent algebraic groups over $\Bbbk$:*

$$\mathsf{Hopf\text{-}alg}^{\mathsf{conil}} \underset{\mathcal{O}(-)}{\overset{Spec(-)}{\rightleftarrows}} \mathsf{pro(UniGrp)}^{\mathsf{op}} ,$$

given by taking the spectrum Spec(−) of the underlying commutative algebra and by taking the global sections of the structure sheaf $\mathcal{O}(-)$.

Remark A.2 We refer to the survey [Vez10] on this subject for more details.

A.1 How to Pro-unipotently Complete?

By the above discussion, it becomes clear that a *pro-unipotent completion* over $\Bbbk$ amounts to the construction of a commutative conilpotent Hopf algebra over $\Bbbk$, which is universal in a suitable sense.

Let us construct it for a finitely generated abstract group G. The first step is to take the group algebra $\Bbbk[G]$ of G over $\Bbbk$, which is naturally a cocommutative Hopf algebra. Now the goal is to dualize it in such a way that it gives a commutative conilpotent Hopf algebra. Notice that, under finiteness assumptions, complete algebras are dual to conilpotent coalgebras in a sense that is explained below.

Definition A.3 (I-Adic Completion) Let A be an augmented $\Bbbk$-algebra, let I be its augmentation ideal. The *I-adic completion* of A is given by the following limit

$$A_I^\wedge := \lim_{n\in\mathbb{N}^*} A/I^n ,$$

taken in the category of $\Bbbk$-algebras.

This pro-nilpotent $\Bbbk$-algebra can be dualized under some minor finiteness assumptions.

Definition A.4 (Topological Dual) Let A be an augmented $\Bbbk$-algebra, let I be its augmentation ideal and suppose A/I^n is finite dimensional over $\Bbbk$ for all $n \geq 1$. The *topological dual* of A is given by

$$A_I^\vee := \operatorname*{colim}_{n\in\mathbb{N}^*} (A/I^n)^* ,$$

where each $(A/I^n)^*$ is the coalgebra given by the linear dual of A/I^n and where the colimit is taken in the category of coalgebras.

The topological dual $A^\vee$ is a conilpotent coalgebra. Indeed, it is pretty straightforward that $(A/I^n)^*$ is conilpotent for all n, and since the category of conilpotent coalgebras is stable under colimits, so is $A^\vee$. The topological dual functor sends comonoids to monoids, thus the topological dual of a Hopf algebra is a conilpotent Hopf algebra.

Remark A.5 Note that the algebra given by the linear dual of $A_I^\vee$ is precisely the completed algebra $A_I^\wedge$.

Lemma A.6 *Let A be an augmented $\Bbbk$-algebra and let I be its augmentation ideal. If I/I^2 is finite dimensional over $\Bbbk$, then A/I^n is finite dimensional over $\Bbbk$ for all $n \geq 1$.*

Proof By a straightforward *dévissage* argument, A/I^n is finite dimensional for every n if and only if I^n/I^{n+1} is finite dimensional. For every $n \geq 1$, there is a $\Bbbk$-linear surjection

$(I/I^2)^{\otimes n} \twoheadrightarrow I^n/I^{n+1}$ given by the multiplication, which implies that I^n/I^{n+1} is finite dimensional over $\Bbbk$. □

Let I be the augmentation ideal of $\Bbbk[G]$. One computes that $I/I^2 \cong G^{ab} \otimes \Bbbk$, where G^{ab} denotes the abelianization of G. Thus if G is a finitely generated group, I/I^2 is finite dimensional over $\Bbbk$ and $\Bbbk[G]^\wedge_I$ admits a topological dual.

Example A.7 It follows from [Arn69] that $\mathrm{H}_1(\mathrm{Conf}_n(\mathbb{C}); \mathbb{Z})$ is a free $\mathbb{Z}$-module, generated by $\frac{n(n-1)}{2}$-generators. Therefore the abelianization of the pure braid group PB_n is given by

$$\mathrm{PB}_n^{ab} \cong \mathbb{Z}^{\frac{n(n-1)}{2}} .$$

Hence $\Bbbk[\mathrm{PB}_n]$ admits a topological dual for all n in $\mathbb{N}$. △

Definition A.8 (Pro-unipotent Completion) Let G be a finitely generated abstract group. Its *pro-unipotent completion* G^{uni} is the pro-unipotent algebraic group given by

$$G^{\mathrm{uni}} := \mathrm{Spec}(\Bbbk[G]^\vee_I) .$$

Remark A.9 The pro-unipotent group G^{uni} is the initial object in the category of pro-unipotent groups U endowed with a group morphism $G \longrightarrow U(\Bbbk)$, where $U(\Bbbk)$ refer to the $\Bbbk$-points of U.

A.2 *Malcev Completion*

Very closely related to the pro-unipotent completion of an abstract group G is the notion of its Malcev completion.

Definition A.10 (Group-Like Elements) Let A be a Hopf algebra. An element x in A is said to be *group-like* if $\Delta(x) = x \otimes x$ and $\epsilon(x) = 1$. The set of group-like elements of A is denoted by $\mathrm{Grp}(A)$.

Group-like elements form a group, where the multiplication is given by the associative product of A, and where the inverse is given by the antipode map of the Hopf algebra.

Definition A.11 (Malcev Completion) Let G be a finitely generated abstract group. Its *Malcev completion* $\widehat{G}(\Bbbk)$ is the abstract group

$$\widehat{G}(\Bbbk) := \mathrm{Grp}(\Bbbk[G]^\wedge_I)$$

given by the group-like elements of the I-adic completion of $\Bbbk[G]$.

The Malcev completion of a group G corresponds to the $\Bbbk$-points of its pro-unipotent completion.

Proposition A.12 *Let G be a finitely generated abstract group. There is a group isomorphism*

$$G^{uni}(\Bbbk) \cong \widehat{G}(\Bbbk) .$$

Proof There is an inclusion

$$G^{\mathrm{uni}}(\Bbbk) := \mathrm{Hom}_{\mathsf{Alg}_\Bbbk}(\Bbbk[G]_I^\vee, \Bbbk) \subset \mathrm{Hom}_{\mathsf{Vect}_\Bbbk}(\Bbbk[G]_I^\vee, \Bbbk) \cong \Bbbk[G]_I^\wedge .$$

Thus the $\Bbbk$-points of the pro-unipotent completion of G will be given by certain kind of element in $\Bbbk[G]_I^\wedge$. One can check that x in $\Bbbk[G]_I^\wedge$ is a group-like element if and only if x in $\mathrm{Hom}_{\mathsf{Vect}_\Bbbk}(\Bbbk[G]_I^\vee, \Bbbk)$ is a morphism of $\Bbbk$-algebras. □

Remark A.13 Let B be a commutative $\Bbbk$-algebra. One can check that

$$G^{\mathrm{uni}}(B) = \mathrm{Hom}_{\mathsf{Alg}_\Bbbk}(\Bbbk[G]_I^\vee, B) \cong \mathrm{Grp}(\Bbbk[G]_I^\wedge \widehat{\otimes} B) ,$$

that is, the B-points of G^{uni} can be computed as the group-like elements of the complete Hopf algebra

$$\Bbbk[G]_I^\wedge \widehat{\otimes} B := \lim_{n \in \mathbb{N}^*} \left(\Bbbk[G]/I^n \otimes B\right) .$$

So a "generalized" Malcev completion allows us to reconstruct the whole pro-unipotent algebraic group G^{uni}.

Remark A.14 The Malcev completion of a group G is a pro-nilpotent uniquely divisible group. Which we will refer to, by a slight abuse of language, as an *abstract pro-unipotent $\Bbbk$-group*.

Proposition A.15 *The Malcev completion defines a strong monoidal functor*

$$\widehat{(-)}(\Bbbk) : \left(\mathsf{Grp}^{\mathsf{f.g}}, \times, \{e\}\right) \longrightarrow \left(\mathsf{p.u\text{-}Grp}_\Bbbk^{\mathsf{f.g}}, \times, \{e\}\right) ,$$

from the category of finitely generated abstract groups to the category of finitely generated abstract pro-unipotent $\Bbbk$-groups.

Proof It is straightforward to check that the various steps of its construction commute with the cartesian product. □

Remark A.16 The inclusion of abstract pro-unipotent $\Bbbk$-groups into all groups is *not* fully faithful in general, except when $\Bbbk = \mathbb{Q}$. See [Qui69, Appendix A] for more details.

A.3 Malcev Completion of Groupoids

There is a generalization of this procedure from groups to groupoids. Given a finitely generated groupoid $\mathcal{G}$, one can produce a category $\Bbbk[\mathcal{G}]$ in the following way:

(1) The set of objects of $\Bbbk[\mathcal{G}]$ is the same as the set of objects of $\mathcal{G}$.
(2) Given two object g, g' in $\mathcal{G}$,

$$\mathrm{Hom}_{\Bbbk[\mathcal{G}]}(g, g') := \Bbbk[\mathrm{Hom}_{\mathcal{G}}(g, g')] .$$

Notice that for every object g in $\Bbbk[\mathcal{G}]$, its endomorphisms $\mathrm{End}_{\Bbbk[\mathcal{G}]}(g)$ form a Hopf algebra. In general, $\mathrm{Hom}_{\Bbbk[\mathcal{G}]}(g, g')$ has a cocommutative coalgebra structure compatible with the composition of morphisms. The category $\Bbbk[\mathcal{G}]$ forms a *Hopf groupoid*. See [Fre17a, Chapter 9] for more on this notion.

For any g, g' in $\Bbbk[\mathcal{G}]$, we can complete the $\mathrm{Hom}_{\Bbbk[\mathcal{G}]}(g, g')$ with respect to its augmentation ideal and get a category $\Bbbk[\mathcal{G}]^\wedge$ enriched in cocommutative coalgebras, which forms a *complete Hopf groupoid*. There is an analogue group-like element functor which produces a groupoid from a Hopf groupoid.

The groupoid obtained $\widehat{\mathcal{G}}(\Bbbk)$ given by the group-like elements of $\Bbbk[\mathcal{G}]^\wedge$ is called the *Malcev completion* of $\mathcal{G}$. Notice that for each object g in $\widehat{\mathcal{G}}(\Bbbk)$, the set of automorphisms of g is the $\Bbbk$-points of a pro-unipotent group. In a slight abuse of terminology, we will refer to it as a *pro-unipotent* $\Bbbk$*-groupoid.*

The Malcev completion functor forms a strong monoidal endofunctor

$$\widehat{(-)}(\Bbbk) : \left(\mathsf{Grpd}^{\mathsf{f.g}}, \times, \{e\}\right) \longrightarrow \left(\mathsf{p.u\text{-}Grpd}_{\Bbbk}^{\mathsf{f.g}}, \times, \{e\}\right) ,$$

from the category of finitely generated groupoids to the category of finitely generated pro-unipotent $\Bbbk$-groupoids.

A.4 How to Compute the Malcev Completion?

Over a field of characteristic zero, a pro-unipotent group is completely determined by its Lie algebra. Therefore, one way to compute the Malcev completion of a G is by first computing its Lie algebra and then integrating it.

Definition A.17 (Primitive Elements) Let A be a Hopf algebra. An element x in A is said to be *primitive* if $\Delta(x) = 1 \otimes x + x \otimes 1$. The set of primitive elements of A is denoted by $\mathrm{Prim}(A)$.

Remark A.18 Since $\Bbbk$ is a field of characteristic different than 2, the condition $\Delta(x) = 1 \otimes x + x \otimes 1$ implies that any primitive element is in the augmentation ideal of A. Indeed, if x is a primitive element, we have that

$$\epsilon(x) = (\epsilon \otimes \epsilon) \circ \Delta(x) = (\epsilon \otimes \epsilon)(1 \otimes x + x \otimes 1) = 2\epsilon(x) ,$$

hence $\epsilon(x) = 0$.

The primitive elements of a Hopf algebra form a Lie algebra, where the bracket is given by the skew-symmetrization of the associative product in A.

Definition A.19 (Exponential Map) Let A be a Hopf algebra and let I be its augmentation ideal. Let

$$\widehat{I} := \lim_{n \in \mathbb{N}^*} I/I^n$$

denote the augmentation ideal of $A_I^\wedge$. The *exponential map* is given by

$$\begin{aligned} \exp : \widehat{I} &\longrightarrow A_I^\wedge \\ x &\longmapsto \sum_{n \geq 0} \frac{x^n}{n!} . \end{aligned}$$

Proposition A.20 *Let G be a finitely generated group. The exponential map defines an isomorphism of groups*

$$exp : \big(Prim(\mathbb{k}[G]_I^\wedge), BCH, 0\big) \xrightarrow{\cong} \widehat{G}(\mathbb{k}) ,$$

where the group law BCH is given by the Baker–Campbell–Hausdorff formula. The inverse of this bijection is given by the logarithm map.

Proof Let x be an primitive element. Its image under the exponential map is a group-like element in $\mathrm{Grp}(\mathbb{k}[G]_I^\wedge)$:

$$\Delta(\exp(x)) = \exp(1 \otimes x + x \otimes 1) = \exp(x) \otimes \exp(x) .$$

Furthermore, the exponential map is a morphism of groups when one considers the group structure on $\mathrm{Prim}(\mathbb{k}[G]_I^\wedge)$ given by the Baker–Campbell–Hausdorff formula, as it is the universal formula such that

$$\exp(\mathrm{BCH}(x, y)) = \exp(x).\exp(y) .$$

Let g be a group-like element. Notice that $\epsilon(g) = 1$ implies that $(g - 1)$ is in the augmentation ideal. Thus one can define the logarithm on group-like elements as

$$\log(g) = \log(1 - (g - 1)) = \sum_{n \geq 1} \frac{(g - 1)^n}{n} .$$

One can check that it is a morphism of groups, which is the inverse of the exponential map. □

This description of $\widehat{G}(\Bbbk)$ gives a canonical morphism of groups

$$\begin{aligned} G &\longrightarrow \widehat{G}(\Bbbk) \\ g &\longmapsto \exp\big(\log(g)\big)\ , \end{aligned}$$

since g is group-like element of $\Bbbk[G]_I^\wedge$.

Remark A.21 In particular, the element g^λ, for λ in $\Bbbk$, makes sense in $\widehat{G}(\Bbbk)$. Indeed, one can define it as $g^\lambda := \exp\big(\lambda.\log(g)\big)$.

Example A.22

(1) One has that $\widehat{\mathbb{Z}}(\Bbbk) \cong \Bbbk$ and that $\widehat{\mathbb{Z}/m\mathbb{Z}}(\Bbbk) = 0$.
(2) In general, if G is an abelian group, then $\widehat{G}(\Bbbk) \cong G \otimes_{\mathbb{Z}} \Bbbk$.
(3) The Malcev completion $\widehat{\mathbb{F}}_n(\Bbbk)$ of the free group $\mathbb{F}_n$ generated by $x_1, \cdots, x_n$ is generated by the expressions $\{x_i^\lambda\}$ for λ is in $\Bbbk$ and $1 \leq i \leq n$.
(4) Using Lemma 2.3, one can compute that $\widehat{\mathrm{PB}_3}(\Bbbk) \cong \widehat{\mathbb{F}}_2(\Bbbk) \times \Bbbk$. △

Appendix B: Operads in Cocartesian Categories

Let C be a category with finite colimits. We view is as a symmetric monoidal category: the monoidal product is the coproduct $\sqcup$ and the unit is the initial object $\emptyset$. The main goal of this appendix is to show how to construct operadic-like objects in $(C, \sqcup, \emptyset)$ from presheaves on certain categories of finite sets with values in C. This formalism of presheaves just encodes a family of objects in C with types of insertion-coproduct morphisms. We deal with three cases: operads, operadic modules, and moperads.

B.1 The Case of Operads

Let Fin_* be the category of finite pointed sets, together with maps that respect the base point. This category is canonically equivalent to the category of finite sets with *partially defined maps*. We consider two distinguished classes of morphisms in Fin_* (viewed as finite sets with partially defined maps):

- *Active* morphisms are the totally defined map.
- *Inert* morphisms are the partially defined bijections.

Example B.1 Let $J \subset K$ be an inclusion of finite sets. The partially defined map $\mathrm{id}_J^K : K \longrightarrow J$ which is given by the identity on J is a partially defined bijection. We will refer to this class of maps as *partially defined equalities*. △

The pair (*Inert*, *Active*) is an orthogonal factorization system, meaning in particular that every morphism factors (uniquely up to a unique isomorphism) as an inert morphism followed by an active one. If, moreover, we require that the inert map is a partially defined equality, then the factorization becomes strictly unique. Now observe that active morphisms are generated by:

- bijections;
- and contracting maps: if I, J are set and $i \in I$, the contracting map $c_i : I\backslash\{i\} \sqcup J \to I$ is defined as $c_i(j) = j$ if $j \notin J$ and $c_i(j) = i$ if $j \in J$. Note that when J is empty, the contracting map is injective.

Theorem B.2 *Any functor $Fin_*^{op} \longrightarrow C$ induces an operad in $(C, \amalg, \emptyset)$.*

Sketch of Proof Let $\mathcal{O} : \mathrm{Fin}_*^{op} \to C$ be a functor. We first restrict it to Fin_{bij}, the category of finite sets with (totally defined) bijections; we thus get a species of structure.

We define the partial compositions as

$$\circ_i := \mathcal{O}(c_i) \amalg \mathcal{O}(\mathrm{id}_J^K) : \mathcal{O}(I) \amalg \mathcal{O}(J) \to \mathcal{O}\big((I\backslash\{i\}) \sqcup J\big)\,,$$

where $K = (I\backslash\{i\}) \sqcup J$. One can check that equivariance, associativity and unitality relations defining an operad are implied by relations among generating morphisms of Fin_*^{op} (these generating morphisms being bijections, contracting maps, and partial equalities).

□

Suppose that C carries a symmetric monoidal structure $\otimes$ such that the identity functor is a symmetric colax monoidal functor $(C, \amalg, \emptyset) \to (C, \otimes, \mathbb{1})$. It might not send operads to operads in general. Nevertheless, if each $\circ_i$ factors through $\mathcal{O}(I) \otimes \mathcal{O}(J)$, then $\mathcal{O}$ becomes an operad in $(C, \otimes, \mathbb{1})$.

Example B.3 Here are two examples of such a symmetric monoidal category:

- The category of unitial associative $\Bbbk$-algebras, with monoidal product being the tensor product;
- The category of Lie $\Bbbk$-algebras, with monoidal product being the direct sum (that is the cartesian monoidal structure).

In both cases, the condition for a map $A_1 \amalg A_2 \to B$ to factor through the monoidal product of A_1 and A_2 is that the images of $A_1 \to B$ and $A_2 \to B$ commute. △

This explains why the insertion-coproduct morphisms we consider in the main body of the paper define operads in Lie/associative $\Bbbk$-algebras:

- Insertion-coproduct morphisms define operads for the cocartesian monoidal structure;
- The image of insertion morphisms commutes with the image of coproduct morphisms, making the operad structure descend to the monoidal structure of interest.

B.2 The Case of Operadic Modules

Let $\mathrm{Fin}_{*,\not *}$ be the following category:

- Objects are finite sets with or without pointing;
- Morphisms are maps preserving the pointing.

In other words, every object of $\mathrm{Fin}_{*,\not *}$ is isomorphic to $\{1, \ldots, n\}$ or $\{*, 1, \ldots, n\}$, where $*$ is the base point, for some $n \geq 0$. For every finite set I there are two objects, denoted by I and I^+, which are given by the finite set without or with a base point. There are three types of morphisms: pointed maps between pointed sets, maps between sets without pointing, and maps from a non-pointed set to a pointed one. Concretely, we have that:

- Morphisms from I to J are totally defined maps $I \to J$;
- Morphisms from I^+ to J^+ are partially defined maps $I \to J$;
- Morphisms from I to J^+ are partially defined maps $I \to J$;
- There are no morphisms from I^+ to J.

As we explained in Appendix B.1, partially defined maps admit an orthogonal factorization system made of inert maps and active maps. Active maps are precisely totally defined maps, and these are generated by bijections and contracting maps.

Remark B.4 There is a more pedantic (but useful!) way of defining $\mathrm{Fin}_{*,\not *}$. Consider the functor $F : \Delta^1 \to \mathrm{Cat}$ representing the functor $(-)^+ : \mathrm{Fin} \to \mathrm{Fin}_*$ sending I to I^+, where Fin is the category of finite sets with totally defined maps. Then $\mathrm{Fin}_{*,\not *} \to \Delta^1$ is the cocartesian fibration associated to F.

Theorem B.5 *Any functor* $\mathrm{Fin}_{*,\not *}^{op} \longrightarrow C$ *induces an operad together with a right operadic module over it in* $(C, \amalg, \emptyset)$.

Sketch of Proof It follows from Remark B.4 that a functor $\mathrm{Fin}_{*,\not *}^{op} \to C$ amounts to the data of a triple $(\mathcal{M}, \mathcal{O}, f)$ where

- $\mathcal{O}$ is a functor $\mathrm{Fin}_*^{op} \to C$, which by Theorem B.2 induces an operad in $(C, \amalg, \emptyset)$;
- $\mathcal{M}$ is a functor $\mathrm{Fin}^{op} \to C$;
- $f : \mathcal{O}((-)^+) \Rightarrow \mathcal{M}$ is a natural transformation.

Since totally defined maps are generated by bijections and contracting maps, the functor $\mathcal{M}$ induces a species of structure when we restrict it to bijections. Furthermore, it comes equipped with the following maps

$$\mathcal{M}(c_i) : \mathcal{M}(I) \longrightarrow \mathcal{M}\big((I \backslash \{i\}) \sqcup J\big) ,$$

induced by the contracting maps. On the other hand, we consider the following composition of morphisms:

$$\mathcal{O}(J) \xrightarrow{O(\mathrm{id}_J^K)} \mathcal{O}\big((I \backslash \{i\}) \sqcup J\big) \xrightarrow{f} \mathcal{M}\big((I \backslash \{i\}) \sqcup J\big) ,$$

where $K = (I\backslash\{i\}) \sqcup J$. These two aforementioned maps assemble into the following structure maps

$$\circ_i : \mathcal{M}(I) \amalg \mathcal{O}(J) \to \mathcal{M}\big((I\backslash\{i\}) \sqcup J\big) .$$

Once again, one can check that all the axioms of a right module over an operad are satisfied, by using all the relations amongst the generating morphisms of the source category. □

Remark B.6 Like in Appendix B.1, the structure defined on the cocartesian monoidal structure can descend to other types of monoidal structures under certain conditions. In particular, the elliptic insertion-coproduct morphisms that we consider in the main body of the paper define a right module structure for the cocartesian monoidal structure in Lie/associative $\Bbbk$-algebras. Since the image of insertion morphisms commutes with the image of coproduct morphisms, they induce a right module structure in the monoidal structure of interest.

B.3 The Case of Moperads

Let $\mathrm{bFin}_{*,*}$ be the following category:

- Objects are finite sets with one or two pointings,
- Morphisms are maps preserving the pointings.

In other words, every object of $\mathrm{bFin}_{*,*}$ is isomorphic to either a pointed set $\{0 = *, 1, \ldots, n\}$ or a double-pointed set $\{*, 0, 1, \ldots, m\}$, where $*$ is the first base point and 0 is the second base point. For every finite set I there are two objects, denoted by I^+ and I^{++}, which are given by the finite set with a base point or with two base points. There are three types of morphisms:

- Morphisms from I^+ to J^+ are partially defined maps $I \to J$;
- Morphisms from I^{++} to J^+ are partially defined maps $I \to J$;
- Morphisms from I^{++} to J^{++} are partially defined pointed maps $I^+ \to J^+$;
- There are no morphisms from I^+ to J^{++}.

Example B.7 If I, J are finite sets, there are pointed-contraction maps $c_0^1 : I^+ \sqcup J^+ \to I^+$ and $c_0^2 : I^+ \sqcup J^+ \to J^+$. The first map c_0^1 is defined by sending I^+ to itself and J^+ to the base point of I^+; the second map c_0^2 is defined by sending J^+ to itself and I^+ to the base point of J^+.

Theorem B.8 *Any functor* $\mathrm{bFin}_{*,*}^{op} \longrightarrow C$ *induces an operad together with a moperad over it in* $(C, \amalg, \emptyset)$.

Sketch of a Proof A functor $\mathrm{bFin}^{op}_{*,*} \to C$ amounts to the data of a triple $(\mathcal{M}, \mathcal{O}, f)$ where

- $\mathcal{O}$ is a functor $\mathrm{Fin}^{op}_{*} \to C$, which by Theorem B.2 induces an operad in $(C, \amalg, \emptyset)$;
- $\mathcal{M}$ is a functor $\mathrm{pFin}^{op}_{*} \to C$ from the opposite category of pointed sets with partially defined pointed maps to C;
- $g : \mathcal{O}(-) \Rightarrow \mathcal{M}((-)^{+})$ is a natural transformation.

The functor $\mathcal{M}$ induces a species of structure by considering its precomposition with the functor $(-)^{+}$ and restricting it to bijections.

The right modules structure over $\mathcal{O}(-)$ is constructed as follows. On one hand, we consider the image of the contracting maps c_i by the functor $(-)^{+}$:

$$\mathcal{M}((c_i)^{+}) : \mathcal{M}(I^{+}) \longrightarrow \mathcal{M}\big(((I\backslash\{i\}) \sqcup J)^{+}\big) ,$$

induced by the contracting maps. On the other hand, we consider the following composition of morphisms:

$$\mathcal{O}(J) \xrightarrow{\mathcal{O}(\mathrm{id}^{K}_{J})} \mathcal{O}\big((I\backslash\{i\}) \sqcup J\big) \xrightarrow{g} \mathcal{M}\big(((I\backslash\{i\}) \sqcup J)^{+}\big) ,$$

where $K = (I\backslash\{i\}) \sqcup J$. These two aforementioned maps assemble into the following structure maps

$$\circ_i : \mathcal{M}(I^{+}) \amalg \mathcal{O}(J) \longrightarrow \mathcal{M}\big(((I\backslash\{i\}) \sqcup J)^{+}\big) .$$

One can check that all the axioms of a right module over an operad are satisfied, by using all the relations amongst the generating morphisms of the source category.

The monoidal structure of $\mathcal{M}$ is constructed using the pointed-contraction maps. It is given by

$$\circ_0 := \mathcal{M}(c^{1}_{0}) \amalg \mathcal{M}(c^{2}_{0}) : \mathcal{M}(I^{+}) \amalg \mathcal{M}(J^{+}) \longrightarrow \mathcal{M}\big(I^{+} \sqcup J^{+}\big) .$$

Using the relations that the pointed-contraction maps satisfy, one can show that they define a monoid structure and that furthermore it is compatible with the right module structure over $\mathcal{O}$, thus forming a moperad. □

Remark B.9 Like in Appendix B.1, the structure defined on the cocartesian monoidal structure can descend to other types of monoidal structures under certain conditions. In particular, the cyclotomic insertion-coproduct morphisms that we consider in the main body of the paper define a moperad structure for the cocartesian monoidal structure in Lie/associative $\Bbbk$-algebras. Since the image of insertion morphisms commutes with the image of coproduct morphisms, they induce a moperad structure in the monoidal structure of interest.

Acknowledgments This survey emerged from a series of lectures given in June 2021 by DC at the "workshop on higher structures and operadic calculus". Both authors thank the organizers and the participants for their enthusiasm, as well as Adrien Brochier for his comments and suggestions.

DC has received funding from the European Research Council (ERC) under the European Union's Horizon 2020 research and innovation programme (grant agreement No 768679).

References

[Arn69] V. I. Arnol'd. The cohomology ring of the colored braid group. *Math. Notes*, 5:138–140, 1969.

[AT10] Anton Alekseev and Charles Torossian. Kontsevich deformation quantization and flat connections. *Comm. Math. Phys.*, 300(1):47–64, 2010.

[Bar98] Dror Bar-Natan. On associators and the Grothendieck-Teichmüller group. I. *Sel. Math., New Ser.*, 4(2):183–212, 1998.

[BLL98] F. Bergeron, G. Labelle, and P. Leroux. *Combinatorial species and tree-like structures. Transl. from the French by Margaret Readdy*, volume 67 of *Encycl. Math. Appl.* Cambridge: Cambridge University Press, 1998.

[BMR95] Michel Broué, Gunter Malle, and Raphaël Rouquier. On complex reflection groups and their associated braid groups. In *Representations of groups (Banff, AB, 1994)*, volume 16 of *CMS Conf. Proc.*, pages 1–13. Amer. Math. Soc., Providence, RI, 1995.

[BR24] Olivia Borghi and Marcy Robertson. Lecture notes on modular infinity operads and Grothendieck–Teichmüller theory. In *this book*. 2024.

[Bro12] Francis Brown. Mixed Tate motives over $\mathbb{Z}$. *Ann. of Math. (2)*, 175(2):949–976, 2012.

[Bro13] Adrien Brochier. Cyclotomic associators and finite type invariants for tangles in the solid torus. *Algebraic and Geometric Topology*, 13(6):3365–3409, Oct 2013.

[Bro24] Adrien Brochier. A combinatorial construction of higher genus associators, 2024, in preparation.

[CEE09] Damien Calaque, Benjamin Enriquez, and Pavel Etingof. Universal KZB equations: the elliptic case. In *Algebra, arithmetic, and geometry. In honor of Yu. I. Manin on the occasion of his 70th birthday. Vol. I*, pages 165–266. Boston, MA: Birkhäuser, 2009.

[CG20a] Damien Calaque and Martin Gonzalez. A moperadic approach to cyclotomic associators, 2020, preprint arXiv:2004.00572v1 [math.QA].

[CG20b] Damien Calaque and Martin Gonzalez. On the universal ellipsitomic KZB connection. *Selecta Math. (N.S.)*, 26(5):Paper No. 73, 59, 2020.

[CG23] Damien Calaque and Martin Gonzalez. Ellipsitomic associators. *Mémoires de la Société Mathématique de France*, 179, 2023.

[Chi12] Michael Ching. A note on the composition product of symmetric sequences. *J. Homotopy Relat. Struct.*, 7(2):237–254, 2012.

[CIW20] Ricardo Campos, Najib Idrissi, and Thomas Willwacher. Configuration spaces of surfaces, 2020, preprint arXiv:1911.12281v2 [math.QA].

[DCNTY19] Kenny De Commer, Sergey Neshveyev, Lars Tuset, and Makoto Yamashita. Ribbon braided module categories, quantum symmetric pairs and Knizhnik-Zamolodchikov equations. *Comm. Math. Phys.*, 367(3):717–769, 2019.

[DG05] Pierre Deligne and Alexander B. Goncharov. Groupes fondamentaux motiviques de Tate mixte. *Ann. Sci. École Norm. Sup. (4)*, 38(1):1–56, 2005.

[Dri91] V. G. Drinfel'd. On quasitriangular quasi-Hopf algebras and a group closely connected with $\mathrm{Gal}(\overline{\mathbb{Q}}/\mathbb{Q})$. *Leningr. Math. J.*, 2(4):829–860, 1991.

[Enr07] Benjamin Enriquez. Quasi-reflection algebras and cyclotomic associators. *Sel. Math., New Ser.*, 13(3):391–463, 2007.

[Enr14a] Benjamin Enriquez. Elliptic associators. *Sel. Math., New Ser.*, 20(2):491–584, 2014.

[Enr14b] Benjamin Enriquez. Flat connections on configuration spaces and braid groups of surfaces. *Adv. Math.*, 252:204–226, 2014.

[Enr16] Benjamin Enriquez. Analogues elliptiques des nombres multizétas. *Bull. Soc. Math. France*, 144(3):395–427, 2016.

[Fre17a] Benoit Fresse. *Homotopy of operads and Grothendieck-Teichmüller groups. Part 1: The algebraic theory and its topological background*, volume 217. Providence, RI: American Mathematical Society (AMS), 2017.

[Fre17b] Benoit Fresse. *Homotopy of operads and Grothendieck-Teichmüller groups. Part 2: The applications of (rational) homotopy theory methods*, volume 217. Providence, RI: American Mathematical Society (AMS), 2017.

[Fur11] Hidekazu Furusho. Double shuffle relation for associators. *Ann. of Math. (2)*, 174(1):341–360, 2011.

[Gon98] A. B. Goncharov. Multiple polylogarithms, cyclotomy and modular complexes. *Math. Res. Lett.*, 5(4):497–516, 1998.

[Gon01a] A. B. Goncharov. The dihedral Lie algebras and Galois symmetries of $\pi_1^{(l)}(\mathbb{P}^1 - (\{0, \infty\} \cup \mu_N))$. *Duke Math. J.*, 110(3):397–487, 2001.

[Gon01b] A. B. Goncharov. Multiple polylogarithms and mixed tate motives, 2001, preprint arXiv:math/0103059v4 [math.AG].

[Gon01c] Alexander B. Goncharov. Multiple ζ-values, Galois groups, and geometry of modular varieties. In *European Congress of Mathematics, Vol. I (Barcelona, 2000)*, volume 201 of *Progr. Math.*, pages 361–392. Birkhäuser, Basel, 2001.

[Gon20] Martin Gonzalez. Surface Drinfeld torsors I : Higher genus associators, 2020, preprint arXiv:2004.07303v1 [math.QA].

[Hai98] Richard M. Hain. The Hodge de Rham theory of relative Malcev completion. *Ann. Sci. Éc. Norm. Supér. (4)*, 31(1):47–92, 1998.

[HM20] Richard Hain and Makoto Matsumoto. Universal mixed elliptic motives. *J. Inst. Math. Jussieu*, 19(3):663–766, 2020.

[Hoe12] Eduardo Hoefel. Some elementary operadic homotopy equivalences. In *Topics in noncommutative geometry*, volume 16 of *Clay Math. Proc.*, pages 67–74. Amer. Math. Soc., Providence, RI, 2012.

[Hor17] Geoffroy Horel. Profinite completion of operads and the Grothendieck-Teichmüller group. *Adv. Math.*, 321:326–390, 2017.

[Hum12] Philippe Humbert. *Intégrale de Kontsevich elliptique et enchevêtrements en genre supérieur*. Phd thesis, Université de Strasbourg, December 2012, tel-00762209v2.

[Iha94] Yasutaka Ihara. On the embedding of $\mathrm{Gal}(\overline{\mathbb{Q}}/\mathbb{Q})$ into $\widehat{\mathrm{GT}}$. (Appendix by Michel Emsalem and Pierre Lochak: The action of the absolute Galois group on the moduli space of spheres with four marked points). In *The Grothendieck theory of dessins d'enfants*, pages 289–321, appendix 307–321. Cambridge: Cambridge University Press, 1994.

[Joy81] Andre Joyal. Une théorie combinatoire des séries formelles. *Adv. Math.*, 42:1–82, 1981.

[JS93] André Joyal and Ross Street. Braided tensor categories. *Adv. Math.*, 102(1):20–78, 1993.

[Kon99a] Maxim Kontsevich. Operads and motives in deformation quantization. *Lett. Math. Phys.*, 48(1):35–72, 1999. Moshé Flato (1937–1998).

[Kon99b] Maxim Kontsevich. Operads and motives in deformation quantization. *Lett. Math. Phys.*, 48(1):35–72, 1999. Moshé Flato (1937–1998).

[KZ01] Maxim Kontsevich and Don Zagier. Periods. In *Mathematics unlimited—2001 and beyond*, pages 771–808. Springer, Berlin, 2001.

[LM96] Thang Tu Quoc Le and Jun Murakami. Kontsevich's integral for the Kauffman polynomial. *Nagoya Math. J.*, 142:39–65, 1996.

[Maa19] Mohamad Maassarani. On orbit configuration spaces associated to finite subgroups of $\mathrm{PSL}_2(\mathbb{C})$. *Bull. Soc. Math. Fr.*, 147(1):123–157, 2019.

[May72] J.P. May. *The geometry of iterated loop spaces*. Springer-Verlag, Berlin, 1972. Lectures Notes in Mathematics, Vol. 271.

[MV19] Yuri I. Manin and Bruno Vallette. Monoidal structures on the categories of quadratic data. *arXiv e-prints*, page arXiv:1902.03778, February 2019, 1902.03778.

[Qui69] Daniel Quillen. Rational homotopy theory. *Ann. Math. (2)*, 90:205–295, 1969.
[Rac02] Georges Racinet. Doubles mélanges des polylogarithmes multiples aux racines de l'unité. *Publ. Math. Inst. Hautes Études Sci.*, 95:185–231, 2002.
[RW14] Carlo A. Rossi and Thomas Willwacher. P. Etingof's conjecture about Drinfeld associators, 2014, preprint arXiv:1404.2047v1 [math.QA].
[ŠW11] Pavol Ševera and Thomas Willwacher. Equivalence of formalities of the little discs operad. *Duke Math. J.*, 160(1):175–206, 2011.
[Tam02] Dmitry E. Tamarkin. Action of the Grothendieck-Teichmüller group on the operad of Gerstenhaber algebras, 2002, preprint arXiv:math/0202039 [math.QA].
[Tam03] Dmitry E. Tamarkin. Formality of chain operad of little discs. *Lett. Math. Phys.*, 66(1-2):65–72, 2003.
[Vez10] Alberto Vezzani. The pro-unipotent completion, 2010, preprint available on the author's webpage.
[Wil16] Thomas Willwacher. The homotopy braces formality morphism. *Duke Math. J.*, 165(10):1815–1964, 2016.
[Zag94] Don Zagier. Values of zeta functions and their applications. In *First European Congress of Mathematics, Vol. II (Paris, 1992)*, volume 120 of *Progr. Math.*, pages 497–512. Birkhäuser, Basel, 1994.

Lecture Notes on Modular Infinity Operads and Grothendieck-Teichmüller Theory

Olivia Borghi and Marcy Robertson

Contents

1 Introduction

These notes represent the transcript of three, 90 minute lectures given by the second author at the CRM in Barcelona in 2021 as part of the "Higher Structures and Operadic Calculus" workshop. The goal of the series was to introduce and motivate modular ∞-operads via their application to what is often called "Grothendieck-Teichmüller" theory. We therefore

O. Borghi · M. Robertson (✉)
School of Mathematics and Statistics, The University of Melbourne, Melbourne, VIC, Australia
e-mail: oborghi@student.unimelb.edu.au; marcy.robertson@unimelb.edu.au

B. Vallette (ed.), *Higher Structures and Operadic Calculus*, Advanced Courses in Mathematics - CRM Barcelona, https://doi.org/10.1007/978-3-031-77779-0_5

start by trying to answer

What is "Grothendieck-Teichmüller" theory?

The absolute Galois group of $\mathbb{Q}$, Gal($\mathbb{Q}$), is the (topological) group of automorphisms of the separable closure $\bar{\mathbb{Q}}$ which fix $\mathbb{Q}$. This is a *profinite group* (Definition 4.2), which means that, to write down an element g of the group Gal($\mathbb{Q}$), one must describe the image of g in *each* of the finite quotients of Gal($\mathbb{Q}$). It turns out, however, we do not know all the finite quotients of Gal($\mathbb{Q}$) (see, for example, [Ser08]).

The "Grothendieck" part of Grothendieck-Teichmüller theory is an idea, laid out in *Esquisse d'un Programme* [Gro97], to use the actions of Gal($\mathbb{Q}$) on the geometric fundamental groups of the stacks $\mathcal{M}_{g,n}$ to gain insight into Gal($\mathbb{Q}$). This idea is inspired by a theorem of Belyi [Bel79] which says that there is a faithful action of Gal($\mathbb{Q}$) on $\pi_1^{et}(\mathcal{M}_{0,4})$ and therefore on general $\pi_1^{et}(\mathcal{M}_{g,n})$.

Grothendieck suggestion that one could consider the collection of fundamental groups, $\pi_1^{et}(\mathcal{M}_{g,n})$, together with the natural maps between them, as a single object called the "Teichmüller tower". He conjectured that the automorphisms of this tower could not only be explicitly described, but that the group of such automorphisms may even be equivalent to Gal($\mathbb{Q}$). In addition, Grothendieck suggested that the ideal tower should be constructed from the genus zero and genus one components of the tower (often called the "two-level principle"). A non-comprehensive list of survey articles about this includes: [Nak97], [Pop22], and [LS94]. It seems then a good question to ask is:

What is the "Teichmüller tower"?

The geometric fundamental group of $\mathcal{M}_{g,n}$ is also the profinite completion of the *mapping class group* of a genus g surface with n marked points, $\Gamma_{g,n}$.[1] Using this fact, Hatcher et al. [HLS00] proposed a model for the Teichmüller tower as the collection of all profinite mapping class groups $\widehat{\Gamma}_{g,n}$ together with all homomorphisms $\widehat{\Gamma}_{h,m} \hookrightarrow \widehat{\Gamma}_{g,n}$ induced by the inclusion of "nice" subsurfaces. Here, Σ' is a nice subsurface of Σ, if Σ' is obtained by cutting along a set of disjoint simple closed curves on Σ. The natural embedding $\Sigma' \hookrightarrow \Sigma$ induces a map on the mapping class groups via the inclusion of Dehn twists (cf. Sect. 4.3).

Any good model for our ideal Teichmüller tower would, in particular, have the property that the Galois action on each of the $\widehat{\Gamma}_{g,n}$ necessarily commutes with all the maps in the tower. In practice, this is done by studying the actions of a family of more explicitly defined profinite groups on a proposed model of the tower. The most famous example is the Grothendieck-Teichmüller group, $\widehat{\mathsf{GT}}$, introduced by Drinfeld, we have described [Dri90, Section 4] and [Iha94]. There are several other interesting groups related to $\widehat{\mathsf{GT}}$, but in these lectures we will focus on the group $\widehat{\mathsf{NS}}$, introduced by Nakamura and Schneps, which is known to act on the tower which includes all the higher genus mapping class groups $\widehat{\Gamma}_{g,n}$ [NS00, Theorem 1.3]. The advantage in studying actions of $\widehat{\mathsf{GT}}$ and $\widehat{\mathsf{NS}}$ on

[1] In brief, a point in the moduli space $\mathcal{M}_{g,n}$ is an isomorphism class of Riemann surfaces and a loop (up to homotopy) is an orientation preserving diffeomorphism of the basepoint surface, up to those homotopic to the identity. Full details can be found in [Oda97].

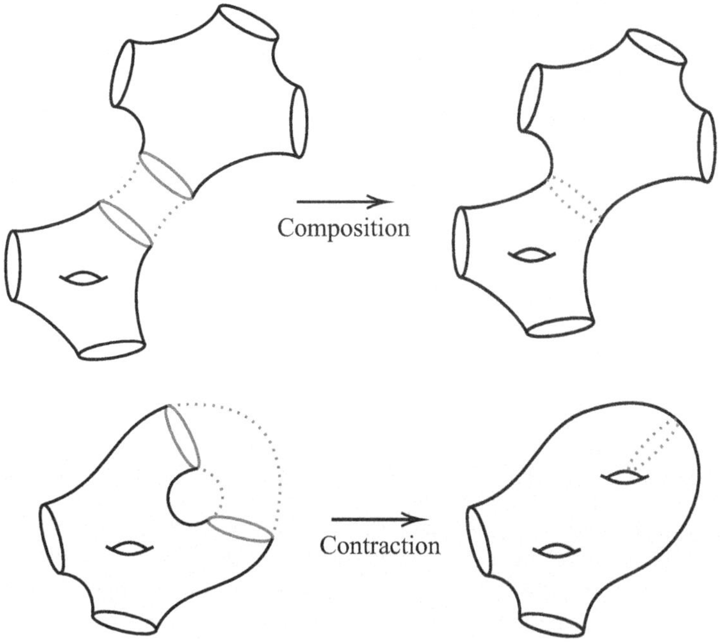

Fig. 1 Examples of the types of "gluings" parameterized by modular operads

the Teichmüller tower is that they have explicit presentations (Definition 4.17 and 4.19) and live "in between" the absolute Galois group and the mapping class groups in a very explicit way. In particular, Ihara [Iha94] showed that the image of the action of $\mathrm{Gal}(\mathbb{Q})$ on the geometric fundamental group $\pi_1^{et}(\mathcal{M}_{0,4})$ lies in $\widehat{\mathsf{GT}}$. Nakamura and Schneps show that their group $\widehat{\mathsf{NS}}$ is a subgroup of $\widehat{\mathsf{GT}}$ which still contains $\mathrm{Gal}(\mathbb{Q})$ [NS00, Theorem 1.2]. In other words,

$$\mathrm{Gal}(\mathbb{Q}) \hookrightarrow \widehat{\mathsf{NS}} \hookrightarrow \widehat{\mathsf{GT}}.$$

The main goal of this lecture series is to explain an "operadic" interpretation of the Teichmüller tower in Hatcher et al. [HLS00]. In this approach, we consider the mapping class groups of surfaces of genus g with n boundary components, Γ_n^g, rather than marked points, $\Gamma_{g,n}$. With this small change, we can assemble the collection of spaces $B\Gamma_n^g$ into a *modular operad* (Definition 2.4) with operations defined by gluing along the boundary components. In this case, the nice subsurface inclusions in the tower of Hatcher, Lochak and Schneps are equivalent to modular operad operations as in Fig. 1. Topologically, this is only a minor change, as there is a short exact sequence

$$0 \longrightarrow \mathbb{Z}^n \longrightarrow \Gamma_n^g \longrightarrow \Gamma_{g,n} \longrightarrow 0$$

where the map $\Gamma_n^g \to \Gamma_{g,n}$ collapses boundary components to points.

To argue that the modular operad $B\Gamma = \{B\Gamma_n^g\}$ models the Teichmüller tower, we start by replacing the mapping class groups Γ_n^g by homotopy equivalent groupoids $\mathbf{S}(g, n)$. The collection of spaces $B\mathbf{S}(g, n)$ assembles into a modular operad, which we call the *modular operad of seamed surfaces*. The genus zero part of this modular operad contains an operad, $B\mathcal{S}$, which we call the *genus zero surface operad* [BdBHR19, Definition 6.5]. Boavida, Horel and the second author show that there is a faithful action of Gal($\mathbb{Q}$) on the profinite completion of the genus zero surface operad, following from the fact that there is an action of $\widehat{\mathsf{GT}}$ on $\widehat{B\mathcal{S}}$ [BdBHR19, Proposition 10.6].

In our third lecture, we demonstrate that there is a relatively straightforward action of $\widehat{\mathsf{GT}}$ on each groupoid $\mathbf{S}(0, n)$, and this action is faithful when n is at least 3. The difficulty, which is beyond the scope of these lectures, is to show that this action of $\widehat{\mathsf{GT}}$, and thus the action of Gal($\mathbb{Q}$), is compatible with the operad structure. This compatibility is established in [BdBHR19] by showing:

Theorem *The profinite Grothendieck-Teichmüller group* $\widehat{\mathsf{GT}}$ *is isomorphic to the group of homotopy automorphisms of the profinite completion of the genus zero surface operad.*

In [BR], Bonatto and the second author show that the group $\widehat{\mathsf{NS}}$ acts on the profinite completion of the modular operad of seamed surfaces in such a way that, when restricting to genus zero, we recover the action of $\widehat{\mathsf{GT}}$. We can show quite directly that there is an action of $\widehat{\mathsf{NS}}$ on each groupoid $\mathbf{S}(g, n)$, but the difficulty is showing that these $\widehat{\mathsf{NS}}$-actions are compatible with the modular operad structure maps. This is overcome by showing that $\widehat{\mathsf{NS}}$ is isomorphic to the group of homotopy automorphisms of the profinite completion of modular operad of seamed surfaces. However, unlike in the genus zero case, we make use an operadic version of the "two-level principle":

Theorem *The group* $\widehat{\mathsf{NS}}$ *acts on the group of homotopy automorphisms of the genus one truncation of the profinite completion of the modular operad of seamed surfaces.*

In this context, a *genus one truncation* of $B\mathbf{S}$ is a modular operad $B\mathbf{S}_{\leq 1}$ in which $B\mathbf{S}(g, n) = \emptyset$ if $g \geq 2$. This translates into an operadic version of the two-level principle because one can show that there is a homotopy equivalence

$$\mathbb{R}\operatorname{End}(\widehat{B\mathbf{S}}) \simeq \mathbb{R}\operatorname{End}(\widehat{B\mathbf{S}}_{\leq 1}),$$

where $\mathbb{R}\operatorname{End}(-)$ denotes the derived endomorphism space of modular dendroidal spaces (Definition 3.1).

What our brief discussion on truncation illustrates is one of the many homotopical difficulties which arise due to profinite completion. In particular, the profinite completion of a (modular) operad in spaces is no longer a (modular) operad. In very special cases, however, the profinite completion of a (modular) operad can be considered as a (modular) operad whose operations hold "up to homotopy". In order to describe these "up to homotopy", or ∞, modular operads the second author, together with Hackney and Yau, developed a Segal model for modular ∞-operads which provides a good setting for working with the profinite completion of modular operads.

In the first two lectures of this series we introduce this homotopical background. In the first lecture we describe a category $\mathbf{U}$, whose objects are undirected, connected graphs with loose ends (Definition 2.7). Morphisms are given by 'blowing up' vertices of the source into "subgraphs" of the target in a way that reflects iterated operations in a modular operad (Definition 2.13). This graphical category models (discrete, coloured) modular operads in a very explicit way.

Theorem (Theorem 2.37) *There exists an equivalence of categories*

$$\mathbf{ModOp} \underset{N}{\overset{\longleftarrow}{\longrightarrow}} (\mathbf{Set}^{\mathbf{U}^{op}})_{Segal}.$$

The category on the right-hand side is a category of modular dendroidal sets (Definition 2.27) satisfying a strict Segal condition (Definition 2.31).

The second lecture builds on the first, weakening the Segal condition of Theorem 2.37 to provide a model for modular ∞-operads (Definition 3.16). To do this we consider space-valued $\mathbf{U}$-presheaves $X : \mathbf{U}^{op} \to \mathbf{sSet}$ and say that (Definition 3.6):

Definition A modular dendroidal space $X \in \mathbf{sSet}^{\mathbf{U}^{op}}$ is *Segal* if for all $G \in \mathbf{U}$, the Segal map

$$\mathbb{R}\,\mathrm{Map}(\mathbf{U}[G], X) \longrightarrow \mathbb{R}\,\mathrm{Map}(\mathbf{Sc}[G], X)$$

is a weak equivalence.

Informally, what this definition says is that a Segal modular operad is a space-valued $\mathbf{U}$-presheaf X in which the value of X at a graph G is determined, up to homotopy, by the value of X at each of the vertices of G. Finally, we describe profinite completion of modular operads and demonstrate why this operation results in a Segal modular operad.

1.1 Structure and Intentions of These Notes

These notes represent the transcript of three, 90 minute lectures and, as such, fall unfortunately short of being a complete survey and introduction to either Grothendieck-Teichmüller theory or (modular) ∞-operads. For a more comprehensive overview of Grothendieck-Teichmüller theory there is an excellent collection [SL97] edited by Lochak and Schneps and a rather recent survey article by Pop [Pop22]. The theory of modular ∞-operads is somewhat new, initiated in [HRY20a, HRY20b], and still in development, but there are several more comprehensive resources for an introduction to ∞-operads. We recommend the lecture series of Moerdijk [Moe10] as well as the book [HM22].

Throughout this series we take for granted that the reader will be familiar with the theory of operads. We give a definition of (coloured) modular operads in the first lecture but assume throughout that the reader, while maybe not familiar with all the details, is aware of the fact that there are adjunctions

$$\mathbf{Operad} \rightleftarrows \mathbf{Cyc} \rightleftarrows \mathbf{ModOp}.$$

where the right adjoint functors are "forgetful" functors. In addition, the open problems we give in Sect. 2.5 assume the reader is familiar with the other "operad-like" objects in the literature such as properads, dioperads and wheeled properads. These results are scattered throughout the literature, but a good first introduction is the survey article [Mar08].

We avoid delving too deep into the homotopical properties of Segal modular operads and suppress many arguments which require the use of Quillen model categories and/or ∞-categories. Our goal for these lectures is to give the reader some basic understanding of modular ∞-operads and their applications without having to understand the full theory. The only real homotopical prerequisite is that the reader have some understanding of *derived mapping spaces* or *homotopy function complexes* which we denote by $\mathbb{R}\,\mathrm{Map}(X, Y)$, throughout. A good standard reference is [Hir03, Chapter 9].

In the original lecture series we gave several exercises and a series of open problems to which we are reasonably confident there is an answer. Most of these open problems are in Sects. 2.5 and 3.3, though some are scattered throughout the notes.

2 Lecture 1: Graphs and Modular Operads

Modular operads are a generalization of operads which allow one to encode algebraic structures that come equipped with a "bilinear form" or "contraction operation". They were introduced by Getzler and Kapronov [GK95] where they gave the canonical example of a modular operad built from the (compactified) moduli spaces of genus g curves, $\bar{\mathcal{M}}_{g,n}$.

This first lecture introduces a very general definition of coloured modular operads in which we allow for an involutive set of colours (Definition 2.4). This differs slightly from the usual definition of a coloured modular operad in the literature ([HVZn10], [DM18], [KW17], etc) where the colour set has trivial involution. Interesting examples of operads with involutive colour sets include [DCH21] where the authors study coloured *cyclic operads* with involutive set of colours and the examples of modular operads coloured by involutive groupoids in [Pet13]. In the case of one-coloured, or monochrome, modular operads the involution on colour sets is trivial and thus all the reader's favorite modular operads are still examples of our definition. As we will discuss in the open problems section (Sect. 2.5), an advantage of considering involutive colour sets is that coloured cyclic operads, coloured operads, and coloured dioperads can all be considered as special types of coloured modular operads. Similarly, we can consider wheeled properads as a special case of modular operads with involutive colour sets.

The main theorem in this lecture is a so-called *nerve theorem* (Theorem 2.37). This is an extension of the classical theorem which says that the inclusion of the simplex category Δ into the category of small categories induces a fully faithful functor from **Cat** into the category of simplicial sets $\mathbf{sSet}^{\Delta^{op}}$ whose essential image consists of the Segal objects, i.e. those $X \in \mathbf{sSet}^{\Delta^{op}}$ with

$$X_n \cong X_1 \times_{X_0} X_1 \times_{X_0} \cdots \times_{X_0} X_1$$

for all $n \geq 2$.

2.1 Cyclic Operads

Modular operads are cyclic operads with contraction operations. As such, we first introduce the notion of a coloured cyclic operad:

Definition 2.1 Let $\mathfrak{C}$ be a non-empty set. A $\mathfrak{C}$-coloured *cyclic operad* **P** consists of:

(1) An involutive set $\mathfrak{C}$ which we call the set of colours and denote by col(**P**). We will write $c \mapsto c^{\dagger}$ for the action of the involution on an element $c \in \mathfrak{C}$.
(2) A collection of sets $\mathbf{P} = \{\mathbf{P}(c_1, \ldots, c_n)\}$, in which, for each $c_1, \ldots, c_n \in \mathfrak{C}$, the set $\mathbf{P}(c_1, \ldots, c_n)$ in equipped with a right Σ_n-action

$$\mathbf{P}(c_1, \ldots, c_n) \xrightarrow{\sigma^*} \mathbf{P}(c_{\sigma(1)}, \ldots c_{\sigma(n)}) \ .$$

(3) A set of distinguished unit elements $id_c \in \mathbf{P}(c^{\dagger}, c)$, one for each $c \in \mathfrak{C}$.
(4) A family of associative, unital and equivariant composition operations

$$\mathbf{P}(c_1, \ldots, c_n) \times \mathbf{P}(d_1, \ldots, d_m) \xrightarrow{\quad \circ_{ij} \quad} \mathbf{P}(c_1, \ldots, \hat{c}_i, \ldots, d_1, \ldots \hat{d}_j, \ldots, d_m$$

whenever $c_i = d_j^{\dagger}$, $(i, j) \in [1, n] \times [1, m]$.

A morphism of coloured cyclic operads $f : \mathbf{P} \to \mathbf{Q}$ consists of an involutive function $f : \mathrm{col}(\mathbf{P}) \to \mathrm{col}(\mathbf{Q})$ together with a family of Σ_n-equivariant maps

$$\mathbf{P}(c_1, \ldots, c_n) \xrightarrow{f_{\underline{c}}} \mathbf{Q}(f(c_1), \ldots, f(c_n))$$

for every list $c_1, \ldots, c_n \in \mathrm{col}(\mathbf{P})$ which commute with composition and identities. We will denote the category of all coloured cyclic operads by **Cyc**.

Remark 2.2

(1) In this first lecture we are only discussing discrete cyclic and modular operads, but one may extend the above to define coloured cyclic operads enriched in any closed, symmetric monoidal category $\mathbf{E} = (\mathbf{E}, \otimes, 1)$. In this case, we denote the category of cyclic operads by **Cyc**(**E**).
(2) In order to avoid introducing too much notation, we have opted not to include the full list of axioms for coloured cyclic operads here. A good reference for these axioms is Definition 2.3 of [DCH21]

Informally, operations of a cyclic operad can be pictured as *simply connected* graphs with n-free edges in the *boundary* whose vertices are *decorated* by elements of the underlying collection $\mathbf{P} = \{\mathbf{P}(c_1, \ldots, c_n)\}$.[2] Two examples of such decorated graphs are depicted in Fig. 2. The operation on the left is a decoration of a *star*–a simply connected graph with a single vertex v. The element $p_v \in \mathbf{P}(c_1, c_2, c_3, c_4)$. Similarly, the operation

[2] A precise definition of decoration can be found in [HRY20b, Definition 2.7].

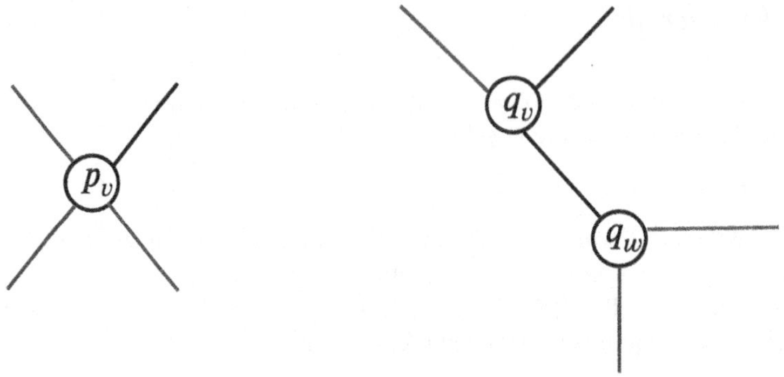

Fig. 2 Operations of a cyclic operad pictured as decorated graphs

on the right is depicting an operation obtained by the composition of two stars. In a moment we will introduce a specific way to label the loose ends of our graphs and make precise the graphical interpretation of when two ends can be composed, but for this informal discussion we just use colours. This informal depiction will be made precise via Theorem 2.37.

If we restrict to one-coloured, or monochrome, cyclic operads the involution is trivial, and as such the familiar examples of one-coloured cyclic operads are still objects in **Cyc**(**E**).

Example 2.3 An $*$-*autonomous category* C is a closed, symmetric monoidal category in which every object x has a dual $x^\dagger$, satisfying the property that $x \cong (x^\dagger)^\dagger$ [Bar79]. When the double dual relation is strict, $x = (x^\dagger)^\dagger$, $*$-autonomous categories are examples of coloured cyclic operads (Example 2.9 of [DCH21]). To our knowledge, it is an open problem to show that all $*$-autonomous categories are coloured cyclic operads.

2.2 Modular Operads

Definition 2.4 A $\mathfrak{C}$-coloured *modular operad* is a $\mathfrak{C}$-coloured cyclic operad which also has a family of contraction operations

$$\mathbf{P}(c_1, \ldots, c_n) \xrightarrow{\xi_{ij}} \mathbf{P}(c_1, \ldots, \hat{c}_i, \ldots, \hat{c}_j, \ldots c_n),$$

whenever $c_i = c_j^\dagger$, $0 \leq i < j \leq n$.

Remark 2.5 Modular operads, as we have defined them in this first lecture, are called "compact symmetric multicategories" in [JK09].

We require the composition and contraction operations to satisfy a series of commutativity, associativity, unitality and equivariance constraints. We do not list the full list of axioms here, as doing so requires one to be extremely careful with re-indexing and

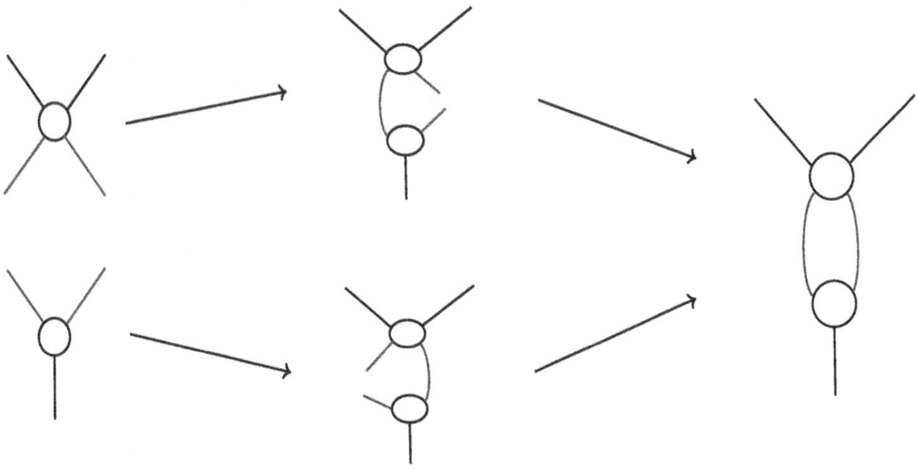

Fig. 3 An illustration of the compatibility of composition and contraction operations

requires the introduction of a significant amount of notion. We do not think this comes at much of a cost because, as we will show, the applications in these notes will often use the identification in Theorem 2.37. For the curious reader, many of the axioms needed are illustrated beautifully in Definition 1.24 of [Ray21] and we also recommend [JY, Section 13.4], where a complete set of axioms is given for coloured modular operads with non-involutive colour sets. To give one example, Fig. 3 illustrates compatibility of composition and contraction operations. The top of the diagram depicts first composing the two stars along the red edges and then contracting the two purple edges. The bottom of the diagram depicts first composing along the purple edges and then contracting the red edges.

A morphism of coloured modular operads $f : \mathbf{P} \to \mathbf{Q}$ consists of an involutive function $f : \mathrm{col}(\mathbf{P}) \to \mathrm{col}(\mathbf{Q})$ together with a family of Σ_n-equivariant maps

$$\mathbf{P}(c_1, \ldots, c_n) \xrightarrow{f_{\underline{c}}} \mathbf{Q}(f(c_1), \ldots, f(c_n))$$

which commute with composition, contractions, and identities. We denote the category of all modular operads by $\mathbf{ModOp}(\mathbf{E})$, suppressing the $\mathbf{E}$ when $\mathbf{E} = \mathbf{Set}$.

Returning to our informal description, operations of modular operads can be depicted as decorated connected graphs. In a modular operad operations might have more than one *internal edge*, such as those depicted in Fig. 4. In this interpretation, simply connected operations, such as those depicted on the right in Fig. 2 are obtained by cyclic operad composition. Contraction, depicted graphically, is obtained by identifying two boundary edges to form a new internal edge. These depictions will be given a precise meaning in terms of *graphical maps* shortly.

As we remarked earlier, when we restrict to one-coloured, or monochrome, modular operads the involution on colour sets is trivial and, as such, the known examples of one-coloured modular operads meet the requirements of our definition.

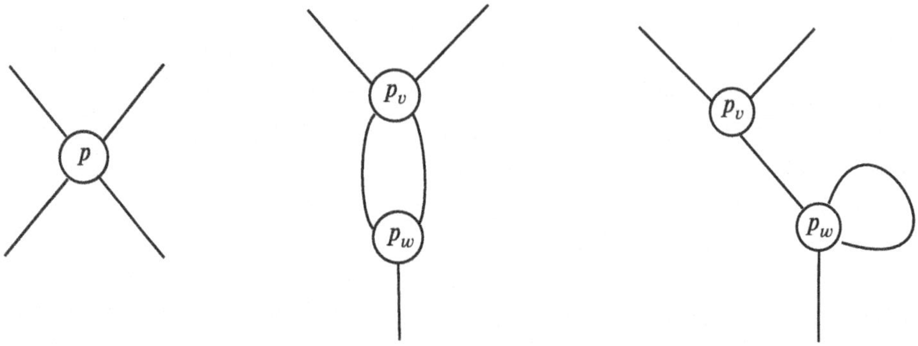

Fig. 4 A graphical interpretation of operations in a modular operad

Example 2.6 A *compact closed category* is a $*$-autonomous category in which the dualizing functor is monoidal [Kel72]. Raynor mentions in [Ray21, Example 1.27] that *involutive* compact closed categories are examples of modular operads. To our knowledge a full proof of this fact does not yet exist in the literature.

2.3 Graphs

We are now ready to introduce a category **U** whose objects are connected, undirected graphs.[3] The category **U** models modular operads in the sense that there is an equivalence of categories

$$\mathbf{ModOp} \cong (\mathbf{Set}^{\mathbf{U}^{op}})_{Segal},$$

where the right-hand side is a subcategory of **U**-presheaves which satisfy a *strict Segal condition*. This equivalence of categories justifies our depiction of cyclic and modular operad operations with decorated graphs.

In these lectures graphs are allowed to have "loose ends" meaning that it is not necessary for both ends (or either end) of an edge to touch a vertex. A typical example is in Fig. 5. To make this precise will use a combinatorial definition of graphs, *Feynman graphs*, due to Joyal and Kock [JK09]. This model for graphs has the advantage that it is extremely easy to write down and the drawback in that it does not fully capture all the graphs we need for defining modular operads (see Remark 2.9 below and Remark 1.1 in [HRY20a]). We note that there are several other combinatorial definitions of graphs in the literature, all of which can be shown to be equivalent to the definition we use here by combining Propositions 15.2, 15.6, and 15.8 of [BB17].

[3] The category is called **U** because "u" stands for undirected.

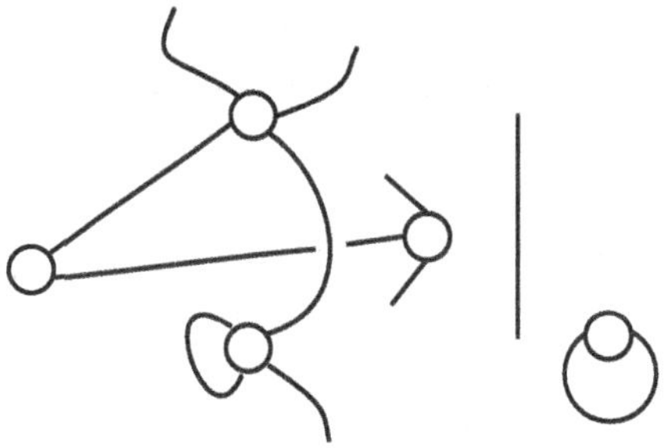

Fig. 5 An example of a graph

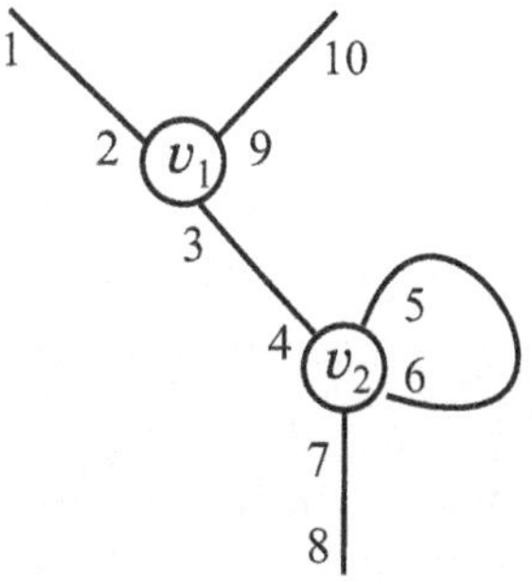

Fig. 6 An example of a graph with half edges labelled

Edges of our graphs are all comprised of two *distinct* "half edges" which you can picture as a copy of the interval (0, 1) equipped with a chosen orientation. We denote the set of half edges by A.[4] This set is equipped with a free involution i which identifies a pair of half edges with the opposite orientation; edges are the orbits of this involution. Half edges are attached to vertices via a partial function $t : A \to V$, where V denotes the set of vertices. Not all half edges will be attached to a vertex, so to distinguish the set of those edges which are adjacent to vertices, we write $D \subseteq A$ for those half edges which are in the *domain* of the function t.

Definition 2.7 A *graph G* consists of:

- a diagram of finite sets

$$i \circlearrowright A \xleftarrow{s} D \xrightarrow{t} V$$

  where

  - i is a fixed point-free involution and
  - s is a monomorphism.

In Fig. 6 the graph depicted has the set of arcs $A = \{1, 2, \ldots, 9, 10\}$ and vertices $V = \{v_1, v_2\}$. The arcs adjacent to the vertices are $D = \{2, 3, 9, 4, 5, 6, 7\}$ and the involution on A interchanges $2n$ and $2n - 1$ for $n = 1, 2, 3, 4, 5$.

[4] Half edges are also often called *arcs* in the literature, hence the use of the letter A for this set.

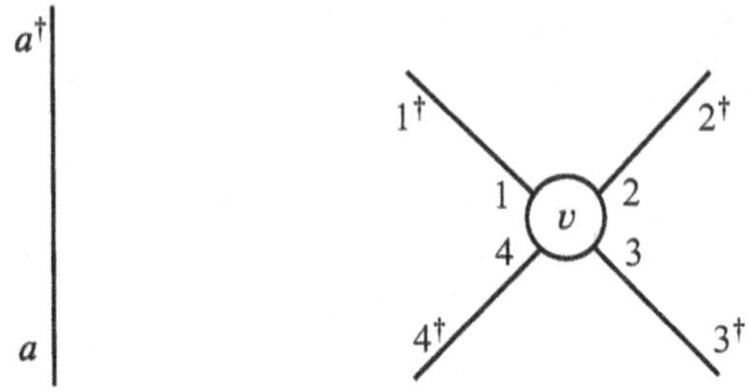

Fig. 7 The exceptional edge $\updownarrow$ and the 4-star $\star_4$

Fig. 8 A loop with 2 vertices and a nodeless loop

The involution on half-edges determines the *edges* of a graph as follows. If we write the action of the involution as $i : a \mapsto a^\dagger$, an *edge* of the graph G is an i-orbit $[a, a^\dagger]$. We write $E(G) = A/i$ for the set of edges of the graph G. An *internal edge* is an edge of the form $[b, b^\dagger]$ where both b and $b^\dagger$ are in D. In other words an internal edge is an edge $[b, b^\dagger]$ where both b and $b^\dagger$ are adjacent to vertices. In Fig. 6 the two internal edges are given by the orbits $e_1 = [3, 4]$ and $e_2 = [5, 6]$. The *boundary* of a graph is the set of half edges not adjacent to a vertex, $\partial(G) = A \setminus D$. The *neighbourhood* of a vertex $v \in V(G)$ consists of the half-edges which are adjacent to the vertex, $\text{nb}(v) = t^{-1}(v) \subseteq D$.

Example 2.8 There are several graphs which warrant special names.

(1) The exceptional *edge*, $G = \updownarrow$ has exactly two arcs $A = \{a, a^\dagger\}$ and $D = V = \emptyset$. The boundary of the edge $\partial(\updownarrow) = A$.
(2) For $n \geq 0$, we write $G = \star_n$ for the *n-star*. This graph has $A = \{1, 1^\dagger, \dots, n, n^\dagger\}$, $D = \{1, \dots, n\}$ and $V = \{v\}$. The boundary of $G = \star_n$ is the set $\{1^\dagger, \dots, n^\dagger\}$ and the neighbourhood of the vertex v is $\{1, \dots, n\}$. The 4-star is depicted in Fig. 7

Many of our graphs have empty boundary such as the loop with n-vertices depicted in Fig. 8.

Remark 2.9 The one graph that we *cannot* describe using our chosen formalism of graphs is the *nodeless loop*, depicted on the right of Fig. 8. As we explain in Remark 1.1 in [HRY20a], if one attempts to describe the nodeless loop in our chosen formalism, we end up with a graph which cannot be distinguished from the *edge*. The technical point will not play a further role in this lecture series, but it plays an essential role in defining the monad for modular operads in [HRY20b].

Every graph G has an associated star, $\star_G$, determined by its boundary. The graph $\star_G$ is the one-vertex graph with $A = \partial(G) \sqcup \partial(G)^\dagger$ and $D = \partial(G)^\dagger$. By definition, $\partial(\star_G) = A \setminus D = \partial(G)$ and the neighbourhood of the unique vertex is $D = \partial(G)^\dagger$. Similarly, every vertex of a graph G has an associated star, $\star_v$, with $V(G) = \{v\}$, $D = \text{nb}(v)$, and $A = \text{nb}(v) \sqcup \text{nb}(v)^\dagger$. The boundary of $\partial(\star_v) = \text{nb}(v)^\dagger$.

Exercise 2.10 Draw a graph G with $A = \{1, 2, 3, 4, 5, 6, 7, 8\}$, $D = \{1, 2, 3, 4, 5, 6, 7, 8\}$, $V = \{v_1, v_2, v_3, v_4\}$ and $i(n) = n - 1$ for $n = 2, 4, 6, 8$. For the graph G you have drawn, write down $\star_G$ and $\star_v$ for each $v \in V(G)$.

2.3.1 Morphisms of Graphs

Our definition of graphical morphisms is designed to capture the units, composition, and contraction operations of modular operads. The first definition of graphical map we give in this lecture series is not, necessarily, the best or most practical definition of graphical map. Instead, we have chosen to present the material in such a way as to motivate how one might arrive at this definition: start with the most "obvious" description of a map between graphs and then modify morphisms until we get to our ideal definition.

A graph G (Definition 2.7) is a diagram in the category of finite sets in the shape of

$$\mathcal{I} := \quad i \circlearrowright \bullet \xleftarrow{s} \bullet \xrightarrow{t} \bullet$$

where the arrow s is sent to a monomorphism and the generating endomorphism i is a free involution. A morphism between two graphs should preserve some structure of the graphs, e.g., a vertex with four adjacent edges should not map to a vertex with three adjacent edges. This leads to the first guess for a definition of graphical map: a graphical map should be maps in the functor category $\mathbf{FinSet}^{\mathcal{I}}$ which preserve the local structure of graphs.

Definition 2.11 Let G and G' be two connected graphs. A natural transformation $\phi : G \to G'$ is called an *embedding* if the right-hand square of:

$$\begin{array}{ccccc} i \circlearrowright A & \xleftarrow{s} & D & \xrightarrow{t} & V \\ \downarrow{\scriptstyle \phi_A} & & \downarrow{\scriptstyle \phi_D} & & \downarrow{\scriptstyle \phi_V} \\ i' \circlearrowright A' & \xleftarrow{s'} & D' & \xrightarrow{t'} & V' \end{array}$$

is a pullback *and* the map $V \to V'$ a monomorphism.

The pullback condition of Definition 2.11 makes sure that the local information i.e., the neighborhoods of vertices, is preserved.[5] The requirement that $V \to V'$ be a monomorphism *almost* implies that every graphical map is an inclusion of a *subgraph*. In particular, for every vertex v of a graph G there is a canonical embedding $\star_v \longrightarrow G$:

$$\begin{array}{ccccc} \mathrm{nb}(v) \sqcup \mathrm{nb}(v)^{\dagger} & \xleftarrow{s} & \mathrm{nb}(v) & \xrightarrow{t} & \{v\} \\ \downarrow & & \downarrow & & \downarrow \\ A & \xleftarrow{s'} & D & \xrightarrow{t'} & V. \end{array} \qquad (\bigstar)$$

[5] A natural transformation $\phi : G \to G'$ which only satisfies the pullback condition of Definition 2.11 is called ètale in [JK09]. Ètale maps play a key role in the description of "graphical species" in [JK09] and [Ray21], but are not quite what we want. See, for example, Remark 2.4 [Hac24] for more details on this fine point.

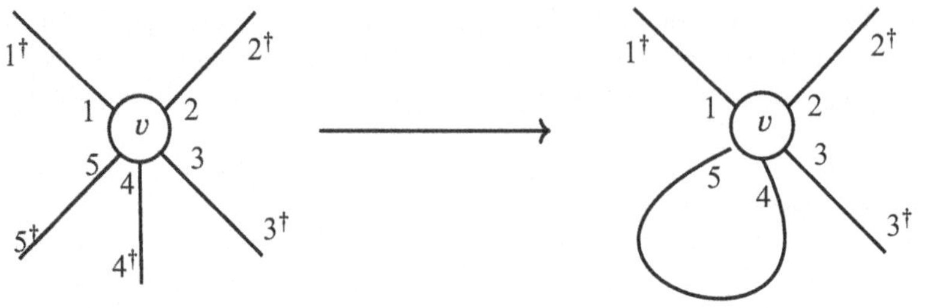

Fig. 9 Embedding a 5-star into a contracted 5-star

The left-hand map in this diagram is just the inclusion $\mathrm{nb}(v) \to D \to A$ on the first component, while the second component (which is forced by compatibility with the involution) sends $a^\dagger$ to ia. Similarly, every edge of a graph G corresponds to an embedding $\updownarrow \longrightarrow G$.

Exercise 2.12 Write down the explicit natural transformation for an edge inclusion $\updownarrow \longrightarrow G$.

Our definition of embedding, however, is more general than a subgraph inclusion, because an embedding is not necessarily injective on the set of half edges. For example, Fig. 9 depicts an embedding $\phi : \star_5 \longrightarrow G$,

$$\begin{array}{ccccc} \{1,2,3,4,5,1^\dagger,2^\dagger,3^\dagger,4^\dagger,5^\dagger\} & \xleftarrow{s} & \{1,2,3,4,5\} & \xrightarrow{t} & \{v\} \\ \downarrow & & \downarrow & & \downarrow \\ \{1,2,3,4,5,1^\dagger,2^\dagger,3^\dagger\} & \xleftarrow{s'} & \{1,2,3,4,5\} & \xrightarrow{t'} & \{v\}. \end{array}$$

The graph G in Fig. 9 is called a *contracted star*. Explicitly, G is the graph with one vertex v, set of half edges $A = \{1, 2, 3, 4, 5, 1^\dagger, 2^\dagger, 3^\dagger\}$ and $D = \{1, 2, 3, 4, 5\}$. The involution $i(n) = n^\dagger$, $n = 1, 2, 3$ and $i(4) = 5$ (and $i(5) = 4$).

Let $\mathrm{Emb}(G)$ denote the set of isomorphism classes of embeddings into G. We are now ready to define graphical maps:

Definition 2.13 A graphical map $\varphi : G \to G'$ consists of:

- a map of involutive sets $\varphi_0 : A_G \to A_{G'}$;
- a function $\varphi_1 : V_G \to \mathrm{Emb}(G')$ satisfying the following conditions:
  - The embeddings $\varphi_1(v)$ do not *overlap* at vertices—no vertex in G' is contained in two graphs $\varphi_1(v)$ and $\varphi_1(v')$;
  - For each v, we have a (necessarily unique) bijection making the diagram:

$$\begin{array}{ccc} \mathrm{nb}(v) & \xrightarrow{i} & A \\ {\scriptstyle\cong}\downarrow & & \downarrow{\scriptstyle\varphi_0} \\ \partial(\varphi_1(v)) & \longrightarrow & A' \end{array}$$

commute, where the top map i is the restriction of the involution on A.

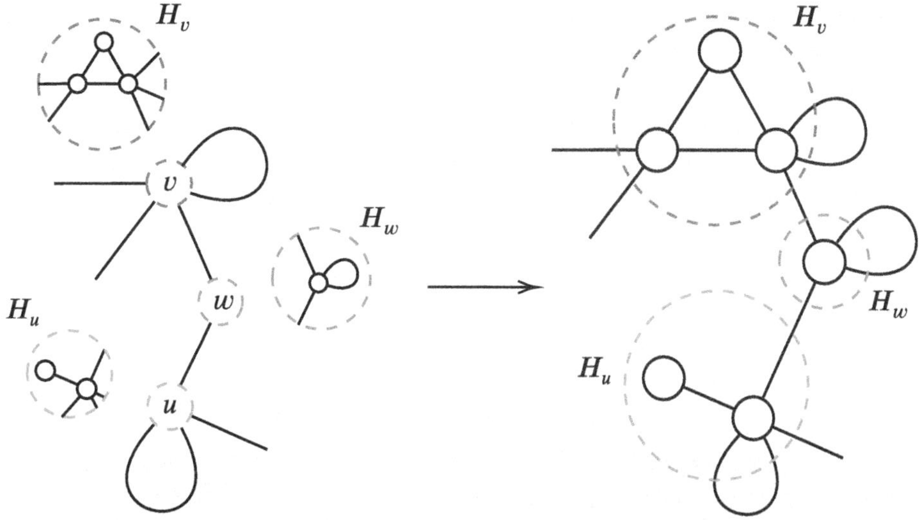

Fig. 10 An example of a graphical map

– If $\partial(G) = \emptyset$, then there exists a v in V so that $\varphi_1(v) \neq \updownarrow$.

The first two conditions on $\varphi_1 : V \to \mathrm{Emb}(G')$ imply that a map $\varphi : G \to G'$ is obtained by "blowing up", or replacing, the vertices of G with another graph H_v.[6] The requirements on the embedding

$$H_v \xhookrightarrow{\varphi_1(v)} G'$$

guarantee that we replace the vertex v by a graph H_v in such a way that we have an isomorphism $\star_{H_v} \cong \star_v$. The third condition is about avoiding the collapse into a *nodeless loop*. In Fig. 10, we have circled the subgraph $H_v = \varphi_1(v)$ in blue.

Exercise 2.14 Write down an explicit description of the graphical map depicted in Fig. 10.

Definition 2.15 The graphical category **U** is the category whose objects are connected Feynman graphs (Definition 2.7). The morphisms are the graphical maps from Definition 2.13.

As we mentioned at the start of this section, the definition of graphical map we have given is a bit cumbersome. Luckily, one can show that all graphical maps can be described (up to isomorphism) as the composite of three elementary classes of graphical maps: inner coface maps, outer coface maps and codegeneracies (Theorem 2.7 [HRY20a]). We have already seen examples of outer coface maps: outer coface maps are embeddings.

[6] The notion of "blowing up" a vertex can be made precise using the language of *graph substitution* which is described for Feynman graphs in Construction 1.18 [HRY20a].

Definition 2.16 An *outer coface map* is either an *embedding* $d_e : G \to G'$ in which G' has precisely *one* more internal edge than G or an *edge inclusion* $\updownarrow \to \star_n$.

Example 2.17 Below we have depicted the outer coface map $d_e : G \to G'$ where G' has the additional inner edge $e = [a, b]$ (highlighted in red). On half edges the graphical map is $(d_e)_0(a^\dagger) = b$, $(d_e)_0(b^\dagger) = a$ and is the identity elsewhere. An outer coface map will not change any of the vertices of G, which is explicitly written as $(d_e)_1(v) = \star_v$ and $(d_e)_1(w) = \star_w$.

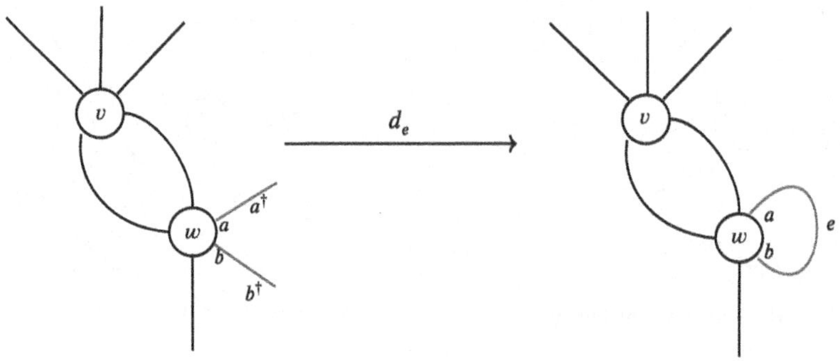

Definition 2.18 An *inner coface map* $d_v : G \to G'$ is a graphical map defined by blowing-up a single vertex v in G by a graph which has precisely *one* internal edge.

Example 2.19 The picture below depicts two possible inner coface maps defined at the vertex v.

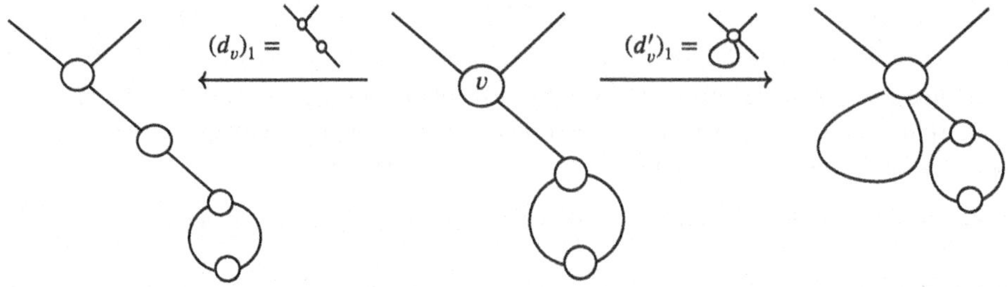

Codegeneracy maps "delete" arity 2 vertices (a vertex v with $|\operatorname{nb}(v)| = 2$).

Definition 2.20 A *codegeneracy map* $s_v : G \to G'$ is a graphical map defined by "blowing-up" a vertex v in G by $\updownarrow$.

Example 2.21 In the graphical map below, the vertex v has arity two. The codegeneracy map $s_v : G \to G'$ is the identity on half edges and $(s_w)_1 = \star_w$ at all vertices except $w = v$ where $(s_v)_1 = \updownarrow$.

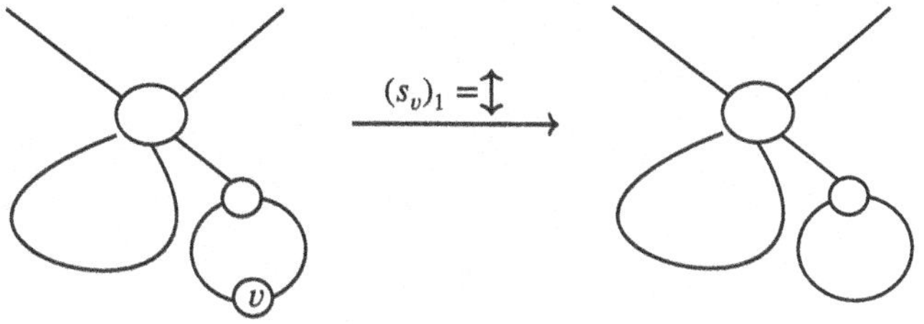

Remark 2.22 In practice, when working with graphical maps it is often enough to study these elementary maps (and isomorphisms). We also note that there is an equivalent, purely combinatorial, definition of graphical maps which does not require reference to graph substitution (see Theorem 1 [Hac24]).

2.4 *Modular Dendroidal Sets and the Nerve Theorem*

As we have (hopefully) motivated with pictures, the graphical category **U** is closely related to modular operads. In particular, every object of **U** freely generates a modular operad (Definition 2.17 [HRY20b]).

Definition 2.23 The modular operad $\langle G\rangle$ generated by a graph G is the free modular operad whose:

- set of colours is the set of half edges A;
- a collection of Σ_n-sets is $E(a_1,\ldots,a_n) = \begin{cases}\{v\} \text{ if } (a_1,\ldots,a_n) = \partial(\star_v)\\ \emptyset \text{ otherwise.}\end{cases}$
- $\langle G\rangle = F(E)$

Example 2.24 To describe the modular operad $\langle G\rangle$ generated by the graph G in Fig. 11 we recall that $\partial(\star_v) = (a_1^\dagger, a_2^\dagger, c_2, c_1)$ and $\partial(\star_w) = (b_1^\dagger, b_2^\dagger, b_3^\dagger, c_4, c_3)$. The underlying collection of $\langle G\rangle$ consists of two one-point sets:

$$E(a_1^\dagger, a_2^\dagger, c_2, c_1) = \{v\},\ E(b_1^\dagger, b_2^\dagger, b_3^\dagger, c_4, c_3) = \{w\}.$$

The graph G provides gluing instructions.

Proposition 2.25 (Proposition 2.25 [HRY20b]) *The assignment $G \mapsto \langle G\rangle$ defines a faithful functor* **U** $\to$ **ModOp** *which is injective on isomorphism classes of objects.*

Exercise 2.26 Write $\mathbf{ModOp}_{\mathfrak{C}}$ for the category of modular operads with $\mathfrak{C}$-colours. Show that $\mathbf{ModOp}_{\emptyset}$ is equivalent to the category of sets and that $\langle \star_0\rangle$ is a generator. Here $G = \star_0$ is the modular operad freely generated by an isolated vertex. See Example 2.20 [HRY20b] for a hint.

We note that the functor

$$J : \mathbf{U} \longrightarrow \mathbf{ModOp}$$

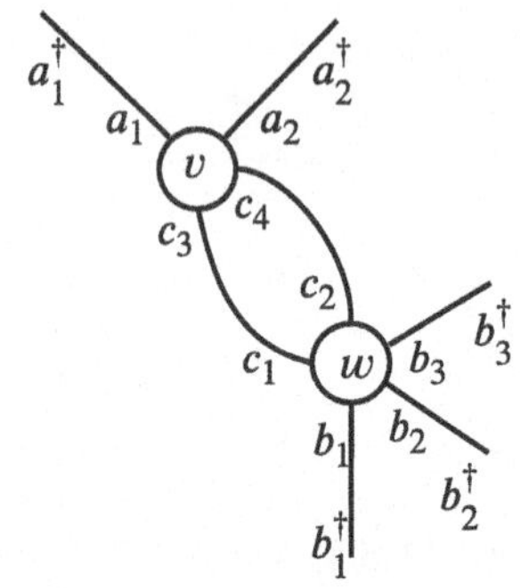

Fig. 11 Two examples of graphs

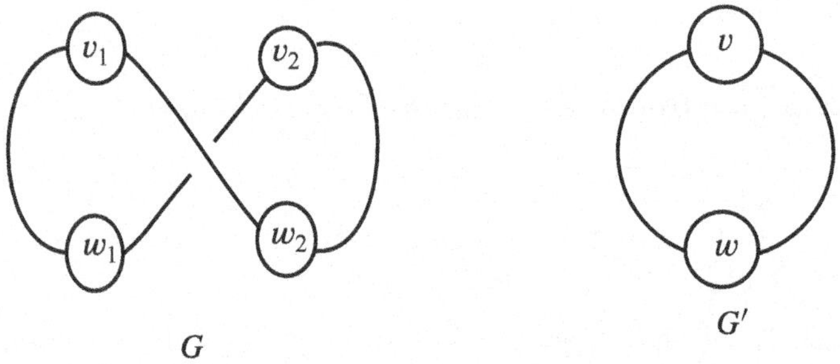

Fig. 12 Example: The embedding of **U** into **Mod** is not full

is not full. To see this one can consider the graphs G and G' in Fig. 12. There is a map of modular operads from $\langle G \rangle$ to $\langle G' \rangle$ which sends each v_i to v and each w_j to w but there is no graphical map $G \to G'$ which has this behavior.

2.4.1 Modular Dendroidal Sets

Using the functor $\mathbf{U} \to \mathbf{ModOp}$ we can define the *nerve* of a modular operad:

$$\mathbf{ModOp} \xrightarrow{N} \mathbf{Set}^{\mathbf{U}^{op}}.$$

The category of presheaves $\mathbf{Set}^{\mathbf{U}^{op}}$ is called the category of *modular dendroidal sets*.[7] Our goal for the remainder of this first lecture is to define a subcategory of modular dendroidal sets which satisfy a strict Segal condition.

[7] In [HRY20a], [HRY20b], we avoid naming the category of set-valued **U**-presheaves largely because we used the term "graphical sets" in [HRY15] when modeling ∞-properads. The name modular dendroidal sets, suggested to us by Ieke Moerdijk, follows the convention in the literature by using the term "modular" or "cyclic" modify the term "operad".

Definition 2.27 The category of *modular dendroidal sets* $\mathbf{Set}^{\mathbf{U}^{op}}$ is the category whose objects are functors $X : \mathbf{U}^{op} \to \mathbf{Set}$. Morphisms in $\mathbf{Set}^{\mathbf{U}^{op}}$ are natural transformations.

In these notes we follow some notation conventions from the theory of dendroidal sets [Moe10], [MW09], and [MW07]. Given an $X \in \mathbf{Set}^{\mathbf{U}^{op}}$ we will write X_G for the evaluation of the presheaf X at a graph $G \in \mathbf{U}$. Similarly, for every morphism $\varphi : G \to G'$ in $\mathbf{U}$, there's map $\varphi^* : X_{G'} \to X_G$ in $\mathbf{Set}^{\mathbf{U}^{op}}$.

There are several examples of modular dendroidal sets which warrant special notation. For any graph G, the *representable* presheaf

$$\mathbf{U}[G] := \mathbf{U}(-; G)$$

is given by

$$\mathbf{U}[G]_H := \mathbf{U}(H, G),$$

where H ranges over all graphs $H \in \mathbf{U}$. We think of an element $x \in X_G$ as a *decoration* of shape G, similar to those we depicted in Fig. 4. The Yoneda Lemma tells us that "the set of all decorations of G" can be identified with the set of maps out of the representable $\mathbf{U}[G]$ as there is a bijection

$$X_G = \mathbf{Set}^{\mathbf{U}^{op}}(\mathbf{U}[G], X).$$

For any graph G with at least two vertices, we define a presheaf X_G^1 which captures the "local decoration data" of G. If G is a graph with at least two vertices, each internal edge between vertices v and w produces a diagram of embeddings:

$$\begin{array}{ccccc} \star_v & \longleftarrow & \updownarrow & \longrightarrow & \star_w \\ & \searrow & \downarrow & \swarrow & \\ & & G & & \end{array}$$

in $\mathbf{U}$. The maps in this diagram consist of edge and star inclusions from ($\bigstar$).

Definition 2.28 Let X be a modular dendroidal set and G be a graph with at least 2 vertices. The *corolla ribbon* or *spine* of X at G is defined by:

$$X_G^1 = \lim_{\star_v \leftarrow \updownarrow \rightarrow \star_w} \left(\begin{array}{ccc} X_{\star_v} & & X_{\star_w} \\ & \searrow \quad \swarrow & \\ & X_{\updownarrow}. & \end{array} \right)$$

Here, the limit ranges over all edge inclusions and stars of G.

Example 2.29 In the graph G in Fig. 13 the internal edges $e_1 = [c_3, c_1]$, $e_2 = [c_4, c_2]$ and $e_3 = [b_2, b_3]$ are highlighted in red. To simplify the edge inclusion diagrams we will

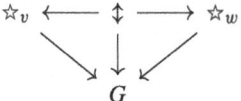

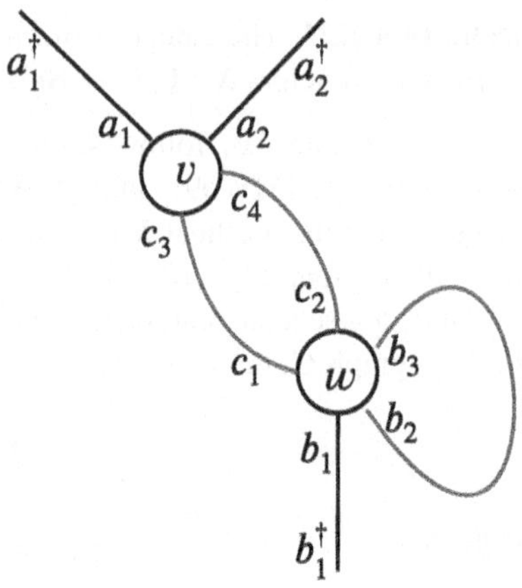

Fig. 13 The edges comprised of $\{c_4, c_2, c_3, c_4, b_3, b_2\}$ form the internal edges of the graph

write v for $\star_v$ and w for $\star_w$:

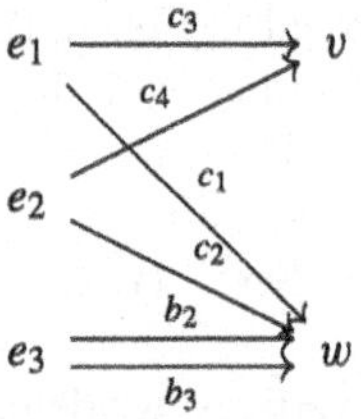

The presheaf X_G^1 is therefore the limit over the diagram:

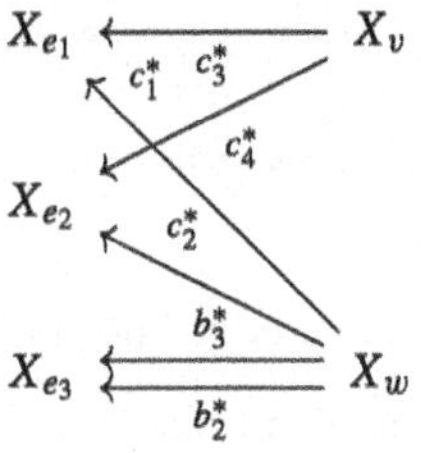

Exercise 2.30 In the case when $X_\updownarrow = *$, show that $X_G^1 = \prod_{v \in V(G)} X_{\star_v}$.

Definition 2.31 For a graph G with at least two vertices the Segal map is the map:

$$X_G \longrightarrow X_G^1 \subseteq \prod_{v \in V(G)} X_{\star_v}$$

induced by the embeddings $\star_v \hookrightarrow G$.

The intuition is that the Segal map says compares the "decorations of the graph G" with the "decorations at each vertex".

Definition 2.32 A modular dendroidal set $X \in \mathbf{Set}^{\mathbf{U}^{op}}$ is strictly *Segal* if the Segal map is a bijection for each G in $\mathbf{U}$.

Modular operads can be identified with the (strictly) Segal modular dendroidal sets via the following construction.

Definition 2.33 Let $\mathbf{P}$ be a discrete modular operad and let G be any graph in $\mathbf{U}$. The modular *nerve* functor

$$N : \mathbf{ModOp} \longrightarrow \mathbf{Set}^{\mathbf{U}^{op}}$$

is defined by

$$N\mathbf{P}_G = \mathbf{ModOp}(\langle G\rangle, \mathbf{P}).$$

Exercise 2.34 Given a graph $G \in \mathbf{U}$, we now have two ways to assign an object in $\mathbf{Set}^{\mathbf{U}^{op}}$ to G: we can take the representable presheaf $\mathbf{U}[G]$ or we can take the nerve of the free modular operad $\langle G\rangle$, $N\langle G\rangle$. The representable $\mathbf{U}[G]$ is a sub-object of $N\langle G\rangle$ (since $J : \mathbf{U} \to \mathbf{ModOp}$ is faithful) but they nearly never coincide.

(1) Let G be the loop with one node and show $\mathbf{U}[G] \subset N\langle G\rangle$.
(2) Show that we have $\mathbf{U}[\star_0] = N\langle\star_0\rangle$.

For any graph G, we picture the set $N\mathbf{P}_G$ as the set of $\mathbf{P}$-decorations of the graph G (Fig. 4). In particular, if $\mathbf{P}$ is a $\mathfrak{C}$-coloured modular operad, then the set

$$N\mathbf{P}_{\updownarrow} := \mathbf{ModOp}(\langle\updownarrow\rangle, \mathbf{P}) = \mathfrak{C}.$$

Exercise 2.35 For any n, check that the set

$$N\mathbf{P}_{\star_n} := \mathbf{ModOp}(\langle\star_n\rangle, \mathbf{P}) = \coprod_{(c_1,\dots,c_n)\in\mathfrak{C}^n} \mathbf{P}(c_1, \dots c_n).$$

Note that the symmetric group actions on $\mathbf{P}(c_1, \dots c_n)$ are captured by the isomorphisms of the graph $\star_n$.

The following theorem says that the strictly Segal modular dendroidal sets are precisely those which live in the essential image of the nerve functor. We will not give the full proof here, but to provide some intuition for the idea, consider that given a modular operad $\mathbf{P}$, the fact that N is a functor means that we have morphisms such as

$$N\mathbf{P}_{\star_{n_1}} \times N\mathbf{P}_{\star_{n_2}} := \mathbf{P}(c_1, \dots c_{n_1}) \times \mathbf{P}(c_1, \dots c_{n_2}) \xrightarrow{\circ_{ij}} \mathbf{P}(c_1, \dots, \hat{c}_i, \dots, \hat{c}_j, \dots c_{n_1+n_2}) =: N\mathbf{P}_{\star_G}$$

given by composition operations. Here, the graph G is the graph with a single internal edge given by $e = [c_i, c_j^\dagger]$. Similar maps are given for contraction operations. The key to a proof of Theorem 2.37, is then showing that compositions of contraction operations of $\mathbf{P}$

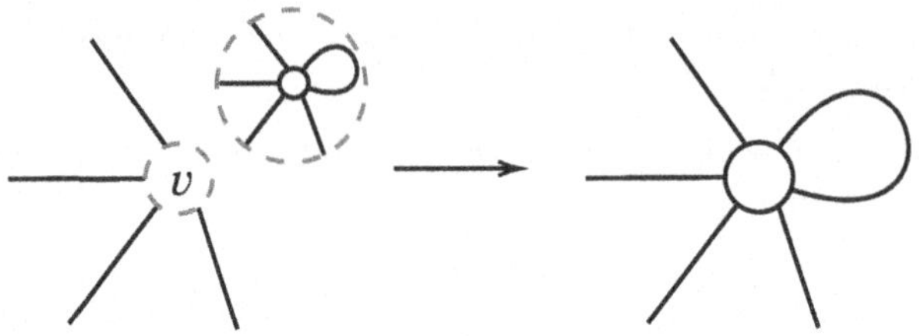

Fig. 14 An example of graph substitution

are precisely the composite with the dashed arrow in

$$\prod_{v\in V(G)} N\mathbf{P}_{\star_v} \overset{Segal}{\underset{\dashrightarrow}{\longleftarrow}} N\mathbf{P}_G \longrightarrow N\mathbf{P}_{\star_G}$$

with the natural maps $N\mathbf{P}_G \to N\mathbf{P}_{\star_G}$ induced by the graphical maps $\star_G \to G$.

Example 2.36 The graphical map $\star_{n-2} \to G$ obtained by substituting G for the vertex v is depicted in Fig. 14 induces the map $N\mathbf{P}_G \to N\mathbf{P}_{\star_{n-2}}$ in the diagram:

$$\begin{array}{ccc} N\mathbf{P}_{\star_n} & \overset{\xi}{\dashrightarrow} & N\mathbf{P}_{\star_{n-2}} \\ \downarrow & \nearrow & \\ N\mathbf{P}_G. & & \end{array}$$

The dashed map is precisely the application of the nerve functor to the contraction operation in $\mathbf{P}$.

Theorem 2.37 ([HRY20b, Theorem 3.6]) *The nerve functor*

$$N : \mathbf{ModOp} \longrightarrow \mathbf{Set}^{\mathbf{U}^{op}}$$

is fully faithful. Moreover, for any $X \in \mathbf{Set}^{\mathbf{U}^{op}}$*, the following statements are equivalent:*

(1) There exists a modular operad $\mathbf{P}$ *and an isomorphism* $X \cong N\mathbf{P}$.
(2) X *satisfies the strict Segal condition.*

Remark 2.38 There are several related constructions and results in the literature. In [Ray18, Ray21], Raynor presents a larger graphical category (the category of graphical species) together with a monad for modular operads which allows her to avoid issues with the nodeless loop in Remark 2.9. This has the formal advantage that her graphical category embeds fully in the category of modular operads, but the practical disadvantage that the resulting corresponding construction of modular ∞-operads [Ray21, Section 8.4] is somewhat opaque.

We also note that our definition of Segal objects (eg: Definition 2.31 and Definition 3.6) have been generalized by several authors. In particular, the definition of Segal objects in

Example 3.11 of Chu and Haugseng [CH21] and Berger's unital hypermoment categories [Ber22] agrees with the one we have given here.

2.5 Further Directions

In the next lecture we will describe how weakening the Segal condition gives us a model for cyclic and modular ∞-operads. At the time of these lectures, there are many open questions one would want to see answered before we can say that we have a comprehensive understanding of what a modular ∞-operad should be. The following are a few open problems.

In Definitions 2.16 and 2.18, we defined the notion of (inner and outer) coface maps of $\mathbf{U}$. Given a coface map δ with codomain G, one can define the **horn** $\Lambda^{\delta}[G]$ which is a sub-object of the representable object $\mathbf{U}[G]$. A *strict inner Kan* graphical set is a presheaf $X \in \mathbf{Sets}^{\mathbf{U}^{op}}$ such that every diagram

$$\begin{array}{ccc} \Lambda^{\delta}[G] & \longrightarrow & X \\ \downarrow & \nearrow & \\ \mathbf{U}[G] & & \end{array}$$

with δ an inner coface map admits a unique filler. Michelle Strumila shows in her PhD thesis that:

Theorem 2.39 (Strumila) *The nerve functor*

$$N : \mathbf{ModOp} \longrightarrow \mathbf{Set}^{\mathbf{U}^{op}}$$

is fully faithful. Moreover, the following statements are equivalent for $X \in \mathbf{Set}^{\mathbf{U}^{op}}$.

(1) There exists a modular operad $\mathbf{P}$ and an isomorphism $X \cong N\mathbf{P}$.
(2) X satisfies the strict Segal condition.
(3) X is strict inner Kan.

If one relaxes the inner Kan condition you arrive at a model for *quasi-modular operads*.

Open Problem 2.40 Following the example of dendroidal sets [CM11] find a model category structure in which the weak inner Kan graphical sets are the fibrant objects.

The involution on colour sets in Definition 2.4 allows us to consider wheeled properads as a subcategory of **ModOp**. In [Hac24], Hackney makes this explicit at the level of graphical categories, identifying a slice category $\mathbf{U}/_{o}$ with the graphical category for wheeled properads defined in [HRY18] and [HRY15]. In particular, he shows that the adjunction

$$\mathbf{WPrd} \rightleftarrows \mathbf{ModOp}$$

can be well understood via graphical presheaves. Similar adjunctions between modular operads, cyclic operads, and operads can all be described via adjunctions of graphical categories.

Open Problem 2.41 Assuming a solution to Problem 2.40, use the adjunctions of graphical categories described in [Hac24] to define a model category structure on the category of graphical sets from [HRY15] and [HRY18] in which the quasi-wheeled properads are the fibrant objects.

3 Lecture 2: A Weak Segal Model for Modular ∞-Operads

For the remainder of this lecture series we will simplify Definition 2.4 and focus on the *one-coloured* modular operads. Just to refresh our memory, a symmetric sequence $\mathbf{P} = \{\mathbf{P}(n)\}$ consists of a sequence of sets $\mathbf{P}(n)$ each of which is equipped with a right Σ_n-action.[8] A *modular operad* $\mathbf{P}$ consists of a symmetric sequence $\mathbf{P} = \{\mathbf{P}(n)\}$ together with:

(1) A distinguished *unit* element $1 \in \mathbf{P}(1)$;
(2) A family of *compositions*

$$\mathbf{P}(n) \times \mathbf{P}(m) \xrightarrow{\circ_{ij}} \mathbf{P}(n+m-2);$$

(3) A family of *contraction* operations

$$\mathbf{P}(n) \xrightarrow{\xi_{ij}} \mathbf{P}(n-2).$$

Moreover, we require the compositions and contractions satisfy a series of axioms (eg: [BM23, Definition A1]) assuring that compositions are associative, unital and equivariant, contractions are associative and equivariant and that the two operations are compatible.

Theorem 2.37 tells us is that, given a discrete modular operad $\mathbf{P}$, we can construct a set-valued presheaf $N\mathbf{P} \in (\mathbf{Set}^{\mathbf{U}^{op}})_{Segal}$ where

$$N\mathbf{P}_{\star_n} = \mathbf{P}(n)$$

in which the Segal maps

$$N\mathbf{P}_G \longrightarrow \prod_{v \in V(G)} N\mathbf{P}_{\star_v}$$

are bijections. In other words, modular operad compositions and contractions of $\mathbf{P}$ are modeled by graphical maps

[8] We follow the convention of calling a one-coloured collection a symmetric sequence.

$$\prod_{v \in V(G)} N\mathbf{P}_{\star_v} \underset{Segal}{\overset{\cdots\cdots\rightarrow}{\longleftarrow}} N\mathbf{P}_G \longrightarrow N\mathbf{P}_{\star_G}.$$

If our goal is to now model modular ∞-operads, i.e. modular operads where operations are defined "up to coherent homotopy". This means we will want to replace our Segal map with a homotopy equivalence. In this second lecture we will introduce space-valued presheaves $\mathbf{sSet}^{\mathbf{U}^{op}}$ and describe a corresponding notion of a *weak* Segal condition on modular dendroidal spaces. At the end of this lecture, we include some brief notes about variations on the graphical category $\mathbf{U}$ which can give genus graded modular ∞-operads, cyclic ∞-operads, etc.

3.1 Modular Dendroidal Spaces

Throughout, we write **sSet** for category of simplicial sets equipped with the standard Kan-Quillen model structure. We often abuse terminology and refer to simplicial sets as "spaces".

Definition 3.1 The category of *modular dendroidal spaces* is the category of space-valued $\mathbf{U}$-presheaves denoted by $\mathbf{sSet}^{\mathbf{U}^{op}}$.

As in the previous lecture, for any $X \in \mathbf{sSet}^{\mathbf{U}^{op}}$ we write X_G for the evaluation of X at a graph $G \in \mathbf{U}$. We consider the representable presheaf in $\mathbf{Set}^{\mathbf{U}^{op}}$:

$$\mathbf{U}[G]_H := \mathbf{U}(H, G)$$

as an object in $\mathbf{sSet}^{\mathbf{U}^{op}}$ via the inclusion $\mathbf{Set}^{\mathbf{U}^{op}} \hookrightarrow \mathbf{sSet}^{\mathbf{U}^{op}}$.

Exercise 3.2 The Yoneda Lemma says that a map $x : \mathbf{U}[G] \to X$ in $\mathbf{Set}^{\mathbf{U}^{op}}$ is equivalent to an element $x \in X_G$. Show that every $X \in \mathbf{Set}^{\mathbf{U}^{op}}$ is, up to isomorphism, a colimit of representables

$$X \cong \operatorname{colim} \mathbf{U}[G]$$

where the colimit is indexed by the maps $\mathbf{U}[G] \to X$.

3.1.1 Segal Cores

In the previous lecture, we introduced the Segal maps via a limit construction (Definition 2.28). To describe the weak Segal maps, we will use a dual construction called the *Segal core*.

To understand the Segal core construction, it can be useful to revisit the definition of a graph. Recall that a graph G as a diagram in the category of finite sets in the shape of

$$\mathcal{I} := \quad i \circlearrowright \bullet \xleftarrow{s} \bullet \xrightarrow{t} \bullet$$

where the arrow s is sent to a monomorphism and the generating endomorphism i is a free involution. If a graph G has at least one vertex, we can choose an orientation for each internal edge, and present G as a coequalizer in $\mathbf{FinSet}^{\mathcal{I}}$:[9]

$$\coprod_{e\in \mathrm{iE}} \updownarrow \rightrightarrows \coprod_{v\in V} \star_v \longrightarrow G.$$

where iE represents the internal edges of G.

Exercise 3.3 Write the graph G from Fig. 13 as a coequalizer:

$$\{e_1, e_2, e_3\} \rightrightarrows \star_v \cup \star_w \longrightarrow G.$$

Definition 3.4 The *Segal core* of a graph G is the coequalizer in $\mathbf{Set}^{\mathbf{U}^{op}}$:

$$\coprod_{e\in \mathrm{iE}} \mathbf{U}[\updownarrow] \rightrightarrows \coprod_{v\in V} \mathbf{U}[\star_v] \longrightarrow \mathbf{Sc}[G]$$

in $\mathbf{Set}^{\mathbf{U}^{op}}$.

The Segal core comes with a natural map $\mathbf{Sc}[G] \to \mathbf{U}[G]$ induced by the embeddings $\star_v \to G$. In the case that $G = \updownarrow$ we declare that the map $\mathbf{Sc}[G] \to \mathbf{U}[G]$ to be the identity map on $\mathbf{U}[G]$.

Exercise 3.5 Check that the Segal core is precisely the colimit so that

$$\mathbf{Set}^{\mathbf{U}^{op}}(\mathbf{Sc}[G], X) = X^1_G = \lim_{\star_v \leftarrow \updownarrow \rightarrow \star_w} \left(\begin{array}{ccc} X_{\star_v} & & X_{\star_w} \\ & \searrow \quad \swarrow & \\ & X_{\updownarrow} & \end{array} \right).$$

As we saw in Exercise 2.30, whenever $X_{\updownarrow} = \mathbf{Set}^{\mathbf{U}^{op}}(\mathbf{U}[\updownarrow], X) = *$, we can identify

$$\mathbf{Set}^{\mathbf{U}^{op}}(\mathbf{Sc}[G], X) = \prod_{v\in V(G)} X_{\star_v}.$$

This implies that the Segal map from Definition 2.31 is equivalent to the map of sets

$$X_G = \mathbf{Set}^{\mathbf{U}^{op}}(\mathbf{U}[G], X) \longrightarrow \mathbf{Set}^{\mathbf{U}^{op}}(\mathbf{Sc}[G], X) = \prod_{v\in V(G)} X_{\star_v}.$$

As with the representable presheaves $\mathbf{U}[G]$, we consider the Segal core $\mathbf{Sc}[G]$ as an object in $\mathbf{sSet}^{\mathbf{U}^{op}}$ via the inclusion $\mathbf{Set}^{\mathbf{U}^{op}} \hookrightarrow \mathbf{sSet}^{\mathbf{U}^{op}}$. This leads to the following definition:

[9] Note this is *not* a coequalizer in $\mathbf{U}$ as these objects don't exist in $\mathbf{U}$ (they are not connected).

Definition 3.6 A modular dendroidal space $X \in \mathbf{sSet}^{\mathbf{U}^{op}}$ is *Segal* if:

- $X_{\updownarrow} = *$;
- for all $G \in \mathbf{U}$, the Segal map

$$\mathbb{R}\,\mathrm{Map}(\mathbf{U}[G], X) \longrightarrow \mathbb{R}\,\mathrm{Map}(\mathbf{Sc}[G], X)$$

  is a weak equivalence.

Here $\mathbb{R}\,\mathrm{Map}(X, Y)$ is the derived mapping space, which is well-defined as long as we can equip the category $\mathbf{sSet}^{\mathbf{U}^{op}}$ with a class of weak equivalences. The category $\mathbf{sSet}^{\mathbf{U}^{op}}$ admits several model category structures including the *projective model structure* and a *Reedy model structure*. In either case, weak equivalences are defined entrywise. In other words, $f : X \to Y$ in $\mathbf{sSet}^{\mathbf{U}^{op}}$ is a weak equivalence if $f^* : X_G \to Y_G$ is a weak equivalence of $\mathbf{sSet}$ for every $G \in \mathbf{U}$.

Remark 3.7 The assumption that $X_{\updownarrow} = *$ is not required for Definition 3.6 but is required in Theorem 3.17. We have included the assumption here for consistency throughout these notes.

Definition 3.6 is a perfectly fine definition for modular dendroidal Segal spaces. In our intended applications, however, we will want to demonstrate that a specific presheaf $X \in \mathbf{sSet}^{\mathbf{U}^{op}}$ is Segal and this simplifies significantly when one uses the Reedy model structure on the category $\mathbf{sSet}^{\mathbf{U}^{op}}$.

3.2 Generalized Reedy Categories

The notion of a generalized Reedy category was introduced in [BM11, Definition 1.1].[10]

Definition 3.8 Let $\mathbb{R}$ be a small category. A dualizable generalized Reedy structure on $\mathbb{R}$ consists of two wide subcategories

$$\mathbb{R}^+ \quad \text{and} \quad \mathbb{R}^-$$

together with a *degree function* on objects $\mathrm{ob}(\mathbb{R}) \to \mathbb{N}$ satisfying:

(1) non-invertible morphisms in $\mathbb{R}^+$ (respectively $\mathbb{R}^-$) raise (respectively lower degree). Isomorphisms preserve degree.
(2) $\mathbb{R}^+ \bigcap \mathbb{R}^- = \mathrm{Iso}(\mathbb{R})$
(3) Every morphism f factors as $f = gh$ such that $g \in \mathbb{R}^+$ and $h \in \mathbb{R}^-$. Moreover, this factorization is unique up to isomorphism.
(4) If $\theta f = f$ for any isomorphism θ and $f \in \mathbb{R}^-$ then θ is an identity.

[10] As an interesting historical note, we noticed while preparing these notes that the initial results on generalized Reedy categories were presented at the CRM program on Homotopy Theory and Higher Categories in 2008.

(5) If $f\theta = f$ for any isomorphism θ and $f \in \mathbb{R}^+$ then θ is an identity.

The subcategory $\mathbb{R}^+$ is commonly called the 'direct category' and $\mathbb{R}^-$ the 'inverse category.' A category $\mathbb{R}$ that satisfies axioms (1) – (4) is a generalised Reedy category. If, in addition, $\mathbb{R}$ satisfies axiom (5) then $\mathbb{R}$ is said to be dualizable, which implies that $\mathbb{R}^{op}$ is also a generalised Reedy category.

Example 3.9 The simplicial category Δ is a Reedy category in which every isomorphism is an identity.

Example 3.10 Other examples of generalized Reedy categories include the *dendroidal category* Ω, finite sets, and pointed finite sets.

The main use of Reedy categories is that one can use latching and matching objects to lift morphisms from $\mathbb{R}$ to the diagram category $\mathbf{E}^{\mathbb{R}}$ by induction on the degree of objects. For any object $r \in \mathbb{R}$, the category $\mathbb{R}^+(r)$ is the full subcategory of $\mathbb{R}^+ \downarrow r$ consisting of non-invertible maps with target r. Similarly, the category $\mathbb{R}^-(r)$ is the full subcategory of $r \downarrow \mathbb{R}^-$ consisting of the non-invertible maps $\alpha : r \to s$.

Definition 3.11 Let X be a diagram in $\mathbf{E}^{\mathbb{R}}$

- The latching object $L_r X = \operatorname{colim}_{\mathbb{R}^+(r)} X$;
- The matching object $M_r X = \lim_{\mathbb{R}^-(r)} X$.

If $\mathbf{E}$ is a cofibrantly generated model category. We say that a morphism $f : X \to Y$ in $\mathbf{E}^{\mathbb{R}}$ is:

- a Reedy cofibration if $X_r \bigcup_{L_r X} L_r Y \to Y_r$ is a cofibration in $\mathbf{E}^{\mathrm{Aut}(r)}$ for all $r \in \mathbb{R}$;
- a Reedy weak equivalence if $X_r \to Y_r$ in $\mathbf{E}^{\mathrm{Aut}(r)}$ for all $r \in \mathbb{R}$;
- a Reedy fibration if $X_r \to M_r X \times_{M_r Y} Y_r$ in $\mathbf{E}^{\mathrm{Aut}(r)}$ for all $r \in \mathbb{R}$.

Theorem 3.12 ([BM11]) *If $\mathbb{R}$ is a dualizable generalized Reedy category and $\mathbf{E}$ is a nice enough model category, then $\mathbf{E}^{\mathbb{R}^{op}}$ admits a cofibrantly generated model category structure with level-wise weak equivalences.*

3.2.1 The Reedy Structure on U

The graphical category $\mathbf{U}$ has many nice factorization properties (eg. Remark 1.8 [HRY20b]) and is, in particular, a generalized Reedy category. We define the *degree of a graph G* to be

$$\deg(G) = |V| + |iE|.$$

Theorem 3.13 ([HRY20a, Theorem 2.22]) *The graphical category $\mathbf{U}$ is a (dualizable) generalised Reedy category. The wide subcategory $\mathbf{U}^-$ is generated by the codegeneracy maps (Definition 2.20) and the wide subcategory $\mathbf{U}^+$ is generated by the inner and outer coface maps (Definitions 2.18 and 2.16).*

Applying Theorem 1.6 of [BM11], we have the following corollary:

Corollary 3.14 *The diagram category* $\mathbf{sSet}^{\mathbf{U}^{op}}$ *admits a model category structure with the Reedy fibrations, Reedy cofibrations, and entrywise weak equivalences.*

Exercise 3.15 (Hard-ish) In Proposition 3.5 of [HRY20a] we show that Segal cores are cofibrant in the Reedy model structure on $\mathbf{sSet}^{\mathbf{U}^{op}}$. Give an example of a graph G in which the Segal core of G fails to be cofibrant in the projective model structure.

The advantage of a Reedy model structure on $\mathbf{sSet}^{\mathbf{U}^{op}}$ is that homotopy limits of Reedy fibrant diagrams are just limits. Revisiting our Definition 3.6 we now have:

Definition 3.16 A modular dendroidal space $X \in \mathbf{sSet}^{\mathbf{U}^{op}}$ is *Segal* if:

(1) $X_{\updownarrow} = *$;
(2) X is Reedy fibrant;
(3) for all $G \in \mathbf{U}$, the Segal map

$$X_G = \mathbb{R}\,\mathrm{Map}(\mathbf{U}[G], X) \longrightarrow \mathbb{R}\,\mathrm{Map}(\mathbf{Sc}[G], X) = \prod_{v \in V(G)} X_{\star_v}$$

is a weak equivalence in $\mathbf{sSet}$.

In the final lecture of this series, we will give our motivating example of a modular ∞-operad. We note, however, that given any (one-coloured) modular operad in $\mathbf{sSet}$, $\mathbf{P}$, the nerve $N\mathbf{P}$ is a Segal modular operad in the sense of Definition 3.6. Moreover, if $N\mathbf{P}$ is Reedy fibrant, then the Segal map

$$N\mathbf{P}_G \to \prod_{v \in V} N\mathbf{P}_{\star_v}$$

is an isomorphism for every G. Thus, up to fibrant replacement, every (one-coloured) modular operad gives rise to a modular ∞-operad. We conclude our description of modular ∞-operads by pointing out that there is a classification of modular dendroidal Segal spaces as the fibrant objects in a localization of the Reedy model category structure on graphical spaces.

Theorem 3.17 ([HRY20a, Theorem 3.8; Proposition 3.19]) *The category* $\mathbf{sSet}^{\mathbf{U}^{op}}$ *admits a cofibrantly generated model structure whose fibrant objects are the Segal modular operads.*

Remark 3.18 While we have discussed space-valued presheaves everything contained in this second lecture about modular dendroidal spaces still makes sense for presheaves in any Cartesian monoidal model category $\mathbf{C}$.

3.3 Variations on the Graphical Category U and Open Problems

The definition of modular operad often comes equipped with an additional "genus" grading (eg: [GK95], [GSNPR05], [DSVV24]). A *graded* modular operad consists of a bi-graded sequence $\mathbf{P} = \{\mathbf{P}(g, n)\}$, in which each $\mathbf{P}(g, n)$ is equipped with a right action of the

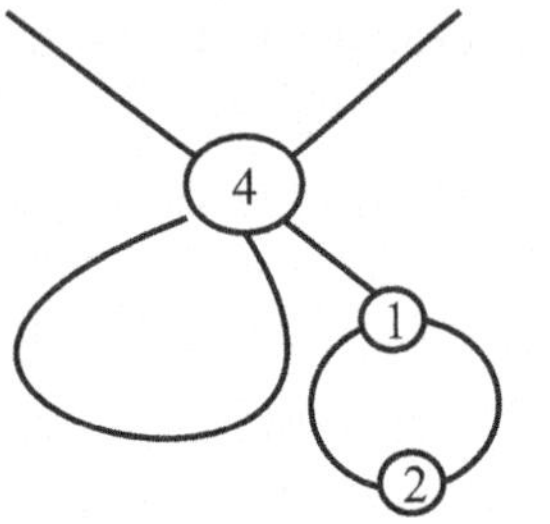

Fig. 15 An example of a graph with genus

symmetric group Σ_n, together with units, composition and contraction maps. Often one also requires that the underlying collection of $\mathbf{P} = \{\mathbf{P}(g, n)\}$ satisfy a *stability condition*, i.e.

$$\mathbf{P}(g, n) = \emptyset \quad \text{whenever} \quad 2g + n - 2 \leq 0.$$

One can define the corresponding genus graded version of the graphical category **U** and a corresponding stable version of modular ∞-operads.

Definition 3.19 Let G be a graph:

(1) A genus function for G is a function $g : V(G) \to \mathbb{N}$.
(2) The total genus of a pair $(G, g : V \to \mathbb{N})$ is given by:

$$g(G) = \beta_1(G) + \Sigma_{v \in V} g(v)$$

where $\beta_1(G)$ is the Betti number of G.
(3) A pair (G, g) is called **stable** if G is connected and for every vertex v:

$$2g(v) + |\operatorname{nb}(v)| - 2 > 0.$$

For example, in Fig. 15, the graph G has genus

$$g(G) = \beta_1(G) + \Sigma_{v \in V} g(v) = 2 + 4 + 1 + 2 = 9.$$

Given an embedding $f : H \to G$, then we can define the *genus of f*

$$g(f) := \beta_1(H) + \Sigma_{v \in V(H)} g(f(v)).$$

Remark 3.20 The exceptional edge admits only one genus function, and $g(\updownarrow) = \beta_1(\updownarrow) = 0$. This graph trivially satisfies the stability condition.

Definition 3.21 The stable graphical category $\mathbf{U}_{st}$ has:

- Objects: stable graphs (G, g)
- Morphisms: $(G, g) \to (G', g')$ are graphical maps $\varphi : G \to G'$ which make the diagram:

$$\begin{array}{ccc} V(G) & \xrightarrow{\quad\varphi_1\quad} & \mathrm{Emb}(G') \\ & {\scriptstyle g}\searrow \qquad \swarrow{\scriptstyle g'} & \\ & \mathbb{N} & \end{array}$$

commute.

The stability condition ensures that in the stable graphical category $\mathbf{U}_{st}$ there are no codegeneracy maps, because a genus 0 vertex with $|\operatorname{nb}(v)| = 2$ cannot be stable and thus the substitution of the edge into an arity 2 vertex is not in our category. This makes the following theorem an immediate corollary to Theorem 3.13.

Theorem 3.22 $\mathbf{U}_{st}$ *is a generalized Reedy category.*

We can therefore define a stable version of a modular ∞-operad as follows:

Definition 3.23 There is model structure on $\mathbf{sSet}^{\mathbf{U}_{st}^{op}}$ in which $X \in \mathbf{sSet}^{\mathbf{U}_{st}^{op}}$ is fibrant if:

- $X_{\updownarrow} = *$;
- X is Reedy fibrant;
- for all $G \in \mathbf{U}$, the Segal map

$$X_{(G,g)} = \mathbb{R}\,\mathbf{Map}(\mathbf{U}[(G,g)], X) \longrightarrow \mathbb{R}\,\mathbf{Map}(\mathbf{Sc}[(G,g)], X)$$

is a weak equivalence.

3.3.1 Cyclic Operads

As we mentioned in the first lecture, modular operads are cyclic operads with contraction operations. In the literature there are actually various notions of cyclic operad, and we define various subcategories of our graphical category $\mathbf{U}$ which correspond to the reader's desired notion of cyclic operad. In particular, there is nested sequence of subcategories:

$$\mathbf{U}_{cyc} \subset \mathbf{U}_0 \subset \mathbf{U}$$

defined as follows:

(1) The category $\mathbf{U}_0$ is the full subcategory of $\mathbf{U}$ whose objects are all *simply connected* graphs. The category $\mathbf{U}_0$ corresponds to *augmented cyclic operads*.
(2) The category $\mathbf{U}_{cyc}$ is the full subcategory of $\mathbf{U}$ whose objects are all *simply connected* graphs with *non-empty boundary*. The category $\mathbf{U}_{cyc}$ corresponds to *cyclic operads*.

Exercise 3.24 Show that $\mathbf{U}_0$ and $\mathbf{U}_{cyc}$ are sieves of $\mathbf{U}$. In other words if $\varphi : G \to T$ is in $\mathbf{U}$ with $T \in \mathbf{U}_0$ (respectively, $\mathbf{U}_{cyc}$) then $G \in \mathbf{U}_0$ (respectively, $\mathbf{U}_{cyc}$).

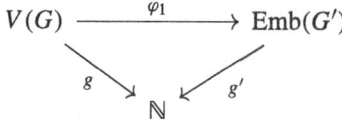

We also note that the category $\mathbf{U}_{cyc}$ is itself related to other categories in the literature:

(1) In [Wal21] Walde introduces a category Ω_{cyc}, which is a non-symmetric version of $\mathbf{U}_{cyc}$. That is: $\mathbf{U}_{cyc}$ is equivalent to a category $\mathbf{U}'_{cyc}$ in which every object has a specified cyclic ordering and Ω_{cyc} is the wide subcategory of $\mathbf{U}'_{cyc}$ where maps preserve the ordering.
(2) There is another category of Segal cyclic operads Ξ in [HRY19]. This is a graphical category in which models coloured cyclic operads where the involution on colour sets is always trivial. In practice, this category has the same objects as $\mathbf{U}_{cyc}$ but slightly different morphisms.

There are well understood (and very useful) adjunctions between operads cyclic operads and modular operads. One would hope that these same relationships hold between ∞-versions of all these objects. This inspires the following open problems.

Open Problem 3.25 Show there are Quillen adjunctions

$$\mathbf{sSet}^{\Xi^{op}} \rightleftarrows \mathbf{sSet}^{\mathbf{U}_{cyc}^{op}} \rightleftarrows \mathbf{sSet}^{\Omega^{op}}.$$

Is the adjunction

$$\mathbf{sSet}^{\Xi^{op}} \rightleftarrows \mathbf{sSet}^{\mathbf{U}_{cyc}^{op}}$$

a Quillen equivalence ? The interested reader may want to look at [HRY19, Proposition 8.5] where we establish a Quillen adjunction

$$\mathbf{sSet}^{\Xi^{op}} \rightleftarrows \mathbf{sSet}^{\Omega^{op}}$$

for inspiration.

Open Problem 3.26 Work of Barwick [Bar10], Hirschhorn and Voliĉ [HV19] characterizes when $F : \mathbb{R} \to \mathbb{S}$ between *strict* Reedy categories result in Quillen adjunctions between diagram categories. Is there a similar characterization for generalized Reedy categories?

4 Lecture 3: Lego-Teichmüller Theory and Modular Operads

In this final lecture, we introduce the genus graded modular operad built from surfaces which, after profinite completion, is related to the ideal Teichmüller towers in our introduction. Throughout this final lecture we will often make use of the fact that we have adjunctions

$$\mathbf{Operad} \rightleftarrows \mathbf{Cyc} \underset{\tau_0}{\rightleftarrows} \mathbf{ModOp}. \tag{4.1}$$

The adjunction between operads and cyclic operads is explicitly described in Section 3 of [DCH21]. The map τ_0 : **ModOp** $\to$ **Cyc** is "truncation at genus 0" or, equivalently, forgetting all contraction operations. This functor is actually a special case of a family of adjunctions

$$\mathbf{ModOp}_k \underset{\longrightarrow}{\overset{\tau_k}{\longleftarrow}} \mathbf{ModOp}$$

where the map τ_k is truncation at genus k, meaning we forget all operations of genus $\geq k$. These functors are a straightforward generalization of those in [GSNPR05, Section 4.1; 8.4] or [War22, Section 2.6].

4.1 Profinite Completion of Modular Operads in Groupoids

In a Cartesian monoidal category **E**, an *inverse system*, $\mathcal{I}$, is a collection of objects $\{X_i\}_{i\in I}$ in **E**, together with maps $\phi_{ij} : X_i \to X_j$ for $i \geq j$, such that:

(1) $\phi_{ii} : X_i \to X_i$ is the identity id_{X_i};
(2) $\phi_{ij} \circ \phi_{jk} = \phi_{ik}$ for $i \geq j \geq k$.

The limit over $\mathcal{I}$ is then given by

$$\varprojlim X_i = \left\{(x_i) \in \prod_{i\in I} X_i \mid \phi_{ij}(x_i) = x_j \ \forall i \geq j\right\}.$$

Example 4.1 A group G has an associated inverse system of finite index subgroups $\{G/N_i\}_{i\in\mathcal{I}}$ where the N_i runs over all normal subgroups of G and the maps

$$\phi_{ij} : G/N_i \to G/N_j$$

are the natural projections.

Definition 4.2 Given a finite group G, the *profinite completion* of G is the limit

$$\widehat{G} = \varprojlim G/N$$

inverse system of the finite index subgroups.

Example 4.3 Let $G = \mathbb{Z}$ which has finite index subgroups $\mathbb{Z}/n\mathbb{Z}$. The profinite completion is the group of *profinite integers* $\widehat{\mathbb{Z}} = \varprojlim \mathbb{Z}/n\mathbb{Z}$. Elements of $\widehat{\mathbb{Z}}$ are given by a sequence $(a_n)_{n\in\mathbb{N}}$ with $a_n \in \mathbb{Z}/n\mathbb{Z}$, and the structure maps for $n \geq m$ are the projections

$$\phi_{nm} : \mathbb{Z}/n\mathbb{Z} \to \mathbb{Z}/m\mathbb{Z}$$

which means that $a_n \equiv a_m \bmod m$ whenever $m \mid n$.

Since profinite completion is a limit, there is a natural map $G \mapsto \widehat{G}$. These maps assemble into a functor $\widehat{(-)} : \mathbf{Grp} \to \widehat{\mathbf{Grp}}$ from the category of groups to the category of profinite groups. This functor is the left adjoint in an adjunction

$$\mathbf{Grp} \underset{|-|}{\overset{\widehat{(-)}}{\rightleftarrows}} \widehat{\mathbf{Grp}}. \tag{4.2}$$

The right adjoint sends a profinite group to its underlying discrete group.

We can generalize profinite completion of groups to define profinite completion of groupoids. In [Hor17], Horel extends the adjunction (4.2) to the category of *groupoids*:

$$\mathbf{Gpd} \underset{|-|}{\overset{\widehat{(-)}}{\rightleftarrows}} \widehat{\mathbf{Gpd}}. \tag{4.3}$$

The category of profinite groupoids $\widehat{\mathbf{Gpd}}$ admits a model category structure [Hor17, Theorem 4.12] and the adjunction (4.3) is a Quillen adjunction. In addition, profinite completion of groupoids is monoidal in the following sense:

Proposition 4.4 ([Hor17, Proposition 4.23]) *Let* $\mathbf{C}$ *and* $\mathbf{D}$ *be two groupoids with finitely many objects. Then the map*

$$\widehat{\mathbf{C} \times \mathbf{D}} \to \widehat{\mathbf{C}} \times \widehat{\mathbf{D}}$$

induced by the projection maps $\widehat{\mathbf{C} \times \mathbf{D}} \to \widehat{\mathbf{C}}$ *and* $\widehat{\mathbf{C} \times \mathbf{D}} \to \widehat{\mathbf{D}}$ *is an isomorphism.*

Proposition 4.4 enables us to define the *profinite completion* of (cyclic and modular) operads in groupoids, so long as every entry of the (cyclic and modular) operad is a groupoid with a finite set of objects.

Definition 4.5 Let $\mathbf{P} = \{\mathbf{P}(n)\}$ be a modular operad in groupoids, in which each groupoid $\mathbf{P}(n)$ has a finite set of objects. The profinite completion of $\mathbf{P}$ is the modular operad

$$\widehat{\mathbf{P}} := \{\widehat{\mathbf{P}}(n)\}$$

where the profinite completion functor is applied entrywise. Composition operations are defined via the dashed lines:

$$\begin{array}{ccc} \widehat{\mathbf{P}}(n) \times \widehat{\mathbf{P}}(m) & \overset{\circ_{ij}}{\dashrightarrow} & \widehat{\mathbf{P}}(n+m-2) \\ \cong\downarrow & \nearrow_{\widehat{\circ_{ij}}} & \\ \widehat{\mathbf{P}(n) \times \mathbf{P}(m)}. & & \end{array}$$

Here the map $\widehat{\circ}_{ij}$ is the result of applying the profinite completion functor in (4.3) to the $\circ_{ij}$-composition maps of **P**. The contraction operations

$$\widehat{\xi}_{ij} : \widehat{\mathbf{P}}(n) \to \widehat{\mathbf{P}}(n-2)$$

result of applying the profinite completion functor (4.3) to the ξ_{ij}-composition maps of **P**.

Remark 4.6 In Definition 4.5 we have restricted ourselves to one-coloured modular operads, but this is not necessary. Definition 4.5 holds for any $\mathfrak{C}$-coloured or genus-graded modular operad in groupoids, so long as each entry of modular operad only has finitely many objects.

The nerve theorem from the first lecture (Theorem 2.37) generalizes to groupoid-valued **U**-presheaves. We can therefore use the nerve functor to identify a modular operad $\widehat{\mathbf{P}}$ in $\widehat{\mathbf{Gpd}}$ with a presheaf $N\widehat{\mathbf{P}} : \mathbf{U}^{op} \to \widehat{\mathbf{Gpd}}$ in which every graph G gives an isomorphism of profinite groupoids

$$N\widehat{\mathbf{P}}_G \longrightarrow \prod_{v \in V} N\widehat{\mathbf{P}}_{\star_v}.$$

4.2 Profinite Completion of Modular Operads in Spaces

To every groupoid **G** we can associate a space by taking the classifying space $B\mathbf{G}$. This fits into an adjunction

$$\mathbf{Gpd} \underset{B}{\overset{\pi_0}{\rightleftarrows}} \mathbf{sSet}.$$

Both the classifying space and fundamental groupoid functors preserves products and thus we can lift this to an adjunction

$$\mathbf{ModOp(Gpd)} \underset{B}{\overset{\pi_0}{\rightleftarrows}} \mathbf{ModOp(sSet)}.$$

A *profinite space* is a simplicial object in profinite sets. The category of profinite spaces is denoted $\widehat{\mathbf{sSet}}$ and is equipped with the model category structure from Quick [Qui08] (See also the discussion in Section 3 of [BdBHR19]). The profinite completion functor of spaces

$$\mathbf{sSet} \underset{|-|}{\overset{\widehat{(-)}}{\rightleftarrows}} \widehat{\mathbf{sSet}}. \tag{4.4}$$

is not as well-behaved as the profinite completion of groupoids. In particular, it is rarely the case that $\widehat{X \times Y} \simeq \widehat{X} \times \widehat{Y}$. In [BdBHR19], we establish a criteria which allows us to profinitely complete a small family of modular operads.

Definition 4.7 A discrete group G is said to be *good* if for any finite abelian group M equipped with a G-action the map $G \to \widehat{G}$ induces an isomorphism

$$H^i(\widehat{G}, M) \to H^i(G, M).$$

Proposition 4.8 (**[BdBHR19, Proposition 3.9]**) *Let X and Y be two connected spaces whose homotopy groups are good. Then the map*

$$\widehat{X \times Y} \longrightarrow \widehat{X} \times \widehat{Y}$$

is a weak equivalence of profinite spaces.

In the case that every space of $B\mathbf{P} = \{B\mathbf{P}(n)\}$ satisfies the conditions of the proposition then

$$N\widehat{B\mathbf{P}}_G \to \prod_{v \in V} N\widehat{B\mathbf{P}}_{\star_v}$$

is a weak equivalence for all graphs and $N\widehat{B\mathbf{P}}$ is a modular ∞-operad [BdBHR19, Proposition 5.1] and [BR].

Remark 4.9 In a recent paper by Blom and Moerdijk [BM22] they provide a more complete characterization of profinte topological operads. In particular, they provide a fibrantly generated model structure on the category of dendroidal profinite sets which characterizes profinte operads.

Open Problem 4.10 Extend the work of Blom and Moerdijk to characterize profinite cyclic and modular operads.

4.3 Operads and Mapping Class Groups

Let Σ be a surface of genus g with n-boundary components. We say that such a surface of *type* (g, n). The boundary components of Σ will always be equipped with an *ordering* $\rho : \mathbb{Z}/n\mathbb{Z} \xrightarrow{\cong} \pi_0(\partial \Sigma)$. Moreover, we require that each boundary, ∂_i, be equipped with a *collar*. In other words, for $1 \leq i \leq n$ and some fixed $\varepsilon > 0$, there is an embedding $\phi_i : S^1 \times [0, \varepsilon) \to \Sigma$ onto a neighborhood of ∂_i such that $\phi_i(S^1 \times \{0\}) = \partial_i$.

Definition 4.11 The *mapping class group* of a surface of type (g, n) is the group of isotopy classes of orientation preserving diffeomorphisms which fix collars pointwise:

$$\Gamma_n^g = \pi_0(\mathrm{Diff}^+(\Sigma, \partial \Sigma)).$$

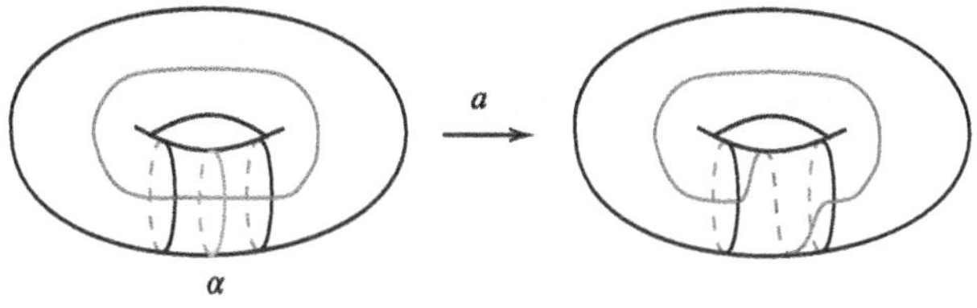

Fig. 16 A Dehn twist around a curve α

A theorem of Hatcher and Thurston [HT80] shows that the mapping class group Γ_g^n has a finite presentation:

$$\Gamma_g^n = \langle b_1, \ldots, b_n, a_1, \ldots, a_k \mid (C), (B), (D), (L) \rangle .$$

The generators of Γ_g^n are Dehn twists along a chosen set of simple closed curves on a surface of type (g, n). Given a curve (i.e. embedded circle) α on a surface, a *Dehn twist* is a diffeomorphism which acts on a neighborhood of α, $N_\alpha := S^1 \times [0, 1]$, by a full twist:

$$a(\theta, t) = (\theta + 2\pi t, t).$$

See Fig. 16 for an example. For the purposes of these lecture notes, we will write a Roman letter a to represent a Dehn twist around α as an element of the mapping class group.

For a surface of type (g, n), the generators of the mapping class consist of Dehn twists around each boundary, $b_1, \ldots, b_n$, as well as Dehn twists, $a_1, \ldots, a_k$, for each curve in a *pants decomposition* of the surface.

Definition 4.12 Let Σ be a surface of type (g, n). A *pants decomposition* of Σ is a finite collection of disjoint simple curves (modulo isotopy) which cut Σ into surfaces of type $(0, 3)$.

An example of a pants decomposition is depicted in Fig. 17a where we have depicted the curves of the decomposition in blue.

Remark 4.13 Though they play a crucial role in the proofs of some of the theorems we state below, we will not explicitly use the relations from the presentation of the mapping class group in this lecture. We refer the curious reader to [HT80] or [HLS00, Theorem 1] for a full description.

Shortly, we will describe how the Grothendieck-Teichmüller and Nakumara-Schneps groups act on the (profinite) mapping class groups. To do this, it is useful to add slightly more structure to the pants decomposition of a given surface. For a surface Σ of type (g, n), a *quilt* is a choice of two distinct points on each curve and boundary component of a pants decomposition of Σ, together with a set of disjoint lines between these points. The disjoint lines, or *seams*, cut each pair of pants in a decomposition into two hexagonal patches. In Fig. 17b the curves providing the pants decomposition are in blue and the seams of the quilt are in orange. A *quilted pants decomposition* of a surface Σ is a pants decomposition of Σ together with a choice of quilt.

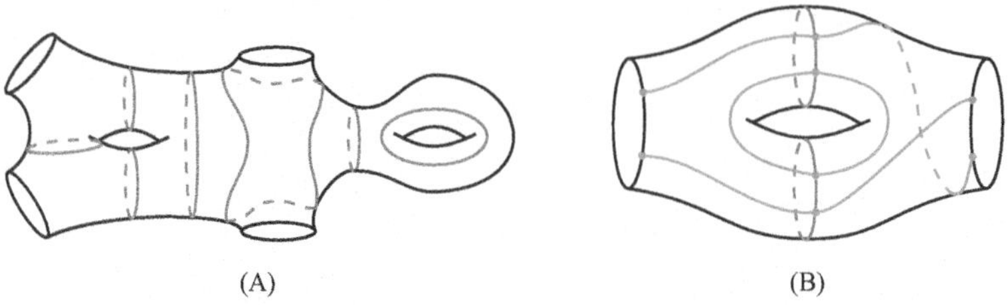

Fig. 17 (**a**) An example of a pants decomposition. (**b**) An example of a quilted pants decomposition

4.3.1 The Modular Operad of Quilted Surfaces

The reader may have noticed by now, that a pants decomposition of a surface Σ is equivalent to giving a prescription for building Σ via modular operad operations such as those pictured in Fig. 1. The second author and L. Bonatto have modified the surface operad of Tillmann [Til00] and Wahl [Wah04] to define a modular operad in groupoids whose objects are surfaces of type (g, n) built from a *standard quilted pair of pants*. Our choice of the standard quilted pair of pants is pictured on the left hand side of Fig. 18b.

Definition 4.14 Define a collection of genus graded groupoids $\mathbf{S}(g, n)$:

- $\mathbf{S}(0, 0) = \mathbf{S}(1, 0) = \emptyset$;
- $\mathbf{S}(0, 2)$ is the groupoid whose only object is homotopic to the circle S^1 (the "thin cylinder") and whose morphisms are given by the integers.[11]
- For all other $g, n \geq 0$, $\mathbf{S}(g, n)$ is the groupoid whose objects are surfaces of type (g, n) which are built from the standard pair of pants. Morphisms are isotopy classes of orientation preserving diffeomorphisms which fix the boundary collars pointwise.

The symmetric group Σ_n acts freely on the groupoid $\mathbf{S}(g, n)$ by permuting the labels of boundaries. Moreover, we can define composition and contraction functors

$$\mathbf{S}(g, n) \times \mathbf{S}(h, m) \xrightarrow{\circ_{ij}} \mathbf{S}(g + h, n + k - 2)$$

$$\mathbf{S}(g, n) \xrightarrow{\xi_{ij}} \mathbf{S}(g + 1, n - 2)$$

at the level of objects by gluing of surfaces. On morphisms these functors act by inclusion of Dehn twists. The following proposition appears in [BR]:

Proposition 4.15 *The composition and contraction operations are well-defined, associative, equivariant and unital operations and thus*

$$\mathbf{S} = \{\mathbf{S}(g, n)\}$$

[11] The groupoid $\mathbf{S}(0, 2)$ is equivalent to the groupoid $\mathcal{S}_{0,1,1}$ in [Wah04, 3.1.1].

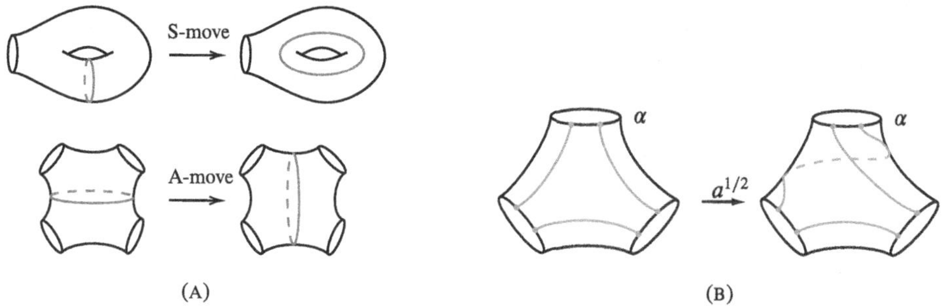

Fig. 18 (**a**) Elementary morphisms in $\mathbf{S}(g, n)$. (**b**) A half Dehn twist of the standard pair of pants

assembles into a genus graded modular operad in groupoids.

The groupoids $\mathbf{S}(g, n)$ are homotopy approximations of the mapping class groups in the sense that:

$$B\mathbf{S}(g, n) \simeq B\Gamma_n^g.$$

Aside from this point, the usefulness of the $\mathbf{S}(g, n)$ is that they are *finite*[12] groupoid approximations of a contractible 2-dimensional simplicial complex called the Seamed Hatcher-Thurston complex, denoted by $\mathcal{SHT}$ in [NS00] (see also the related $\mathcal{HT}$ complex in [HLS00] and [HT80]). In short, we select a finite number of the points of $\mathcal{SHT}$ to be the objects in our groupoids. Morphisms in $\mathbf{S}(g, n)$ are then 1-cells (or composites of 1-cells) from the complex $\mathcal{SHT}$. A consequence of this fact is that morphisms in $\mathbf{S}(g, n)$ are generated by three types of elementary morphisms: half-Dehn twists, A-moves ("associative moves"), and S-moves ("simple moves"). Examples of these diffeomorphisms are depicted in Fig. 18.

Remark 4.16 The half-Dehn twists are diffeomorphisms which change the quilt on a pair of pants relative to a specific curve. It is beyond the scope of these lectures to explicitly describe the effect of an A-move or S-move on the quilting of a surface, but the important part is that effect of an A-move or S-move is strictly defined. In particular, an A-move or S-move which takes a curve α to a curve β, changes a quilt by a uniquely determined half-Dehn twist around α, $a^{n/2}$, $n \in \mathbb{Z}$. Full details will appear in [BR], but see also the notion of a "quilt adjustment" in [NS00].

The 2-cells of the $\mathcal{SHT}$ complex can then be used to show that our generating morphisms in $\mathbf{S}(g, n)$ satisfy the following local relations:

(3A) The loops β_1, β_2 and β_3 in Fig. 19a represent all possible pants decompositions of a surface of type $(0, 4)$. If we let A_{b_i,b_j} denote the A-move which takes curve $\beta_i \to \beta_j$, then $A_{b_1,b_2}A_{b_2,b_3}A_{b_3,b_1} = 1$ in $\mathbf{S}(0, 4)$.

[12] We are using the term finite here in the sense that our groupoids $\mathbf{S}(g, n)$ all have finitely many objects.

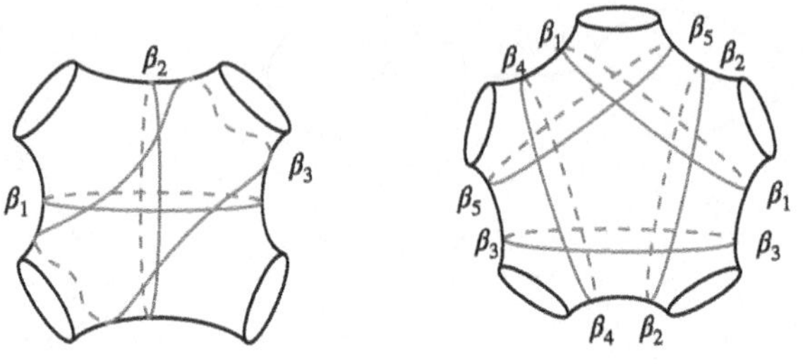

(A) 3A (B) 5A

Fig. 19 (**a**) 3A and (b) 5A

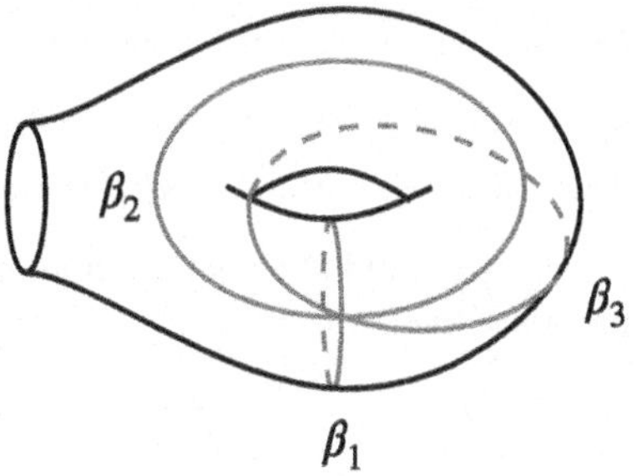

Fig. 20 3S

(5A) Similar to above, the loops $\beta_1, \ldots, \beta_5$ in Fig. 19b pairwise comprise all possible pants decompositions of a surface of type (0, 5). Then the equation:

$$A_{b_3,b_4} A_{b_1,b_2} A_{b_4,b_5} A_{b_2,b_3} A_{b_5,b_1} = 1$$

holds in $\mathbf{S}(0, 5)$

(3S) The curves β_1, β_2 and β_3 in Fig. 20 represent all possible pants decompositions of a surface of type (1, 1). If we write S_{b_i,b_j} for an S-move mapping a curve $\beta_i \to \beta_j$, then the equation

$$S_{b_1,b_2} S_{b_2,b_3} S_{b_3,b_1} = 1$$

holds in $\mathbf{S}(1, 1)$.

(6AS) Considering all possible pants decompositions of a surface of type (1, 2) (Fig. 21) the equation

$$A_{a_3,e_3} S_{a_1,a_2} A_{e_3,e_2} A_{e_2,e_1} S_{a_2,a_3} A_{e_1,a_1} = 1$$

in $\mathbf{S}(1, 2)$.

(C) Any two moves $\alpha_1 \to \alpha_2$ and $\beta_1 \to \beta_2$ (either S-moves or A-moves) supported in disjoint subsurfaces commute.

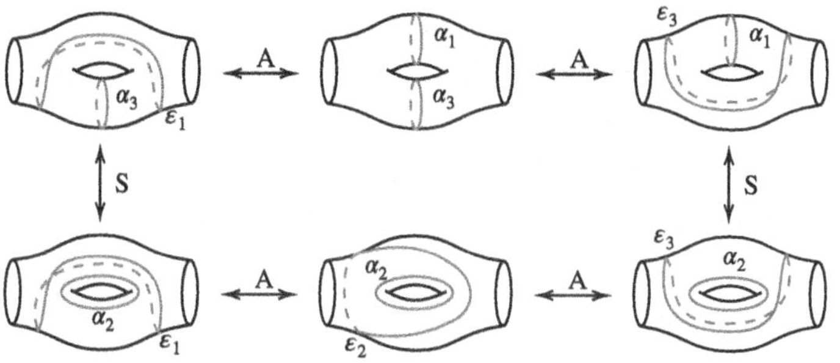

Fig. 21 3AS

4.4 The Grothendieck-Teichmüller and Nakumara-Schneps Groups

Recall from our introduction that the goal of this series is to convince the reader that the modular operad of seamed surfaces, $\mathbf{BS}$, gives a reasonable model for the Teichmüller tower (after completion). In order to argue this, we need to show that (1) the absolute Galois group acts on our proposed model and (2) that this action commutes with the modular operad structure. In practice, we don't really know how to do this without passing through some intermediate profinte groups. Here, we introduce two such groups: Grothendieck-Teichmüller and Nakumara-Schneps groups.

Let $\widehat{F_2}$ denote the profinite completion of the free group on two letters $F_2 = \langle x, y \rangle$. If the reader still finds the abstract definition of profinite group confusing, it can be helpful to think of an element $f \in \widehat{F_2}$ as a (possibly infinite) word in x and y. Any homomorphism of profinite groups

$$\widehat{F_2} \longrightarrow \widehat{G}$$

will necessarily be determined by where it sends the generators, $(x, y) \mapsto (a, b)$, and we will write $f(a, b)$ for the image of a word $f \in \widehat{F_2}$ in $\widehat{G}$. For example, we will write $f(y, x)$ for the image of any $f \in \widehat{F_2}$ under the map $\widehat{F_2} \to \widehat{F_2}$ given by $(x, y) \mapsto (y, x)$.

Definition 4.17 The Grothendieck-Teichmüller group $\widehat{\mathsf{GT}}$ is the group of pairs

$$(\lambda, f) \in \widehat{\mathbb{Z}}^* \times \widehat{F}_2'$$

which satisfy the property that

$$x \mapsto x^\lambda \quad \text{and} \quad y \mapsto f^{-1} y^\lambda f$$

induces an automorphism of $\widehat{F_2}$. Moreover, we require the pair (λ, f) satisfy the following axioms:

(I) $f(x, y) f(y, x) = 1$,
(II) $f(x, y) x^m f(z, x) z^m f(y, z) y^m = 1$ where $xyz = 1$ and $m = (\lambda - 1)/2$,

(III) $f(b_3, b_4)f(b_5, b_1)f(b_2, b_3)f(b_4, b_5)f(b_1, b_2) = 1$ in $\widehat{\Gamma}_{0,5}$ where b_i is a *Dehn twist* along a loop β_i depicted in Fig. 19b.

The profinite *Grothendieck-Teichmüller group*, $\widehat{\mathsf{GT}}$, is closely related to the absolute Galois group. In particular, a theorem of Ihara says:

Theorem 4.18 ([Iha94]) *There is an injection* $Gal(\mathbb{Q}) \hookrightarrow \widehat{\mathsf{GT}}$.

A related group, defined by Nakumara and Schneps in [NS00] is defined by adding a "higher genus" relation to $\widehat{\mathsf{GT}}$.

Definition 4.19 Let $\widehat{\mathsf{NS}}$ denote the group of pairs

$$(\lambda, f) \in \widehat{\mathbb{Z}}^* \times \widehat{F}_2'$$

which satisfy the property that

$$x \mapsto x^\lambda \quad \text{and} \quad y \mapsto f^{-1}y^\lambda f$$

induce an automorphism of $\widehat{F}_2$. Moreover, we require pairs (λ, f) satisfy relations $(I) - (III)$ of $\widehat{\mathsf{GT}}$ and:

(IV) $f(e_1, a_1)a_3^{-8\rho_2} f(a_2^2, a_3^2)(a_3a_2a_3)^{2m} f(e_2, e_1)e_2^{2m} f(e_3, e_2)a_2^{-2m}(a_1a_2a_1)^{2m}f(a_1^2, a_2^2)$ $a_1^{8\rho_2} f(a_3, e_3) = 1$ where a_1, a_2, a_3, e_1, e_2 are Dehn twists in $\Gamma_{1,2}$ corresponding to the curves in Fig. 21.

Nakumara and Schneps show that $\widehat{\mathsf{NS}}$ is a subgroup of $\widehat{\mathsf{GT}}$ and that Ihara's injection $\mathrm{Gal}(\mathbb{Q}) \hookrightarrow \widehat{\mathsf{GT}}$ also maps $\mathrm{Gal}(\mathbb{Q})$ into $\widehat{\mathsf{NS}}$ (Theorem 1.2 [NS00]).

Remark 4.20 At this time, it is not known if $\widehat{\mathsf{NS}}$ is a *proper* subgroup of $\widehat{\mathsf{GT}}$.

The goal for the remainder of this lecture is to investigate the Galois actions on our proposed model for the Teichmüller tower by studying actions of $\widehat{\mathsf{NS}}$ and $\widehat{\mathsf{GT}}$.

4.4.1 The Genus Zero Case

The Grothendieck-Teichmüller group is closely related to the *genus zero* component of our tower. Let $\mathbf{S}_0 = \{\mathbf{S}(0, n)\}$ for the restriction of $\mathbf{S}$ to genus 0. In other words, $\mathbf{S}_0$ is the *underlying cyclic operad* of $\mathbf{S}$ via the adjunction in (4.1). The underlying *operad* $\mathcal{S} = \{\mathcal{S}(0, n+1)\}$ is obtained from the cyclic operad $\mathbf{S}_0 = \{\mathbf{S}(0, n)\}$ by marking one boundary of each surface as the distinguished output of the surface. Operad composition is then defined by gluing the *marked* boundary component of a surface in $\mathcal{S}(0, m+1)$ to the ith free boundary component of $\mathcal{S}(0, n+1)$:

$$\mathcal{S}(0, n+1) \times \mathcal{S}(0, m+1) \xrightarrow{\circ_i} \mathcal{S}(0, n+m+1).$$

Remark 4.21 As we mentioned, the modular operad $\mathbf{S} = \{\mathbf{S}(g, n)\}$ is an extension of the surface operad of Tillmann [Til00] and Wahl [Wah04, Section 3.1]. Our genus zero operad

$\mathcal{S} = \{\mathcal{S}(0, n+1)\}$ is precisely the genus zero part of Tillmann's operad [BdBHR19, Definition 6.5]. This operad is equivalent to the operad of parenthesized ribbon braids $\mathsf{PaRB} = \{\mathsf{PaRB}(n)\}$ and closely related to the operad of framed discs, in the sense that there are homotopy equivalences

$$B\mathcal{S}(0, n+1) \simeq B(\mathsf{PaRB}(n)) \simeq \mathsf{fD}(n).$$

Each groupoid $\mathcal{S}(0, n+1)$ in our operad $\mathcal{S}$ has finitely many objects and thus, after applying the profinite completion functor entrywise, we obtain an operad in profinite groupoids

$$\widehat{\mathcal{S}} = \{\widehat{\mathcal{S}}(0, n+1).\}$$

We write End_0 for the set of operad endomorphisms which fix objects. Proposition 7.3 and Proposition 8.1 of [BdBHR19] combine to show:

Proposition 4.22 *There is an isomorphism*

$$\widehat{\mathsf{GT}} \cong \mathrm{End}_0(\widehat{\mathcal{S}}).$$

Putting this together with the nerve theorem from the first lecture (Theorem 2.37), we identify the group $\widehat{\mathsf{GT}}$ with the group of (path components of) self maps of the ∞-operad $N\widehat{B\mathcal{S}}$ [BdBHR19, Theorem 8.4]:

Theorem 4.23 *There is an isomorphism*

$$\widehat{\mathsf{GT}} \cong \pi_0 \mathbb{R}\,\mathrm{End}(N\widehat{B\mathcal{S}}).$$

Remark 4.24 In [BdBHR19], we actually show $\widehat{\mathsf{GT}} \cong \mathrm{End}_0(\widehat{\mathsf{PaRB}})$. We have been a bit loose with the translation, because one can show that the operads PaRB and $\mathcal{S}$ are equivalent. This presentation just translates a bit easier to the higher genus case.

4.4.2 The $\widehat{\mathsf{GT}}$ Action

The proof of Proposition 4.22 is outside of the scope of these lectures. However, we can describe the arity-wise action of $\widehat{\mathsf{GT}}$

$$\widehat{\mathsf{GT}} \longrightarrow \mathrm{End}_0(\widehat{\mathbf{S}}(0, n))$$

on the profinte cyclic operad rather easily by translating the action of $\widehat{\mathsf{GT}}$ on the $\mathcal{SHT}$ complex from [NS00] to our groupoids.

Recall that an object Σ in one of the groupoids $\mathbf{S}(0, n)$ is equivalent to fixing a surface of type $(0, n)$ together with an "atomic" quilted pants decomposition. For each $(\lambda, f) \in \widehat{\mathsf{GT}}$, we define an $F_{(\lambda, f)} : \mathbf{S}(0, n) \to \mathbf{S}(0, n)$ which is the identity on objects and acts on elementary morphisms by:

$$(\lambda, f) \longrightarrow \begin{cases} a_{\frac{1}{2}} \mapsto a_{\frac{1}{2}}^{\lambda} \\ A_{\alpha,b} \mapsto A_{\alpha,b} \cdot f(a,b) a^{n(\lambda-1)/2}. \end{cases} \tag{4.5}$$

Here $a_{\frac{1}{2}}$ is a half Dehn twist around the boundary components or any curve $\{\alpha_i\}$ in the pants decomposition of the relevant object. The integer n which arises in the action on an A-move $A_{\alpha,\beta} : \Sigma \to \Sigma'$ can be calculated based on the interaction of the A-move with the quilt on Σ. We wont discuss how this integer is computed in these notes, but a similar computation is done in Section 7 of [BdBHR19]. Full details will appear in [BR].

Remark 4.25 Note that there are no S-moves in the genus zero groupoids $\mathbf{S}(0, n)$.

To check that the action of $\widehat{\mathsf{GT}}$ in (4.5) is well-defined, we need to check that the maps $F_{(\lambda,f)}$ commute with the defining relations of $\widehat{\mathsf{GT}}$. In general, this is a bit involved, but we can sketch how one shows that the map $F_{(\lambda,f)}$ commutes with relation (III) from Definition 4.17. We fix a surface Σ of type $(0, 5)$. Then the action of $\widehat{\mathsf{GT}}$ on the automorphism of Σ given by the composite of A-moves

$$A_{b_3,b_4} A_{b_1,b_2} A_{b_4,b_5} A_{b_2,b_3} A_{b_5,b_1}$$

becomes:

$$\begin{aligned} &A_{b_3,b_4} A_{b_1,b_2} A_{b_4,b_5} A_{b_2,b_3} A_{b_5,b_1} \\ &\quad \mapsto A_{b_3,b_4} f(b_1, b_2) b_1^{n(\lambda-1)/2} \ldots A_{b_5,b_1} f(b_5, b_1) b_5^{n(\lambda-1)/2}. \end{aligned} \tag{4.6}$$

A quick computation shows that for this particular action that $n = 0$ (A proof is similar to [NS00, Proposition 8.3]. Full details in this setting will appear in [BR]). The action commutes with categorical composition, and so Eq. (4.6), simplifies to

$$A_{b_3,b_4} f(b_3, b_4) \ldots A_{b_5,b_1} f(b_5, b_1) = (A_{b_3,b_4} \cdots A_{b_5,b_1}) \cdot (f(b_3, b_4) \ldots f(b_5, b_1)) = 1. \tag{4.7}$$

But now relation $(5A)$ between morphisms in $\mathbf{S}(0, 5)$, reduces Eq. (4.7) to

$$f(b_3, b_4) f(b_1, b_2) f(b_4, b_5) f(b_2, b_3) f(b_5, b_1) = 1.$$

This is precisely relation (III) in the definition of $\widehat{\mathsf{GT}}$.

The other relations follow a similar pattern. The difficult part is showing that the $\widehat{\mathsf{GT}}$-action commutes with the cyclic operad structure maps. As we did in [BdBHR19], we can overcome this by showing:

Proposition 4.26 *There is an isomorphism*

$$\widehat{\mathsf{GT}} \cong \mathrm{End}_0(\widehat{\mathbf{S}}_0).$$

4.4.3 The Higher Genus Action

The Nakamura-Schneps group $\widehat{\mathsf{NS}}$ acts on the full modular operad $\mathbf{S}$ in such a way that the restriction to genus zero is precisely the action of $\widehat{\mathsf{GT}}$ on $\mathbf{S}_0$ we have just described. In [BR] we show:

Proposition 4.27 ([BR]) *The profinte group* $\widehat{\mathsf{NS}}$ *acts on* $\mathrm{End}_0(\widehat{\mathbf{S}})$.

As in the genus zero case, we can described the arity wise action

$$\widehat{\mathsf{NS}} \longrightarrow \mathrm{End}_0(\widehat{\mathbf{S}}(g,n)).$$

Given a $(\lambda, f) \in \widehat{\mathsf{NS}}$ we wish to define a functor $F_{(\lambda,f)} : \mathbf{S}(g,n) \to \mathbf{S}(g,n)$ which fixes objects and acts on elementary morphisms via

$$(\lambda, f) \xrightarrow{F_{(\lambda,f)}} \begin{cases} a_{\frac{1}{2}} \mapsto a_{\frac{1}{2}}^{\lambda} \\ A_{\alpha,b} \mapsto A_{\alpha,b} \cdot f(a,b)a^{n(\lambda-1)/2} \\ S_{\alpha,b} \mapsto S_{\alpha,b} \cdot (aba)^{\lambda-1} b^{n(\lambda-1)/2-8\rho_2} f(a^2,b^2)a^{8\rho_2}. \end{cases}$$

As before the, if $S_{a,b} : \Sigma \to \Sigma'$ is our S-move, the integer n is dependent on the quilt of Σ. The integer ρ_2 is the Kummer 1-cocycle with respect to the roots of 2 (See Section 5 of [NS00] for full details).

To check that the action we have given is well-defined, it remains to verify that it is compatible with the defining relations of our group $\widehat{\mathsf{NS}}$. For example, if we consider the $\widehat{\mathsf{NS}}$ action on $\mathbf{S}(1,2)$ then we know that all A and S moves necessarily satisfy the relation (6AS):

$$A_{\alpha_3,\epsilon_3} S_{\alpha_1,\alpha_2} A_{\epsilon_3,\epsilon_2} A_{\epsilon_2,\epsilon_1} S_{\alpha_2,\alpha_3} A_{\epsilon_1,\alpha_1} = 1. \tag{4.8}$$

Acting by (λ, f) gives the equation

$$\begin{aligned} &A_{\alpha_3,\epsilon_3} f(a_3,e_3) a_3^{n(\lambda-1)/2} S_{\alpha_1,\alpha_2} (a_1a_2a_1)^{\lambda-1} a_2^{n(\lambda-1)/2-8\rho_2} \\ &f(a_1^2,a_2^2) a_1^{8\rho_2} \dots A_{\epsilon_1,\alpha_1} f(e_1,a_1) e_1^{n(\lambda-1)/2}. \end{aligned} \tag{4.9}$$

The integer n is computed based on the quilting (see [NS00, Proposition 8.3]) and the action, coming from a group homomorphism, commutes with the categorical compositions. This yields the relation (IV) from Definition 4.19:

$$\begin{aligned} &f(e_1,a_1)a_3^{-8\rho_2} f(a_2^2,a_3^2)(a_3a_2a_3)^{2m} f(e_2,e_1)e_2^{2m} f(e_3,e_2)a_2^{-2m} \\ &(a_1a_2a_1)^{2m} f(a_1^2,a_2^2)a_1^{8\rho_2} f(a_3,e_3) = 1 \end{aligned}$$

4.4.4 An Operadic Two Level Principle

The groupoids $\mathbf{S}(g, n)$ are only homotopy approximations of Γ_g^n in the sense that

$$B\mathbf{S}(g, n) \simeq B\Gamma_g^n.$$

Therefore, in order to complete our description of Teichmüller tower we want to see the action of $\widehat{\mathsf{NS}}$ on the profinite completion of the modular operad $B\mathbf{S}$. Applying the profinite completion functor (4.4) entrywise results in a sequence of profinite spaces

$$\widehat{B\mathbf{S}} = \{\widehat{B\mathbf{S}(g, n)}, \}$$

but, unfortunately, these profinite spaces do not form a modular ∞-operad. This is because we do not know if the mapping class groups Γ_g^n are good groups (Definition 4.7) for $g \geq 2$ and thus we cannot apply Proposition 4.8 to get a family of weak composition maps. One can show, however, that we have a modular dendroidal space:

$$N\widehat{B\mathbf{S}} : \mathbf{U}^{op} \to \widehat{\mathbf{sSet}}.$$

The truncation of the modular operad $\mathbf{S}$ at genus 1 to defines a modular $B\mathbf{S}_1$ with

$$B\mathbf{S}_1(g, n) = \begin{cases} B\mathbf{S}(g, n) \text{ if } g \leq 1 \\ \emptyset \text{ otherwise.} \end{cases}$$

In this case, applying the profinite completion functor entrywise results in a modular ∞-operad, $N\widehat{B\mathbf{S}}_1$.

The modular operad $\mathbf{S}$ is generated by a single object (our standard pair of pants) and morphisms in genus zero and one. It follows that one can show:

Proposition 4.28 *There is an isomorphism of profinite groups:*

$$\mathrm{End}_0(\widehat{\mathbf{S}}) \cong \mathrm{End}_0(\widehat{\mathbf{S}}_1).$$

The classifying space functor $B : \mathbf{Gpd} \to \mathbf{sSet}$ is homotopically fully faithful, meaning that for any two groupoids $\mathbf{C}$ and $\mathbf{D}$:

$$\mathbb{R}\,\mathrm{Map}(\mathbf{C}, \mathbf{D}) \cong \mathbb{R}\,\mathrm{Map}(B\mathbf{C}, B\mathbf{D}).$$

Combining this with the observation that the truncation functor

$$\mathbf{ModOp}_1(\widehat{\mathbf{Gpd}}) \underset{\tau_1}{\overset{\longrightarrow}{\longleftarrow}} \mathbf{ModOp}(\widehat{\mathbf{Gpd}})$$

is part of a Quillen adjunction leads us to our final theorem of this lecture series:

Theorem 4.29 ([BR]) *There is an action of the profinite group* $\widehat{\mathsf{NS}}$ *on the profinite modular* ∞*-operad* $N\widehat{B\mathsf{S}}$.

Acknowledgments The work presented in these lectures covers joint work of the second author with Luciana Basualdo Bonatto, Pedro Boavida de Brito, Philip Hackney, Geoffroy Horel, and Donald Yau. We would like to thank all of them for their work, comments, and suggestions and take full responsibility for any typos, errors or bad writing contained in these notes.

Over the years this work has also benefited immensely from feedback and comments from Ezra Getzler, Joachim Kock, Ieke Moerdijk, Sophie Raynor, Marco Robalo, Leila Schneps, Michelle Strumila and many others. We would like to specifically acknowledge Luciana Basualdo Bonatto, Philip Hackney and Sophie Raynor who provided feedback on these lectures as they were being prepared and the graduate students at the University of Melbourne who were subjected to the first version(s) of these lectures. A special thank-you goes to Santiago Nahuel Martinez who made many of the beautiful figures in these notes.

Lastly, we would like to thank the organizers and participants of the workshop on "Higher Structures and Operadic Calculus". We are particularly grateful for the efforts made to make this event possible during a (hopefully!) once-in-a-lifetime pandemic and for making it possible for the authors to participate remotely from the USA and Australia, respectively.

References

[Bar79] Michael Barr, *$*$-autonomous categories*, Lecture Notes in Mathematics, vol. 752, Springer, Berlin, 1979, With an appendix by Po Hsiang Chu. MR550878

[Bar10] Clark Barwick, *On left and right model categories and left and right Bousfield localizations*, Homology Homotopy Appl. **12** (2010), no. 2, 245–320. MR2771591

[BB17] M. A. Batanin and C. Berger, *Homotopy theory for algebras over polynomial monads*, Theory Appl. Categ. **32** (2017), Paper No. 6, 148–253. MR3607212

[BdBHR19] Pedro Boavida de Brito, Geoffroy Horel, and Marcy Robertson, *Operads of genus zero curves and the Grothendieck-Teichmüller group*, Geom. Topol. **23** (2019), no. 1, 299–346. MR3921321

[Bel79] G. V. Belyĭ, *Galois extensions of a maximal cyclotomic field*, Izv. Akad. Nauk SSSR Ser. Mat. **43** (1979), no. 2, 267–276, 479. MR534593

[Ber22] Clemens Berger, *Moment categories and operads*, Theory Appl. Categ. **38** (2022), Paper No. 39, 1485–1537. MR4541944

[BM11] Clemens Berger and Ieke Moerdijk, *On an extension of the notion of Reedy category*, Math. Z. **269** (2011), no. 3–4, 977–1004. MR2860274

[BM22] Thomas Blom and Ieke Moerdijk, *Profinite ∞-operads*, Adv. Math. **408** (2022), Paper No. 108601, 50. MR4462939

[BM23] Michael Batanin and Martin Markl, *Koszul duality for operadic categories*, Compositionality **5** (2023), no. 4, 56. MR4611064

[BR] Luciana Basualdo Bonatto and Marcy Robertson, *A modular operad of seamed surfaces and the grothendieck-teichmüller group*, in preparation.

[CH21] Hongyi Chu and Rune Haugseng, *Homotopy-coherent algebra via Segal conditions*, Adv. Math. **385** (2021), Paper No. 107733, 95. MR4256131

[CM11] Denis-Charles Cisinski and Ieke Moerdijk, *Dendroidal sets as models for homotopy operads*, J. Topol. **4** (2011), no. 2, 257–299. MR2805991

[DCH21] Gabriel C. Drummond-Cole and Philip Hackney, *Dwyer-Kan homotopy theory for cyclic operads*, Proc. Edinb. Math. Soc. (2) **64** (2021), no. 1, 29–58. MR4249838

[DM18] Martin Doubek and Martin Markl, *Open-closed modular operads, the Cardy condition and string field theory*, J. Noncommut. Geom. **12** (2018), no. 4, 1359–1424. MR3896229

[Dri90] V. G. Drinfel'd, *On quasitriangular quasi-Hopf algebras and on a group that is closely connected with* $\mathrm{Gal}(\overline{\mathbf{Q}}/\mathbf{Q})$, Algebra i Analiz **2** (1990), no. 4, 149–181. MR1080203

[DSVV24] Vladimir Dotsenko, Sergey Shadrin, Arkady Vaintrob, and Bruno Vallette, *Deformation theory of cohomological field theories*, J. Reine Angew. Math. **809** (2024), 91–157. MR4726567

[GK95] E. Getzler and M. M. Kapranov, *Cyclic operads and cyclic homology*, Geometry, topology, & physics, Conf. Proc. Lecture Notes Geom. Topology, IV, Int. Press, Cambridge, MA, 1995, pp. 167–201. MR1358617

[Gro97] Alexandre Grothendieck, *Esquisse d'un programme*, Geometric Galois actions, 1, London Math. Soc. Lecture Note Ser., vol. 242, Cambridge Univ. Press, Cambridge, 1997, With an English translation on pp. 243–283, pp. 5–48. MR1483107

[GSNPR05] F. Guillén Santos, V. Navarro, P. Pascual, and A. Roig, *Moduli spaces and formal operads*, Duke Math. J. **129** (2005), no. 2, 291–335. MR2165544

[Hac24] Philip Hackney, *Categories of graphs for operadic structures*, Math. Proc. Cambridge Philos. Soc. **176** (2024), no. 1, 155–212. MR4680484

[Hir03] Philip S. Hirschhorn, *Model categories and their localizations*, Mathematical Surveys and Monographs, vol. 99, American Mathematical Society, Providence, RI, 2003. MR1944041

[HLS00] Allen Hatcher, Pierre Lochak, and Leila Schneps, *On the Teichmüller tower of mapping class groups*, J. Reine Angew. Math. **521** (2000), 1–24. MR1752293

[HM22] Gijs Heuts and Ieke Moerdijk, *Simplicial and dendroidal homotopy theory*, Ergebnisse der Mathematik und ihrer Grenzgebiete. 3. Folge. A Series of Modern Surveys in Mathematics [Results in Mathematics and Related Areas. 3rd Series. A Series of Modern Surveys in Mathematics], vol. 75, Springer, Cham, [2022] ©2022. MR4485749

[Hor17] Geoffroy Horel, *Profinite completion of operads and the Grothendieck-Teichmüller group*, Adv. Math. **321** (2017), 326–390. MR3715714

[HRY15] Philip Hackney, Marcy Robertson, and Donald Yau, *Infinity properads and infinity wheeled properads*, Lecture Notes in Mathematics, vol. 2147, Springer, Cham, 2015. MR3408444

[HRY18] Philip Hackney, Marcy Robertson, and Donald Yau, *On factorizations of graphical maps*, Homology Homotopy Appl. **20** (2018), no. 2, 217–238. MR3812464

[HRY19] Philip Hackney, Marcy Robertson, and Donald Yau, *Higher cyclic operads*, Algebr. Geom. Topol. **19** (2019), no. 2, 863–940. MR3924179

[HRY20a] Philip Hackney, Marcy Robertson, and Donald Yau, *A graphical category for higher modular operads*, Adv. Math. **365** (2020), 107044, 61. MR4064770

[HRY20b] Philip Hackney, Marcy Robertson, and Donald Yau, *Modular operads and the nerve theorem*, Adv. Math. **370** (2020), 107206, 39. MR4099828

[HT80] A. Hatcher and W. Thurston, *A presentation for the mapping class group of a closed orientable surface*, Topology **19** (1980), no. 3, 221–237. MR579573

[HV19] Philip S. Hirschhorn and Ismar Volić, *Functors between Reedy model categories of diagrams*, North-West. Eur. J. Math. **5** (2019), 21–67, i. MR3978001

[HVZn10] Eric Harrelson, Alexander A. Voronov, and J. Javier Zúñiga, *Open-closed moduli spaces and related algebraic structures*, Lett. Math. Phys. **94** (2010), no. 1, 1–26. MR2720252

[Iha94] Yasutaka Ihara, *On the embedding of* $\mathrm{Gal}(\overline{\mathbf{Q}}/\mathbf{Q})$ *into* $\widehat{\mathrm{GT}}$, The Grothendieck theory of dessins d'enfants (Luminy, 1993), London Math. Soc. Lecture Note Ser., vol. 200, Cambridge Univ. Press, Cambridge, 1994, With an appendix: the action of the absolute Galois group on the moduli space of spheres with four marked points by Michel Emsalem and Pierre Lochak, pp. 289–321. MR1305402

[JK09] André Joyal and Joachim Kock, *Feynman graphs, and nerve theorem for compact symmetric multicategories (extended abstract)*, 2009.

[JY] Mark Johnson and Donald Yau, *Boardman-Vogt resolutions of generalized props*, Book draft, available on https://u.osu.edu/yau.22/main/.

[Kel72] G. M. Kelly, *Many-variable functorial calculus. I*, Coherence in categories, 1972, pp. 66–105. Lecture Notes in Math., Vol. 281. MR0340371

[KW17] Ralph M. Kaufmann and Benjamin C. Ward, *Feynman categories*, Astérisque (2017), no. 387, vii+161. MR3636409

[LS94] Pierre Lochak and Leila Schneps, *The Grothendieck-Teichmüller group and automorphisms of braid groups*, The Grothendieck theory of dessins d'enfants (Luminy, 1993), London Math. Soc. Lecture Note Ser., vol. 200, Cambridge Univ. Press, Cambridge, 1994, pp. 323–358. MR1305403

[Mar08] Martin Markl, *Operads and PROPs*, Handbook of algebra. Vol. 5, Handb. Algebr., vol. 5, Elsevier/North-Holland, Amsterdam, 2008, pp. 87–140. MR2523450

[Moe10] Ieke Moerdijk, *Lectures on dendroidal sets*, Simplicial methods for operads and algebraic geometry, Adv. Courses Math. CRM Barcelona, Birkhäuser/Springer Basel AG, Basel, 2010, Notes written by Javier J. Gutiérrez, pp. 1–118. MR2778589

[MW07] Ieke Moerdijk and Ittay Weiss, *Dendroidal sets*, Algebr. Geom. Topol. **7** (2007), 1441–1470. MR2366165

[MW09] I. Moerdijk and I. Weiss, *On inner Kan complexes in the category of dendroidal sets*, Adv. Math. **221** (2009), no. 2, 343–389. MR2508925

[Nak97] Hiroaki Nakamura, *Galois rigidity of profinite fundamental groups [translation of Sūgaku* **47** *(1995), no. 1, 1–17; MR1362515 (98d:14027)]*, vol. 10, 1997, Sugaku Expositions, pp. 195–215. MR1600655

[NS00] Hiroaki Nakamura and Leila Schneps, *On a subgroup of the Grothendieck-Teichmüller group acting on the tower of profinite Teichmüller modular groups*, Invent. Math. **141** (2000), no. 3, 503–560. MR1779619

[Oda97] Takayuki Oda, *Etale homotopy type of the moduli spaces of algebraic curves*, Geometric Galois actions, 1, London Math. Soc. Lecture Note Ser., vol. 242, Cambridge Univ. Press, Cambridge, 1997, pp. 85–95. MR1483111

[Pet13] Dan Petersen, *The operad structure of admissible G-covers*, Algebra Number Theory **7** (2013), no. 8, 1953–1975. MR3134040

[Pop22] Florian Pop, *Little survey on I/OM and its variants and their relation to (variants of)* $\widehat{GT}$*—old & new*, Topology Appl. **313** (2022), Paper No. 107993, 15. MR4423101

[Qui08] Gereon Quick, *Profinite homotopy theory*, Doc. Math. **13** (2008), 585–612. MR2466189

[Ray18] Sophie Raynor, *Compact symmetric multicategories and the problem of loops.*, Ph.D. thesis, 2018.

[Ray21] Sophie Raynor, *Graphical combinatorics and a distributive law for modular operads*, Adv. Math. **392** (2021), Paper No. 108011, 87. MR4316667

[Ser08] Jean-Pierre Serre, *Topics in Galois theory*, second ed., Research Notes in Mathematics, vol. 1, A K Peters, Ltd., Wellesley, MA, 2008, With notes by Henri Darmon. MR2363329

[SL97] Leila Schneps and Pierre Lochak (eds.), *Geometric Galois actions. 1*, London Mathematical Society Lecture Note Series, vol. 242, Cambridge University Press, Cambridge, 1997, Around Grothendieck's "Esquisse d'un programme". MR1483106

[Til00] Ulrike Tillmann, *Higher genus surface operad detects infinite loop spaces*, Math. Ann. **317** (2000), no. 3, 613–628. MR1776120

[Wah04] Nathalie Wahl, *Infinite loop space structure(s) on the stable mapping class group*, Topology **43** (2004), no. 2, 343–368. MR2052967

[Wal21] Tashi Walde, *2-Segal spaces as invertible infinity-operads*, Algebr. Geom. Topol. **21** (2021), no. 1, 211–246. MR4224740

[War22] Benjamin C. Ward, *Toward a minimal model for* $H_*(\overline{\mathcal{M}})$, J. Homotopy Relat. Struct. **17** (2022), no. 4, 465–492. MR4514122

The manufacturer's authorised representative in the EU is Springer Nature Customer Service Centre GmbH, Europaplatz 3, 69115 Heidelberg, Germany. If you have any concerns regarding our products, please contact ProductSafety@springernature.com

Printed and bound by CPI Group (UK) Ltd, Croydon, CR0 4YY
29/03/2026
02080433-0001